Functional Analysis and Applied Optimization in Banach Spaces

Fabio Botelho

Functional Analysis and Applied Optimization in Banach Spaces

Applications to Non-Convex Variational Models

With Contributions by Anderson Ferreira and Alexandre Molter

Fabio Botelho
Department of Mathematics and Statistics
Federal University of Pelotas
Pelotas, RS-Brazil

ISBN 978-3-319-38206-7 ISBN 978-3-319-06074-3 (eBook)
DOI 10.1007/978-3-319-06074-3
Springer Cham Heidelberg New York Dordrecht London

Mathematics Subject Classification: 46N10, 46E15, 46N50, 49J40, 49K20

Softcover reprint of the hardcover 1st edition 2014

Printed on acid-free paper

Springer is part of Springer Science+Business Media (www.springer.com)

Preface

The first objective of this work is to present, to some extent, a deep introduction to the basic concepts on real and functional analysis.

In principle, the text is written for applied mathematicians and postgraduate students in applied mathematics, with interest in applications of functional analysis, calculus of variations, and optimization to problems in physics and engineering.

However, engineers, physicists, and other professionals in related areas may find the text very interesting by the possibility of background development towards graduate-level mathematics applicable in their respective work fields.

We have proven almost all results presented. The proofs are rigorous, but we believe are almost all very clear and relatively easy to read, even at the most complex text parts.

The material presented in Parts I and II concerns standard real and functional analysis. Hence in these two parts the results in general are not new, with the exception of some sections on domains of class $\hat{C}_1$ and relating Sobolev spaces and some sections about Lagrange multiplier results and the basic theorem about relaxation for the scalar case, where we show a different proof concerning the original one in the book *Convex Analysis and Variational Problems* (indeed such a book is the theoretical base of the present work) by Ekeland and Témam's.

About the basic part, specifically Chaps. 1–3 correspond to standard functional analysis. In Chaps. 4–6 we present basic and advanced concepts in measure and integration which will be relevant in subsequent results (in fact perhaps a little more than the minimum necessary). Moreover, Chaps. 7 and 8 correspond to a basic exposition on Sobolev spaces and again, the fundamental results presented are relevant for subsequent developments. In Chaps. 9–11 we introduce some basic and more advanced concepts on calculus of variations, convex analysis, and optimization.

Finally, the applications presented in Chaps. 12–23 correspond to the work of the present author along the last years, and almost all results including the applications of duality for micro-magnetism, composites in elasticity, and conductivity and phase transitions are extensions and natural developments of prior ones presented in the author's Ph.D. thesis at Virginia Tech, USA, and the previous book *Topics on Functional Analysis, Calculus of Variations and Duality* published by Academic

Publications. The present book overlaps to some extent with the previous one just on a part concerning standard mathematics. The applications in the present one are almost all new developments.

Anyway, a key feature of the present work is that while all problems studied here are nonlinear with corresponding non-convex variational formulation, it has been almost always possible to develop convex (in fact concave) dual variational formulations, which in general are more amenable to numerical computations.

The section on relaxation for the vectorial case, as its title suggests, presents duality principles that are valid even for vectorial problems. It is worth noting that such results were used in this text to develop concave dual variational formulations in situations such as for conductivity in composites and vectorial examples in phase transitions. In Chap. 15 we present the generalized method of lines, a numerical procedure in which the solution of the partial differential equation in question is written on lines as functions of boundary conditions and boundary shape. In Chap. 22 we develop some examples concerning the Navier–Stokes system.

Summary of Main Results

The main results of this work are summarized as follows.

Duality Applied to Elasticity

Chapter 12 develops duality for a model in finite elasticity. The dual formulations obtained allow the matrix of stresses to be nonpositive or nonnegative definite. This is, in some sense, an extension of earlier results (which establish the complementary energy as a perfect global optimization duality principle only if the stress tensor is positive definite at the equilibrium point). The results are based on standard tools of convex analysis and the concept of the Legendre transform.

Duality Applied to a Plate Model

Chapter 13 develops dual variational formulations for the two-dimensional equations of the nonlinear elastic Kirchhoff–Love plate model. We obtain a convex dual variational formulation which allows nonpositive definite membrane forces. In the third section, similar to the triality criterion introduced in [36], we obtain sufficient conditions of optimality for the present case. Again the results are based on the fundamental tools of convex analysis and the Legendre transform, which can easily be analytically expressed for the model in question.

Duality Applied to Ginzburg–Landau-Type Equations

Chapters 14–16 are concerned with existence theory and the development of dual variational formulations for Ginzburg–Landau-type equations. Since the primal formulations are non-convex, we use specific results for distance between two convex functions to obtain the dual approaches. Note that we obtain a convex dual formulation for the simpler real case. For such a formulation optimality conditions are also established.

Duality Applied to Multi-well Variational Problems

The main focus of Chaps. 17 and 18 is the development of dual variational formulations for multi-well optimization problems in phase transitions, conductivity, and elasticity. The primal formulation may not have minimizers in the classical sense. In this case, the solution through the dual formulation is a weak limit of minimizing sequences for the original problem.

Duality for a Model in Quantum Mechanics

In Chap. 19 we develop a duality principle and computation for a class of nonlinear eigenvalue problems found in quantum mechanics models. We present numerical results for one- and two-dimensional problems. We highlight that this chapter is coauthored by myself and my colleague Professor Anderson Ferreira.

Duality Applied to the Optimal Design in Elasticity

The first part of Chap. 20 develops a dual variational formulation for the optimal design of a plate of variable thickness. The design variable, namely the plate thickness, is supposed to minimize the plate deformation work due to a given external load. The second part is concerned with the optimal design for a two-phase problem in elasticity. In this case, we are looking for the mixture of two constituents that minimizes the structural internal work. In both applications the dual formulations were obtained through basic tools of convex analysis. Finally, we highlight that this chapter is coauthored by myself and my colleague Professor Alexandre Molter.

Duality Applied to Micro-magnetism

The main focus of Chap. 21 is the development of dual variational formulations for functionals related to ferromagnetism models. We develop duality principles for the so-called hard and full (semi-linear) uniaxial cases. It is important to emphasize that the dual formulations here presented are convex and are useful to compute the average behavior of minimizing sequences, specially as the primal formulation has no minimizers in the classical sense. Once more the results are obtained through standard tools of convex analysis.

Duality Applied to Fluid Mechanics

In Chap. 22 we develop approximate solutions for the incompressible Navier–Stokes system through the generalized method of lines. We also obtain a linear system whose solution solves the steady-state incompressible Euler equations.

Duality Applied to the Optimal Control and Optimal Design of a Beam Model

Chapter 23 develops duality for the optimal control and design of a beam model. We emphasize the dual formulation is useful to obtain numerical results. Finally, numerical examples of optimal design are provided, concerning the maximization of buckling load and fundamental frequency, respectively.

Acknowledgments

This monograph is based on my Ph.D. thesis at Virginia Tech and I am especially grateful to Professor Robert C. Rogers for his excellent work as advisor. I would like to thank the Department of Mathematics for its constant support and this opportunity of studying mathematics in an advanced level. I am also grateful to all the professors who have been teaching me during the last years for their valuable work. Among the professors, I particularly thank Martin Day (calculus of variations), James Thomson (real analysis), and George Hagedorn (functional analysis) for the excellent lectured courses. Finally, special thanks to all my professors at ITA (Instituto Tecnológico de Aeronáutica, SP-Brasil), my undergraduate and masters school, and Virginia Tech,

USA, two wonderful institutions in the American continent, forever in my heart and memories. Specifically about ITA, among many others, I would like to express my gratitude to Professors Leo H. Amaral and Tânia Rabello and my master thesis advisor Antônio Marmo de Oliveira also for their valuable work.

Pelotas, Brazil

Fabio Botelho

Contents

Acronyms

$\mathbb{N}$	The set of natural numbers
$\mathbb{R}$	The real set
$\mathbb{R}^n$	The n-dimensional real set
∞	Infinity symbol
$\overline{\mathbb{R}}$	$\mathbb{R} \cup \{+\infty\}$
Ω	A subset of $\mathbb{R}^n$
$\partial\Omega$	The boundary of Ω
$\mathbb{C}$	The complex set
U, V, Y	Banach spaces
U^*	Dual space of U
$\langle \cdot, \cdot \rangle_U : U \times U^* \to \mathbb{R}$	Duality pairing between U and U^*
H	Hilbert space
$(\cdot, \cdot)_H$	Inner product in a Hilbert space H
$A : U \to Y$	Operator whose domain is U and co-domain is Y
$A^* : Y^* \to U^*$	Adjoint operator relating A
$D(A)$	Domain of operator A
$N(A)$	Null space of operator A
$R(A)$	Range of operator A
$F : U \to \mathbb{R}$	Functional whose domain is U
$F^* : U^* \to \mathbb{R}$	Fenchel conjugate functional of F
$F'(u), \delta F(u), \frac{\partial F(u)}{\partial u}$	Notation for the Gâteaux derivative of F, at $u \in U$
$\delta F(u; \varphi)$	The Gâteaux variation of F at $u \in U$ relating the direction φ
$\partial F(u)$	The set of sub-gradients of F at $u \in U$
$\lvert \cdot \rvert$	Absolute value for real or complex numbers
$\lVert \cdot \rVert_U$	Norm, for vectors in a normed space U
$d : U \times U \to [0, +\infty)$	Metric in a metric space U
$\emptyset$	The empty set
$S \Rightarrow T$	S implies T
$S \Leftrightarrow T$	S if and only if T
$\forall$	For all

$\exists$	Exists symbol
$C^m(\Omega)$	Space of real functions on Ω that are continuously differentiable up to order m, $0 \leq m \leq \infty$
$C_c^m(\Omega)$	Set of functions in $C^m(\Omega)$ with compact support in Ω
$\mathscr{D}(\Omega), C_c^\infty(\Omega)$	Set of functions in $C^\infty(\Omega)$ with compact support in Ω
$L^p(\Omega)$	Space of measurable functions whose p-power of their absolute values is finite integrable
$W^{m,p}(\Omega)$	Sobolev space $\{u \in L^p(\Omega),\ D^\alpha u \in L^p(\Omega),\ \lvert\alpha\rvert \leq m,\ 1 \leq p \leq \infty\}$, where derivatives are in the distributional sense
$L^p(\Omega;\mathbb{R}^n)$	n-Dimensional L^p
$W^{m,p}(\Omega;\mathbb{R}^n)$	n-Dimensional $W^{m,p}$

Part I
Basic Functional Analysis

Chapter 1
Topological Vector Spaces

1.1 Introduction

The main objective of this chapter is to present an outline of the basic tools of analysis necessary to develop the subsequent chapters. We assume the reader has a background in linear algebra and elementary real analysis at an undergraduate level. The main references for this chapter are the excellent books on functional analysis: Rudin [58], Bachman and Narici [6], and Reed and Simon [52]. All proofs are developed in details.

1.2 Vector Spaces

We denote by $\mathbb{F}$ a scalar field. In practice this is either $\mathbb{R}$ or $\mathbb{C}$, the set of real or complex numbers.

Definition 1.2.1 (Vector Spaces). A vector space over $\mathbb{F}$ is a set which we will denote by U whose elements are called vectors, for which are defined two operations, namely, addition denoted by $(+) : U \times U \to U$ and scalar multiplication denoted by $(\cdot) : \mathbb{F} \times U \to U$, so that the following relations are valid:

1. $u+v = v+u, \forall u, v \in U$,
2. $u+(v+w) = (u+v)+w, \forall u, v, w \in U$,
3. there exists a vector denoted by θ such that $u+\theta = u$, $\forall u \in U$,
4. for each u $\in U$, there exists a unique vector denoted by $-u$ such that $u+(-u) = \theta$,
5. $\alpha \cdot (\beta \cdot u) = (\alpha \cdot \beta) \cdot u, \forall \alpha, \beta \in \mathbb{F}$, $u \in U$,
6. $\alpha \cdot (u+v) = \alpha \cdot u + \alpha \cdot v, \forall \alpha \in \mathbb{F}$, $u, v \in U$,
7. $(\alpha+\beta) \cdot u = \alpha \cdot u + \beta \cdot u, \forall \alpha, \beta \in \mathbb{F}$, $u \in U$,
8. $1 \cdot u = u, \forall u \in U$.

F. Botelho, *Functional Analysis and Applied Optimization in Banach Spaces: Applications to Non-Convex Variational Models*, DOI 10.1007/978-3-319-06074-3_1,

Remark 1.2.2. From now on we may drop the dot $(\cdot)$ in scalar multiplications and denote $\alpha \cdot u$ simply as αu.

Definition 1.2.3 (Vector Subspace). Let U be a vector space. A set $V \subset U$ is said to be a vector subspace of U if V is also a vector space with the same operations as those of U. If $V \neq U$, we say that V is a proper subspace of U.

Definition 1.2.4 (Finite-Dimensional Space). A vector space is said to be of finite dimension if there exists fixed $u_1, u_2, \ldots, u_n \in U$ such that for each $u \in U$ there are corresponding $\alpha_1, \ldots, \alpha_n \in \mathbb{F}$ for which

$$u = \sum_{i=1}^{n} \alpha_i u_i. \tag{1.1}$$

Definition 1.2.5 (Topological Spaces). A set U is said to be a topological space if it is possible to define a collection σ of subsets of U called a topology in U, for which the following properties are valid:

1. $U \in \sigma$,
2. $\emptyset \in \sigma$,
3. if $A \in \sigma$ and $B \in \sigma$, then $A \cap B \in \sigma$,
4. arbitrary unions of elements in σ also belong to σ.

Any $A \in \sigma$ is said to be an open set.

Remark 1.2.6. When necessary, to clarify the notation, we shall denote the vector space U endowed with the topology σ by (U, σ).

Definition 1.2.7 (Closed Sets). Let U be a topological space. A set $A \subset U$ is said to be closed if $U \setminus A$ is open. We also denote $U \setminus A = A^c = \{u \in U \mid u \notin A\}$.

Remark 1.2.8. For any sets $A, B \subset U$ we denote

$$A \setminus B = \{u \in A \mid u \notin B\}.$$

Also, when the meaning is clear we may denote $A \setminus B$ by $A - B$.

Proposition 1.2.9. *For closed sets we have the following properties:*

1. *U and $\emptyset$ are closed,*
2. *if A and B are closed sets, then $A \cup B$ is closed,*
3. *arbitrary intersections of closed sets are closed.*

Proof.

1. Since $\emptyset$ is open and $U = \emptyset^c$, by Definition 1.2.7, U is closed. Similarly, since U is open and $\emptyset = U \setminus U = U^c$, $\emptyset$ is closed.
2. A, B closed implies that A^c and B^c are open, and by Definition 1.2.5, $A^c \cup B^c$ is open, so that $A \cap B = (A^c \cup B^c)^c$ is closed.

3. Consider $A = \cap_{\lambda \in L} A_\lambda$, where L is a collection of indices and A_λ is closed, $\forall \lambda \in L$. We may write $A = (\cup_{\lambda \in L} A_\lambda^c)^c$ and since A_λ^c is open $\forall \lambda \in L$ we have, by Definition 1.2.5, that A is closed.

Definition 1.2.10 (Closure). Given $A \subset U$ we define the closure of A, denoted by $\bar{A}$, as the intersection of all closed sets that contain A.

Remark 1.2.11. From Proposition 1.2.9 item 3 we have that $\bar{A}$ is the smallest closed set that contains A, in the sense that if C is closed and $A \subset C$, then $\bar{A} \subset C$.

Definition 1.2.12 (Interior). Given $A \subset U$ we define its interior, denoted by A°, as the union of all open sets contained in A.

Remark 1.2.13. It is not difficult to prove that if A is open, then $A = A^\circ$.

Definition 1.2.14 (Neighborhood). Given $u_0 \in U$ we say that $\mathscr{V}$ is a neighborhood of u_0 if such a set is open and contains u_0. We denote such neighborhoods by $\mathscr{V}_{u_0}$.

Proposition 1.2.15. *If $A \subset U$ is a set such that for each $u \in A$ there exists a neighborhood $\mathscr{V}_u \ni u$ such that $\mathscr{V}_u \subset A$, then A is open.*

Proof. This follows from the fact that $A = \cup_{u \in A} \mathscr{V}_u$ and any arbitrary union of open sets is open.

Definition 1.2.16 (Function). Let U and V be two topological spaces. We say that $f : U \to V$ is a function if f is a collection of pairs $(u, v) \in U \times V$ such that for each $u \in U$ there exists only one $v \in V$ such that $(u, v) \in f$.

Definition 1.2.17 (Continuity at a Point). A function $f : U \to V$ is continuous at $u \in U$ if for each neighborhood $\mathscr{V}_{f(u)} \subset V$ of $f(u)$, there exists a neighborhood $\mathscr{V}_u \subset U$ of u such that $f(\mathscr{V}_u) \subset \mathscr{V}_{f(u)}$.

Definition 1.2.18 (Continuous Function). A function $f : U \to V$ is continuous if it is continuous at each $u \in U$.

Proposition 1.2.19. *A function $f : U \to V$ is continuous if and only if $f^{-1}(\mathscr{V})$ is open for each open $\mathscr{V} \subset V$, where*

$$f^{-1}(\mathscr{V}) = \{u \in U \mid f(u) \in \mathscr{V}\}. \tag{1.2}$$

Proof. Suppose $f^{-1}(\mathscr{V})$ is open whenever $\mathscr{V} \subset V$ is open. Pick $u \in U$ and any open $\mathscr{V}$ such that $f(u) \in \mathscr{V}$. Since $u \in f^{-1}(\mathscr{V})$ and $f(f^{-1}(\mathscr{V})) \subset \mathscr{V}$, we have that f is continuous at $u \in U$. Since $u \in U$ is arbitrary we have that f is continuous. Conversely, suppose f is continuous and pick $\mathscr{V} \subset V$ open. If $f^{-1}(\mathscr{V}) = \emptyset$, we are done, since $\emptyset$ is open. Thus, suppose $u \in f^{-1}(\mathscr{V})$, since f is continuous, there exists $\mathscr{V}_u$ a neighborhood of u such that $f(\mathscr{V}_u) \subset \mathscr{V}$. This means $\mathscr{V}_u \subset f^{-1}(\mathscr{V})$ and therefore, from Proposition 1.2.15, $f^{-1}(\mathscr{V})$ is open.

Definition 1.2.20. We say that (U,σ) is a Hausdorff topological space if, given u_1, $u_2 \in U$, $u_1 \neq u_2$, there exists $\mathscr{V}_1$, $\mathscr{V}_2 \in \sigma$ such that

$$u_1 \in \mathscr{V}_1 \ , \ u_2 \in \mathscr{V}_2 \text{ and } \mathscr{V}_1 \cap \mathscr{V}_2 = \emptyset. \tag{1.3}$$

Definition 1.2.21 (Base). A collection $\sigma' \subset \sigma$ is said to be a base for σ if every element of σ may be represented as a union of elements of σ'.

Definition 1.2.22 (Local Base). A collection $\hat{\sigma}$ of neighborhoods of a point $u \in U$ is said to be a local base at u if each neighborhood of u contains a member of $\hat{\sigma}$.

Definition 1.2.23 (Topological Vector Space). A vector space endowed with a topology, denoted by (U,σ), is said to be a topological vector space if and only if

1. every single point of U is a closed set,
2. the vector space operations (addition and scalar multiplication) are continuous with respect to σ.

More specifically, addition is continuous if given $u, v \in U$ and $\mathscr{V} \in \sigma$ such that $u+v \in \mathscr{V}$, then there exists $\mathscr{V}_u \ni u$ and $\mathscr{V}_v \ni v$ such that $\mathscr{V}_u + \mathscr{V}_v \subset \mathscr{V}$. On the other hand, scalar multiplication is continuous if given $\alpha \in \mathbb{F}$, $u \in U$ and $\mathscr{V} \ni \alpha \cdot u$, there exists $\delta > 0$ and $\mathscr{V}_u \ni u$ such that $\forall \beta \in \mathbb{F}$ satisfying $|\beta - \alpha| < \delta$ we have $\beta \mathscr{V}_u \subset \mathscr{V}$.

Given (U,σ), let us associate with each $u_0 \in U$ and $\alpha_0 \in \mathbb{F}$ ($\alpha_0 \neq 0$) the functions $T_{u_0} : U \to U$ and $M_{\alpha_0} : U \to U$ defined by

$$T_{u_0}(u) = u_0 + u \tag{1.4}$$

and

$$M_{\alpha_0}(u) = \alpha_0 \cdot u. \tag{1.5}$$

The continuity of such functions is a straightforward consequence of the continuity of vector space operations (addition and scalar multiplication). It is clear that the respective inverse maps, namely T_{-u_0} and M_{1/α_0}, are also continuous. So if $\mathscr{V}$ is open, then $u_0 + \mathscr{V}$, that is, $(T_{-u_0})^{-1}(\mathscr{V}) = T_{u_0}(\mathscr{V}) = u_0 + \mathscr{V}$ is open. By analogy $\alpha_0 \mathscr{V}$ is open. Thus σ is completely determined by a local base, so that the term local base will be understood henceforth as a local base at θ. So to summarize, a local base of a topological vector space is a collection Ω of neighborhoods of θ, such that each neighborhood of θ contains a member of Ω.

Now we present some simple results.

Proposition 1.2.24. *If $A \subset U$ is open, then $\forall u \in A$, there exists a neighborhood $\mathscr{V}$ of θ such that $u + \mathscr{V} \subset A$.*

Proof. Just take $\mathscr{V} = A - u$.

Proposition 1.2.25. *Given a topological vector space (U,σ), any element of σ may be expressed as a union of translates of members of Ω, so that the local base Ω generates the topology σ.*

Proof. Let $A \subset U$ open and $u \in A$. $\mathcal{V} = A - u$ is a neighborhood of θ and by definition of local base, there exists a set $\mathcal{V}_{\Omega u} \subset \mathcal{V}$ such that $\mathcal{V}_{\Omega_u} \in \Omega$. Thus, we may write

$$A = \cup_{u \in A}(u + \mathcal{V}_{\Omega_u}). \tag{1.6}$$

1.3 Some Properties of Topological Vector Spaces

In this section we study some fundamental properties of topological vector spaces. We start with the following proposition.

Proposition 1.3.1. *Any topological vector space U is a Hausdorff space.*

Proof. Pick $u_0, u_1 \in U$ such that $u_0 \neq u_1$. Thus $\mathcal{V} = U \setminus \{u_1 - u_0\}$ is an open neighborhood of zero. As $\theta + \theta = \theta$, by the continuity of addition, there exist $\mathcal{V}_1$ and $\mathcal{V}_2$ neighborhoods of θ such that

$$\mathcal{V}_1 + \mathcal{V}_2 \subset \mathcal{V} \tag{1.7}$$

define $\mathcal{U} = \mathcal{V}_1 \cap \mathcal{V}_2 \cap (-\mathcal{V}_1) \cap (-\mathcal{V}_2)$, thus $\mathcal{U} = -\mathcal{U}$ (symmetric) and $\mathcal{U} + \mathcal{U} \subset \mathcal{V}$ and hence

$$u_0 + \mathcal{U} + \mathcal{U} \subset u_0 + \mathcal{V} \subset U \setminus \{u_1\} \tag{1.8}$$

so that

$$u_0 + v_1 + v_2 \neq u_1, \quad \forall v_1, v_2 \in \mathcal{U}, \tag{1.9}$$

or

$$u_0 + v_1 \neq u_1 - v_2, \quad \forall v_1, v_2 \in \mathcal{U}, \tag{1.10}$$

and since $\mathcal{U} = -\mathcal{U}$

$$(u_0 + \mathcal{U}) \cap (u_1 + \mathcal{U}) = \emptyset. \tag{1.11}$$

Definition 1.3.2 (Bounded Sets). A set $A \subset U$ is said to be bounded if to each neighborhood of zero $\mathcal{V}$ there corresponds a number $s > 0$ such that $A \subset t\mathcal{V}$ for each $t > s$.

Definition 1.3.3 (Convex Sets). A set $A \subset U$ such that

$$\text{if } u, v \in A \text{ then } \lambda u + (1 - \lambda)v \in A, \quad \forall \lambda \in [0, 1], \tag{1.12}$$

is said to be convex.

Definition 1.3.4 (Locally Convex Spaces). A topological vector space U is said to be locally convex if there is a local base Ω whose elements are convex.

Definition 1.3.5 (Balanced Sets). A set $A \subset U$ is said to be balanced if $\alpha A \subset A$, $\forall \alpha \in \mathbb{F}$ such that $|\alpha| \leq 1$.

Theorem 1.3.6. *In a topological vector space U we have:*

1. *every neighborhood of zero contains a balanced neighborhood of zero,*
2. *every convex neighborhood of zero contains a balanced convex neighborhood of zero.*

Proof.

1. Suppose $\mathscr{U}$ is a neighborhood of zero. From the continuity of scalar multiplication, there exist $\mathscr{V}$ (neighborhood of zero) and $\delta > 0$, such that $\alpha\mathscr{V} \subset \mathscr{U}$ whenever $|\alpha| < \delta$. Define $\mathscr{W} = \cup_{|\alpha|<\delta}\alpha\mathscr{V}$; thus $\mathscr{W} \subset \mathscr{U}$ is a balanced neighborhood of zero.
2. Suppose $\mathscr{U}$ is a convex neighborhood of zero in U. Define

$$A = \{\cap \alpha\mathscr{U} \mid \alpha \in \mathbb{C},\ |\alpha| = 1\}. \tag{1.13}$$

As $0 \cdot \theta = \theta$ (where $\theta \in U$ denotes the zero vector) from the continuity of scalar multiplication there exists $\delta > 0$ and there is a neighborhood of zero $\mathscr{V}$ such that if $|\beta| < \delta$, then $\beta\mathscr{V} \subset \mathscr{U}$. Define $\mathscr{W}$ as the union of all such $\beta\mathscr{V}$. Thus $\mathscr{W}$ is balanced and $\alpha^{-1}\mathscr{W} = \mathscr{W}$ as $|\alpha| = 1$, so that $\mathscr{W} = \alpha\mathscr{W} \subset \alpha\mathscr{U}$, and hence $\mathscr{W} \subset A$, which implies that the interior A° is a neighborhood of zero. Also $A^\circ \subset \mathscr{U}$. Since A is an intersection of convex sets, it is convex and so is A°. Now we will show that A° is balanced and complete the proof. For this, it suffices to prove that A is balanced. Choose r and β such that $0 \leq r \leq 1$ and $|\beta| = 1$. Then

$$r\beta A = \cap_{|\alpha|=1} r\beta\alpha\mathscr{U} = \cap_{|\alpha|=1} r\alpha\mathscr{U}. \tag{1.14}$$

Since $\alpha\mathscr{U}$ is a convex set that contains zero, we obtain $r\alpha\mathscr{U} \subset \alpha\mathscr{U}$, so that $r\beta A \subset A$, which completes the proof.

Proposition 1.3.7. *Let U be a topological vector space and $\mathscr{V}$ a neighborhood of zero in U. Given $u \in U$, there exists $r \in \mathbb{R}^+$ such that $\beta u \in \mathscr{V}$, $\forall \beta$ such that $|\beta| < r$.*

Proof. Observe that $u + \mathscr{V}$ is a neighborhood of $1 \cdot u$, and then by the continuity of scalar multiplication, there exists $\mathscr{W}$ neighborhood of u and $r > 0$ such that

$$\beta\mathscr{W} \subset u + \mathscr{V}, \forall \beta \text{ such that } |\beta - 1| < r, \tag{1.15}$$

so that

$$\beta u \in u + \mathscr{V}, \tag{1.16}$$

or

$$(\beta - 1)u \in \mathscr{V}, \text{ where } |\beta - 1| < r, \tag{1.17}$$

and thus

$$\hat{\beta} u \in \mathscr{V}, \forall \hat{\beta} \text{ such that } |\hat{\beta}| < r, \tag{1.18}$$

which completes the proof.

Corollary 1.3.8. *Let $\mathscr{V}$ be a neighborhood of zero in U; if $\{r_n\}$ is a sequence such that $r_n > 0$, $\forall n \in \mathbb{N}$, and $\lim_{n\to\infty} r_n = \infty$, then $U \subset \cup_{n=1}^{\infty} r_n \mathscr{V}$.*

Proof. Let $u \in U$, then $\alpha u \in \mathscr{V}$ for any α sufficiently small, from the last proposition $u \in \frac{1}{\alpha}\mathscr{V}$. As $r_n \to \infty$ we have that $r_n > \frac{1}{\alpha}$ for n sufficiently big, so that $u \in r_n \mathscr{V}$, which completes the proof.

Proposition 1.3.9. *Suppose $\{\delta_n\}$ is a sequence such that $\delta_n \to 0$, $\delta_n < \delta_{n-1}$, $\forall n \in \mathbb{N}$ and $\mathscr{V}$ a bounded neighborhood of zero in U, then $\{\delta_n \mathscr{V}\}$ is a local base for U.*

Proof. Let $\mathscr{U}$ be a neighborhood of zero; as $\mathscr{V}$ is bounded, there exists $t_0 \in \mathbb{R}^+$ such that $\mathscr{V} \subset t\mathscr{U}$ for any $t > t_0$. As $\lim_{n\to\infty} \delta_n = 0$, there exists $n_0 \in \mathbb{N}$ such that if $n \geq n_0$, then $\delta_n < \frac{1}{t_0}$, so that $\delta_n \mathscr{V} \subset \mathscr{U}, \forall n$ such that $n \geq n_0$.

Definition 1.3.10 (Convergence in Topological Vector Spaces). Let U be a topological vector space. We say $\{u_n\}$ converges to $u_0 \in U$, if for each neighborhood $\mathscr{V}$ of u_0, then there exists $N \in \mathbb{N}$ such that

$$u_n \in \mathscr{V}, \forall n \geq N.$$

1.4 Compactness in Topological Vector Spaces

We start this section with the definition of open covering.

Definition 1.4.1 (Open Covering). Given $B \subset U$ we say that $\{\mathscr{O}_\alpha, \quad \alpha \in A\}$ is a covering of B if $B \subset \cup_{\alpha \in A} \mathscr{O}_\alpha$. If $\mathscr{O}_\alpha$ is open $\forall \alpha \in A$, then $\{\mathscr{O}_\alpha\}$ is said to be an open covering of B.

Definition 1.4.2 (Compact Sets). A set $B \subset U$ is said to be compact if each open covering of B has a finite subcovering. More explicitly, if $B \subset \cup_{\alpha \in A} \mathscr{O}_\alpha$, where $\mathscr{O}_\alpha$ is open $\forall \alpha \in A$, then there exist $\alpha_1, \ldots, \alpha_n \in A$ such that $B \subset \mathscr{O}_{\alpha_1} \cup \ldots \cup \mathscr{O}_{\alpha_n}$, for some n, a finite positive integer.

Proposition 1.4.3. *A compact subset of a Hausdorff space is closed.*

Proof. Let U be a Hausdorff space and consider $A \subset U$, A compact. Given $x \in A$ and $y \in A^c$, there exist open sets $\mathscr{O}_x$ and $\mathscr{O}_y^x$ such that $x \in \mathscr{O}_x$, $y \in \mathscr{O}_y^x$, and $\mathscr{O}_x \cap \mathscr{O}_y^x = \emptyset$. It is clear that $A \subset \cup_{x \in A} \mathscr{O}_x$, and since A is compact, we may find $\{x_1, x_2, \ldots, x_n\}$ such that $A \subset \cup_{i=1}^n \mathscr{O}_{x_i}$. For the selected $y \in A^c$ we have $y \in \cap_{i=1}^n \mathscr{O}_y^{x_i}$ and $(\cap_{i=1}^n \mathscr{O}_y^{x_i}) \cap (\cup_{i=1}^n \mathscr{O}_{x_i}) = \emptyset$. Since $\cap_{i=1}^n \mathscr{O}_y^{x_i}$ is open and y is an arbitrary point of A^c we have that A^c is open, so that A is closed, which completes the proof.

The next result is very useful.

Theorem 1.4.4. *Let $\{K_\alpha,\ \alpha \in L\}$ be a collection of compact subsets of a Hausdorff topological vector space U, such that the intersection of every finite subcollection (of $\{K_\alpha,\ \alpha \in L\}$) is nonempty.*

Under such hypotheses

$$\cap_{\alpha \in L} K_\alpha \neq \emptyset.$$

Proof. Fix $\alpha_0 \in L$. Suppose, to obtain contradiction, that

$$\cap_{\alpha \in L} K_\alpha = \emptyset.$$

That is,

$$K_{\alpha_0} \cap [\cap_{\alpha \in L}^{\alpha \neq \alpha_0} K_\alpha] = \emptyset.$$

Thus,

$$\cap_{\alpha \in L}^{\alpha \neq \alpha_0} K_\alpha \subset K_{\alpha_0}^c,$$

so that

$$K_{\alpha_0} \subset [\cap_{\alpha \in L}^{\alpha \neq \alpha_0} K_\alpha]^c,$$

$$K_{\alpha_0} \subset [\cup_{\alpha \in L}^{\alpha \neq \alpha_0} K_\alpha^c].$$

However, K_{α_0} is compact and K_α^c is open, $\forall \alpha \in L$.

Hence, there exist $\alpha_1, \ldots, \alpha_n \in L$ such that

$$K_{\alpha_0} \subset \cup_{i=1}^n K_{\alpha_i}^c.$$

From this we may infer that

$$K_{\alpha_0} \cap [\cap_{i=1}^n K_{\alpha_i}] = \emptyset,$$

which contradicts the hypotheses.

The proof is complete.

Proposition 1.4.5. *A closed subset of a compact space U is compact.*

Proof. Consider $\{\mathcal{O}_\alpha, \alpha \in L\}$ an open cover of A. Thus $\{A^c,\ \mathcal{O}_\alpha,\ \alpha \in L\}$ is a cover of U. As U is compact, there exist $\alpha_1, \alpha_2, \ldots, \alpha_n$ such that $A^c \cup (\cup_{i=1}^n \mathcal{O}_{\alpha_i}) \supset U$, so that $\{\mathcal{O}_{\alpha_i},\ i \in \{1, \ldots, n\}\}$ covers A, so that A is compact. The proof is complete.

Definition 1.4.6 (Countably Compact Sets). A set A is said to be countably compact if every infinite subset of A has a limit point in A.

Proposition 1.4.7. *Every compact subset of a topological space U is countably compact.*

Proof. Let B an infinite subset of A compact and suppose B has no limit point. Choose $\{x_1, x_2, \ldots\} \subset B$ and define $F = \{x_1, x_2, x_3, \ldots\}$. It is clear that F has no limit point. Thus, for each $n \in \mathbb{N}$, there exist $\mathcal{O}_n$ open such that $\mathcal{O}_n \cap F = \{x_n\}$. Also, for each $x \in A - F$, there exist $\mathcal{O}_x$ such that $x \in \mathcal{O}_x$ and $\mathcal{O}_x \cap F = \emptyset$. Thus $\{\mathcal{O}_x,\ x \in A - F;\ \mathcal{O}_1, \mathcal{O}_2, \ldots\}$ is an open cover of A without a finite subcover, which contradicts the fact that A is compact.

1.5 Normed and Metric Spaces

The idea here is to prepare a route for the study of Banach spaces defined below. We start with the definition of norm.

Definition 1.5.1 (Norm). A vector space U is said to be a normed space, if it is possible to define a function $\|\cdot\|_U : U \to \mathbb{R}^+ = [0,+\infty)$, called a norm, which satisfies the following properties:

1. $\|u\|_U > 0$, if $u \neq \theta$ and $\|u\|_U = 0 \Leftrightarrow u = \theta$,
2. $\|u+v\|_U \leq \|u\|_U + \|v\|_U, \forall\ u,v \in U$,
3. $\|\alpha u\|_U = |\alpha|\|u\|_U, \forall u \in U, \alpha \in \mathbb{F}$.

Now we present the definition of metric.

Definition 1.5.2 (Metric Space). A vector space U is said to be a metric space if it is possible to define a function $d : U \times U \to \mathbb{R}^+$, called a metric on U, such that

1. $0 \leq d(u,v),\ \ \forall u,v \in U$,
2. $d(u,v) = 0 \Leftrightarrow u = v$,
3. $d(u,v) = d(v,u),\ \ \forall u,v \in U$,
4. $d(u,w) \leq d(u,v) + d(v,w), \forall u,v,w \in U$.

A metric can be defined through a norm, that is,

$$d(u,v) = \|u-v\|_U. \tag{1.19}$$

In this case we say that the metric is induced by the norm.

The set $B_r(u) = \{v \in U \mid d(u,v) < r\}$ is called the open ball with center at u and radius r. A metric $d : U \times U \to \mathbb{R}^+$ is said to be invariant if

$$d(u+w, v+w) = d(u,v), \forall u,v,w \in U. \tag{1.20}$$

The following are some basic definitions concerning metric and normed spaces:

Definition 1.5.3 (Convergent Sequences). Given a metric space U, we say that $\{u_n\} \subset U$ converges to $u_0 \in U$ as $n \to \infty$, if for each $\varepsilon > 0$, there exists $n_0 \in \mathbb{N}$, such that if $n \geq n_0$, then $d(u_n, u_0) < \varepsilon$. In this case we write $u_n \to u_0$ as $n \to +\infty$.

Definition 1.5.4 (Cauchy Sequence). $\{u_n\} \subset U$ is said to be a Cauchy sequence if for each $\varepsilon > 0$ there exists $n_0 \in \mathbb{N}$ such that $d(u_n, u_m) < \varepsilon, \forall m, n \geq n_0$

Definition 1.5.5 (Completeness). A metric space U is said to be complete if each Cauchy sequence related to $d : U \times U \to \mathbb{R}^+$ converges to an element of U.

Definition 1.5.6 (Limit Point). Let (U,d) be a metric space and let $E \subset U$. We say that $v \in U$ is a limit point of E if for each $r > 0$ there exists $w \in B_r(v) \cap E$ such that $w \neq v$.

Definition 1.5.7 (Interior Point, Topology for (U,d)). Let (U,d) be a metric space and let $E \subset U$. We say that $u \in E$ is interior point if there exists $r > 0$ such that $B_r(u) \subset E$. We may define a topology for a metric space (U,d) by declaring as open all set $E \subset U$ such that all its points are interior. Such a topology is said to be induced by the metric d.

Definition 1.5.8. Let (U,d) be a metric space. The set σ of all open sets, defined through the last definition, is indeed a topology for (U,d).

Proof.

1. Obviously $\emptyset$ and U are open sets.
2. Assume A and B are open sets and define $C = A \cap B$. Let $u \in C = A \cap B$; thus, from $u \in A$, there exists $r_1 > 0$ such that $B_{r_1}(u) \subset A$. Similarly from $u \in B$ there exists $r_2 > 0$ such that $B_{r_2}(u) \subset B$.
   Define $r = \min\{r_1, r_2\}$. Thus, $B_r(u) \subset A \cap B = C$, so that u is an interior point of C. Since $u \in C$ is arbitrary, we may conclude that C is open.
3. Suppose $\{A_\alpha,\ \alpha \in L\}$ is a collection of open sets. Define $E = \cup_{\alpha \in L} A_\alpha$, and we shall show that E is open.
   Choose $u \in E = \cup_{\alpha \in L} A_\alpha$. Thus there exists $\alpha_0 \in L$ such that $u \in A_{\alpha_0}$. Since A_{α_0} is open there exists $r > 0$ such that $B_r(u) \subset A_{\alpha_0} \subset \cup_{\alpha \in L} A_\alpha = E$. Hence u is an interior point of E, since $u \in E$ is arbitrary, we may conclude that $E = \cup_{\alpha \in L} A_\alpha$ is open.

The proof is complete.

Definition 1.5.9. Let (U,d) be a metric space and let $E \subset U$. We define E' as the set of all the limit points of E.

Theorem 1.5.10. *Let (U,d) be a metric space and let $E \subset U$. Then E is closed if and only if $E' \subset E$.*

Proof. Suppose $E' \subset E$. Let $u \in E^c$; thus $u \notin E$ and $u \notin E'$. Therefore there exists $r > 0$ such that $B_r(u) \cap E = \emptyset$, so that $B_r(u) \subset E^c$. Therefore u is an interior point of E^c. Since $u \in E^c$ is arbitrary, we may infer that E^c is open, so that $E = (E^c)^c$ is closed.

Conversely, suppose that E is closed, that is, E^c is open.

If $E' = \emptyset$, we are done.

Thus assume $E' \neq \emptyset$ and choose $u \in E'$. Thus, for each $r > 0$, there exists $v \in B_r(u) \cap E$ such that $v \neq u$. Thus $B_r(u) \not\subset E^c, \forall r > 0$ so that u is not a interior point of E^c. Since E^c is open, we have that $u \notin E^c$ so that $u \in E$. We have thus obtained, $u \in E, \forall u \in E'$, so that $E' \subset E$.

The proof is complete.

Remark 1.5.11. From this last result, we may conclude that in a metric space, $E \subset U$ is closed if and only if $E' \subset E$.

Definition 1.5.12 (Banach Spaces). A normed vector space U is said to be a Banach space if each Cauchy sequence related to the metric induced by the norm converges to an element of U.

Remark 1.5.13. We say that a topology σ is compatible with a metric d if any $A \subset \sigma$ is represented by unions and/or finite intersections of open balls. In this case we say that $d : U \times U \to \mathbb{R}^+$ induces the topology σ.

Definition 1.5.14 (Metrizable Spaces). A topological vector space (U, σ) is said to be metrizable if σ is compatible with some metric d.

Definition 1.5.15 (Normable Spaces). A topological vector space (U, σ) is said to be normable if the induced metric (by this norm) is compatible with σ.

1.6 Compactness in Metric Spaces

Definition 1.6.1 (Diameter of a Set). Let (U,d) be a metric space and $A \subset U$. We define the diameter of A, denoted by $\text{diam}(A)$ by

$$\text{diam}(A) = \sup\{d(u,v) \mid u,v \in A\}.$$

Definition 1.6.2. Let (U,d) be a metric space. We say that $\{F_k\} \subset U$ is a nested sequence of sets if

$$F_1 \supset F_2 \supset F_3 \supset \ldots .$$

Theorem 1.6.3. *If (U,d) is a complete metric space, then every nested sequence of nonempty closed sets $\{F_k\}$ such that*

$$\lim_{k \to +\infty} diam(F_k) = 0$$

has nonempty intersection, that is,

$$\cap_{k=1}^{\infty} F_k \neq \emptyset.$$

Proof. Suppose $\{F_k\}$ is a nested sequence and $\lim_{k \to \infty} \text{diam}(F_k) = 0$. For each $n \in \mathbb{N}$, select $u_n \in F_n$. Suppose given $\varepsilon > 0$. Since

$$\lim_{n \to \infty} \text{diam}(F_n) = 0,$$

there exists $N \in \mathbb{N}$ such that if $n \geq N$, then

$$\text{diam}(F_n) < \varepsilon.$$

Thus if $m, n > N$ we have $u_m, u_n \in F_N$ so that

$$d(u_n, u_m) < \varepsilon.$$

Hence $\{u_n\}$ is a Cauchy sequence. Being U complete, there exists $u \in U$ such that

$$u_n \to u \text{ as } n \to \infty.$$

Choose $m \in \mathbb{N}$. We have that $u_n \in F_m, \forall n > m$, so that

$$u \in \bar{F}_m = F_m.$$

Since $m \in \mathbb{N}$ is arbitrary we obtain

$$u \in \cap_{m=1}^{\infty} F_m.$$

The proof is complete.

Theorem 1.6.4. *Let (U,d) be a metric space. If $A \subset U$ is compact, then it is closed and bounded.*

Proof. We have already proved that A is closed. Suppose, to obtain contradiction, that A is not bounded. Thus for each $K \in \mathbb{N}$ there exists $u, v \in A$ such that

$$d(u,v) > K.$$

Observe that

$$A \subset \cup_{u \in A} B_1(u).$$

Since A is compact there exists $u_1, u_2, \ldots, u_n \in A$ such that

$$A = \subset \cup_{k=1}^{n} B_1(u_k).$$

Define

$$R = \max\{d(u_i, u_j) \mid i, j \in \{1, \ldots, n\}\}.$$

Choose $u, v \in A$ such that

$$d(u,v) > R + 2. \tag{1.21}$$

Observe that there exist $i, j \in \{1, \ldots, n\}$ such that

$$u \in B_1(u_i),\ v \in B_1(u_j).$$

Thus

$$\begin{aligned} d(u,v) &\leq d(u,u_i) + d(u_i,u_j) + d(u_j,v) \\ &\leq 2 + R, \end{aligned} \tag{1.22}$$

which contradicts (1.21). This completes the proof.

Definition 1.6.5 (Relative Compactness). In a metric space (U,d), a set $A \subset U$ is said to be relatively compact if $\overline{A}$ is compact.

Definition 1.6.6 (ε-Nets). Let (U,d) be a metric space. A set $N \subset U$ is sat to be a ε-net with respect to a set $A \subset U$ if for each $u \in A$ there exists $v \in N$ such that

$$d(u,v) < \varepsilon.$$

Definition 1.6.7. Let (U,d) be a metric space. A set $A \subset U$ is said to be totally bounded if for each $\varepsilon > 0$, there exists a finite ε-net with respect to A.

Proposition 1.6.8. *Let (U,d) be a metric space. If $A \subset U$ is totally bounded, then it is bounded.*

Proof. Choose $u, v \in A$. Let $\{u_1, \ldots, u_n\}$ be the 1-net with respect to A. Define

$$R = \max\{d(u_i, u_j) \mid i, j \in \{1, \ldots, n\}\}.$$

Observe that there exist $i, j \in \{1, \ldots, n\}$ such that

$$d(u, u_i) < 1, \; d(v, u_j) < 1.$$

Thus

$$\begin{aligned} d(u,v) &\leq d(u,u_i) + d(u_i,u_j) + d(u_j,v) \\ &\leq R + 2. \end{aligned} \tag{1.23}$$

Since $u, v \in A$ are arbitrary, A is bounded.

Theorem 1.6.9. *Let (U,d) be a metric space. If from each sequence $\{u_n\} \subset A$ we can select a convergent subsequence $\{u_{n_k}\}$, then A is totally bounded.*

Proof. Suppose, to obtain contradiction, that A is not totally bounded. Thus there exists $\varepsilon_0 > 0$ such that there exists no ε_0-net with respect to A. Choose $u_1 \in A$; hence $\{u_1\}$ is not a ε_0-net, that is, there exists $u_2 \in A$ such that

$$d(u_1, u_2) > \varepsilon_0.$$

Again $\{u_1, u_2\}$ is not a ε_0-net for A, so that there exists $u_3 \in A$ such that

$$d(u_1, u_3) > \varepsilon_0 \text{ and } d(u_2, u_3) > \varepsilon_0.$$

Proceeding in this fashion we can obtain a sequence $\{u_n\}$ such that

$$d(u_n, u_m) > \varepsilon_0, \text{ if } m \neq n. \tag{1.24}$$

Clearly we cannot extract a convergent subsequence of $\{u_n\}$; otherwise such a subsequence would be Cauchy contradicting (1.24). The proof is complete.

Definition 1.6.10 (Sequentially Compact Sets). Let (U,d) be a metric space. A set $A \subset U$ is said to be sequentially compact if for each sequence $\{u_n\} \subset A$, there exist a subsequence $\{u_{n_k}\}$ and $u \in A$ such that

$$u_{n_k} \to u, \text{ as } k \to \infty.$$

Theorem 1.6.11. *A subset A of a metric space (U,d) is compact if and only if it is sequentially compact.*

Proof. Suppose A is compact. By Proposition 1.4.7 A is countably compact. Let $\{u_n\} \subset A$ be a sequence. We have two situations to consider:

1. $\{u_n\}$ has infinitely many equal terms, that is, in this case we have
$$u_{n_1} = u_{n_2} = \ldots = u_{n_k} = \ldots = u \in A.$$
Thus the result follows trivially.
2. $\{u_n\}$ has infinitely many distinct terms. In such a case, being A countably compact, $\{u_n\}$ has a limit point in A, so that there exist a subsequence $\{u_{n_k}\}$ and $u \in A$ such that
$$u_{n_k} \to u, \text{ as } k \to \infty.$$

In both cases we may find a subsequence converging to some $u \in A$.

Thus A is sequentially compact.

Conversely suppose A is sequentially compact, and suppose $\{G_\alpha,\ \alpha \in L\}$ is an open cover of A. For each $u \in A$ define
$$\delta(u) = \sup\{r \mid B_r(u) \subset G_\alpha, \text{ for some } \alpha \in L\}.$$

First we prove that $\delta(u) > 0, \forall u \in A$. Choose $u \in A$. Since $A \subset \cup_{\alpha \in L} G_\alpha$, there exists $\alpha_0 \in L$ such that $u \in G_{\alpha_0}$. Being G_{α_0} open, there exists $r_0 > 0$ such that $B_{r_0}(u) \subset G_{\alpha_0}$.

Thus,
$$\delta(u) \geq r_0 > 0.$$

Now define δ_0 by
$$\delta_0 = \inf\{\delta(u) \mid u \in A\}.$$

Therefore, there exists a sequence $\{u_n\} \subset A$ such that
$$\delta(u_n) \to \delta_0 \text{ as } n \to \infty.$$

Since A is sequentially compact, we may obtain a subsequence $\{u_{n_k}\}$ and $u_0 \in A$ such that
$$\delta(u_{n_k}) \to \delta_0 \text{ and } u_{n_k} \to u_0,$$
as $k \to \infty$. Therefore, we may find $K_0 \in \mathbb{N}$ such that if $k > K_0$, then
$$d(u_{n_k}, u_0) < \frac{\delta(u_0)}{4}. \tag{1.25}$$

We claim that
$$\delta(u_{n_k}) \geq \frac{\delta(u_0)}{4}, \text{ if } k > K_0.$$

To prove the claim, suppose
$$z \in B_{\frac{\delta(u_0)}{4}}(u_{n_k}), \forall k > K_0,$$

(observe that in particular from (1.25)

$$u_0 \in B_{\frac{\delta(u_0)}{4}}(u_{n_k}), \forall k > K_0).$$

Since

$$\frac{\delta(u_0)}{2} < \delta(u_0),$$

there exists some $\alpha_1 \in L$ such that

$$B_{\frac{\delta(u_0)}{2}}(u_0) \subset G_{\alpha_1}.$$

However, since

$$d(u_{n_k}, u_0) < \frac{\delta(u_0)}{4}, \text{ if } k > K_0,$$

we obtain

$$B_{\frac{\delta(u_0)}{2}}(u_0) \supset B_{\frac{\delta(u_0)}{4}}(u_{n_k}), \text{ if } k > K_0,$$

so that

$$\delta(u_{n_k}) \geq \frac{\delta(u_0)}{4}, \forall k > K_0.$$

Therefore

$$\lim_{k\to\infty} \delta(u_{n_k}) = \delta_0 \geq \frac{\delta(u_0)}{4}.$$

Choose $\varepsilon > 0$ such that

$$\delta_0 > \varepsilon > 0.$$

From the last theorem since A is sequentially compact, it is totally bounded. For the $\varepsilon > 0$ chosen above, consider an ε-net contained in A (the fact that the ε-net may be chosen contained in A is also a consequence of the last theorem) and denote it by N that is,

$$N = \{v_1, \ldots, v_n\} \in A.$$

Since $\delta_0 > \varepsilon$, there exists

$$\alpha_1, \ldots, \alpha_n \in L$$

such that

$$B_\varepsilon(v_i) \subset G_{\alpha_i}, \forall i \in \{1, \ldots, n\},$$

considering that

$$\delta(v_i) > \delta_0 > \varepsilon > 0, \forall i \in \{1, \ldots, n\}.$$

For $u \in A$, since N is an ε-net we have

$$u \in \cup_{i=1}^n B_\varepsilon(v_i) \subset \cup_{i=1}^n G_{\alpha_i}.$$

Since $u \in U$ is arbitrary we obtain

$$A \subset \cup_{i=1}^n G_{\alpha_i}.$$

Thus

$$\{G_{\alpha_1}, \ldots, G_{\alpha_n}\}$$

is a finite subcover for A of

$$\{G_\alpha, \ \alpha \in L\}.$$

Hence, A is compact.

The proof is complete.

Theorem 1.6.12. *Let (U,d) be a metric space. Thus $A \subset U$ is relatively compact if and only if for each sequence in A, we may select a convergent subsequence.*

Proof. Suppose A is relatively compact. Thus $\overline{A}$ is compact so that from the last theorem, $\overline{A}$ is sequentially compact.

Thus from each sequence in $\overline{A}$ we may select a subsequence which converges to some element of $\overline{A}$. In particular, for each sequence in $A \subset \overline{A}$, we may select a subsequence that converges to some element of $\overline{A}$.

Conversely, suppose that for each sequence in A, we may select a convergent subsequence. It suffices to prove that $\overline{A}$ is sequentially compact. Let $\{v_n\}$ be a sequence in $\overline{A}$. Since A is dense in $\overline{A}$, there exists a sequence $\{u_n\} \subset A$ such that

$$d(u_n, v_n) < \frac{1}{n}.$$

From the hypothesis we may obtain a subsequence $\{u_{n_k}\}$ and $u_0 \in \overline{A}$ such that

$$u_{n_k} \to u_0, \text{ as } k \to \infty.$$

Thus,

$$v_{n_k} \to u_0 \in \overline{A}, \text{ as } k \to \infty.$$

Therefore $\overline{A}$ is sequentially compact so that it is compact.

Theorem 1.6.13. *Let (U,d) be a metric space.*

1. *If $A \subset U$ is relatively compact, then it is totally bounded.*
2. *If (U,d) is a complete metric space and $A \subset U$ is totally bounded, then A is relatively compact.*

Proof.

1. Suppose $A \subset U$ is relatively compact. From the last theorem, from each sequence in A, we can extract a convergent subsequence. From Theorem 1.6.9, A is totally bounded.
2. Let (U,d) be a metric space and let A be a totally bounded subset of U.
   Let $\{u_n\}$ be a sequence in A. Since A is totally bounded for each $k \in \mathbb{N}$ we find a ε_k-net where $\varepsilon_k = 1/k$, denoted by N_k where

$$N_k = \{v_1^{(k)}, v_2^{(k)}, \ldots, v_{n_k}^{(k)}\}.$$

In particular for $k = 1$ $\{u_n\}$ is contained in the 1-net N_1. Thus at least one ball of radius 1 of N_1 contains infinitely many points of $\{u_n\}$. Let us select a subsequence $\{u_{n_k}^{(1)}\}_{k\in\mathbb{N}}$ of this infinite set (which is contained in a ball of radius 1). Similarly, we may select a subsequence here just partially relabeled $\{u_{n_l}^{(2)}\}_{l\in\mathbb{N}}$ of $\{u_{n_k}^{(1)}\}$ which is contained in one of the balls of the $\frac{1}{2}$-net. Proceeding in this fashion for each $k \in \mathbb{N}$ we may find a subsequence denoted by $\{u_{n_m}^{(k)}\}_{m\in\mathbb{N}}$ of the original sequence contained in a ball of radius $1/k$.
Now consider the diagonal sequence denoted by $\{u_{n_k}^{(k)}\}_{k\in\mathbb{N}} = \{z_k\}$. Thus

$$d(z_n, z_m) < \frac{2}{k}, \text{ if } m, n > k,$$

that is, $\{z_k\}$ is a Cauchy sequence, and since (U, d) is complete, there exists $u \in U$ such that

$$z_k \to u \text{ as } k \to \infty.$$

From Theorem 1.6.12, A is relatively compact.

The proof is complete.

1.7 The Arzela–Ascoli Theorem

In this section we present a classical result in analysis, namely the Arzela–Ascoli theorem.

Definition 1.7.1 (Equicontinuity). Let $\mathscr{F}$ be a collection of complex functions defined on a metric space (U, d). We say that $\mathscr{F}$ is equicontinuous if for each $\varepsilon > 0$, there exists $\delta > 0$ such that if $u, v \in U$ and $d(u, v) < \delta$, then

$$|f(u) - f(v)| < \varepsilon, \forall f \in \mathscr{F}.$$

Furthermore, we say that $\mathscr{F}$ is point-wise bounded if for each $u \in U$ there exists $M(u) \in \mathbb{R}$ such that

$$|f(u)| < M(u), \forall f \in \mathscr{F}.$$

Theorem 1.7.2 (Arzela–Ascoli). *Suppose $\mathscr{F}$ is a point-wise bounded equicontinuous collection of complex functions defined on a metric space (U, d). Also suppose that U has a countable dense subset E. Thus, each sequence $\{f_n\} \subset \mathscr{F}$ has a subsequence that converges uniformly on every compact subset of U.*

Proof. Let $\{u_n\}$ be a countable dense set in (U, d). By hypothesis, $\{f_n(u_1)\}$ is a bounded sequence; therefore, it has a convergent subsequence, which is denoted by $\{f_{n_k}(u_1)\}$. Let us denote

$$f_{n_k}(u_1) = \tilde{f}_{1,k}(u_1), \forall k \in \mathbb{N}.$$

Thus there exists $g_1 \in \mathbb{C}$ such that

$$\tilde{f}_{1,k}(u_1) \to g_1, \text{ as } k \to \infty.$$

Observe that $\{f_{n_k}(u_2)\}$ is also bounded and also it has a convergent subsequence, which similarly as above we will denote by $\{\tilde{f}_{2,k}(u_2)\}$. Again there exists $g_2 \in \mathbb{C}$ such that

$$\tilde{f}_{2,k}(u_1) \to g_1, \text{ as } k \to \infty.$$

$$\tilde{f}_{2,k}(u_2) \to g_2, \text{ as } k \to \infty.$$

Proceeding in this fashion for each $m \in \mathbb{N}$ we may obtain $\{\tilde{f}_{m,k}\}$ such that

$$\tilde{f}_{m,k}(u_j) \to g_j, \text{ as } k \to \infty, \forall j \in \{1,\ldots,m\},$$

where the set $\{g_1, g_2, \ldots, g_m\}$ is obtained as above. Consider the diagonal sequence

$$\{\tilde{f}_{k,k}\},$$

and observe that the sequence

$$\{\tilde{f}_{k,k}(u_m)\}_{k>m}$$

is such that

$$\tilde{f}_{k,k}(u_m) \to g_m \in \mathbb{C}, \text{ as } k \to \infty, \forall m \in \mathbb{N}.$$

Therefore we may conclude that from $\{f_n\}$ we may extract a subsequence also denoted by

$$\{f_{n_k}\} = \{\tilde{f}_{k,k}\}$$

which is convergent in

$$E = \{u_n\}_{n \in \mathbb{N}}.$$

Now suppose $K \subset U$, being K compact. Suppose given $\varepsilon > 0$. From the equicontinuity hypothesis there exists $\delta > 0$ such that if $u, v \in U$ and $d(u,v) < \delta$ we have

$$|f_{n_k}(u) - f_{n_k}(v)| < \frac{\varepsilon}{3}, \forall k \in \mathbb{N}.$$

Observe that

$$K \subset \cup_{u \in K} B_{\frac{\delta}{2}}(u),$$

and being K compact we may find $\{\tilde{u}_1, \ldots, \tilde{u}_M\}$ such that

$$K \subset \cup_{j=1}^{M} B_{\frac{\delta}{2}}(\tilde{u}_j).$$

Since E is dense in U, there exists

$$v_j \in B_{\frac{\delta}{2}}(\tilde{u}_j) \cap E, \forall j \in \{1,\ldots,M\}.$$

Fixing $j \in \{1,\ldots,M\}$, from $v_j \in E$ we obtain that

$$\lim_{k\to\infty} f_{n_k}(v_j)$$

exists as $k \to \infty$. Hence there exists $K_{0_j} \in \mathbb{N}$ such that if $k,l > K_{0_j}$, then

$$|f_{n_k}(v_j) - f_{n_l}(v_j)| < \frac{\varepsilon}{3}.$$

Pick $u \in K$; thus

$$u \in B_{\frac{\delta}{2}}(\tilde{u}_{\hat{j}})$$

for some $\hat{j} \in \{1,\ldots,M\}$, so that

$$d(u, v_{\hat{j}}) < \delta.$$

Therefore if

$$k,l > \max\{K_{0_1},\ldots,K_{0_M}\},$$

then

$$\begin{aligned} |f_{n_k}(u) - f_{n_l}(u)| &\leq |f_{n_k}(u) - f_{n_k}(v_{\hat{j}})| + |f_{n_k}(v_{\hat{j}}) - f_{n_l}(v_{\hat{j}})| \\ &\quad + |f_{n_l}(v_{\hat{j}}) - f_{n_l}(u)| \\ &\leq \frac{\varepsilon}{3} + \frac{\varepsilon}{3} + \frac{\varepsilon}{3} = \varepsilon. \end{aligned} \quad (1.26)$$

Since $u \in K$ is arbitrary, we conclude that $\{f_{n_k}\}$ is uniformly Cauchy on K.

The proof is complete.

1.8 Linear Mappings

Given U,V topological vector spaces, a function (mapping) $f : U \to V$, $A \subset U$, and $B \subset V$, we define

$$f(A) = \{f(u) \mid u \in A\}, \quad (1.27)$$

and the inverse image of B, denoted $f^{-1}(B)$ as

$$f^{-1}(B) = \{u \in U \mid f(u) \subset B\}. \quad (1.28)$$

Definition 1.8.1 (Linear Functions). A function $f : U \to V$ is said to be linear if

$$f(\alpha u + \beta v) = \alpha f(u) + \beta f(v), \forall u,v \in U,\ \alpha,\beta \in \mathbb{F}. \quad (1.29)$$

Definition 1.8.2 (Null Space and Range). Given $f : U \to V$, we define the null space and the range of f, denoted by $N(f)$ and $R(f)$, respectively, as

$$N(f) = \{u \in U \mid f(u) = \theta\} \tag{1.30}$$

and

$$R(f) = \{v \in V \mid \exists u \in U \text{ such that } f(u) = v\}. \tag{1.31}$$

Note that if f is linear, then $N(f)$ and $R(f)$ are subspaces of U and V, respectively.

Proposition 1.8.3. *Let U, V be topological vector spaces. If $f : U \to V$ is linear and continuous at θ, then it is continuous everywhere.*

Proof. Since f is linear, we have $f(\theta) = \theta$. Since f is continuous at θ, given $\mathscr{V} \subset V$ a neighborhood of zero, there exists $\mathscr{U} \subset U$ neighborhood of zero, such that

$$f(\mathscr{U}) \subset \mathscr{V}. \tag{1.32}$$

Thus

$$v - u \in \mathscr{U} \Rightarrow f(v-u) = f(v) - f(u) \in \mathscr{V}, \tag{1.33}$$

or

$$v \in u + \mathscr{U} \Rightarrow f(v) \in f(u) + \mathscr{V}, \tag{1.34}$$

which means that f is continuous at u. Since u is arbitrary, f is continuous everywhere.

1.9 Linearity and Continuity

Definition 1.9.1 (Bounded Functions). A function $f : U \to V$ is said to be bounded if it maps bounded sets into bounded sets.

Proposition 1.9.2. *A set E is bounded if and only if the following condition is satisfied: whenever $\{u_n\} \subset E$ and $\{\alpha_n\} \subset \mathbb{F}$ are such that $\alpha_n \to 0$ as $n \to \infty$ we have $\alpha_n u_n \to \theta$ as $n \to \infty$.*

Proof. Suppose E is bounded. Let $\mathscr{U}$ be a balanced neighborhood of θ in U and then $E \subset t\mathscr{U}$ for some t. For $\{u_n\} \subset E$, as $\alpha_n \to 0$, there exists N such that if $n > N$, then $t < \frac{1}{|\alpha_n|}$. Since $t^{-1}E \subset \mathscr{U}$ and $\mathscr{U}$ is balanced, we have that $\alpha_n u_n \in \mathscr{U}, \forall n > N$, and thus $\alpha_n u_n \to \theta$. Conversely, if E is not bounded, there is a neighborhood $\mathscr{V}$ of θ and $\{r_n\}$ such that $r_n \to \infty$ and E is not contained in $r_n\mathscr{V}$, that is, we can choose u_n such that $r_n^{-1}u_n$ is not in $\mathscr{V}$, $\forall n \in \mathbb{N}$, so that $\{r_n^{-1}u_n\}$ does not converge to θ.

Proposition 1.9.3. *Let $f : U \to V$ be a linear function. Consider the following statements:*

1. f is continuous,
2. f is bounded,
3. if $u_n \to \theta$, then $\{f(u_n)\}$ is bounded,
4. if $u_n \to \theta$, then $f(u_n) \to \theta$.

Then,

- 1 *implies* 2,
- 2 *implies* 3,
- *if U is metrizable, then* 3 *implies* 4, *which implies* 1.

Proof.

1. 1 implies 2: Suppose f is continuous, for $\mathscr{W} \subset V$ neighborhood of zero, there exists a neighborhood of zero in U, denoted by $\mathscr{V}$, such that
$$f(\mathscr{V}) \subset \mathscr{W}. \tag{1.35}$$
If E is bounded, there exists $t_0 \in \mathbb{R}^+$ such that $E \subset t\mathscr{V}, \forall t \geq t_0$, so that
$$f(E) \subset f(t\mathscr{V}) = tf(\mathscr{V}) \subset t\mathscr{W}, \quad \forall t \geq t_0, \tag{1.36}$$
and thus f is bounded.
2. 2 implies 3: Suppose $u_n \to \theta$ and let $\mathscr{W}$ be a neighborhood of zero. Then, there exists $N \in \mathbb{N}$ such that if $n \geq N$, then $u_n \in \mathscr{V} \subset \mathscr{W}$ where $\mathscr{V}$ is a balanced neighborhood of zero. On the other hand, for $n < N$, there exists K_n such that $u_n \in K_n\mathscr{V}$. Define $K = \max\{1, K_1, \ldots, K_n\}$. Then, $u_n \in K\mathscr{V}, \forall n \in \mathbb{N}$ and hence $\{u_n\}$ is bounded. Finally from 2, we have that $\{f(u_n)\}$ is bounded.
3. 3 implies 4: Suppose U is metrizable and let $u_n \to \theta$. Given $K \in \mathbb{N}$, there exists $n_K \in \mathbb{N}$ such that if $n > n_K$, then $d(u_n, \theta) < \frac{1}{K^2}$. Define $\gamma_n = 1$ if $n < n_1$ and $\gamma_n = K$, if $n_K \leq n < n_{K+1}$ so that
$$d(\gamma_n u_n, \theta) = d(K u_n, \theta) \leq K d(u_n, \theta) < K^{-1}. \tag{1.37}$$
Thus since 2 implies 3 we have that $\{f(\gamma_n u_n)\}$ is bounded so that, by Proposition 1.9.2, $f(u_n) = \gamma_n^{-1} f(\gamma_n u_n) \to \theta$ as $n \to \infty$.
4. 4 implies 1: suppose 1 fails. Thus there exists a neighborhood of zero $\mathscr{W} \subset V$ such that $f^{-1}(\mathscr{W})$ contains no neighborhood of zero in U. Particularly, we can select $\{u_n\}$ such that $u_n \in B_{1/n}(\theta)$ and $f(u_n)$ not in $\mathscr{W}$ so that $\{f(u_n)\}$ does not converge to zero. Thus 4 fails.

1.10 Continuity of Operators on Banach Spaces

Let U, V be Banach spaces. We call a function $A : U \to V$ an operator.

Proposition 1.10.1. *Let U, V be Banach spaces. A linear operator $A : U \to V$ is continuous if and only if there exists $K \in \mathbb{R}^+$ such that*
$$\|A(u)\|_V < K\|u\|_U, \forall u \in U.$$

Proof. Suppose A is linear and continuous. From Proposition 1.9.3,

$$\text{if } \{u_n\} \subset U \text{ is such that } u_n \to \theta \text{ then } A(u_n) \to \theta. \tag{1.38}$$

We claim that for each $\varepsilon > 0$ there exists $\delta > 0$ such that if $\|u\|_U < \delta$, then $\|A(u)\|_V < \varepsilon$.

Suppose, to obtain contradiction, that the claim is false.

Thus there exists $\varepsilon_0 > 0$ such that for each $n \in \mathbb{N}$ there exists $u_n \in U$ such that $\|u_n\|_U \leq \frac{1}{n}$ and $\|A(u_n)\|_V \geq \varepsilon_0$.

Therefore $u_n \to \theta$ and $A(u_n)$ does not converge to θ, which contradicts (1.38).

Thus the claim holds.

In particular, for $\varepsilon = 1$, there exists $\delta > 0$ such that if $\|u\|_U < \delta$, then $\|A(u)\|_V < 1$. Thus given an arbitrary not relabeled $u \in U$, $u \neq \theta$, for

$$w = \frac{\delta u}{2\|u\|_U}$$

we have

$$\|A(w)\|_V = \frac{\delta\|A(u)\|_V}{2\|u\|_U} < 1,$$

that is

$$\|A(u)\|_V < \frac{2\|u\|_U}{\delta}, \forall u \in U.$$

Defining

$$K = \frac{2}{\delta}$$

the first part of the proof is complete. Reciprocally, suppose there exists $K > 0$ such that

$$\|A(u)\|_V < K\|u\|_U, \forall u \in U.$$

Hence $u_n \to \theta$ implies $\|A(u_n)\|_V \to \theta$, so that from Proposition 1.9.3, A is continuous.

The proof is complete.

1.11 Some Classical Results on Banach Spaces

In this section we present some important results in Banach spaces. We start with the following theorem.

Theorem 1.11.1. *Let U and V be Banach spaces and let $A : U \to V$ be a linear operator. Then A is bounded if and only if the set $C \subset U$ has at least one interior point, where*

$$C = A^{-1}[\{v \in V \mid \|v\|_V \leq 1\}].$$

Proof. Suppose there exists $u_0 \in U$ in the interior of C. Thus, there exists $r > 0$ such that

$$B_r(u_0) = \{u \in U \mid \|u - u_0\|_U < r\} \subset C.$$

Fix $u \in U$ such that $\|u\|_U < r$. Thus, we have

$$\|A(u)\|_V \leq \|A(u + u_0)\|_V + \|A(u_0)\|_V.$$

Observe also that

$$\|(u + u_0) - u_0\|_U < r,$$

so that $u + u_0 \in B_r(u_0) \subset C$ and thus

$$\|A(u + u_0)\|_V \leq 1$$

and hence

$$\|A(u)\|_V \leq 1 + \|A(u_0)\|_V, \tag{1.39}$$

$\forall u \in U$ such that $\|u\|_U < r$. Fix an arbitrary not relabeled $u \in U$ such that $u \neq \theta$. From (1.39)

$$w = \frac{u}{\|u\|_U} \frac{r}{2}$$

is such that

$$\|A(w)\|_V = \frac{\|A(u)\|_V}{\|u\|_U} \frac{r}{2} \leq 1 + \|A(u_0)\|_V,$$

so that

$$\|A(u)\|_V \leq (1 + \|A(u_0)\|_V)\|u\|_U \frac{2}{r}.$$

Since $u \in U$ is arbitrary, A is bounded.

Reciprocally, suppose A is bounded. Thus

$$\|A(u)\|_V \leq K\|u\|_U, \forall u \in U,$$

for some $K > 0$. In particular

$$D = \left\{u \in U \mid \|u\|_U \leq \frac{1}{K}\right\} \subset C.$$

The proof is complete.

Definition 1.11.2. A set S in a metric space U is said to be nowhere dense if $\overline{S}$ has an empty interior.

Theorem 1.11.3 (Baire Category Theorem). *A complete metric space is never the union of a countable number of nowhere dense sets.*

Proof. Suppose, to obtain contradiction, that U is a complete metric space and

$$U = \cup_{n=1}^{\infty} A_n,$$

where each A_n is nowhere dense. Since A_1 is nowhere dense, there exist $u_1 \in U$ which is not in $\bar{A}_1$; otherwise we would have $U = \bar{A}_1$, which is not possible since U is open. Furthermore, $\bar{A}_1^c$ is open, so that we may obtain $u_1 \in A_1^c$ and $0 < r_1 < 1$ such that

$$B_1 = B_{r_1}(u_1)$$

satisfies

$$B_1 \cap A_1 = \emptyset.$$

Since A_2 is nowhere dense we have B_1 is not contained in $\bar{A}_2$. Therefore we may select $u_2 \in B_1 \setminus \bar{A}_2$ and since $B_1 \setminus \bar{A}_2$ is open, there exists $0 < r_2 < 1/2$ such that

$$\bar{B}_2 = \bar{B}_{r_2}(u_2) \subset B_1 \setminus \bar{A}_2,$$

that is,

$$B_2 \cap A_2 = \emptyset.$$

Proceeding inductively in this fashion, for each $n \in \mathbb{N}$, we may obtain $u_n \in B_{n-1} \setminus \bar{A}_n$ such that we may choose an open ball $B_n = B_{r_n}(u_n)$ such that

$$\bar{B}_n \subset B_{n-1},$$

$$B_n \cap A_n = \emptyset,$$

and

$$0 < r_n < 2^{1-n}.$$

Observe that $\{u_n\}$ is a Cauchy sequence, considering that if $m, n > N$, then $u_n, u_m \in B_N$, so that

$$d(u_n, u_m) < 2(2^{1-N}).$$

Define

$$u = \lim_{n \to \infty} u_n.$$

Since

$$u_n \in B_N, \forall n > N,$$

we get

$$u \in \bar{B}_N \subset B_{N-1}.$$

Therefore u is not in $A_{N-1}, \forall N > 1$, which means u is not in $\cup_{n=1}^{\infty} A_n = U$, a contradiction.

The proof is complete.

Theorem 1.11.4 (The Principle of Uniform Boundedness). *Let U be a Banach space. Let $\mathscr{F}$ be a family of linear bounded operators from U into a normed linear space V. Suppose for each $u \in U$ there exists a $K_u \in \mathbb{R}$ such that*

$$\|T(u)\|_V < K_u, \forall T \in \mathscr{F}.$$

Then, there exists $K \in \mathbb{R}$ such that

$$\|T\| < K, \forall T \in \mathscr{F}.$$

Proof. Define

$$B_n = \{u \in U \mid \|T(u)\|_V \leq n, \forall T \in \mathscr{F}\}.$$

By the hypotheses, given $u \in U$, $u \in B_n$ for all n is sufficiently big. Thus,

$$U = \cup_{n=1}^{\infty} B_n.$$

Moreover each B_n is closed. By the Baire category theorem there exists $n_0 \in \mathbb{N}$ such that B_{n_0} has nonempty interior. That is, there exists $u_0 \in U$ and $r > 0$ such that

$$B_r(u_0) \subset B_{n_0}.$$

Thus, fixing an arbitrary $T \in \mathscr{F}$, we have

$$\|T(u)\|_V \leq n_0, \forall u \in B_r(u_0).$$

Thus if $\|u\|_U < r$ then $\|(u+u_0) - u_0\|_U < r$, so that

$$\|T(u+u_0)\|_V \leq n_0,$$

that is,

$$\|T(u)\|_V - \|T(u_0)\|_V \leq n_0.$$

Thus,

$$\|T(u)\|_V \leq 2n_0, \text{ if } \|u\|_U < r. \tag{1.40}$$

For $u \in U$ arbitrary, $u \neq \theta$, define

$$w = \frac{ru}{2\|u\|_U},$$

from (1.40) we obtain

$$\|T(w)\|_V = \frac{r\|T(u)\|_V}{2\|u\|_U} \leq 2n_0,$$

so that

$$\|T(u)\|_V \leq \frac{4n_0\|u\|_U}{r}, \forall u \in U.$$

Hence

$$\|T\| \leq \frac{4n_0}{r}, \forall T \in \mathscr{F}.$$

The proof is complete.

Theorem 1.11.5 (The Open Mapping Theorem). *Let U and V be Banach spaces and let $A : U \to V$ be a bounded onto linear operator. Thus, if $\mathcal{O} \subset U$ is open, then $A(\mathcal{O})$ is open in V.*

Proof. First we will prove that given $r > 0$, there exists $r' > 0$ such that

$$A(B_r(\theta)) \supset B^V_{r'}(\theta). \tag{1.41}$$

Here $B^V_{r'}(\theta)$ denotes a ball in V of radius r' with center in θ. Since A is onto

$$V = \cup_{n=1}^{\infty} A(nB_1(\theta)).$$

By the Baire category theorem, there exists $n_0 \in \mathbb{N}$ such that the closure of $A(n_0B_1(\theta))$ has nonempty interior, so that $\overline{A(B_1(\theta))}$ has nonempty interior. We will show that there exists $r' > 0$ such that

$$B^V_{r'}(\theta) \subset \overline{A(B_1(\theta))}.$$

Observe that there exists $y_0 \in V$ and $r_1 > 0$ such that

$$B^V_{r_1}(y_0) \subset \overline{A(B_1(\theta))}. \tag{1.42}$$

Define $u_0 \in B_1(\theta)$ which satisfies $A(u_0) = y_0$. We claim that

$$\overline{A(B_{r_2}(\theta))} \supset B^V_{r_1}(\theta),$$

where $r_2 = 1 + \|u_0\|_U$. To prove the claim, pick

$$y \in A(B_1(\theta))$$

thus there exists $u \in U$ such that $\|u\|_U < 1$ and $A(u) = y$. Therefore

$$A(u) = A(u - u_0 + u_0) = A(u - u_0) + A(u_0).$$

But observe that

$$\begin{aligned} \|u - u_0\|_U &\leq \|u\|_U + \|u_0\|_U \\ &< 1 + \|u_0\|_U \\ &= r_2, \end{aligned} \tag{1.43}$$

so that

$$A(u - u_0) \in A(B_{r_2}(\theta)).$$

This means

$$y = A(u) \in A(u_0) + A(B_{r_2}(\theta)),$$

and hence

$$A(B_1(\theta)) \subset A(u_0) + A(B_{r_2}(\theta)).$$

That is, from this and (1.42), we obtain

$$A(u_0)+\overline{A(B_{r_2}(\theta))}\supset\overline{A(B_1(\theta))}\supset B^V_{r_1}(y_0)=A(u_0)+B^V_{r_1}(\theta),$$

and therefore

$$\overline{A(B_{r_2}(\theta))}\supset B^V_{r_1}(\theta).$$

Since

$$A(B_{r_2}(\theta))=r_2A(B_1(\theta)),$$

we have, for some not relabeled $r_1>0$, that

$$\overline{A(B_1(\theta))}\supset B^V_{r_1}(\theta).$$

Thus it suffices to show that

$$\overline{A(B_1(\theta))}\subset A(B_2(\theta)),$$

to prove (1.41). Let $y\in\overline{A(B_1(\theta))}$; since A is continuous, we may select $u_1\in B_1(\theta)$ such that

$$y-A(u_1)\in B^V_{r_1/2}(\theta)\subset\overline{A(B_{1/2}(\theta))}.$$

Now select $u_2\in B_{1/2}(\theta)$ so that

$$y-A(u_1)-A(u_2)\in B^V_{r_1/4}(\theta).$$

By induction, we may obtain

$$u_n\in B_{2^{1-n}}(\theta),$$

such that

$$y-\sum_{j=1}^{n}A(u_j)\in B^V_{r_1/2^n}(\theta).$$

Define

$$u=\sum_{n=1}^{\infty}u_n,$$

we have that $u\in B_2(\theta)$, so that

$$y=\sum_{n=1}^{\infty}A(u_n)=A(u)\in A(B_2(\theta)).$$

Therefore

$$\overline{A(B_1(\theta))}\subset A(B_2(\theta)).$$

The proof of (1.41) is complete.

To finish the proof of this theorem, assume $\mathcal{O} \subset U$ is open. Let $v_0 \in A(\mathcal{O})$. Let $u_0 \in \mathcal{O}$ be such that $A(u_0) = v_0$. Thus there exists $r > 0$ such that

$$B_r(u_0) \subset \mathcal{O}.$$

From (1.41),

$$A(B_r(\theta)) \supset B^V_{r'}(\theta),$$

for some $r' > 0$. Thus

$$A(\mathcal{O}) \supset A(u_0) + A(B_r(\theta)) \supset v_0 + B^V_{r'}(\theta).$$

This means that v_0 is an interior point of $A(\mathcal{O})$. Since $v_0 \in A(\mathcal{O})$ is arbitrary, we may conclude that $A(\mathcal{O})$ is open.

The proof is complete.

Theorem 1.11.6 (The Inverse Mapping Theorem). *A continuous linear bijection of one Banach space onto another has a continuous inverse.*

Proof. Let $A : U \to V$ satisfying the theorem hypotheses. Since A is open, A^{-1} is continuous.

Definition 1.11.7 (Graph of a Mapping). Let $A : U \to V$ be an operator, where U and V are normed linear spaces. The *graph* of A denoted by $\Gamma(A)$ is defined by

$$\Gamma(A) = \{(u,v) \in U \times V \mid v = A(u)\}.$$

Theorem 1.11.8 (The Closed Graph Theorem). *Let U and V be Banach spaces and let $A : U \to V$ be a linear operator. Then A is bounded if and only if its graph is closed.*

Proof. Suppose $\Gamma(A)$ is closed. Since A is linear, $\Gamma(A)$ is a subspace of $U \oplus V$. Also, being $\Gamma(A)$ closed, it is a Banach space with the norm

$$\|(u,A(u)\| = \|u\|_U + \|A(u)\|_V.$$

Consider the continuous mappings

$$\Pi_1(u,A(u)) = u$$

and

$$\Pi_2(u,A(u)) = A(u).$$

Observe that Π_1 is a bijection, so that by the inverse mapping theorem, Π_1^{-1} is continuous. As

$$A = \Pi_2 \circ \Pi_1^{-1},$$

it follows that A is continuous. The converse is trivial.

1.12 Hilbert Spaces

At this point we introduce an important class of spaces, namely the Hilbert spaces.

Definition 1.12.1. Let H be a vector space. We say that H is a real pre-Hilbert space if there exists a function $(\cdot,\cdot)_H : H \times H \to \mathbb{R}$ such that

1. $(u,v)_H = (v,u)_H, \ \forall u,v \in H,$
2. $(u+v,w)_H = (u,w)_H + (v,w)_H, \ \forall u,v,w \in H,$
3. $(\alpha u,v)_H = \alpha(u,v)_H, \ \forall u,v \in H, \ \alpha \in \mathbb{R},$
4. $(u,u)_H \geq 0, \ \forall u \in H,$ and $(u,u)_H = 0,$ if and only if $u = \theta$.

Remark 1.12.2. The function $(\cdot,\cdot)_H : H \times H \to \mathbb{R}$ is called an inner product.

Proposition 1.12.3 (Cauchy–Schwarz Inequality). *Let H be a pre-Hilbert space. Defining*

$$\|u\|_H = \sqrt{(u,u)_H}, \forall u \in H,$$

we have

$$|(u,v)_H| \leq \|u\|_H \|v\|_H, \forall u,v \subset H.$$

Equality holds if and only if $u = \alpha v$ for some $\alpha \in \mathbb{R}$ or $v = \theta$.

Proof. If $v = \theta$, the inequality is immediate. Assume $v \neq \theta$. Given $\alpha \in \mathbb{R}$ we have

$$\begin{aligned} 0 &\leq (u-\alpha v, u-\alpha v)_H \\ &= (u,u)_H + \alpha^2 (v,v)_H - 2\alpha(u,v)_H \\ &= \|u\|_H^2 + \alpha^2 \|v\|_H^2 - 2\alpha(u,v)_H. \end{aligned} \tag{1.44}$$

In particular, for $\alpha = (u,v)_H / \|v\|_H^2$, we obtain

$$0 \leq \|u\|_H^2 - \frac{(u,v)_H^2}{\|v\|_H^2},$$

that is,

$$|(u,v)_H| \leq \|u\|_H \|v\|_H.$$

The remaining conclusions are left to the reader.

Proposition 1.12.4. *On a pre-Hilbert space H, the function*

$$\| \ \|_H : H \to \mathbb{R}$$

is a norm, where as above

$$\|u\|_H = \sqrt{(u,u)}.$$

Proof. The only nontrivial property to be verified, concerning the definition of norm, is the triangle inequality.

Observe that given $u, v \in H$, from the Cauchy–Schwarz inequality, we have

$$\begin{aligned}
\|u+v\|_H^2 &= (u+v, u+v)_H \\
&= (u,u)_H + (v,v)_H + 2(u,v)_H \\
&\leq (u,u)_H + (v,v)_H + 2|(u,v)_H| \\
&\leq \|u\|_H^2 + \|v\|_H^2 + 2\|u\|_H \|v\|_H \\
&= (\|u\|_H + \|v\|_H)^2.
\end{aligned} \tag{1.45}$$

Therefore

$$\|u+v\|_H \leq \|u\|_H + \|v\|_H, \forall u, v \in H.$$

The proof is complete.

Definition 1.12.5. A pre-Hilbert space H is to be a Hilbert space if it is complete, that is, if any Cauchy sequence in H converges to an element of H.

Definition 1.12.6 (Orthogonal Complement). Let H be a Hilbert space. Considering $M \subset H$ we define its orthogonal complement, denoted by $M^\perp$, by

$$M^\perp = \{u \in H \mid (u,m)_H = 0, \ \forall m \in M\}.$$

Theorem 1.12.7. *Let H be a Hilbert space and M a closed subspace of H and suppose $u \in H$. Under such hypotheses there exists a unique $m_0 \in M$ such that*

$$\|u - m_0\|_H = \min_{m \in M}\{\|u - m\|_H\}.$$

Moreover $n_0 = u - m_0 \in M^\perp$ so that

$$u = m_0 + n_0,$$

where $m_0 \in M$ and $n_0 \in M^\perp$. Finally, such a representation through $M \oplus M^\perp$ is unique.

Proof. Define d by

$$d = \inf_{m \in M}\{\|u - m\|_H\}.$$

Let $\{m_i\} \subset M$ be a sequence such that

$$\|u - m_i\|_H \to d, \ \text{as } i \to \infty.$$

Thus, from the parallelogram law, we have

$$\begin{aligned}
\|m_i - m_j\|_H^2 &= \|m_i - u - (m_j - u)\|_H^2 \\
&= 2\|m_i - u\|_H^2 + 2\|m_j - u\|_H^2 \\
&\quad -2\| -2u + m_i + m_j\|_H^2 \\
&= 2\|m_i - u\|_H^2 + 2\|m_j - u\|_H^2
\end{aligned}$$

$$-4\|-u+(m_i+m_j)/2\|_H^2$$
$$\to 2d^2+2d^2-4d^2=0, \text{ as } i,j\to+\infty. \tag{1.46}$$

Thus $\{m_i\}\subset M$ is a Cauchy sequence. Since M is closed, there exists $m_0\in M$ such that

$$m_i\to m_0, \text{ as } i\to+\infty,$$

so that

$$\|u-m_i\|_H\to\|u-m_0\|_H=d.$$

Define

$$n_0=u-m_0.$$

We will prove that $n_0\in M^{\perp}$.

Pick $m\in M$ and $t\in\mathbb{R}$, and thus we have

$$\begin{aligned} d^2 &\le \|u-(m_0-tm)\|_H^2 \\ &= \|n_0+tm\|_H^2 \\ &= \|n_0\|_H^2+2(n_0,m)_Ht+\|m\|_H^2t^2. \end{aligned} \tag{1.47}$$

Since

$$\|n_0\|_H^2=\|u-m_0\|_H^2=d^2,$$

we obtain

$$2(n_0,m)_Ht+\|m\|_H^2t^2\ge 0,\forall t\in\mathbb{R}$$

so that

$$(n_0,m)_H=0.$$

Being $m\in M$ arbitrary, we obtain

$$n_0\in M^{\perp}.$$

It remains to prove the uniqueness. Let $m\in M$, and thus

$$\begin{aligned} \|u-m\|_H^2 &= \|u-m_0+m_0-m\|_H^2 \\ &= \|u-m_0\|_H^2+\|m-m_0\|_H^2, \end{aligned} \tag{1.48}$$

since

$$(u-m_0,m-m_0)_H=(n_0,m-m_0)_H=0.$$

From (1.48) we obtain

$$\|u-m\|_H^2>\|u-m_0\|_H^2=d^2,$$

if $m\neq m_0$.

Therefore m_0 is unique.

Now suppose

$$u=m_1+n_1,$$

where $m_1 \in M$ and $n_1 \in M^\perp$. As above, for $m \in M$

$$\begin{aligned} \|u-m\|_H^2 &= \|u-m_1+m_1-m\|_H^2 \\ &= \|u-m_1\|_H^2+\|m-m_1\|_H^2, \\ &\geq \|u-m_1\|_H \end{aligned} \tag{1.49}$$

and thus since m_0 such that

$$d=\|u-m_0\|_H$$

is unique, we get

$$m_1=m_0$$

and therefore

$$n_1=u-m_0=n_0.$$

The proof is complete.

Theorem 1.12.8 (The Riesz Lemma). *Let H be a Hilbert space and let $f : H \to \mathbb{R}$ be a continuous linear functional. Then there exists a unique $u_0 \in H$ such that*

$$f(u)=(u,u_0)_H, \forall u \in H.$$

Moreover

$$\|f\|_{H^*}=\|u_0\|_H.$$

Proof. Define N by

$$N=\{u \in H \mid f(u)=0\}.$$

Thus, as f is a continuous and linear, N is a closed subspace of H. If $N=H$, then $f(u)=0=(u,\theta)_H, \forall u \in H$ and the proof would be complete. Thus, assume $N \neq H$. By the last theorem there exists $v \neq \theta$ such that $v \in N^\perp$.

Define

$$u_0=\frac{f(v)}{\|v\|_H^2}v.$$

Thus,if $u \in N$ we have

$$f(u)=0=(u,u_0)_H=0.$$

On the other hand, if $u=\alpha v$ for some $\alpha \in \mathbb{R}$, we have

$$\begin{aligned} f(u) &= \alpha f(v) \\ &= \frac{f(v)(\alpha v,v)_H}{\|v\|_H^2} \\ &= \left(\alpha v, \frac{f(v)v}{\|v\|_H^2}\right)_H \\ &= (\alpha v,u_0)_H. \end{aligned} \tag{1.50}$$

Therefore $f(u)$ equals $(u,u_0)_H$ in the space spanned by N and v. Now we show that this last space (then span of N and v) is in fact H. Just observe that given $u \in H$ we may write

$$u = \left(u - \frac{f(u)v}{f(v)}\right) + \frac{f(u)v}{f(v)}. \tag{1.51}$$

Since

$$u - \frac{f(u)v}{f(v)} \in N$$

we have finished the first part of the proof, that is, we have proven that

$$f(u) = (u, u_0)_H, \forall u \in H.$$

To finish the proof, assume $u_1 \in H$ is such that

$$f(u) = (u, u_1)_H, \forall u \in H.$$

Thus,

$$\begin{aligned}\|u_0 - u_1\|_H^2 &= (u_0 - u_1, u_0 - u_1)_H \\ &= (u_0 - u_1, u_0)_H - (u_0 - u_1, u_1)_H \\ &= f(u_0 - u_1) - f(u_0 - u_1) = 0. \end{aligned} \tag{1.52}$$

Hence $u_1 = u_0$.

Let us now prove that

$$\|f\|_{H^*} = \|u_0\|_H.$$

First observe that

$$\begin{aligned}\|f\|_{H^*} &= \sup\{f(u) \mid u \in H,\ \|u\|_H \leq 1\} \\ &= \sup\{|(u, u_0)_H| \mid u \in H,\ \|u\|_H \leq 1\} \\ &\leq \sup\{\|u\|_H \|u_0\|_H \mid u \in H,\ \|u\|_H \leq 1\} \\ &\leq \|u_0\|_H. \end{aligned} \tag{1.53}$$

On the other hand

$$\begin{aligned}\|f\|_{H^*} &= \sup\{f(u) \mid u \in H,\ \|u\|_H \leq 1\} \\ &\geq f\left(\frac{u_0}{\|u_0\|_H}\right) \\ &= \frac{(u_0, u_0)_H}{\|u_0\|_H} \\ &= \|u_0\|_H. \end{aligned} \tag{1.54}$$

From (1.53) and (1.54)

$$\|f\|_{H^*} = \|u_0\|_H.$$

The proof is complete.

Remark 1.12.9. Similarly as above we may define a Hilbert space H over $\mathbb{C}$, that is, a complex one. In this case the complex inner product $(\cdot,\cdot)_H : H \times H \to \mathbb{C}$ is defined through the following properties:

1. $(u,v)_H = \overline{(v,u)_H}$, $\forall u,v \in H$,
2. $(u+v,w)_H = (u,w)_H + (v,w)_H$, $\forall u,v,w \in H$,
3. $(\alpha u,v)_H = \overline{\alpha}(u,v)_H$, $\forall u,v \in H$, $\alpha \in \mathbb{C}$,
4. $(u,u)_H \geq 0$, $\forall u \in H$, and $(u,u) = 0$, if and only if $u = \theta$.

Observe that in this case we have

$$(u,\alpha v)_H = \alpha(u,v)_H,\ \forall u,v \in H,\ \alpha \in \mathbb{C},$$

where for $\alpha = a + bi \in \mathbb{C}$, we have $\overline{\alpha} = a - bi$. Finally, similar results as those proven above are valid for complex Hilbert spaces.

1.13 Orthonormal Basis

In this section we study separable Hilbert spaces and the related orthonormal bases.

Definition 1.13.1. Let H be a Hilbert space. A set $S \subset H$ is said to be orthonormal if

$$\|u\|_H = 1,$$

and

$$(u,v)_H = 0, \forall u,v \in S, \text{ such that } u \neq v.$$

If S is not properly contained in any other orthonormal set, it is said to be an orthonormal basis for H.

Theorem 1.13.2. *Let H be a Hilbert space and let $\{u_n\}_{n=1}^N$ be an orthonormal set. Then, for all $u \in H$, we have*

$$\|u\|_H^2 = \sum_{n=1}^N |(u,u_n)_H|^2 + \left\| u - \sum_{n=1}^N (u,u_n)_H u_n \right\|_H^2.$$

Proof. Observe that

$$u = \sum_{n=1}^N (u,u_n)_H u_n + \left(u - \sum_{n=1}^N (u,u_n)_H u_n \right).$$

Furthermore, we may easily obtain that

$$\sum_{n=1}^N (u,u_n)_H u_n \text{ and } u - \sum_{n=1}^N (u,u_n)_H u_n$$

are orthogonal vectors so that

$$\begin{aligned}\|u\|_H^2 &= (u,u)_H \\ &= \left\|\sum_{n=1}^N |(u,u_n)_H u_n\right\|_H^2 + \left\|u - \sum_{n=1}^N (u,u_n)_H u_n\right\|_H^2 \\ &= \sum_{n=1}^N |(u,u_n)_H|^2 + \left\|u - \sum_{n=1}^N (u,u_n)_H u_n\right\|_H^2. \end{aligned} \tag{1.55}$$

Corollary 1.13.3 (Bessel Inequality). *Let H be a Hilbert space and let $\{u_n\}_{n=1}^N$ be an orthonormal set. Then, for all $u \in H$, we have*

$$\|u\|_H^2 \geq \sum_{n=1}^N |(u,u_n)_H|^2.$$

Theorem 1.13.4. *Each Hilbert space has an orthonormal basis.*

Proof. Define by C the collection of all orthonormal sets in H. Define an order in C by stating $S_1 \prec S_2$ if $S_1 \subset S_2$. Then, C is partially ordered and obviously nonempty, since

$$v/\|v\|_H \in C, \forall v \in H, v \neq \theta.$$

Now let $\{S_\alpha\}_{\alpha \in L}$ be a linearly ordered subset of C. Clearly, $\cup_{\alpha \in L} S_\alpha$ is an orthonormal set which is an upper bound for $\{S_\alpha\}_{\alpha \in L}$.

Therefore, every linearly ordered subset has an upper bound, so that by Zorn's lemma C has a maximal element, that is, an orthonormal set not properly contained in any other orthonormal set.

This completes the proof.

Theorem 1.13.5. *Let H be a Hilbert space and let $S = \{u_\alpha\}_{\alpha \in L}$ be an orthonormal basis. Then for each $v \in H$ we have*

$$v = \sum_{\alpha \in L} (u_\alpha, v)_H u_\alpha,$$

and

$$\|v\|_H^2 = \sum_{\alpha \in L} |(u_\alpha, v)_H|^2.$$

Proof. Let $L' \subset L$ be a finite subset of L. From Bessel's inequality we have

$$\sum_{\alpha \in L'} |(u_\alpha, v)_H| \leq \|v\|_H^2.$$

From this, we may infer that the set $A_n = \{\alpha \in L \mid |(u_\alpha, v)_H| > 1/n\}$ is finite, so that

$$A = \{\alpha \in L \mid |(u_\alpha, v)_H| > 0\} = \cup_{n=1}^\infty A_n$$

is at most countable.

Thus $(u_\alpha, v)_H \neq 0$ for at most countably many $\alpha' s \in L$, which we order by $\{\alpha_n\}_{n\in\mathbb{N}}$. Since the sequence

$$s_N = \sum_{i=1}^{N} |(u_{\alpha_i}, v)_H|^2,$$

is monotone and bounded, it is converging to some real limit as $N \to \infty$. Define

$$v_n = \sum_{i=1}^{n} (u_{\alpha_i}, v)_H u_{\alpha_i},$$

so that for $n > m$ we have

$$\begin{aligned} \|v_n - v_m\|_H^2 &= \left\| \sum_{i=m+1}^{n} (u_{\alpha_i}, v)_H u_{\alpha_i} \right\|_H^2 \\ &= \sum_{i=m+1}^{n} |(u_{\alpha_i}, v)_H|^2 \\ &= |s_n - s_m|. \end{aligned} \tag{1.56}$$

Hence, $\{v_n\}$ is a Cauchy sequence which converges to some $v' \in H$.

Observe that

$$\begin{aligned} (v - v', u_{\alpha_l})_H &= \lim_{N\to\infty} (v - \sum_{i=1}^{N} (u_{\alpha_i}, v)_H u_{\alpha_i}, u_{\alpha_l})_H \\ &= (v, u_{\alpha_l})_H - (v, u_{\alpha_l})_H \\ &= 0. \end{aligned} \tag{1.57}$$

Also, if $\alpha \neq \alpha_l, \forall l \in \mathbb{N}$, then

$$(v - v', u_\alpha)_H = \lim_{N\to\infty} (v - \sum_{i=1}^{\infty} (u_{\alpha_i}, v)_H u_{\alpha_i}, u_\alpha)_H = 0.$$

Hence

$$v - v' \perp u_\alpha, \ \forall \alpha \in L.$$

If

$$v - v' \neq \theta,$$

then we could obtain an orthonormal set

$$\left\{ u_\alpha, \ \alpha \in L, \frac{v - v'}{\|v - v'\|_H} \right\}$$

which would properly contain the complete orthonormal set

$$\{u_\alpha, \ \alpha \in L\},$$

a contradiction.

Therefore, $v - v' = \theta$, that is,

$$v = \lim_{N \to \infty} \sum_{i=1}^{N} (u_{\alpha_i}, v)_H u_{\alpha_i}.$$

1.13.1 The Gram–Schmidt Orthonormalization

Let H be a Hilbert space and $\{u_n\} \subset H$ be a sequence of linearly independent vectors. Consider the procedure

$$w_1 = u_1, \ v_1 = \frac{w_1}{\|w_1\|_H},$$

$$w_2 = u_2 - (v_1, u_2)_H v_1, \ v_2 = \frac{w_2}{\|w_2\|_H},$$

and inductively,

$$w_n = u_n - \sum_{k=1}^{n-1} (v_k, u_n)_H v_k, \ v_n = \frac{w_n}{\|w_n\|_H}, \forall n \in \mathbb{N}, n > 2.$$

Observe that clearly $\{v_n\}$ is an orthonormal set and for each $m \in \mathbb{N}$, $\{v_k\}_{k=1}^m$ and $\{u_k\}_{k=1}^m$ span the same vector subspace of H.

Such a process of obtaining the orthonormal set $\{v_n\}$ is known as the Gram–Schmidt orthonormalization.

We finish this section with the following theorem.

Theorem 1.13.6. *A Hilbert space H is separable if and only if it has a countable orthonormal basis. If $\dim(H) = N < \infty$, the H is isomorphic to $\mathbb{C}^N$. If $\dim(H) = +\infty$, then H is isomorphic to l^2, where*

$$l^2 = \left\{ \{y_n\} \mid y_n \in \mathbb{C}, \forall n \in \mathbb{N} \ und \ \sum_{n=1}^{\infty} |y_n|^2 < +\infty \right\}.$$

Proof. Suppose H is separable and let $\{u_n\}$ be a countable dense set in H. To obtain an orthonormal basis it suffices to apply the Gram–Schmidt orthonormalization procedure to the greatest linearly independent subset of $\{u_n\}$.

Conversely, if $B = \{v_n\}$ is an orthonormal basis for H, the set of all finite linear combinations of elements of B with rational coefficients are dense in H, so that H is separable.

Moreover, if $dim(H) = +\infty$, consider the isomorphism $F : H \to l^2$ given by

$$F(u) = \{(u_n, u)_H\}_{n\in\mathbb{N}}.$$

Finally, if $dim(H) = N < +\infty$, consider the isomorphism $F : H \to \mathbb{C}^N$ given by

$$F(u) = \{(u_n, u)_H\}_{n=1}^N.$$

The proof is complete.

Chapter 2
The Hahn–Banach Theorems and Weak Topologies

2.1 Introduction

The notion of weak topologies and weak convergence is fundamental in the modern variational analysis. Many important problems are non-convex and have no minimizers in the classical sense. However, the minimizing sequences in reflexive spaces may be weakly convergent, and it is important to evaluate the average behavior of such sequences in many practical applications. Finally, we emphasize the main reference for this chapter is Brezis [16], where more details may be found.

2.2 The Hahn–Banach Theorem

In this chapter U denotes a Banach space, unless otherwise indicated. We start this section by stating and proving the Hahn–Banach theorem for real vector spaces, which is sufficient for our purposes.

Theorem 2.2.1 (The Hahn–Banach Theorem). *Consider a functional $p : U \to \mathbb{R}$ satisfying*

$$p(\lambda u) = \lambda p(u), \forall u \in U, \lambda > 0, \tag{2.1}$$

$$p(u+v) \leq p(u) + p(v), \forall u, v \in U. \tag{2.2}$$

Let $V \subset U$ be a vector subspace and let $g : V \to \mathbb{R}$ be a linear functional such that

$$g(u) \leq p(u), \forall u \in V. \tag{2.3}$$

Then there exists a linear functional $f : U \to \mathbb{R}$ such that

$$g(u) = f(u), \forall u \in V, \tag{2.4}$$

F. Botelho, *Functional Analysis and Applied Optimization in Banach Spaces: Applications to Non-Convex Variational Models*, DOI 10.1007/978-3-319-06074-3_2,

and

$$f(u) \leq p(u), \forall u \in U. \tag{2.5}$$

Proof. Pick $z \in U - V$. Denote by $\tilde{V}$ the space spanned by V and z, that is,

$$\tilde{V} = \{v + \alpha z \mid v \in V \text{ and } \alpha \in \mathbb{R}\}. \tag{2.6}$$

We may define an extension of g to $\tilde{V}$, denoted by $\tilde{g}$, as

$$\tilde{g}(\alpha z + v) = \alpha \tilde{g}(z) + g(v), \tag{2.7}$$

where $\tilde{g}(z)$ will be appropriately defined. Suppose given $v_1, v_2 \in V$, $\alpha > 0$, $\beta > 0$. Then

$$\begin{aligned}
\beta g(v_1) + \alpha g(v_2) &= g(\beta v_1 + \alpha v_2) \\
&= (\alpha + \beta) g(\frac{\beta}{\alpha + \beta} v_1 + \frac{\alpha}{\alpha + \beta} v_2) \\
&\leq (\alpha + \beta) p(\frac{\beta}{\alpha + \beta}(v_1 - \alpha z) + \frac{\alpha}{\alpha + \beta}(v_2 + \beta z)) \\
&\leq \beta p(v_1 - \alpha z) + \alpha p(v_2 + \beta z)
\end{aligned} \tag{2.8}$$

and therefore

$$\frac{1}{\alpha}[-p(v_1 - \alpha z) + g(v_1)] \leq \frac{1}{\beta}[p(v_2 + \beta z) - g(v_2)], \quad \forall v_1, v_2 \in V,\ \alpha, \beta > 0. \tag{2.9}$$

Thus, there exists $a \in \mathbb{R}$ such that

$$\sup_{v \in V, \alpha > 0} [\frac{1}{\alpha}(-p(v - \alpha z) + g(v))] \leq a \leq \inf_{v \in V, \alpha > 0} [\frac{1}{\alpha}(p(v + \alpha z) - g(v))]. \tag{2.10}$$

If we define $\tilde{g}(z) = a$, we obtain $\tilde{g}(u) \leq p(u), \forall u \in \tilde{V}$. Define by $\mathscr{E}$ the set of extensions e of g, which satisfy $e(u) \leq p(u)$ on the subspace where e is defined. We define a partial order in $\mathscr{E}$ by setting $e_1 \prec e_2$ if e_2 is defined in a larger set than e_1 and $e_1 = e_2$ where both are defined. Let $\{e_\alpha\}_{\alpha \in A}$ be a linearly ordered subset of $\mathscr{E}$. Let V_α be the subspace on which e_α is defined. Define e on $\cup_{\alpha \in A} V_\alpha$ by setting $e(u) = e_\alpha$ on V_α. Clearly $e_\alpha \prec e$ so each linearly ordered set of $\mathscr{E}$ has an upper bound. By Zorn's lemma, $\mathscr{E}$ has a maximal element f defined on some set $\tilde{U}$ such that $f(u) \leq p(u), \forall u \in \tilde{U}$. We can conclude that $\tilde{U} = U$; otherwise, if there was an $z_1 \in U - \tilde{U}$, as above, we could have a new extension f_1 to the subspace spanned by z_1 and $\tilde{U}$, contradicting the maximality of f.

Definition 2.2.2 (Topological Dual Space). For a Banach space U, we define its topological dual space as the set of all linear continuous functionals defined on U. We suppose that such dual space of U may be identified with a space denoted by U^* through a bilinear form $\langle \cdot, \cdot \rangle_U : U \times U^* \to \mathbb{R}$ (here we are referring to the standard representations of dual spaces concerning Lebesgue and Sobolev spaces). That is, given $f : U \to \mathbb{R}$ linear continuous functional, there exists $u^* \in U^*$ such that

$$f(u) = \langle u, u^* \rangle_U, \forall u \in U. \tag{2.11}$$

The norm of f, denoted by $\|f\|_{U^*}$, is defined as

$$\|f\|_{U^*} = \sup_{u \in U}\{|\langle u, u^* \rangle_U| \mid \|u\|_U \leq 1\}. \tag{2.12}$$

Corollary 2.2.3. *Let $V \subset U$ be a vector subspace of U and let $g : V \to \mathbb{R}$ be a linear continuous functional of norm*

$$\|g\|_{V^*} = \sup_{u \in V}\{|g(u)| \mid \|u\|_V \leq 1\}. \tag{2.13}$$

Then, there exists an u^* in U^* such that

$$\langle u, u^* \rangle_U = g(u), \forall u \in V, \tag{2.14}$$

and

$$\|u^*\|_{U^*} = \|g\|_{V^*}. \tag{2.15}$$

Proof. Apply Theorem 2.2.1 with $p(u) = \|g\|_{V^*}\|u\|_V$.

Corollary 2.2.4. *Given $u_0 \in U$ there exists $u_0^* \in U^*$ such that*

$$\|u_0^*\|_{U^*} = \|u_0\|_U \text{ and } \langle u_0, u_0^* \rangle_U = \|u_0\|_U^2. \tag{2.16}$$

Proof. Apply Corollary 2.2.3 with $V = \{\alpha u_0 \mid \alpha \in \mathbb{R}\}$ and $g(tu_0) = t\|u_0\|_U^2$ so that $\|g\|_{V^*} = \|u_0\|_U$.

Corollary 2.2.5. *Given $u \in U$ we have*

$$\|u\|_U = \sup_{u^* \in U^*}\{|\langle u, u^* \rangle_U| \mid \|u^*\|_{U^*} \leq 1\}. \tag{2.17}$$

Proof. Suppose $u \neq \theta$. Since

$$|\langle u, u^* \rangle_U| \leq \|u\|_U \|u^*\|_{U^*}, \forall u \in U, u^* \in U^*$$

we have

$$\sup_{u^* \in U^*}\{|\langle u, u^* \rangle_U| \mid \|u^*\|_{U^*} \leq 1\} \leq \|u\|_U. \tag{2.18}$$

However, from last corollary, we have that there exists $u_0^* \in U^*$ such that $\|u_0^*\|_{U^*} = \|u\|_U$ and $\langle u, u_0^* \rangle_U = \|u\|_U^2$. Define $u_1^* = \|u\|_U^{-1} u_0^*$. Then $\|u_1^*\|_U = 1$ and $\langle u, u_1^* \rangle_U = \|u\|_U$.

Definition 2.2.6 (Affine Hyperplane). Let U be a Banach space. An affine hyperplane H is a set of the form

$$H = \{u \in U \mid \langle u, u^* \rangle_U = \alpha\} \tag{2.19}$$

for some $u^* \in U^*$ and $\alpha \in \mathbb{R}$.

Proposition 2.2.7. *A hyperplane H defined as above is closed.*

Proof. The result follows from the continuity of $\langle u, u^* \rangle_U$ as a functional defined in U.

Definition 2.2.8 (Separation). Given $A, B \subset U$ we say that a hyperplane H, defined as above, separates A and B if

$$\langle u, u^* \rangle_U \leq \alpha, \forall u \in A, \text{ and } \langle u, u^* \rangle_U \geq \alpha, \forall u \in B. \tag{2.20}$$

We say that H separates A and B strictly if there exists $\varepsilon > 0$ such that

$$\langle u, u^* \rangle_U \leq \alpha - \varepsilon, \forall u \in A, \text{ and } \langle u, u^* \rangle_U \geq \alpha + \varepsilon, \forall u \in B, \tag{2.21}$$

Theorem 2.2.9 (Hahn–Banach Theorem, Geometric Form). *Consider $A, B \subset U$ two convex disjoint nonempty sets, where A is open. Then there exists a closed hyperplane that separates A and B.*

We need the following lemma.

Lemma 2.2.10. *Consider $C \subset U$ a convex open set such that $\theta \in C$. Given $u \in U$, define*

$$p(u) = \inf\{\alpha > 0, \ \alpha^{-1} u \in C\}. \tag{2.22}$$

Thus, p is such that there exists $M \in \mathbb{R}^+$ satisfying

$$0 \leq p(u) \leq M \|u\|_U, \forall u \in U, \tag{2.23}$$

and

$$C = \{u \in U \mid p(u) < 1\}. \tag{2.24}$$

Also

$$p(u+v) \leq p(u) + p(v), \forall u, v \in U.$$

Proof. Let $r > 0$ be such that $B(\theta, r) \subset C$; thus

$$p(u) \leq \frac{\|u\|_U}{r}, \forall u \in U \tag{2.25}$$

which proves (2.23). Now suppose $u \in C$. Since C is open, $(1+\varepsilon)u \in C$ for ε is sufficiently small. Therefore $p(u) \leq \frac{1}{1+\varepsilon} < 1$. Conversely, if $p(u) < 1$, there exists $0 < \alpha < 1$ such that $\alpha^{-1} u \in C$ and therefore, since C is convex, $u = \alpha(\alpha^{-1} u) + (1 - \alpha)\theta \in C$.

Also, let $u, v \in C$ and $\varepsilon > 0$. Thus $\frac{u}{p(u)+\varepsilon} \in C$ and $\frac{v}{p(v)+\varepsilon} \in C$ so that $\frac{tu}{p(u)+\varepsilon} + \frac{(1-t)v}{p(v)+\varepsilon} \in C, \forall t \in [0,1]$. Particularly for $t = \frac{p(u)+\varepsilon}{p(u)+p(v)+2\varepsilon}$ we obtain $\frac{u+v}{p(u)+p(v)+2\varepsilon} \in C$, which means $p(u+v) \leq p(u)+p(v)+2\varepsilon, \forall \varepsilon > 0$

Lemma 2.2.11. *Consider $C \subset U$ a convex open set and let $u_0 \in U$ be a vector not in C. Then there exists $u^* \in U^*$ such that $\langle u, u^* \rangle_U < \langle u_0, u^* \rangle_U, \forall u \in C$*

Proof. By a translation, we may assume $\theta \in C$. Consider the functional p as in the last lemma. Define $V = \{\alpha u_0 \mid \alpha \in \mathbb{R}\}$. Define g on V by

$$g(tu_0) = t,\ t \in \mathbb{R}. \tag{2.26}$$

We have that $g(u) \leq p(u), \forall u \in V$. From the Hahn–Banach theorem, there exists a linear functional f on U which extends g such that

$$f(u) \leq p(u) \leq M\|u\|_U. \tag{2.27}$$

Here we have used Lemma 2.2.10. In particular, $f(u_0) = 1$ and (also from the last lemma) $f(u) < 1, \forall u \in C$. The existence of u^* satisfying the theorem follows from the continuity of f indicated in (2.27).

Proof of Theorem 2.2.9. Define $C = A + (-B)$ so that C is convex and $\theta \notin C$. From Lemma 2.2.11, there exists $u^* \in U^*$ such that $\langle w, u^* \rangle_U < 0, \forall w \in C$, which means

$$\langle u, u^* \rangle_U < \langle v, u^* \rangle_U, \forall u \in A,\ \ v \in B. \tag{2.28}$$

Thus, there exists $\alpha \in \mathbb{R}$ such that

$$\sup_{u \in A} \langle u, u^* \rangle_U \leq \alpha \leq \inf_{v \in B} \langle v, u^* \rangle_U, \tag{2.29}$$

which completes the proof.

Theorem 2.2.12 (Hahn–Banach Theorem, Second Geometric Form). *Consider $A, B \subset U$ two convex disjoint nonempty sets. Suppose A is closed and B is compact. Then there exists a hyperplane which separates A and B strictly.*

Proof. There exists $\varepsilon > 0$ sufficiently small such that $A_\varepsilon = A + B(0,\varepsilon)$ and $B_\varepsilon = B + B(0,\varepsilon)$ are convex disjoint sets. From Theorem 2.2.9, there exists $u^* \in U^*$ such that $u^* \neq \theta$ and

$$\langle u + \varepsilon w_1, u^* \rangle_U < \langle u + \varepsilon w_2, u^* \rangle_U, \forall u \in A,\ v \in B,\ w_1, w_2 \in B(0,1). \tag{2.30}$$

Thus, there exists $\alpha \in \mathbb{R}$ such that

$$\langle u, u^* \rangle_U + \varepsilon \|u^*\|_{U^*} \leq \alpha \leq \langle v, u^* \rangle_U - \varepsilon \|u^*\|_{U^*}, \forall u \in A,\ \ v \in B. \tag{2.31}$$

Corollary 2.2.13. *Suppose $V \subset U$ is a vector subspace such that $\bar{V} \neq U$. Then there exists $u^* \in U^*$ such that $u^* \neq \theta$ and*

$$\langle u, u^* \rangle_U = 0, \forall u \in V. \tag{2.32}$$

Proof. Consider $u_0 \in U$ such that $u_0 \notin \overline{V}$. Applying Theorem 2.2.9 to $A = \overline{V}$ and $B = \{u_0\}$ we obtain $u^* \in U^*$ and $\alpha \in \mathbb{R}$ such that $u^* \neq \theta$ and

$$\langle u, u^* \rangle_U < \alpha < \langle u_0, u^* \rangle_U, \forall u \in V. \tag{2.33}$$

Since V is a subspace we must have $\langle u, u^* \rangle_U = 0, \forall u \in V$.

2.3 Weak Topologies

Definition 2.3.1 (Weak Neighborhoods and Weak Topologies). For the topological space U and $u_0 \in U$, we define a weak neighborhood of u_0, denoted by $\mathscr{V}_w$ as

$$\mathscr{V}_w = \{u \in U \mid |\langle u - u_0, u_i^* \rangle_U| < \varepsilon, \forall i \in \{1, \ldots, m\}\}, \tag{2.34}$$

for some $m \in \mathbb{N}$, $\varepsilon > 0$, and $u_i^* \in U^*$, $\forall i \in \{1, \ldots, m\}$. Also, we define the weak topology for U, denoted by $\sigma(U, U^*)$, as the set of arbitrary unions and finite intersections of weak neighborhoods in U.

Proposition 2.3.2. *Consider Z a topological vector space and ψ a function of Z into U. Then ψ is continuous as U is endowed with the weak topology, if and only if $u^* \circ \psi$ is continuous, for all $u^* \in U^*$.*

Proof. It is clear that if ψ is continuous with U endowed with the weak topology, then $u^* \circ \psi$ is continuous for all $u^* \in U^*$. Conversely, consider $\mathscr{U}$ a weakly open set in U. We have to show that $\psi^{-1}(\mathscr{U})$ is open in Z. But observe that $\mathscr{U} = \cup_{\lambda \in L} \mathscr{V}_\lambda$, where each $\mathscr{V}_\lambda$ is a weak neighborhood. Thus $\psi^{-1}(\mathscr{U}) = \cup_{\lambda \in L} \psi^{-1}(\mathscr{V}_\lambda)$. The result follows considering that $u^* \circ \psi$ is continuous for all $u^* \in U^*$, so that $\psi^{-1}(\mathscr{V}_\lambda)$ is open, for all $\lambda \in L$.

Proposition 2.3.3. *A Banach space U is Hausdorff as endowed with the weak topology $\sigma(U, U^*)$.*

Proof. Pick $u_1, u_2 \in U$ such that $u_1 \neq u_2$. From the Hahn–Banach theorem, second geometric form, there exists a hyperplane separating $\{u_1\}$ and $\{u_2\}$. That is, there exist $u^* \in U^*$ and $\alpha \in \mathbb{R}$ such that

$$\langle u_1, u^* \rangle_U < \alpha < \langle u_2, u^* \rangle_U. \tag{2.35}$$

Defining

$$\mathscr{V}_{w1} = \{u \in U \mid |\langle u - u_1, u^* \rangle| < \alpha - \langle u_1, u^* \rangle_U\}, \tag{2.36}$$

and

$$\mathscr{V}_{w2} = \{u \in U \mid |\langle u - u_2, u^* \rangle_U| < \langle u_2, u^* \rangle_U - \alpha\}, \tag{2.37}$$

we obtain $u_1 \in \mathscr{V}_{w1}$, $u_2 \in \mathscr{V}_{w2}$ and $\mathscr{V}_{w1} \cap \mathscr{V}_{w2} = \emptyset$.

Remark 2.3.4. If $\{u_n\} \in U$ is such that u_n converges to u in $\sigma(U, U^*)$, then we write $u_n \rightharpoonup u$.

Proposition 2.3.5. *Let U be a Banach space. Considering $\{u_n\} \subset U$ we have*

1. *$u_n \rightharpoonup u$, for $\sigma(U, U^*) \Leftrightarrow \langle u_n, u^* \rangle_U \to \langle u, u^* \rangle_U, \forall u^* \in U^*$,*
2. *if $u_n \to u$ strongly (in norm), then $u_n \rightharpoonup u$ weakly,*
3. *if $u_n \rightharpoonup u$ weakly, then $\{\|u_n\|_U\}$ is bounded and $\|u\|_U \leq \liminf_{n\to\infty} \|u_n\|_U$,*
4. *if $u_n \rightharpoonup u$ weakly and $u_n^* \to u^*$ strongly in U^*, then $\langle u_n, u_n^* \rangle_U \to \langle u, u^* \rangle_U$.*

Proof.

1. The result follows directly from the definition of topology $\sigma(U, U^*)$.
2. This follows from the inequality

$$|\langle u_n, u^* \rangle_U - \langle u, u^* \rangle_U| \leq \|u^*\|_{U^*} \|u_n - u\|_U. \tag{2.38}$$

3. Since for every $u^* \in U^*$ the sequence $\{\langle u_n, u^* \rangle_U\}$ is bounded, from the uniform boundedness principle, we have that there exists $M > 0$ such that $\|u_n\|_U \leq M, \forall n \in \mathbb{N}$. Furthermore, for $u^* \in U^*$, we have

$$|\langle u_n, u^* \rangle_U| \leq \|u^*\|_{U^*} \|u_n\|_U, \tag{2.39}$$

and taking the limit, we obtain

$$|\langle u, u^* \rangle_U| \leq \liminf_{n\to\infty} \|u^*\|_{U^*} \|u_n\|_U. \tag{2.40}$$

Thus

$$\|u\|_U = \sup_{\|u\|_{U^*} \leq 1} |\langle u, u^* \rangle_U| \leq \liminf_{n\to\infty} \|u_n\|_U. \tag{2.41}$$

4. Just observe that

$$\begin{aligned} |\langle u_n, u_n^* \rangle_U - \langle u, u^* \rangle_U| &\leq |\langle u_n, u_n^* - u^* \rangle_U| \\ &\quad + |\langle u - u_n, u^* \rangle_U| \\ &\leq \|u_n^* - u^*\|_{U^*} \|u_n\|_U \\ &\quad + |\langle u_n - u, u^* \rangle_U| \\ &\leq M \|u_n^* - u^*\|_{U^*} \\ &\quad + |\langle u_n - u, u^* \rangle_U|. \end{aligned} \tag{2.42}$$

Theorem 2.3.6. *Consider $A \subset U$ a convex set. Thus A is weakly closed if and only if it is strongly closed.*

Proof. Suppose A is strongly closed. Consider $u_0 \notin A$. By the Hahn–Banach theorem there exists a closed hyperplane which separates u_0 and A strictly. Therefore there exists $\alpha \in \mathbb{R}$ and $u^* \in U^*$ such that

$$\langle u_0, u^* \rangle_U < \alpha < \langle v, u^* \rangle_U, \forall v \in A. \tag{2.43}$$

Define

$$\mathcal{V} = \{u \in U \mid \langle u, u^* \rangle_U < \alpha\}, \tag{2.44}$$

so that $u_0 \in \mathcal{V}$, $\mathcal{V} \subset U - A$. Since $\mathcal{V}$ is open for $\sigma(U, U^*)$ we have that $U - A$ is weakly open; hence A is weakly closed. The converse is obvious.

2.4 The Weak-Star Topology

Definition 2.4.1 (Reflexive Spaces). Let U be a Banach space. We say that U is reflexive if the canonical injection $J : U \to U^{**}$ defined by

$$\langle u, u^* \rangle_U = \langle u^*, J(u) \rangle_{U^*}, \forall u \in U, \ \ u^* \in U^*, \tag{2.45}$$

is onto.

The weak topology for U^* is denoted by $\sigma(U^*, U^{**})$. By analogy, we can define the topology $\sigma(U^*, U)$, which is called the weak-star topology. A standard neighborhood of $u_0^* \in U^*$ for the weak-star topology, which we denoted by $\mathcal{V}_{w^*}$, is given by

$$\mathcal{V}_{w^*} = \{u^* \in U^* \mid |\langle u_i, u^* - u_0^* \rangle_U| < \varepsilon, \forall i \in \{1, \ldots, m\}\} \tag{2.46}$$

for some $\varepsilon > 0$, $m \in \mathbb{N}$, $u_i \in U, \forall i \in \{1, \ldots, m\}$. It is clear that the weak topology for U^* and the weak-star topology coincide if U is reflexive.

Proposition 2.4.2. *Let U be a Banach space. U^* as endowed with the weak-star topology is a Hausdorff space.*

Proof. The proof is similar to that of Proposition 2.3.3.

2.5 Weak-Star Compactness

We start with an important theorem about weak-star compactness.

Theorem 2.5.1 (Banach–Alaoglu Theorem). *The set $B_{U^*} = \{f \in U^* \mid \|f\|_{U^*} \leq 1\}$ is compact for the topology $\sigma(U^*, U)$ (the weak-star topology).*

Proof. For each $u \in U$, we will associate a real number ω_u and denote $\omega = \prod_{u \in U} \omega_u$. We have that $\omega \in \mathbb{R}^U$ and let us consider the projections $P_u : \mathbb{R}^U \to \mathbb{R}$, where $P_u(\omega) = \omega_u$. Consider the weakest topology σ for which the functions P_u $(u \in U)$ are continuous. For U^*, with the topology $\sigma(U^*, U)$, define $\phi : U^* \to \mathbb{R}^U$ by

$$\phi(u^*) = \prod_{u \in U} \langle u, u^* \rangle_U, \forall u^* \in U^*. \tag{2.47}$$

Since for each fixed u the mapping $u^* \to \langle u, u^* \rangle_U$ is weakly star continuous, we see that ϕ is σ continuous, since weak-star convergence and convergence in σ are equivalent in U^*. To prove that ϕ^{-1} is continuous, from Proposition 2.3.2, it suffices to show that the function $\omega \to \langle u, \phi^{-1}(\omega) \rangle_U$ is continuous on $\phi(U^*)$. This is true because $\langle u, \phi^{-1}(\omega) \rangle_U = \omega_u$ on $\phi(U^*)$. On the other hand, it is also clear that $\phi(B_{U^*}) = K$, where

$$\begin{aligned} K = \{ \omega \in \mathbb{R}^U \mid |\omega_u| \leq \|u\|_U, \\ \omega_{u+v} = \omega_u + \omega_v, \; \omega_{\lambda u} = \lambda \omega_u, \forall u, v \in U, \; \lambda \in \mathbb{R} \}. \end{aligned} \tag{2.48}$$

To finish the proof, it is sufficient, from the continuity of ϕ^{-1}, to show that K is compact in $\mathbb{R}^U$, concerning the topology σ. Observe that $K = K_1 \cap K_2$, where

$$K_1 = \{ \omega \in \mathbb{R}^U \mid |\omega_u| \leq \|u\|_U, \forall u \in U \}, \tag{2.49}$$

and

$$K_2 = \{ \omega \in \mathbb{R}^U \mid \omega_{u+v} = \omega_u + \omega_v, \; \omega_{\lambda u} = \lambda \omega_u, \forall u, v \in U, \; \lambda \in \mathbb{R} \}. \tag{2.50}$$

The set $K_3 \equiv \prod_{u \in U} [-\|u\|_U, \|u\|_U]$ is compact as a cartesian product of compact intervals. Since $K_1 \subset K_3$ and K_1 is closed, we have that K_1 is compact (for the topology in question). On the other hand, K_2 is closed, because defining the closed sets $A_{u,v}$ and $B_{\lambda,u}$ as

$$A_{u,v} = \{ \omega \in \mathbb{R}^U \mid \omega_{u+v} - \omega_u - \omega_v = 0 \}, \tag{2.51}$$

and

$$B_{\lambda,u} = \{ \omega \in \mathbb{R}^U | \omega_{\lambda u} - \lambda \omega_u = 0 \} \tag{2.52}$$

we may write

$$K_2 = (\cap_{u,v \in U} A_{u,v}) \cap (\cap_{(\lambda,u) \in \mathbb{R} \times U} B_{\lambda,u}). \tag{2.53}$$

We recall that the K_2 is closed because arbitrary intersections of closed sets are closed. Finally, we have that $K_1 \cap K_2$ is compact, which completes the proof.

Theorem 2.5.2 (Kakutani). *Let U be a Banach space. Then U is reflexive if and only if*

$$B_U = \{u \in U \mid \|u\|_U \leq 1\} \tag{2.54}$$

is compact for the weak topology $\sigma(U, U^*)$.

Proof. Suppose U is reflexive, and then $J(B_U) = B_{U^{**}}$. From the last theorem $B_{U^{**}}$ is compact for the topology $\sigma(U^{**}, U^*)$. Therefore it suffices to verify that $J^{-1} : U^{**} \to U$ is continuous from U^{**} with the topology $\sigma(U^{**}, U^*)$ to U, with the topology $\sigma(U, U^*)$.

From Proposition 2.3.2 it is sufficient to show that the function $u \mapsto \langle J^{-1}u, f\rangle_U$ is continuous for the topology $\sigma(U^{**}, U^*)$, for each $f \in U^*$. Since $\langle J^{-1}u, f\rangle_U = \langle f, u\rangle_{U^*}$ we have completed the first part of the proof. For the second we need two lemmas.

Lemma 2.5.3 (Helly). *Let* U *be a Banach space,* $f_1, \ldots, f_n \in U^*$*, and* $\alpha_1, \ldots, \alpha_n \in \mathbb{R}$*, and then* 1 *and* 2 *are equivalent, where:*

1.

$$\text{Given } \varepsilon > 0, \text{ there exists } u_\varepsilon \in U \text{ such that } \|u_\varepsilon\|_U \leq 1 \text{ and}$$
$$|\langle u_\varepsilon, f_i\rangle_U - \alpha_i| < \varepsilon, \forall i \in \{1, \ldots, n\}.$$

2.

$$\left|\sum_{i=1}^n \beta_i \alpha_i\right| \leq \left\|\sum_{i=1}^n \beta_i f_i\right\|_{U^*}, \forall \beta_1, \ldots, \beta_n \in \mathbb{R}. \tag{2.55}$$

Proof. $1 \Rightarrow 2$: Fix $\beta_1, \ldots, \beta_n \in \mathbb{R}$, $\varepsilon > 0$ and define $S = \sum_{i=1}^n |\beta_i|$. From 1, we have

$$\left|\sum_{i=1}^n \beta_i \langle u_\varepsilon, f_i\rangle_U - \sum_{i=1}^n \beta_i \alpha_i\right| < \varepsilon S \tag{2.56}$$

and therefore

$$\left|\sum_{i=1}^n \beta_i \alpha_i\right| - \left|\sum_{i=1}^n \beta_i \langle u_\varepsilon, f_i\rangle_U\right| < \varepsilon S \tag{2.57}$$

or

$$\left|\sum_{i=1}^n \beta_i \alpha_i\right| < \left\|\sum_{i=1}^n \beta_i f_i\right\|_{U^*} \|u_\varepsilon\|_U + \varepsilon S \leq \left\|\sum_{i=1}^n \beta_i f_i\right\|_{U^*} + \varepsilon S \tag{2.58}$$

so that

$$\left|\sum_{i=1}^n \beta_i \alpha_i\right| \leq \left\|\sum_{i=1}^n \beta_i f_i\right\|_{U^*} \tag{2.59}$$

since ε is arbitrary. Now let us show that $2 \Rightarrow 1$. Define $\alpha = (\alpha_1, \ldots, \alpha_n) \in \mathbb{R}^n$ and consider the function $\varphi(u) = (\langle u, f_1\rangle_U, \ldots, \langle u, f_n\rangle_U)$. Item 1 implies that α belongs to the closure of $\varphi(B_U)$. Let us suppose that α does not belong to the closure of $\varphi(B_U)$ and obtain a contradiction. Thus we can separate α and the closure of $\varphi(B_U)$ strictly, that is, there exists $\beta = (\beta_1, \ldots, \beta_n) \in \mathbb{R}^n$ and $\gamma \in \mathbb{R}$ such that

$$\varphi(u) \cdot \beta < \gamma < \alpha \cdot \beta, \forall u \in B_U \tag{2.60}$$

Taking the supremum in u we contradict 2.

Also we need the lemma.

Lemma 2.5.4. *Let U be a Banach space. Then $J(B_U)$ is dense in $B_{U^{**}}$ for the topology $\sigma(U^{**}, U^*)$.*

Proof. Let $u^{**} \in B_{U^{**}}$ and consider $\mathcal{V}_{u^{**}}$ a neighborhood of u^{**} for the topology $\sigma(U^{**}, U^*)$. It suffices to show that $J(B_U) \cap \mathcal{V}_{u^{**}} \neq \emptyset$. As $\mathcal{V}_{u^{**}}$ is a weak neighborhood, there exists $f_1, \ldots, f_n \in U^*$ and $\varepsilon > 0$ such that

$$\mathcal{V}_{u^{**}} = \{\eta \in U^{**} \mid \langle f_i, \eta - u^{**} \rangle_{U^*} | < \varepsilon, \forall i \in \{1, \ldots, n\}\}. \tag{2.61}$$

Define $\alpha_i = \langle f_i, u^{**} \rangle_{U^*}$ and thus for any given $\beta_1, \ldots, \beta_n \in \mathbb{R}$ we have

$$\left| \sum_{i=1}^{n} \beta_i \alpha_i \right| = \left| \left\langle \sum_{i=1}^{n} \beta_i f_i, u^{**} \right\rangle_{U^*} \right| \leq \left\| \sum_{i=1}^{n} \beta_i f_i \right\|_{U^*}, \tag{2.62}$$

so that from Helly lemma, there exists $u_\varepsilon \in U$ such that $\|u_\varepsilon\|_U \leq 1$ and

$$|\langle u_\varepsilon, f_i \rangle_U - \alpha_i| < \varepsilon, \forall i \in \{1, \ldots, n\} \tag{2.63}$$

or,

$$|\langle f_i, J(u_\varepsilon) - u^{**} \rangle_{U^*}| < \varepsilon, \forall i \in \{1, \ldots, n\} \tag{2.64}$$

and hence

$$J(u_\varepsilon) \in \mathcal{V}_{u^{**}}. \tag{2.65}$$

Now we will complete the proof of Kakutani theorem. Suppose B_U is weakly compact (i.e., compact for the topology $\sigma(U, U^*)$). Observe that $J : U \to U^{**}$ is weakly continuous, that is, it is continuous with U endowed with the topology $\sigma(U, U^*)$ and U^{**} endowed with the topology $\sigma(U^{**}, U^*)$. Thus as B_U is weakly compact, we have that $J(B_U)$ is compact for the topology $\sigma(U^{**}, U^*)$. From the last lemma, $J(B_U)$ is dense $B_{U^{**}}$ for the topology $\sigma(U^{**}, U^*)$. Hence $J(B_U) = B_{U^{**}}$, or $J(U) = U^{**}$, which completes the proof.

Proposition 2.5.5. *Let U be a reflexive Banach space. Let $K \subset U$ be a convex closed bounded set. Then K is weakly compact.*

Proof. From Theorem 2.3.6, K is weakly closed (closed for the topology $\sigma(U, U^*)$). Since K is bounded, there exists $\alpha \in \mathbb{R}^+$ such that $K \subset \alpha B_U$. Since K is weakly closed and $K = K \cap \alpha B_U$, we have that it is weakly compact.

Proposition 2.5.6. *Let U be a reflexive Banach space and $M \subset U$ a closed subspace. Then M with the norm induced by U is reflexive.*

Proof. We can identify two weak topologies in M:

$$\sigma(M,M^*) \text{ and the trace of } \sigma(U,U^*). \tag{2.66}$$

It can be easily verified that these two topologies coincide (through restrictions and extensions of linear forms). From Theorem 2.5.2, it suffices to show that B_M is compact for the topology $\sigma(M,M^*)$. But B_U is compact for $\sigma(U,U^*)$ and $M \subset U$ is closed (strongly) and convex so that it is weakly closed; thus, from the last proposition, B_M is compact for the topology $\sigma(U,U^*)$, and therefore it is compact for $\sigma(M,M^*)$.

2.6 Separable Sets

Definition 2.6.1 (Separable Spaces). A metric space U is said to be separable if there exists a set $K \subset U$ such that K is countable and dense in U.

The next proposition is proved in [16].

Proposition 2.6.2. *Let U be a separable metric space. If $V \subset U$, then V is separable.*

Theorem 2.6.3. *Let U be a Banach space such that U^* is separable. Then U is separable.*

Proof. Consider $\{u_n^*\}$ a countable dense set in U^*. Observe that

$$\|u_n^*\|_{U^*} = \sup\{|\langle u_n^*, u\rangle_U| \mid u \in U \text{ and } \|u\|_U = 1\} \tag{2.67}$$

so that for each $n \in \mathbb{N}$, there exists $u_n \in U$ such that $\|u_n\|_U = 1$ and $\langle u_n^*, u_n\rangle_U \geq \frac{1}{2}\|u_n^*\|_{U^*}$.

Define U_0 as the vector space on $\mathbb{Q}$ spanned by $\{u_n\}$ and U_1 as the vector space on $\mathbb{R}$ spanned by $\{u_n\}$. It is clear that U_0 is dense in U_1 and we will show that U_1 is dense in U, so that U_0 is a dense set in U. Suppose u^* is such that $\langle u, u^*\rangle_U = 0, \forall u \in U_1$. Since $\{u_n^*\}$ is dense in U^*, given $\varepsilon > 0$, there exists $n \in \mathbb{N}$ such that $\|u_n^* - u^*\|_{U^*} < \varepsilon$, so that

$$\begin{aligned}\frac{1}{2}\|u_n^*\|_{U^*} \leq \langle u_n, u_n^*\rangle_U &= \langle u_n, u_n^* - u^*\rangle_U + \langle u_n, u^*\rangle_U \\ &\leq \|u_n^* - u^*\|_{U^*}\|u_n\|_U + 0 < \varepsilon \end{aligned} \tag{2.68}$$

or

$$\|u^*\|_{U^*} \leq \|u_n^* - u^*\|_{U^*} + \|u_n^*\|_{U^*} < \varepsilon + 2\varepsilon = 3\varepsilon. \tag{2.69}$$

Therefore, since ε is arbitrary, $\|u^*\|_{U^*} = 0$, that is, $u^* = \theta$. By Corollary 2.2.13 this completes the proof.

Proposition 2.6.4. *U is reflexive if and only if U^* is reflexive.*

Proof. Suppose U is reflexive; as B_{U^*} is compact for $\sigma(U^*,U)$ and $\sigma(U^*,U) = \sigma(U^*,U^{**})$, we have that B_{U^*} is compact for $\sigma(U^*,U^{**})$, which means that U^* is reflexive.

Suppose U^* is reflexive; from above U^{**} is reflexive. Since $J(U)$ is a closed subspace of U^{**}, from Proposition 2.5.6, J(U) is reflexive. Thus, U is reflexive, since J is an isometry.

Proposition 2.6.5. *Let U be a Banach space. Then U is reflexive and separable if and only if U^* is reflexive and separable.*

Our final result in this section refers to the metrizability of B_{U^*}.

Theorem 2.6.6. *Let U be separable Banach space. Under such hypotheses B_{U^*} is metrizable with respect to the weak-star topology $\sigma(U^*,U)$. Conversely, if B_{U^*} is mertizable in $\sigma(U^*,U)$, then U is separable.*

Proof. Let $\{u_n\}$ be a dense countable set in B_U. For each $u^* \in U^*$ define

$$\|u^*\|_w = \sum_{n=1}^{\infty} \frac{1}{2^n} | \langle u_n, u^* \rangle_U |.$$

It may be easily verified that $\|\cdot\|_w$ is a norm in U^* and

$$\|u^*\|_w \leq \|u^*\|_U.$$

So, we may define a metric in U^* by

$$d(u^*, v^*) = \|u^* - v^*\|_w.$$

Now we shall prove that the topology induced by d coincides with $\sigma(U^*,U)$ in U^*.

Let $u_0^* \in B_{U^*}$ and let V be neighborhood of u_0^* in $\sigma(U^*,U)$.

We need to prove that there exists $r > 0$ such that

$$V_w = \{u^* \in B_{U^*} \mid d(u_0^*, u^*) < r\} \subset V.$$

Observe that for V we may assume the general format

$$V = \{u^* \in U^* \mid |\langle v_i, u^* - u_0^* \rangle_U| < \varepsilon, \forall l \in \{1, ..., k\}\}$$

for some $\varepsilon > 0$ and $v_1, \ldots, v_k \in U$.

There is no loss in generality in assuming

$$\|v_i\|_U \leq 1, \forall i \in \{1, \ldots, k\}.$$

Since $\{u_n\}$ is dense in U, for each $i \in \{1,\ldots,k\}$, there exists $n_i \in \mathbb{N}$ such that

$$\|u_{n_i} - v_i\|_U < \frac{\varepsilon}{4}.$$

Choose $r > 0$ small enough such that

$$2^{n_i} r < \frac{\varepsilon}{2}, \forall i \in \{1,\ldots,k\}.$$

We are going to show that $V_w \subset V$, where

$$V_w = \{u^* \in B_{U^*} \mid d(u_0^*, u^*) < r\} \subset V.$$

Observe that if $u^* \in V_w$, then

$$d(u_0^*, u^*) < r,$$

so that

$$\frac{1}{2^{n_i}} |\langle u_{n_i}, u^* - u_0^* \rangle_U| < r, \forall i \in \{1,\ldots,k\},$$

so that

$$\begin{aligned} |\langle v_i, u^* - u_0^* \rangle_U| &\leq |\langle v_i - u_{n_i}, u^* - u_0^* \rangle_U| + |\langle u_{n_i}, u^* - u_0^* \rangle_U| \\ &\leq (\|u^*\|_{U^*} + \|u_0^*\|_{U^*}) \|v_i - u_{n_i}\|_U + |\langle u_{n_i}, u^* - u_0^* \rangle_U| \\ &< 2\frac{\varepsilon}{4} + \frac{\varepsilon}{2} = \varepsilon. \end{aligned} \tag{2.70}$$

Therefore, $u^* \in V$, so that $V_w \subset V$.

Now let $u_0 \in B_{U^*}$ and fix $r > 0$. We have to obtain a neighborhood $V \in \sigma(U^*U)$ such that

$$V \subset V_w = \{u^* \in B_{U^*} \mid d(u_0^*, u^*) < r\}.$$

We shall define $k \in \mathbb{N}$ and $\varepsilon > 0$ in the next lines so that $V \subset V_w$, where

$$V = \{u^* \in B_{U^*} \mid |\langle u_i, u^* - u_0^* \rangle_U| < \varepsilon, \forall i \in \{1,\ldots,k\}\}.$$

For $u^* \in V_w$ we have

$$\begin{aligned} d(u^*, u_0^*) &= \sum_{n=1}^{k} \frac{1}{2^n} |\langle u_n, u^* - u_0^* \rangle_U| \\ &\quad + \sum_{n=k+1}^{\infty} \frac{1}{2^n} |\langle u_n, u^* - u_0^* \rangle_U| \\ &< \varepsilon + 2 \sum_{n=k+1}^{\infty} \frac{1}{2^n} \\ &= \varepsilon + \frac{1}{2^{k-1}}. \end{aligned} \tag{2.71}$$

Hence, it suffices to take $\varepsilon = r/2$, and k sufficiently big such that

$$\frac{1}{2^{k-1}} < r/2.$$

The first part of the proof is finished.

Conversely, assume B_{U^*} is metrizable in $\sigma(U^*,U)$. We are going to show that U is separable.

Define,

$$\tilde{V}_n = \left\{ u^* \in B_{U^*} \mid d(u^*,\theta) < \frac{1}{n} \right\}.$$

From the first part, we may find V_n a neighborhood of zero in $\sigma(U^*,U)$ such that

$$V_n \subset \tilde{V}_n.$$

Moreover, we may assume that V_n has the form

$$V_n = \{u^* \in B_{U^*} \mid |\langle u, u^* - \theta\rangle_U| < \varepsilon_n, \forall u \in C_n\},$$

where C_n is a finite set.

Define

$$D = \cup_{i=1}^{\infty} C_n.$$

Thus D is countable and we are going to prove that such a set is dense in U.

Suppose $u^* \in U^*$ is such that

$$\langle u, u^*\rangle_U = 0, \forall u \in D.$$

Hence,

$$u^* \in V_n \subset \tilde{V}_n, \forall n \in \mathbb{N},$$

so that $u^* = \theta$.

The proof is complete.

2.7 Uniformly Convex Spaces

Definition 2.7.1 (Uniformly Convex Spaces). A Banach space U is said to be uniformly convex if for each $\varepsilon > 0$, there exists $\delta > 0$ such that:

If $u, v \in U$, $\|u\|_U \leq 1$, $\|v\|_U \leq 1$, and $\|u - v\|_U > \varepsilon$, then $\frac{\|u+v\|_U}{2} < 1 - \delta$.

Theorem 2.7.2 (Milman Pettis). *Every uniformly convex Banach space is reflexive.*

Proof. Let $\eta \in U^{**}$ be such that $\|\eta\|_{U^{**}} = 1$. It suffices to show that $\eta \in J(B_U)$. Since $J(B_U)$ is closed in U^{**}, we have only to show that for each $\varepsilon > 0$ there exists $u \in U$ such that $\|\eta - J(u)\|_{U^{**}} < \varepsilon$.

Thus, suppose given $\varepsilon > 0$. Let $\delta > 0$ be the corresponding constant relating the uniformly convex property.

Choose $f \in U^*$ such that $\|f\|_{U^*} = 1$ and

$$\langle f, \eta \rangle_{U^*} > 1 - \frac{\delta}{2}. \tag{2.72}$$

Define

$$V = \left\{ \zeta \in U^{**} \mid |\langle f, \zeta - \eta \rangle_{U^*}| < \frac{\delta}{2} \right\}.$$

Observe that V is neighborhood of η in $\sigma(U^{**}, U^*)$. Since $J(B_U)$ is dense in $B_{U^{**}}$ concerning the topology $\sigma(U^{**}, U^*)$, we have that $V \cap J(B_U) \neq \emptyset$ and thus there exists $u \in B_U$ such that $J(u) \in V$. Suppose, to obtain contradiction, that

$$\|\eta - J(u)\|_{U^{**}} > \varepsilon.$$

Therefore, defining

$$W = (J(u) + \varepsilon B_{U^{**}})^c,$$

we have that $\eta \in W$, where W is also a weak neighborhood of η in $\sigma(U^{**}, U^*)$, since $B_{U^{**}}$ is closed in $\sigma(U^{**}, U^*)$.

Hence $V \cap W \cap J(B_U) \neq \emptyset$, so that there exists some $v \in B_U$ such that $J(v) \in V \cap W$. Thus, $J(u) \in V$ and $J(v) \in V$, so that

$$|\langle u, f \rangle_U - \langle f, \eta \rangle_{U^*}| < \frac{\delta}{2},$$

and

$$|\langle v, f \rangle_U - \langle f, \eta \rangle_{U^*}| < \frac{\delta}{2}.$$

Hence,

$$\begin{aligned} 2\langle f, \eta \rangle_{U^*} &< \langle u + v, f \rangle_U + \delta \\ &\leq \|u + v\|_U + \delta. \end{aligned} \tag{2.73}$$

From this and (2.72) we obtain

$$\frac{\|u + v\|_U}{2} > 1 - \delta,$$

and thus from the definition of uniform convexity, we obtain

$$\|u - v\|_U \leq \varepsilon. \tag{2.74}$$

On the other hand, since $J(v) \in W$, we have

$$\|J(u) - J(v)\|_{U^{**}} = \|u - v\|_U > \varepsilon,$$

which contradicts (2.74). The proof is complete.

Chapter 3
Topics on Linear Operators

The main references for this chapter are Reed and Simon [52] and Bachman and Narici [6].

3.1 Topologies for Bounded Operators

First we recall that the set of all bounded linear operators, denoted by $\mathscr{L}(U,Y)$, is a Banach space with the norm

$$\|A\| = \sup\{\|Au\|_Y \mid \|u\|_U \leq 1\}.$$

The topology related to the metric induced by this norm is called the uniform operator topology.

Let us introduce now the strong operator topology, which is defined as the weakest topology for which the functions

$$E_u : \mathscr{L}(U,Y) \to Y$$

are continuous where

$$E_u(A) = Au, \forall A \in \mathscr{L}(U,Y).$$

For such a topology a base at origin is given by sets of the form

$$\{A \mid A \in \mathscr{L}(U,Y),\ \|Au_i\|_Y < \varepsilon, \forall i \in \{1,\ldots,n\}\},$$

where $u_1,\ldots,u_n \in U$ and $\varepsilon > 0$.

Observe that a sequence $\{A_n\} \subset \mathscr{L}(U,Y)$ converges to A concerning this last topology if

$$\|A_n u - Au\|_Y \to 0, \text{ as } n \to \infty, \forall u \in U.$$

F. Botelho, *Functional Analysis and Applied Optimization in Banach Spaces: Applications to Non-Convex Variational Models*, DOI 10.1007/978-3-319-06074-3_3,

In the next lines we describe the weak operator topology in $\mathscr{L}(U,Y)$. Such a topology is weakest one such that the functions

$$E_{u,v} : \mathscr{L}(U,Y) \to \mathbb{C}$$

are continuous, where

$$E_{u,v}(A) = \langle Au, v\rangle_Y, \forall A \in \mathscr{L}(U,Y), u \in U,\ v \in Y^*.$$

For such a topology, a base at origin is given by sets of the form

$$\{A \in \mathscr{L}(U,Y) \mid |\langle Au_i, v_j\rangle_Y| < \varepsilon, \forall i \in \{1,\ldots,n\},\ j \in \{1,\ldots,m\}\},$$

where $\varepsilon > 0$, $u_1,\ldots,u_n \in U$, $v_1,\ldots,v_m \in Y^*$.

A sequence $\{A_n\} \subset \mathscr{L}(U,Y)$ converges to $A \in \mathscr{L}(U,Y)$ if

$$|\langle A_n u, v\rangle_Y - \langle Au, v\rangle_Y| \to 0,$$

as $n \to \infty$, $\forall u \in U,\ v \in Y^*$.

3.2 Adjoint Operators

We start this section recalling the definition of adjoint operator.

Definition 3.2.1. Let U,Y be Banach spaces. Given a bounded linear operator $A : U \to Y$ and $v^* \in Y^*$, we have that $T(u) = \langle Au, v^*\rangle_Y$ is such that

$$|T(u)| \leq \|Au\|_Y \cdot \|v^*\| \leq \|A\| \|v^*\|_{Y^*} \|u\|_U.$$

Hence $T(u)$ is a continuous linear functional on U and considering our fundamental representation hypothesis, there exists $u^* \in U^*$ such that

$$T(u) = \langle u, u^*\rangle_U, \forall u \in U.$$

We define A^* by setting $u^* = A^* v^*$, so that

$$T(u) = \langle u, u^*\rangle_U = \langle u, A^* v^*\rangle_U$$

that is,

$$\langle u, A^* v^*\rangle_U = \langle Au, v^*\rangle_Y, \forall u \in U,\ v^* \in Y^*.$$

We call $A^* : Y^* \to U^*$ the adjoint operator relating $A : U \to Y$.

Theorem 3.2.2. *Let U,Y be Banach spaces and let $A : U \to Y$ be a bounded linear operator. Then*

$$\|A\| = \|A^*\|.$$

Proof. Observe that

$$\begin{aligned}
\|A\| &= \sup_{u\in U}\{\|Au\| \mid \|u\|_U = 1\} \\
&= \sup_{u\in U}\left\{\sup_{v^*\in Y^*}\{\langle Au, v^*\rangle_Y \mid \|v^*\|_{Y^*} = 1\}, \|u\|_U = 1\right\} \\
&= \sup_{(u,v^*)\in U\times Y^*}\{\langle Au, v^*\rangle_Y \mid \|v^*\|_{Y^*} = 1, \|u\|_U = 1\} \\
&= \sup_{(u,v^*)\in U\times Y^*}\{\langle u, A^*v^*\rangle_U \mid \|v^*\|_{Y^*} = 1, \|u\|_U = 1\} \\
&= \sup_{v^*\in Y^*}\left\{\sup_{u\in U}\{\langle u, A^*v^*\rangle_U \mid \|u\|_U = 1\}, \|v^*\|_{Y^*} = 1\right\} \\
&= \sup_{v^*\in Y^*}\{\|A^*v^*\|, \|v^*\|_{Y^*} = 1\} \\
&= \|A^*\|. \qquad (3.1)
\end{aligned}$$

In particular, if $U = Y = H$ where H is Hilbert space, we have

Theorem 3.2.3. *Given the bounded linear operators $A, B : H \to H$ we have*

1. $(AB)^* = B^*A^*$,
2. $(A^*)^* = A$,
3. *if A has a bounded inverse A^{-1}, then A^* has a bounded inverse and*

$$(A^*)^{-1} = (A^{-1})^*.$$

4. $\|AA^*\| = \|A\|^2$.

Proof.

1. Observe that

$$(ABu, v)_H = (Bu, A^*v)_H = (u, B^*A^*v)_H, \forall u, v \in H.$$

2. Observe that

$$(u, Av)_H = (A^*u, v)_H = (u, A^{**}v)_H, \forall u, v \in H.$$

3. We have that

$$I = AA^{-1} = A^{-1}A,$$

so that

$$I = I^* = (AA^{-1})^* = (A^{-1})^*A^* = (A^{-1}A)^* = A^*(A^{-1})^*.$$

4. Observe that

$$\|A^*A\| \leq \|A\|\|A^*\| = \|A\|^2,$$

and

$$\|A^*A\| \geq \sup_{u\in U}\{(u, A^*Au)_H \mid \|u\|_H = 1\}$$

$$= \sup_{u \in U}\{(Au, Au)_H \mid \|u\|_U = 1\}$$
$$= \sup_{u \in U}\{\|Au\|_H^2 \mid \|u\|_U = 1\} = \|A\|^2, \tag{3.2}$$

and hence

$$\|A^*A\| = \|A\|^2.$$

Definition 3.2.4. Given $A \in \mathscr{L}(H)$ we say that A is self-adjoint if

$$A = A^*.$$

Theorem 3.2.5. *Let U and Y be Banach spaces and let $A : U \to Y$ be a bounded linear operator. Then*

$$[R(A)]^\perp = N(A^*),$$

where

$$[R(A)]^\perp = \{v^* \in Y^* \mid \langle Au, v^* \rangle_Y = 0, \ \forall u \in U\}.$$

Proof. Let $v^* \in N(A^*)$. Choose $v \in R(A)$. Thus there exists u in U such that $Au = v$ so that

$$\langle v, v^* \rangle_Y = \langle Au, v^* \rangle_Y = \langle u, A^*v^* \rangle_U = 0.$$

Since $v \in R(A)$ is arbitrary we have obtained

$$N(A^*) \subset [R(A)]^\perp.$$

Suppose $v^* \in [R(A)]^\perp$. Choose $u \in U$. Thus,

$$\langle Au, v^* \rangle_Y = 0,$$

so that

$$\langle u, A^*v^* \rangle_U, \forall u \in U.$$

Therefore $A^*v^* = \theta$, that is, $v^* \in N(A^*)$. Since $v^* \in [R(A)]^\perp$ is arbitrary, we get

$$[R(A)]^\perp \subset N(A^*).$$

This completes the proof.

The next result is relevant for subsequent developments.

Lemma 3.1. *Let U, Y be Banach spaces and let $A : U \to Y$ be a bounded linear operator. Suppose also that $R(A) = \{A(u) \ : \ u \in U\}$ is closed. Under such hypotheses, there exists $K > 0$ such that for each $v \in R(A)$ there exists $u_0 \in U$ such that*

$$A(u_0) = v$$

and

$$\|u_0\|_U \leq K \|v\|_Y.$$

Proof. Define $L = N(A) = \{u \in U \, : \, A(u) = \theta\}$ (the null space of A). Consider the space U/L, where

$$U/L = \{\overline{u} \, : \, u \in U\},$$

where

$$\overline{u} = \{u + w \, : \, w \in L\}.$$

Define $\overline{A} : U/L \to R(A)$, by

$$\overline{A}(\overline{u}) = A(u).$$

Observe that $\overline{A}$ is one-to-one, linear, onto, and bounded. Moreover $R(A)$ is closed so that it is a Banach space. Hence by the inverse mapping theorem we have that $\overline{A}$ has a continuous inverse. Thus, for any $v \in R(A)$, there exists $\overline{u} \in U/L$ such that

$$\overline{A}(\overline{u}) = v$$

so that

$$\overline{u} = \overline{A}^{-1}(v),$$

and therefore

$$\|\overline{u}\| \leq \|\overline{A}^{-1}\| \|v\|_Y.$$

Recalling that

$$\|\overline{u}\| = \inf_{w \in L} \{\|u + w\|_U\},$$

we may find $u_0 \in \overline{u}$ such that

$$\|u_0\|_U \leq 2\|\overline{u}\| \leq 2\|\overline{A}^{-1}\| \|v\|_Y,$$

and so that

$$A(u_0) = \overline{A}(\overline{u_0}) = \overline{A}(\overline{u}) = v.$$

Taking $K = 2\|\overline{A}^{-1}\|$ we have completed the proof.

Theorem 3.1. *Let U, Y be Banach spaces and let $A : U \to Y$ be a bound linear operator. Assume $R(A)$ is closed. Under such hypotheses*

$$R(A^*) = [N(A)]^{\perp}.$$

Proof. Let $u^* \in R(A^*)$. Thus there exists $v^* \in Y^*$ such that

$$u^* = A^*(v^*).$$

Let $u \in N(A)$. Hence,

$$\langle u, u^* \rangle_U = \langle u, A^*(v^*) \rangle_U = \langle A(u), v^* \rangle_Y = 0.$$

Since $u \in N(A)$ is arbitrary, we get $u^* \in [N(A)]^{\perp}$, so that

$$R(A^*) \subset [N(A)]^{\perp}.$$

Now suppose $u^* \in [N(A)]^\perp$. Thus

$$\langle u, u^* \rangle_U = 0, \; \forall u \in N(A).$$

Fix $v \in R(A)$. From the Lemma 3.1, there exists $K > 0$ (which does not depend on v) and $u_v \in U$ such that

$$A(u_v) = v$$

and

$$\|u_v\|_U \leq K \|v\|_Y.$$

Define $f : R(A) \to \mathbb{R}$ by

$$f(v) = \langle u_v, u^* \rangle_U.$$

Observe that

$$|f(v)| \leq \|u_v\|_U \|u^*\|_{U^*} \leq K \|v\|_Y \|u^*\|_{U^*},$$

so that f is a bounded linear functional. Hence by a Hahn–Banach theorem corollary there exists $v^* \in Y^*$ such that

$$f(v) = \langle v, v^* \rangle_Y \equiv F(v), \; \forall v \in R(A),$$

that is, F is an extension of f from $R(A)$ to Y.

In particular

$$f(v) = \langle u_v, u^* \rangle_U = \langle v, v^* \rangle_Y = \langle A(u_v), v^* \rangle_Y \; \forall v \in R(A),$$

where $A(u_v) = v$, so that

$$\langle u_v, u^* \rangle_U = \langle A(u_v), v^* \rangle_Y \; \forall v \in R(A).$$

Now let $u \in U$ and define $A(u) = v_0$. Observe that

$$u = (u - u_{v_0}) + u_{v_0},$$

and

$$A(u - u_{v_0}) = A(u) - A(u_{v_0}) = v_0 - v_0 = \theta.$$

Since $u^* \in [N(A)]^\perp$ we get

$$\langle u - u_{v_0}, u^* \rangle_U = 0$$

so that

$$\begin{aligned} \langle u, u^* \rangle_U &= \langle (u - u_{v_0}) + u_{v_0}, u^* \rangle_U \\ &= \langle u_{v_0}, u^* \rangle_U \\ &= \langle A(u_{v_0}), v^* \rangle_Y \\ &= \langle A(u - u_{v_0}) + A(u_{v_0}), v^* \rangle_Y \\ &= \langle A(u), v^* \rangle_Y. \end{aligned} \tag{3.3}$$

Hence,

$$\langle u, u^* \rangle_U = \langle A(u), v^* \rangle_Y, \ \forall u \in U.$$

We may conclude that $u^* = A^*(v^*) \in R(A^*)$. Since $u^* \in [N(A)]^\perp$ is arbitrary we obtain

$$[N(A)]^\perp \subset R(A^*).$$

The proof is complete.

We finish this section with the following result.

Definition 3.2.6. Let U be a Banach space and $S \subset U$. We define the positive conjugate cone of S, denoted by $S^\oplus$ by

$$S^\oplus = \{u^* \in U^* \ : \ \langle u, u^* \rangle_U \geq 0, \ \forall u \in S\}.$$

Similarly, we define the negative cone of S, denoted by $S^\ominus$ by

$$S^\ominus = \{u^* \in U^* \ : \ \langle u, u^* \rangle_U \leq 0, \ \forall u \in S\}.$$

Theorem 3.2.7. *Let U, Y be Banach spaces and $A : U \to Y$ be a bounded linear operator. Let $S \subset U$. Then*

$$[A(S)]^\oplus = (A^*)^{-1}(S^\oplus),$$

where

$$(A^*)^{-1} = \{v^* \in Y^* \ : \ A^* v^* \in S^\oplus\}.$$

Proof. Let $v^* \in [A(S)]^\oplus$ and $u \in S$. Thus,

$$\langle A(u), v^* \rangle_Y \geq 0,$$

so that

$$\langle u, A^*(v^*) \rangle_U \geq 0.$$

Since $u \in S$ is arbitrary, we get

$$v^* \in (A^*)^{-1}(S^\oplus).$$

From this

$$[A(S)]^\oplus \subset (A^*)^{-1}(S^\oplus).$$

Reciprocally, let $v^* \in (A^*)^{-1}(S^\oplus)$. Hence $A^*(v^*) \in S^\oplus$ so that for $u \in S$ we obtain

$$\langle u, A^*(v^*) \rangle_U \geq 0,$$

and therefore

$$\langle A(u), v^* \rangle_Y \geq 0.$$

Since $u \in S$ is arbitrary, we get $v^* \in [A(S)]^{\oplus}$, that is,

$$(A^*)^{-1}(S^{\oplus}) \subset [A(S)]^{\oplus}.$$

The proof is complete.

3.3 Compact Operators

We start this section defining compact operators.

Definition 3.3.1. Let U and Y be Banach spaces. An operator $A \in \mathscr{L}(U,Y)$ (linear and bounded) is said to compact if A takes bounded sets into pre-compact sets. Summarizing, A is compact if for each bounded sequence $\{u_n\} \subset U$, $\{Au_n\}$ has a convergent subsequence in Y.

Theorem 3.3.2. *A compact operator maps weakly convergent sequences into norm convergent sequences.*

Proof. Let $A : U \to Y$ be a compact operator. Suppose

$$u_n \rightharpoonup u \text{ weakly in U.}$$

By the uniform boundedness theorem, $\{\|u_n\|\}$ is bounded. Thus, given $v^* \in Y^*$ we have

$$\begin{aligned} \langle v^*, Au_n \rangle_Y &= \langle A^*v^*, u_n \rangle_U \\ &\to \langle A^*v^*, u \rangle_U \\ &= \langle v^*, Au \rangle_Y. \end{aligned} \tag{3.4}$$

Being $v^* \in Y^*$ arbitrary, we get that

$$Au_n \rightharpoonup Au \text{ weakly in } Y. \tag{3.5}$$

Suppose Au_n does not converge in norm to Au. Thus there exists $\varepsilon > 0$ and a subsequence $\{Au_{n_k}\}$ such that

$$\|Au_{n_k} - Au\|_Y \geq \varepsilon, \forall k \in \mathbb{N}.$$

As $\{u_{n_k}\}$ is bounded and A is compact, $\{Au_{n_k}\}$ has a subsequence converging para $\tilde{v} \neq Au$. But then such a sequence converges weakly to $\tilde{v} \neq Au$, which contradicts (3.5). The proof is complete.

Theorem 3.3.3. *Let H be a separable Hilbert space. Thus each compact operator in $\mathscr{L}(H)$ is the limit in norm of a sequence of finite rank operators.*

Proof. Let A be a compact operator in H. Let $\{\phi_j\}$ an orthonormal basis in H. For each $n \in \mathbb{N}$ define

$$\lambda_n = \sup\{\|A\psi\|_H \mid \psi \in [\phi_1, \ldots, \phi_n]^{\perp} \text{ and } \|\psi\|_H = 1\}.$$

It is clear that $\{\lambda_n\}$ is a nonincreasing sequence that converges to a limit $\lambda \geq 0$. We will show that $\lambda = 0$. Choose a sequence $\{\psi_n\}$ such that

$$\psi_n \in [\phi_1, \ldots, \phi_n]^{\perp},$$

$\|\psi_n\|_H = 1$, and $\|A\psi_n\|_H \geq \lambda/2$. Now we will show that

$$\psi_n \rightharpoonup \theta, \text{ weakly in } H.$$

Let $\psi^* \in H^* = H$,; thus there exists a sequence $\{a_j\} \subset \mathbb{C}$ such that

$$\psi^* = \sum_{j=1}^{\infty} a_j \phi_j.$$

Suppose given $\varepsilon > 0$. We may find $n_0 \in \mathbb{N}$ such that

$$\sum_{j=n_0}^{\infty} |a_j|^2 < \varepsilon.$$

Choose $n > n_0$. Hence there exists $\{b_j\}_{j>n}$ such that

$$\psi_n = \sum_{j=n+1}^{\infty} b_j \phi_j,$$

and

$$\sum_{j=n+1}^{\infty} |b_j|^2 = 1.$$

Therefore

$$\begin{aligned} |(\psi_n, \psi^*)_H| &= \left| \sum_{j=n+1}^{\infty} (\phi_j, \phi_j)_H a_j \cdot b_j \right| \\ &= \left| \sum_{j=n+1}^{\infty} a_j \cdot b_j \right| \\ &\leq \sqrt{\sum_{j=n+1}^{\infty} |a_j|^2} \sqrt{\sum_{j=n+1}^{\infty} |b_j|^2} \\ &\leq \sqrt{\varepsilon}, \end{aligned} \tag{3.6}$$

if $n > n_0$. Since $\varepsilon > 0$ is arbitrary,

$$(\psi_n, \psi^*)_H \to 0, \text{ as } n \to \infty.$$

Since $\psi^* \in H$ is arbitrary, we get

$$\psi_n \rightharpoonup \theta, \text{ weakly in } H.$$

Hence, as A is compact, we have

$$A\psi_n \to \theta \text{ in norm },$$

so that $\lambda = 0$. Finally, we may define $\{A_n\}$ by

$$A_n(u) = A\left(\sum_{j=1}^{n}(u,\phi_j)_H\phi_j\right) = \sum_{j=1}^{n}(u,\phi_j)_H A\phi_j,$$

for each $u \in H$. Thus

$$\|A - A_n\| = \lambda_n \to 0, \text{ as } n \to \infty.$$

The proof is complete.

3.4 The Square Root of a Positive Operator

Definition 3.4.1. Let H be a Hilbert space. A mapping $E : H \to H$ is said to be a projection on $M \subset H$ if for each $z \in H$ we have

$$Ez = x,$$

where $z = x + y$, $x \in M$, and $y \in M^{\perp}$.

Observe that

1. E is linear,
2. E is idempotent, that is, $E^2 = E$,
3. $R(E) = M$,
4. $N(E) = M^{\perp}$.

Also observe that from

$$Ez = x$$

we have

$$\|Ez\|_H^2 = \|x\|_H^2 \leq \|x\|_H^2 + \|y\|_H^2 = \|z\|_H^2,$$

so that

$$\|E\| \leq 1.$$

Definition 3.4.2. Let $A, B \in \mathscr{L}(H)$. We write

$$A \geq \theta$$

if

$$(Au,u)_H \geq 0, \forall u \in H,$$

and in this case we say that A is positive. Finally, we denote

$$A \geq B$$

if

$$A - B \geq \theta.$$

Theorem 3.4.3. *Let A and B be bounded self-adjoint operators such that $A \geq \theta$ and $B \geq \theta$. If $AB = BA$, then*

$$AB \geq \theta.$$

Proof. If $A = \theta$, the result is obvious. Assume $A \neq \theta$ and define the sequence

$$A_1 = \frac{A}{\|A\|}, \quad A_{n+1} = A_n - A_n^2, \forall n \in \mathbb{N}.$$

We claim that

$$\theta \leq A_n \leq I, \forall n \in \mathbb{N}.$$

We prove the claim by induction.

For $n = 1$, it is clear that $A_1 \geq \theta$. And since $\|A_1\| = 1$, we get

$$(A_1 u, u)_H \leq \|A_1\| \|u\|_H \|u\|_H = (Iu, u)_H, \forall u \in H,$$

so that

$$A_1 \leq I.$$

Thus

$$\theta \leq A_1 \leq I.$$

Now suppose $\theta \leq A_n \leq I$. Since A_n is self-adjoint, we have

$$\begin{aligned} (A_n^2(I - A_n)u, u)_H &= ((I - A_n)A_n u, A_n u)_H \\ &= ((I - A_n)v, v)_H \geq 0, \forall u \in H, \end{aligned} \tag{3.7}$$

where $v = A_n u$. Therefore

$$A_n^2(I - A_n) \geq \theta.$$

Similarly, we may obtain

$$A_n(I - A_n)^2 \geq \theta,$$

so that

$$\theta \leq A_n^2(I - A_n) + A_n(I - A_n)^2 = A_n - A_n^2 = A_{n+1}.$$

So, also we have

$$\theta \leq I - A_n + A_n^2 = I - A_{n+1},$$

that is,

$$\theta \leq A_{n+1} \leq I,$$

so that

$$\theta \leq A_n \leq I, \ \forall n \in \mathbb{N}.$$

Observe that

$$\begin{aligned} A_1 &= A_1^2 + A_2 \\ &= A_1^2 + A_2^2 + A_3 \\ &\ldots\ldots\ldots\ldots\ldots\ldots\ldots \\ &= A_1^2 + \ldots + A_n^2 + A_{n+1}. \end{aligned} \tag{3.8}$$

Since $A_{n+1} \geq \theta$, we obtain

$$A_1^2 + A_2^2 + \ldots + A_n^2 = A_1 - A_{n+1} \leq A_1. \tag{3.9}$$

From this, for a fixed $u \in H$, we have

$$\begin{aligned} \sum_{j=1}^{n} \|A_j u\|^2 &= \sum_{j=1}^{n} (A_j u, A_j u)_H \\ &= \sum_{j=1}^{n} (A_j^2 u, u)_H \\ &\leq (A_1 u, u)_H. \end{aligned} \tag{3.10}$$

Since $n \in \mathbb{N}$ is arbitrary, we get

$$\sum_{j=1}^{\infty} \|A_j u\|^2$$

is a converging series, so that

$$\|A_n u\| \to 0,$$

that is,

$$A_n u \to \theta, \text{ as } n \to \infty.$$

From this and (3.9), we get

$$\sum_{j=1}^{n} A_j^2 u = (A_1 - A_{n+1})u \to A_1 u, \text{ as } n \to \infty.$$

Finally, we may write

$$\begin{aligned} (ABu, u)_H &= \|A\|(A_1 Bu, u)_H \\ &= \|A\|(BA_1 u, u)_H \\ &= \|A\|(B \lim_{n..} \sum_j = 1^n A_j^2 u, u)_H \\ &= \|A\| \lim_{n\ldots} \sum_j = 1^n (BA_j^2 u, u)_H \\ &= \|A\| \lim_{n\ldots} \sum_j = 1^n (BA_j u, BA_j u)_H \\ &\geq 0. \end{aligned} \tag{3.11}$$

Hence

$$(ABu, u)_H \geq 0, \forall u \in H.$$

The proof is complete.

Theorem 3.4.4. *Let $\{A_n\}$ be a sequence of self-adjoint commuting operators in $\mathscr{L}(H)$. Let $B \in \mathscr{L}(H)$ be a self-adjoint operator such that*

$$A_iB = BA_i, \forall i \in \mathbb{N}.$$

Suppose also that

$$A_1 \leq A_2 \leq A_3 \leq \ldots \leq A_n \leq \ldots \leq B.$$

Under such hypotheses there exists a self-adjoint, bounded, linear operator A such that

$$A_n \to A \text{ in norm},$$

and

$$A \leq B.$$

Proof. Consider the sequence $\{C_n\}$ where

$$C_n = B - A_n \geq 0, \forall n \in \mathbb{N}.$$

Fix $u \in H$. First, we show that $\{C_n u\}$ converges. Observe that

$$C_iC_j = C_jC_i, \forall i, j \in \mathbb{N}.$$

Also, if $n > m$, then

$$A_n - A_m \geq \theta$$

so that

$$C_m = B - A_m \geq B - A_n = C_n.$$

Therefore, from $C_m \geq \theta$ and $C_m - C_n \geq \theta$, we obtain

$$(C_m - C_n)C_m \geq \theta, \text{ if } n > m$$

and also

$$C_n(C_m - C_n) \geq \theta.$$

Thus,

$$(C_m^2 u, u)_H \geq (C_nC_m u, u)_H \geq (C_n^2 u, u)_H,$$

and we may conclude that

$$(C_n^2 u, u)_H$$

is a monotone nonincreasing sequence of real numbers, bounded below by 0, so that there exists $\alpha \in \mathbb{R}$ such that

$$\lim_{n \to \infty} (C_n^2 u, u)_H = \alpha.$$

Since each C_n is self-adjoint we obtain

$$\begin{aligned}\|(C_n - C_m)u\|_H^2 &= ((C_n - C_m)u, (C_n - C_m)u)_H \\ &= ((C_n - C_m)(C_n - C_m)u, u)_H \\ &= (C_n^2 u, u)_H - 2(C_n C_m u, u) + (C_m^2 u, u)_H \\ &\to \alpha - 2\alpha + \alpha = 0, \end{aligned} \tag{3.12}$$

as

$$m, n \to \infty.$$

Therefore $\{C_n u\}$ is a Cauchy sequence in norm, so that there exists the limit

$$\lim_{n\to\infty} C_n u = \lim_{n\to\infty} (B - A_n)u,$$

and hence there exists

$$\lim_{n\to\infty} A_n u, \forall u \in H.$$

Now define A by

$$Au = \lim_{n\to\infty} A_n u.$$

Since the limit

$$\lim_{n\to\infty} A_n u, \forall u \in H$$

exists we have that

$$\sup_{n\in\mathbb{N}} \{\|A_n u\|_H\}$$

is finite for all $u \in H$. By the principle of uniform boundedness

$$\sup_{n\in\mathbb{N}} \{\|A_n\|\} < \infty$$

so that there exists $K > 0$ such that

$$\|A_n\| \leq K, \forall n \in \mathbb{N}.$$

Therefore

$$\|A_n u\|_H \leq K\|u\|_H,$$

so that

$$\|Au\| = \lim_{n\to\infty} \{\|A_n u\|_H\} \leq K\|u\|_H, \forall u \in H$$

which means that A is bounded. Fixing $u, v \in H$, we have

$$(Au, v)_H = \lim_{n\to\infty} (A_n u, v)_H = \lim_{n\to\infty} (u, A_n v)_H = (u, Av)_H,$$

and thus A is self-adjoint. Finally

$$(A_n u, u)_H \leq (Bu, u)_H, \forall n \in \mathbb{N},$$

so that

$$(Au,u) = \lim_{n\to\infty}(A_nu,u)_H \leq (Bu,u)_H, \forall u \in H.$$

Hence $A \leq B$.

The proof is complete.

Definition 3.4.5. Let $A \in \mathscr{L}(A)$ be a positive operator. The self-adjoint operator $B \in \mathscr{L}(H)$ such that

$$B^2 = A$$

is called the square root of A. If $B \geq \theta$, we denote

$$B = \sqrt{A}.$$

Theorem 3.4.6. *Suppose $A \in \mathscr{L}(H)$ is positive. Then there exists $B \geq \theta$ such that*

$$B^2 = A.$$

Furthermore B commutes with any $C \in \mathscr{L}(H)$ such that commutes with A.

Proof. There is no loss of generality in considering

$$\|A\| \leq 1,$$

which means $\theta \leq A \leq I$, because we may replace A by

$$\frac{A}{\|A\|}$$

so that if

$$C^2 = \frac{A}{\|A\|}$$

then

$$B = \|A\|^{1/2}C.$$

Let

$$B_0 = \theta,$$

and consider the sequence of operators given by

$$B_{n+1} = B_n + \frac{1}{2}(A - B_n^2), \forall n \in \mathbb{N} \cup \{0\}.$$

Since each B_n is polynomial in A, we have that B_n is self-adjoint and commute with any operator with commutes with A. In particular

$$B_iB_j = B_jB_i, \forall i,j \in \mathbb{N}.$$

First we show that

$$B_n \leq I, \forall n \in \mathbb{N} \cup \{0\}.$$

Since $B_0 = \theta$, and $B_1 = \frac{1}{2}A$, the statement holds for $n = 1$. Suppose $B_n \leq I$. Thus

$$\begin{aligned} I - B_{n+1} &= I - B_n - \frac{1}{2}A + \frac{1}{2}B_n^2 \\ &= \frac{1}{2}(I - B_n)^2 + \frac{1}{2}(I - A) \geq \theta \end{aligned} \tag{3.13}$$

so that

$$B_{n+1} \leq I.$$

The induction is complete, that is,

$$B_n \leq I, \forall n \in \mathbb{N}.$$

Now we prove the monotonicity also by induction. Observe that

$$B_0 \leq B_1,$$

and supposing

$$B_{n-1} \leq B_n,$$

we have

$$\begin{aligned} B_{n+1} - B_n &= B_n + \frac{1}{2}(A - B_n^2) - B_{n-1} - \frac{1}{2}(A - B_{n-1}^2) \\ &= B_n - B_{n-1} - \frac{1}{2}(B_n^2 - B_{n-1}^2) \\ &= B_n - B_{n-1} - \frac{1}{2}(B_n + B_{n-1})(B_n - B_{n-1}) \\ &= (I - \frac{1}{2}(B_n + B_{n-1}))(B_n - B_{n-1}) \\ &= \frac{1}{2}((I - B_{n-1}) + (I - B_n))(B_n - B_{n-1}) \geq \theta. \end{aligned}$$

The induction is complete, that is,

$$\theta = B_0 \leq B_1 \leq B_2 \leq \ldots \leq B_n \leq \ldots \leq I.$$

By the last theorem there exists a self-adjoint operator B such that

$$B_n \to B \text{ in norm.}$$

Fixing $u \in H$ we have

$$B_{n+1}u = B_n u + \frac{1}{2}(A - B_n^2)u,$$

so that taking the limit in norm as $n \to \infty$, we get

$$\theta = (A - B^2)u.$$

Being $u \in H$ arbitrary we obtain

$$A = B^2.$$

It is also clear that

$$B \geq \theta$$

The proof is complete.

3.5 About the Spectrum of a Linear Operator

Definition 3.5.1. Let U be a Banach space and let $A \in \mathscr{L}(U)$. A complex number λ is said to be in the resolvent set $\rho(A)$ of A, if

$$\lambda I - A$$

is a bijection with a bounded inverse. We call

$$R_\lambda(A) = (\lambda I - A)^{-1}$$

the resolvent of A in λ.

If $\lambda \notin \rho(A)$, we write

$$\lambda \in \sigma(A) - \mathbb{C} - \rho(A),$$

where $\sigma(A)$ is said to be the spectrum of A.

Definition 3.5.2. Let $A \in \mathscr{L}(U)$.

1. If $u \neq \theta$ and $Au = \lambda u$ for some $\lambda \in \mathbb{C}$, then u is said to be an eigenvector of A and λ the corresponding eigenvalue. If λ is an eigenvalue, then $(\lambda I - A)$ is not injective and therefore $\lambda \in \sigma(A)$.
   The set of eigenvalues is said to be the point spectrum of A.
2. If λ is not an eigenvalue but
$$R(\lambda I - A)$$
   is not dense in U and therefore $\lambda I - A$ is not a bijection, we have that $\lambda \in \sigma(A)$. In this case we say that λ is in the residual spectrum of A, or briefly $\lambda \in Res[\sigma(A)]$.

Theorem 3.5.3. *Let U be a Banach space and suppose that $A \in \mathscr{L}(U)$. Then $\rho(A)$ is an open subset of $\mathbb{C}$ and*

$$F(\lambda) = R_\lambda(A)$$

is an analytic function with values in $\mathscr{L}(U)$ on each connected component of $\rho(A)$. For λ, $\mu \in \sigma(A)$, $R_\lambda(A)$, and $R_\mu(A)$ commute and

$$R_\lambda(A) - R_\mu(A) = (\mu - \lambda) R_\mu(A) R_\lambda(A).$$

Proof. Let $\lambda_0 \in \rho(A)$. We will show that λ_0 is an interior point of $\rho(A)$.

Observe that symbolically we may write

$$\begin{aligned}\frac{1}{\lambda - A} &= \frac{1}{\lambda - \lambda_0 + (\lambda_0 - A)} \\ &= \frac{1}{\lambda_0 - A}\left[\frac{1}{1 - \left(\frac{\lambda_0 - \lambda}{\lambda_0 - A}\right)}\right] \\ &= \frac{1}{\lambda_0 - A}\left(1 + \sum_{n=1}^{\infty}\left(\frac{\lambda_0 - \lambda}{\lambda_0 - A}\right)^n\right). \end{aligned} \tag{3.14}$$

Define

$$\hat{R}_\lambda(A) = R_{\lambda_0}(A)\left\{I + \sum_{n=1}^{\infty}(\lambda - \lambda_0)^n (R_{\lambda_0})^n\right\}. \tag{3.15}$$

Observe that

$$\|(R_{\lambda_0})^n\| \leq \|R_{\lambda_0}\|^n.$$

Thus, the series indicated in (3.15) will converge in norm if

$$|\lambda - \lambda_0| < \|R_{\lambda_0}\|^{-1}. \tag{3.16}$$

Hence, for λ satisfying (3.16), $\hat{R}(A)$ is well defined and we can easily check that

$$(\lambda I - A)\hat{R}_\lambda(A) = I = \hat{R}_\lambda(A)(\lambda I - A).$$

Therefore

$$\hat{R}_\lambda(A) = R_\lambda(A), \text{ if } |\lambda - \lambda_0| < \|R_{\lambda_0}\|^{-1},$$

so that λ_0 is an interior point. Since $\lambda_0 \in \rho(A)$ is arbitrary, we have that $\rho(A)$ is open. Finally, observe that

$$\begin{aligned} R_\lambda(A) - R_\mu(A) &= R_\lambda(A)(\mu I - A)R_\mu(A) - R_\lambda(A)(\lambda I - A)R_\mu(A) \\ &= R_\lambda(A)(\mu I)R_\mu(A) - R_\lambda(A)(\lambda I)R_\mu(A) \\ &= (\mu - \lambda)R_\lambda(A)R_\mu(A). \end{aligned} \tag{3.17}$$

Interchanging the roles of λ and μ we may conclude that R_λ and R_μ commute.

Corollary 3.5.4. *Let U be a Banach space and $A \in \mathcal{L}(U)$. Then the spectrum of A is nonempty.*

Proof. Observe that if

$$\frac{\|A\|}{|\lambda|} < 1$$

we have

$$\begin{aligned} (\lambda I - A)^{-1} &= [\lambda(I - A/\lambda)]^{-1} \\ &= \lambda^{-1}(I - A/\lambda)^{-1} \end{aligned}$$

$$= \lambda^{-1}\left(I + \sum_{n=1}^{\infty}\left(\frac{A}{\lambda}\right)^n\right). \tag{3.18}$$

Therefore we may obtain

$$R_\lambda(A) = \lambda^{-1}\left(I + \sum_{n=1}^{\infty}\left(\frac{A}{\lambda}\right)^n\right).$$

In particular

$$\|R_\lambda(A)\| \to 0, \text{ as } |\lambda| \to \infty. \tag{3.19}$$

Suppose, to obtain contradiction, that

$$\sigma(A) = \emptyset.$$

In such a case $R_\lambda(A)$ would be an entire bounded analytic function. From Liouville's theorem, $R_\lambda(A)$ would be constant, so that from (3.19) we would have

$$R_\lambda(A) = \theta, \forall \lambda \in \mathbb{C},$$

which is a contradiction.

Proposition 3.5.5. *Let H be a Hilbert space and $A \in \mathcal{L}(H)$.*

1. *If $\lambda \in Res[\sigma(A)]$, then $\overline{\lambda} \in P\sigma(A^*)$.*
2. *If $\lambda \in P\sigma(A)$, then $\overline{\lambda} \in P\sigma(A^*) \cup Res[\sigma(A^*)]$.*

Proof.

1. If $\lambda \in Res[\sigma(A)]$, then
$$R(A - \lambda I) \neq H.$$
Therefore there exists $v \in (R(A - \lambda I))^\perp$, $v \neq \theta$ such that
$$(v, (A - \lambda I)u)_H = 0, \forall u \in H$$
that is,
$$((A^* - \overline{\lambda} I)v, u)_H = 0, \forall u \in H$$
so that
$$(A^* - \overline{\lambda} I)v = \theta,$$
which means that $\overline{\lambda} \in P\sigma(A^*)$.
2. Suppose there exists $v \neq \theta$ such that
$$(A - \lambda I)v = \theta,$$
and
$$\overline{\lambda} \not\in P\sigma(A^*).$$

Thus

$$(u,(A-\lambda I)v))_H=0, \forall u \in H,$$

so that

$$((A^*-\overline{\lambda}I)u,v)_H, \forall u \in H.$$

Since

$$(A^*-\overline{\lambda}I)u \neq \theta, \forall u \in H, u \neq \theta,$$

we get $v \in (R(A^*-\overline{\lambda}I))^{\perp}$, so that $R(A^*-\overline{\lambda}I) \neq H$.
Hence $\overline{\lambda} \in Res[\sigma(A^*)]$.

Theorem 3.5.6. *Let $A \in \mathscr{L}(H)$ be a self-adjoint operator, then*

1. $\sigma(A) \subset \mathbb{R}$.
2. Eigenvectors corresponding to distinct eigenvalues of A are orthogonal.

Proof. Let $\mu, \lambda \in \mathbb{R}$. Thus, given $u \in H$ we have

$$\|(A-(\lambda+\mu i))u\|^2 = \|(A-\lambda)u\|^2+\mu^2\|u\|^2,$$

so that

$$\|(A-(\lambda+\mu i))u\|^2 \geq \mu^2\|u\|^2.$$

Therefore if $\mu \neq 0$, $A-(\lambda+\mu i)$ has a bounded inverse on its range, which is closed. If $R(A-(\lambda+\mu i)) \neq H$, then by the last result $(\lambda-\mu i)$ would be in the point spectrum of A, which contradicts the last inequality. Hence, if $\mu \neq 0$, then $\lambda+\mu i \in \rho(A)$. To complete the proof, suppose

$$Au_1 = \lambda_1 u_1,$$

and

$$Au_2 = \lambda_2 u_2,$$

where

$$\lambda_1, \lambda_2 \in \mathbb{R},\ \lambda_1 \neq \lambda_2,\ and\ u_1, u_2 \neq \theta.$$

Thus

$$\begin{aligned}(\lambda_1-\lambda_2)(u_1,u_2)_H &= \lambda_1(u_1,u_2)_H-\lambda_2(u_1,u_2)_H \\ &= (\lambda_1 u_1,u_2)_H-(u_1,\lambda_2 u_2)_H \\ &= (Au_1,u_2)_H-(u_1,Au_2)_H \\ &= (u_1,Au_2)_H-(u_1,Au_2)_H \\ &= 0. \end{aligned} \tag{3.20}$$

Since $\lambda_1-\lambda_2 \neq 0$ we get

$$(u_1,u_2)_H=0.$$

3.6 The Spectral Theorem for Bounded Self-Adjoint Operators

Let H be a complex Hilbert space. Consider $A : H \to H$ a linear bounded operator, that is, $A \in \mathscr{L}(H)$, and suppose also that such an operator is self-adjoint. Define

$$m = \inf_{u \in H}\{(Au,u)_H \mid \|u\|_H = 1\},$$

and

$$M = \sup_{u \in H}\{(Au,u)_H \mid \|u\|_H = 1\}.$$

Remark 3.6.1. It is possible to prove that for a linear self-adjoint operator $A : H \to H$ we have

$$\|A\| = \sup\{|(Au,u)_H| \mid u \in H,\ \|u\|_H = 1\}.$$

This propriety, which prove in the next lines, is crucial for the subsequent results, since, for example, for A, B linear and self-adjoint and $\varepsilon > 0$, we have

$$-\varepsilon I \leq A - B \leq \varepsilon I,$$

we also would have

$$\|A - B\| < \varepsilon.$$

So, we present the following basic result.

Theorem 3.6.2. *Let $A : H \to H$ be a bounded linear self-adjoint operator. Define*

$$\alpha = \max\{|m|, |M|\},$$

where

$$m = \inf_{u \in H}\{(Au,u)_H \mid \|u\|_H = 1\},$$

and

$$M = \sup_{u \in H}\{(Au,u)_H \mid \|u\|_H = 1\}.$$

Then

$$\|A\| = \alpha.$$

Proof. Observe that

$$(A(u+v), u+v)_H = (Au,u)_H + (Av,v)_H + 2(Au,v)_H,$$

and

$$(A(u-v), u-v)_H = (Au,u)_H + (Av,v)_H - 2(Au,v)_H.$$

Thus,

$$4(Au,v) = (A(u+v), u+v)_H - (A(u-v), u-v)_H \leq M\|u+v\|_U^2 - m\|u-v\|_U^2,$$

so that

$$4(Au,v)_H \leq \alpha(\|u+v\|_U^2 + \|u-v\|_U^2).$$

Hence, replacing v by $-v$, we obtain

$$-4(Au,v)_H \leq \alpha(\|u+v\|_U^2 + \|u-v\|_U^2),$$

and therefore

$$4|(Au,v)_H| \leq \alpha(\|u+v\|_U^2 + \|u-v\|_U^2).$$

Replacing v by βv, we get

$$4|(A(u),v)_H| \leq 2\alpha(\|u\|_U^2/\beta + \beta\|v\|_U^2).$$

Minimizing the last expression in $\beta > 0$, for the optimal

$$\beta = \|u\|_U/\|v\|_U,$$

we obtain

$$|(Au,v)_H| \leq \alpha\|u\|_U\|v\|_U, \forall u,v \in U.$$

Thus

$$\|A\| \leq \alpha.$$

On the other hand,

$$|(Au,u)_H| \leq \|A\|\|u\|_U^2,$$

so that

$$|M| \leq \|A\|$$

and

$$|m| \leq \|A\|,$$

so that

$$\alpha \leq \|A\|.$$

The proof is complete.

At this point we start to develop the spectral theory. Define by P the set of all real polynomials defined in $\mathbb{R}$. Define

$$\Phi_1 : P \to \mathscr{L}(H),$$

by

$$\Phi_1(p(\lambda)) = p(A), \forall p \in P.$$

Thus we have

1. $\Phi_1(p_1 + p_2) = p_1(A) + p_2(A)$,
2. $\Phi_1(p_1 \cdot p_2) = p_1(A)p_2(A)$,
3. $\Phi_1(\alpha p) = \alpha p(A), \forall \alpha \in \mathbb{R},\ p \in P$,
4. if $p(\lambda) \geq 0$, on $[m,M]$, then $p(A) \geq \theta$.

We will prove (4):

Consider $p \in P$. Denote the real roots of $p(\lambda)$ less or equal to m by $\alpha_1, \alpha_2, \ldots, \alpha_n$ and denote those that are greater or equal to M by $\beta_1, \beta_2, \ldots, \beta_l$. Finally denote all the remaining roots, real or complex, by

$$\nu_1 + i\mu_1, \ldots, \nu_k + i\mu_k.$$

Observe that if $\mu_i = 0$, then $\nu_i \in (m, M)$. The assumption that $p(\lambda) \geq 0$ on $[m, M]$ implies that any real root in (m, M) must be of even multiplicity.

Since complex roots must occur in conjugate pairs, we have the following representation for $p(\lambda)$:

$$p(\lambda) = a \prod_{i=1}^{n} (\lambda - \alpha_i) \prod_{i=1}^{l} (\beta_i - \lambda) \prod_{i=1}^{k} ((\lambda - \nu_i)^2 + \mu_i^2),$$

where $a \geq 0$. Observe that

$$A - \alpha_i I \geq \theta,$$

since

$$(Au, u)_H \geq m(u, u)_H > \alpha_i (u, u)_H, \forall u \in H,$$

and by analogy

$$\beta_i I - A \geq \theta.$$

On the other hand, since $A - \nu_k I$ is self-adjoint, its square is positive, and hence since the sum of positive operators is positive, we obtain

$$(A - \nu_k I)^2 + \mu_k^2 I \geq \theta.$$

Therefore,

$$p(A) \geq \theta.$$

The idea is now to extend the domain of Φ_1 to the set of upper semicontinuous functions, and such set we will denote by C^{up}.

Observe that if $f \in C^{up}$, there exists a sequence of continuous functions $\{g_n\}$ such that

$$g_n \downarrow f, \text{ pointwise },$$

that is,

$$g_n(\lambda) \downarrow f(\lambda), \forall \lambda \in \mathbb{R}.$$

Considering the Weierstrass Theorem, since $g_n \in C([m, M])$, we may obtain a se quence of polynomials $\{p_n\}$ such that

$$\left\| \left(g_n + \frac{1}{2^n} \right) - p_n \right\|_\infty < \frac{1}{2^n},$$

where the norm $\|\cdot\|_\infty$ refers to $[m,M]$. Thus

$$p_n(\lambda) \downarrow f(\lambda), \text{ on } [m,M].$$

Therefore

$$p_1(A) \geq p_2(A) \geq p_3(A) \geq \ldots \geq p_n(A) \geq \ldots$$

Since $p_n(A)$ is self-adjoint for all $n \in \mathbb{N}$, we have

$$p_j(A)p_k(A) = p_k(A)p_j(A), \forall j,k \in \mathbb{N}.$$

Then the $\lim_{n\to\infty} p_n(A)$ (in norm) exists, and we denote

$$\lim_{n\to\infty} p_n(A) = f(A).$$

Now recall the Dini's theorem.

Theorem 3.6.3 (Dini). *Let $\{g_n\}$ be a sequence of continuous functions defined on a compact set $K \subset \mathbb{R}$. Suppose $g_n \to g$ point-wise and monotonically on K. Under such assumptions the convergence in question is also uniform.*

Now suppose that $\{p_n\}$ and $\{q_n\}$ are sequences of polynomial such that

$$p_n \downarrow f, \text{ and } q_n \downarrow f,$$

we will show that

$$\lim_{n\to\infty} p_n(A) = \lim_{n\to\infty} q_n(A).$$

First observe that being $\{p_n\}$ and $\{q_n\}$ sequences of continuous functions we have that

$$\hat{h}_{nk}(\lambda) = \max\{p_n(\lambda), q_k(\lambda)\}, \forall \lambda \in [m,M]$$

is also continuous, $\forall n,k \in \mathbb{N}$. Now fix $n \in \mathbb{N}$ and define

$$h_k(\lambda) = \max\{p_k(\lambda), q_n(\lambda)\}.$$

Observe that

$$h_k(\lambda) \downarrow q_n(\lambda), \forall \lambda \in \mathbb{R},$$

so that by Dini's theorem

$$h_k \to q_n, \text{ uniformly on } [m,M].$$

It follows that for each $n \in \mathbb{N}$ there exists $k_n \in \mathbb{N}$ such that if $k > k_n$ then

$$h_k(\lambda) - q_n(\lambda) \leq \frac{1}{n}, \forall \lambda \in [m,M].$$

Since

$$p_k(\lambda) \leq h_k(\lambda), \forall \lambda \in [m,M],$$

we obtain

$$p_k(\lambda) - q_n(\lambda) \leq \frac{1}{n}, \forall \lambda \in [m, M].$$

By analogy, we may show that for each $n \in \mathbb{N}$ there exists $\hat{k}_n \in \mathbb{N}$ such that if $k > \hat{k}_n$, then

$$q_k(\lambda) - p_n(\lambda) \leq \frac{1}{n}.$$

From above we obtain

$$\lim_{k \to \infty} p_k(A) \leq q_n(A) + \frac{1}{n}.$$

Since the self-adjoint $q_n(A) + 1/n$ commutes with the

$$\lim_{k \to \infty} p_k(A)$$

we obtain

$$\begin{aligned} \lim_{k \to \infty} p_k(A) &\leq \lim_{n \to \infty} \left(q_n(A) + \frac{1}{n} \right) \\ &\leq \lim_{n \to \infty} q_n(A). \end{aligned} \tag{3.21}$$

Similarly we may obtain

$$\lim_{k \to \infty} q_k(A) \leq \lim_{n \to \infty} p_n(A),$$

so that

$$\lim_{n \to \infty} q_n(A) = \lim_{n \to \infty} p_n(A) = f(A).$$

Hence, we may extend $\Phi_1 : P \to \mathscr{L}(H)$ to $\Phi_2 : C^{up} \to \mathscr{L}(H)$, where C^{up}, as earlier indicated, denotes the set of upper semicontinuous functions, where

$$\Phi_2(f) = f(A).$$

Observe that Φ_2 has the following properties:

1. $\Phi_2(f_1 + f_2) = \Phi_2(f_1) + \Phi_2(f_2)$,
2. $\Phi_2(f_1 \cdot f_2) = f_1(A) f_2(A)$,
3. $\Phi_2(\alpha f) = \alpha \Phi_2(f), \forall \alpha \in \mathbb{R},\ \alpha \geq 0$,
4. if $f_1(\lambda) \geq f_2(\lambda), \forall \lambda \in [m, M]$, then

$$f_1(A) \geq f_2(A).$$

The next step is to extend Φ_2 to $\Phi_3 : C^{up}_{-} \to \mathscr{L}(H)$, where

$$C^{up}_{-} = \{ f - g \mid f, g \in C^{up} \}.$$

For $h = f - g \in C^{up}_{-}$ we define

$$\Phi_3(h) = f(A) - g(A).$$

Now we will show that Φ_3 is well defined. Suppose that $h \in C_-^{up}$ and

$$h = f_1 - g_1 \text{ and } h = f_2 - g_2.$$

Thus

$$f_1 - g_1 = f_2 - g_2,$$

that is

$$f_1 + g_2 = f_2 + g_1,$$

so that from the definition of Φ_2 we obtain

$$f_1(A) + g_2(A) = f_2(A) + g_1(A),$$

that is,

$$f_1(A) - g_1(A) = f_2(A) - g_2(A).$$

Therefore Φ_3 is well defined. Finally observe that for $\alpha < 0$

$$\alpha(f - g) = -\alpha g - (-\alpha)f,$$

where $-\alpha g \in C^{up}$ and $-\alpha f \in C^{up}$. Thus

$$\Phi_3(\alpha f) = \alpha f(A) = \alpha \Phi_3(f), \forall \alpha \in \mathbb{R}.$$

3.6.1 The Spectral Theorem

Consider the upper semicontinuous function

$$h_\mu(\lambda) = \begin{cases} 1, \text{ if } \lambda \leq \mu, \\ 0, \text{ if } \lambda > \mu. \end{cases} \tag{3.22}$$

Denote

$$E(\mu) = \Phi_3(h_\mu) = h_\mu(A).$$

Observe that

$$h_\mu(\lambda)h_\mu(\lambda) = h_\mu(\lambda), \forall \lambda \in \mathbb{R},$$

so that

$$[E(\mu)]^2 = E(\mu), \forall \mu \in \mathbb{R}.$$

Therefore

$$\{E(\mu) \mid \mu \in \mathbb{R}\}$$

is a family of orthogonal projections. Also observe that if $\nu \geq \mu$, we have

$$h_\nu(\lambda)h_\mu(\lambda) = h_\mu(\lambda)h_\nu(\lambda) = h_\mu(\lambda),$$

so that

$$E(\nu)E(\mu) = E(\mu)E(\nu) = E(\mu), \forall \nu \geq \mu.$$

If $\mu < m$, then $h_\mu(\lambda) = 0$, on $[m, M]$, so that

$$E(\mu) = 0, \text{ if } \mu < m.$$

Similarly, if $\mu \geq M$, then $h_\mu(\lambda) = 1$, on $[m, M]$, so that

$$E(\mu) = I, \text{ if } \mu \geq M.$$

Next we show that the family $\{E(\mu)\}$ is strongly continuous from the right. First we will establish a sequence of polynomials $\{p_n\}$ such that

$$p_n \downarrow h_\mu$$

and

$$p_n(\lambda) \geq h_{\mu+\frac{1}{n}}(\lambda), \text{ on } [m, M].$$

Observe that for any fixed n there exists a sequence of polynomials $\{p_j^n\}$ such that

$$p_j^n \downarrow h_{\mu+1/n}, \text{ point-wise.}$$

Consider the monotone sequence

$$g_n(\lambda) = \min\{p_s^r(\lambda) \mid r, s \in \{1, \ldots, n\}\}.$$

Thus

$$g_n(\lambda) \geq h_{\mu+\frac{1}{n}}(\lambda), \forall \lambda \in \mathbb{R},$$

and we obtain

$$\lim_{n\to\infty} g_n(\lambda) \geq \lim_{n\to\infty} h_{\mu+\frac{1}{n}}(\lambda) = h_\mu(\lambda).$$

On the other hand

$$g_n(\lambda) \leq p_n^r(\lambda), \forall \lambda \in \mathbb{R}, \forall r \in \{1, \ldots, n\},$$

so that

$$\lim_{n\to\infty} g_n(\lambda) \leq \lim_{n\to\infty} p_n^r(\lambda).$$

Therefore

$$\begin{aligned} \lim_{n\to\infty} g_n(\lambda) &\leq \lim_{r\to\infty} \lim_{n\to\infty} p_n^r(\lambda) \\ &= h_\mu(\lambda). \end{aligned} \tag{3.23}$$

Thus

$$\lim_{n\to\infty} g_n(\lambda) = h_\mu(\lambda).$$

Observe that g_n are not necessarily polynomials. To set a sequence of polynomials, observe that we may obtain a sequence $\{p_n\}$ of polynomials such that

$$|g_n(\lambda) + 1/n - p_n(\lambda)| < \frac{1}{2^n}, \forall \lambda \in [m, M],\ n \in \mathbb{N},$$

so that

$$p_n(\lambda) \geq g_n(\lambda) + 1/n - 1/2^n \geq g_n(\lambda) \geq h_{\mu+1/n}(\lambda).$$

Thus

$$p_n(A) \to E(\mu),$$

and

$$p_n(A) \geq h_{\mu+\frac{1}{n}}(A) = E(\mu + 1/n) \geq E(\mu).$$

Therefore we may write

$$E(\mu) = \lim_{n\to\infty} p_n(A) \geq \lim_{n\to\infty} E(\mu + 1/n) \geq E(\mu).$$

Thus

$$\lim_{n\to\infty} E(\mu + 1/n) = E(\mu).$$

From this we may easily obtain the strong continuity from the right.

For $\mu \leq \nu$ we have

$$\begin{aligned}\mu(h_\nu(\lambda) - h_\mu(\lambda)) &\leq \lambda(h_\nu(\lambda) - h_\mu(\lambda)) \\ &\leq \nu(h_\nu(\lambda) - h_\mu(\lambda)).\end{aligned} \tag{3.24}$$

To verify this observe that if $\lambda < \mu$ or $\lambda > \nu$, then all terms involved in the above inequalities are zero. On the other hand if

$$\mu \leq \lambda \leq \nu$$

then

$$h_\nu(\lambda) - h_\mu(\lambda) = 1,$$

so that in any case (3.24) holds. From the monotonicity property we have

$$\begin{aligned}\mu(E(\nu) - E(\mu)) &\leq A(E(\nu) - E(\mu)) \\ &\leq \nu(E(\nu) - E(\mu)).\end{aligned} \tag{3.25}$$

Now choose $a, b \in \mathbb{R}$ such that

$$a < m \text{ and } b \geq M.$$

Suppose given $\varepsilon > 0$. Choose a partition P_0 of $[a,b]$, that is,

$$P_0 = \{a = \lambda_0, \lambda_1, \ldots, \lambda_n = b\},$$

such that

$$\max_{k \in \{1,\ldots,n\}} \{|\lambda_k - \lambda_{k-1}|\} < \varepsilon.$$

Hence

$$\begin{aligned} \lambda_{k-1}(E(\lambda_k) - E(\lambda_{k-1})) &\leq A(E(\lambda_k) - E(\lambda_{k-1})) \\ &\leq \lambda_k(E(\lambda_k) - E(\lambda_{k-1})). \end{aligned} \tag{3.26}$$

Summing up on k and recalling that

$$\sum_{k=1}^{n} E(\lambda_k) - E(\lambda_{k-1}) = I,$$

we obtain

$$\begin{aligned} \sum_{k=1}^{n} \lambda_{k-1}(E(\lambda_k) - E(\lambda_{k-1})) &\leq A \\ &\leq \sum_{k=1}^{n} \lambda_k(E(\lambda_k) - E(\lambda_{k-1})). \end{aligned} \tag{3.27}$$

Let $\lambda_k^0 \in [\lambda_{k-1}, \lambda_k]$. Since $(\lambda_k - \lambda_k^0) \leq (\lambda_k - \lambda_{k-1})$ from (3.26) we obtain

$$\begin{aligned} A - \sum_{k=1}^{n} \lambda_k^0(E(\lambda_k) - E(\lambda_{k-1})) &\leq \varepsilon \sum_{k=1}^{n} (E(\lambda_k) - E(\lambda_{k-1})) \\ &= \varepsilon I. \end{aligned} \tag{3.28}$$

By analogy

$$-\varepsilon I \leq A - \sum_{k=1}^{n} \lambda_k^0(E(\lambda_k) - E(\lambda_{k-1})). \tag{3.29}$$

Since

$$A - \sum_{k=1}^{n} \lambda_k^0(E(\lambda_k) - E(\lambda_{k-1}))$$

is self adjoint we obtain

$$\|A - \sum_{k=1}^{n} \lambda_k^0(E(\lambda_k) - E(\lambda_{k-1}))\| < \varepsilon.$$

Being $\varepsilon > 0$ arbitrary, we may write

$$A = \int_a^b \lambda dE(\lambda),$$

that is,

$$A = \int_{m^-}^{M} \lambda dE(\lambda).$$

3.7 The Spectral Decomposition of Unitary Transformations

Definition 3.7.1. Let H be a Hilbert space. A transformation $U : H \to H$ is said to be unitary if

$$(Uu, Uv)_H = (u, v)_H, \forall u, v \in H.$$

Observe that in this case

$$U^*U = UU^* = I,$$

so that

$$U^{-1} = U^*.$$

Theorem 3.7.2. *Every unitary transformation U has a spectral decomposition*

$$U = \int_{0^-}^{2\pi} e^{i\phi} dE(\phi),$$

where $\{E(\phi)\}$ is a spectral family on $[0, 2\pi]$. Furthermore $E(\phi)$ is continuous at 0 and it is the limit of polynomials in U and U^{-1}.

We present just a sketch of the proof. For the trigonometric polynomials

$$p(e^{i\phi}) = \sum_{k=-n}^{n} c_k e^{ik\phi},$$

consider the transformation

$$p(U) = \sum_{k=-n}^{n} c_k U^k,$$

where $c_k \in \mathbb{C}, \forall k \in \{-n, \ldots, 0, \ldots, n\}$.

Observe that

$$\overline{p(e^{i\phi})} = \sum_{k=-n}^{n} \overline{c}_k e^{-ik\phi},$$

so that the corresponding operator is

$$p(U)^* = \sum_{k=-n}^{n} \overline{c}_k U^{-k} = \sum_{k=-n}^{n} \overline{c}_k (U^*)^k.$$

Also if

$$p(e^{i\phi}) \geq 0$$

there exists a polynomial q such that

$$p(e^{i\phi}) = |q(e^{i\phi})|^2 = \overline{q(e^{i\phi})}q(e^{i\phi}),$$

so that

$$p(U) = [q(U)]^* q(U).$$

Therefore

$$(p(U)v, v)_H = (q(U)^* q(U)v, v)_H = (q(U)v, q(U)v)_H \geq 0, \forall v \in H,$$

which means

$$p(U) \geq 0.$$

Define the function $h_\mu(\phi)$ by

$$h_\mu(\phi) = \begin{cases} 1, \text{ if } 2k\pi < \phi \leq 2k\pi + \mu, \\ 0, \text{ if } 2k\pi + \mu < \phi \leq 2(k+1)\pi, \end{cases} \tag{3.30}$$

for each $k \in \{0, \pm 1, \pm 2, \pm 3, \ldots\}$. Define $E(\mu) = h_\mu(U)$. Observe that the family $\{E(\mu)\}$ are projections and in particular

$$E(0) = 0,$$

$$E(2\pi) = I$$

and if $\mu \leq \nu$, since

$$h_\mu(\phi) \leq h_\nu(\phi),$$

we have

$$E(\mu) \leq E(\nu).$$

Suppose given $\varepsilon > 0$. Let P_0 be a partition of $[0, 2\pi]$, that is,

$$P_0 = \{0 = \phi_0, \phi_1, \ldots, \phi_n = 2\pi\}$$

such that

$$\max_{j \in \{1, \ldots, n\}} \{|\phi_j - \phi_{j-1}|\} < \varepsilon.$$

For fixed $\phi \in [0, 2\pi]$, let $j \in \{1, \ldots, n\}$ be such that

$$\phi \in [\phi_{j-1}, \phi_j].$$

$$\begin{aligned} |e^{i\phi} - \sum_{k=1}^{n} e^{i\phi_k}(h_{\phi_k}(\phi) - h_{\phi_{k-1}}(\phi))| &= |e^{i\phi} - e^{i\phi_j}| \\ &\leq |\phi - \phi_j| < \varepsilon. \end{aligned} \tag{3.31}$$

Thus,

$$0 \leq |e^{i\phi} - \sum_{k=1}^{n} e^{i\phi_k}(h_{\phi_k}(\phi) - h_{\phi_{k-1}}(\phi))|^2 \leq \varepsilon^2$$

so that, for the corresponding operators

$$0 \leq [U - \sum_{k=1}^{n} e^{i\phi_k}(E(\phi_k) - E(\phi_{k-1}))]^*[U - \sum_{k=1}^{n} e^{i\phi_k}(E(\phi_k) - E(\phi_{k-1}))]$$
$$\leq \varepsilon^2 I$$

and hence

$$\left\| U - \sum_{k=1}^{n} e^{i\phi_k}(E(\phi_k) - E(\phi_{k-1})\right\| < \varepsilon.$$

Being $\varepsilon > 0$ arbitrary, we may infer that

$$U = \int_0^{2\pi} e^{i\phi} dE(\phi).$$

3.8 Unbounded Operators

3.8.1 Introduction

Let H be a Hilbert space. Let $A : D(A) \to H$ be an operator, where unless indicated $D(A)$ is a dense subset of H. We consider in this section the special case where A is unbounded.

Definition 3.8.1. Given $A : D \to H$ we define the graph of A, denoted by $\Gamma(A)$, by

$$\Gamma(A) = \{(u, Au) \mid u \in D\}.$$

Definition 3.8.2. An operator $A : D \to H$ is said to be closed if $\Gamma(A)$ is closed.

Definition 3.8.3. Let $A_1 : D_1 \to H$ and $A_2 : D_2 \to H$ operators. We write $A_2 \supset A_1$ if $D_2 \supset D_1$ and

$$A_2 u = A_1 u, \forall u \in D_1.$$

In this case we say that A_2 is an extension of A_1.

Definition 3.8.4. A linear operator $A : D \to H$ is said to be closable if it has a linear closed extension. The smallest closed extension of A is denoted by $\overline{A}$ and is called the closure of A.

Proposition 3.8.5. *Let $A : D \to H$ be a linear operator. If A is closable, then*

$$\Gamma(\overline{A}) = \overline{\Gamma(A)}.$$

Proof. Suppose B is a closed extension of A. Then

$$\overline{\Gamma(A)} \subset \overline{\Gamma(B)} = \Gamma(B),$$

so that if $(\theta,\phi) \in \overline{\Gamma(A)}$, then $(\theta,\phi) \in \Gamma(B)$, and hence $\phi = \theta$. Define the operator C by

$$D(C) = \{\psi \mid (\psi,\phi) \in \overline{\Gamma(A)} \text{ for some } \phi\},$$

and $C(\psi) = \phi$, where ϕ is the unique point such that $(\psi,\phi) \in \overline{\Gamma(A)}$. Hence

$$\Gamma(C) = \overline{\Gamma(A)} \subset \Gamma(B),$$

so that

$$A \subset C.$$

However $C \subset B$ and since B is an arbitrary closed extension of A we have

$$C = \overline{A}$$

so that

$$\Gamma(C) = \Gamma(\overline{A}) = \overline{\Gamma(A)}.$$

Definition 3.8.6. Let $A : D \to H$ be a linear operator where D is dense in H. Define $D(A^*)$ by

$$D(A^*) = \{\phi \in H \mid (A\psi,\phi)_H = (\psi,\eta)_H,\ \forall \psi \in D \text{ for some } \eta \in H\}.$$

In this case we denote

$$A^*\phi = \eta.$$

A^* defined in this way is called the adjoint operator related to A.

Observe that by the Riesz lemma, $\phi \in D(A^*)$ if and only if there exists $K > 0$ such that

$$|(A\psi,\phi)_H| \leq K\|\psi\|_H, \forall \psi \in D.$$

Also note that if

$$A \subset B \text{ then } B^* \subset A^*.$$

Finally, as D is dense in H, then

$$\eta = A^*(\phi)$$

is uniquely defined. However the domain of A^* may not be dense, and in some situations we may have $D(A^*) = \{\theta\}$.

If $D(A^*)$ is dense, we define

$$A^{**} = (A^*)^*.$$

Theorem 3.8.7. *Let $A : D \to H$ a linear operator, being D dense in H. Then*

1. *A^* is closed,*
2. *A is closable if and only if $D(A^*)$ is dense and in this case*

$$\overline{A} = A^{**},$$

3. *If A is closable, then $(\overline{A})^* = A^*$.*

Proof.

1. We define the operator $V : H \times H \to H \times H$ by

$$V(\phi, \psi) = (-\psi, \phi).$$

Let $E \subset H \times H$ be a subspace. Thus, if $(\phi_1, \psi_1) \in V(E^\perp)$, then there exists $(\phi, \psi) \in E^\perp$ such that

$$V(\phi, \psi) = (-\psi, \phi) = (\phi_1, \psi_1).$$

Hence

$$\psi = -\phi_1 \text{ and } \phi = \psi_1,$$

so that for $(\psi_1, -\phi_1) \in E^\perp$ and $(w_1, w_2) \in E$ we have

$$((\psi_1, -\phi_1), (w_1, w_2))_{H \times H} = 0 = (\psi_1, w_1)_H + (-\phi_1, w_2)_H.$$

Thus

$$(\phi_1, -w_2)_H + (\psi_1, w_1)_H = 0,$$

and therefore

$$((\phi_1, \psi_1), (-w_2, w_1))_{H \times H} = 0,$$

that is,

$$((\phi_1, \psi_1), V(w_1, w_2))_{H \times H} = 0, \forall (w_1, w_2) \in E.$$

This means that

$$(\phi_1, \psi_1) \in (V(E))^\perp,$$

so that

$$V(E^\perp) \subset (V(E))^\perp.$$

It is easily verified that the implications from which the last inclusion results are in fact equivalences, so that

$$V(E^\perp) = (V(E))^\perp.$$

Suppose $(\phi, \eta) \in H \times H$. Thus, $(\phi, \eta) \in V(\Gamma(A))^\perp$ if and only if

$$((\phi, \eta), (-A\psi, \psi))_{H \times H} = 0, \forall \psi \in D,$$

which holds if and only if

$$(\phi, A\psi)_H = (\eta, \psi)_H, \forall \psi \in D,$$

that is, if and only if

$$(\phi, \eta) \in \Gamma(A^*).$$

Thus

$$\Gamma(A^*) = V(\Gamma(A))^{\perp}.$$

Since $(V(\Gamma(A))^{\perp}$ is closed, A^* is closed.

2. Observe that $\Gamma(A)$ is a linear subset of $H \times H$ so that

$$\begin{aligned}\overline{\Gamma(A)} &= [\Gamma(A)^{\perp}]^{\perp} \\ &= V^2[\Gamma(A)^{\perp}]^{\perp} \\ &= [V[V(\Gamma(A))^{\perp}]]^{\perp} \\ &= [V(\Gamma(A^*)]^{\perp} \end{aligned} \tag{3.32}$$

so that from the proof of item 1, if A^* is densely defined, we get

$$\overline{\Gamma(A)} = \Gamma[(A^*)^*].$$

Conversely, suppose $D(A^*)$ is not dense. Thus there exists $\psi \in [D(A^*)]^{\perp}$ such that $\psi \neq \theta$. Let $(\phi, A^*\phi) \in \Gamma(A^*)$. Hence

$$((\psi, \theta), (\phi, A^*\phi))_{H \times H} = (\psi, \phi)_H = 0,$$

so that

$$(\psi, \theta) \in [\Gamma(A^*)]^{\perp}.$$

Therefore $V[\Gamma(A^*)]^{\perp}$ is not the graph of a linear operator. Since $\overline{\Gamma(A)} = V[\Gamma(A^*)]^{\perp}$ A is not closable.

3. Observe that if A is closable, then

$$A^* = \overline{(A^*)} = A^{***} = (\overline{A})^*.$$

3.9 Symmetric and Self-Adjoint Operators

Definition 3.9.1. Let $A : D \to H$ be a linear operator, where D is dense in H. A is said to be symmetric if $A \subset A^*$, that is, if $D \subset D(A^*)$ and

$$A^*\phi = A\phi, \forall \phi \subset D.$$

Equivalently, A is symmetric if and only if

$$(A\phi, \psi)_H = (\phi, A\psi)_H, \forall \phi, \psi \in D.$$

Definition 3.9.2. Let $A : D \to H$ be a linear operator. We say that A is self-adjoint if $A = A^*$, that is, if A is symmetric and $D = D(A^*)$.

Definition 3.9.3. Let $A : D \to H$ be a symmetric operator. We say that A is essentially self-adjoint if its closure $\overline{A}$ is self-adjoint. If A is closed, a subset $E \subset D$ is said to be a core for A if $\overline{A|_E} = A$.

Theorem 3.9.4. *Let $A : D \to H$ be a symmetric operator. Then the following statements are equivalent:*

1. *A is self-adjoint,*
2. *A is closed and $N(A^* \pm iI) = \{\theta\}$,*
3. *$R(A \pm iI) = H$.*

Proof.

- 1 implies 2:
  Suppose A is self-adjoint, let $\phi \in D = D(A^*)$ be such that
  $$A\phi = i\phi$$
  so that
  $$A^*\phi = i\phi.$$
  Observe that
  $$\begin{aligned} -i(\phi,\phi)_H &= (i\phi,\phi)_H \\ &= (A\phi,\phi)_H \\ &= (\phi,A\phi)_H \\ &= (\phi,i\phi)_H \\ &= i(\phi,\phi)_H, \end{aligned} \tag{3.33}$$
  so that $(\phi,\phi)_H = 0$, that is, $\phi = \theta$. Thus
  $$N(A - iI) = \{\theta\}.$$
  Similarly we prove that $N(A + iI) = \{\theta\}$. Finally, since $\overline{A^*} = A^* = A$, we get that $A = A^*$ is closed.
- 2 implies 3:
  Suppose 2 holds. Thus the equation
  $$A^*\phi = -i\phi$$
  has no nontrivial solution. We will prove that $R(A - iI)$ is dense in H. If $\psi \in R(A - iI)^\perp$, then
  $$((A - iI)\phi, \psi)_H = 0, \forall \phi \in D,$$
  so that $\psi \in D(A^*)$ and

$$(A - iI)^* \psi = (A^* + iI)\psi = \theta,$$

and hence by above $\psi = \theta$. Now we will prove that $R(A - iI)$ is closed and conclude that

$$R(A - iI) = H.$$

Given $\phi \in D$ we have

$$\|(A - iI)\phi\|_H^2 = \|A\phi\|_H^2 + \|\phi\|_H^2. \tag{3.34}$$

Let $\psi_0 \in H$ be a limit point of $R(A - iI)$. Thus we may find $\{\phi_n\} \subset D$ such that

$$(A - iI)\phi_n \to \psi_0.$$

From (3.34)

$$\|\phi_n - \phi_m\|_H \leq \|(A - iI)(\phi_n - \phi_m)\|_H, \forall m, n \in \mathbb{N}$$

so that $\{\phi_n\}$ is a Cauchy sequence, therefore converging to some $\phi_0 \in H$. Also from (3.34)

$$\|A\phi_n - A\phi_m\|_H \leq \|(A - iI)(\phi_n - \phi_m)\|_H, \forall m, n \in \mathbb{N}$$

so that $\{A\phi_n\}$ is a Cauchy sequence, hence also a converging one. Since A is closed, we get $\phi_0 \in D$ and

$$(A - iI)\phi_0 = \psi_0.$$

Therefore $R(A - iI)$ is closed, so that

$$R(A - iI) = H.$$

Similarly

$$R(A + iI) = H.$$

- 3 implies 1: Let $\phi \in D(A^*)$. Since $R(A - iI) = H$, there is an $\eta \in D$ such that

$$(A - iI)\eta = (A^* - iI)\phi,$$

and since $D \subset D(A^*)$ we obtain $\phi - \eta \in D(A^*)$ and

$$(A^* - iI)(\phi - \eta) = \theta.$$

Since $R(A + iI) = H$ we have $N(A^* - iI) = \{\theta\}$. Therefore $\phi = \eta$, so that $D(A^*) = D$. The proof is complete.

3.9.1 The Spectral Theorem Using Cayley Transform

In this section H is a complex Hilbert space. We suppose A is defined on a dense subspace of H, being A self-adjoint but possibly unbounded. We have shown that $(A+i)$ and $(A-i)$ are onto H and it is possible to prove that

$$U=(A-i)(A+i)^{-1},$$

exists on all H and it is unitary. Furthermore, on the domain of A,

$$A=i(I+U)(I-U)^{-1}.$$

The operator U is called the Cayley transform of A. We have already proven that

$$U=\int_0^{2\pi} e^{i\phi}\,dF(\phi),$$

where $\{F(\phi)\}$ is a monotone family of orthogonal projections, strongly continuous from the right and we may consider it such that

$$F(\phi)=\begin{cases} 0, \text{ if } \phi\leq 0, \\ I, \text{ if } \phi\geq 2\pi. \end{cases} \tag{3.35}$$

Since $F(\phi)=0$, for all $\phi\leq 0$ and

$$F(0)=F(0^+)$$

we obtain

$$F(0^+)=0=F(0^-),$$

that is, $F(\phi)$ is continuous at $\phi=0$. We claim that F is continuous at $\phi=2\pi$. Observe that $F(2\pi)=F(2\pi^+)$ so that we need only to show that

$$F(2\pi^-)=F(2\pi).$$

Suppose

$$F(2\pi)-F(2\pi^-)\neq\theta.$$

Thus, there exists some $u,v\in H$ such that

$$(F(2\pi)-F(2(\pi^-)))u=v\neq\theta.$$

Therefore

$$F(\phi)v=F(\phi)[(F(2\pi)-F(2\pi^-))u],$$

so that

$$F(\phi)v=\begin{cases} 0, \text{ if } \phi<2\pi, \\ v, \text{ if } \phi\geq 2\pi. \end{cases} \tag{3.36}$$

Observe that

$$U - I = \int_0^{2\pi} (e^{i\phi} - 1) dF(\phi),$$

and

$$U^* - I = \int_0^{2\pi} (e^{-i\phi} - 1) dF(\phi).$$

Let $\{\phi_n\}$ be a partition of $[0, 2\pi]$. From the monotonicity of $[0, 2\pi]$ and pairwise orthogonality of

$$\{F(\phi_n) - F(\phi_{n-1})\}$$

we can show that (this is not proved in details here)

$$(U^* - I)(U - I) = \int_0^{2\pi} (e^{-i\phi} - 1)(e^{i\phi} - 1) dF(\phi),$$

so that, given $z \in H$, we have

$$((U^* - I)(U - I)z, z)_H = \int_0^{2\pi} |e^{i\phi} - 1|^2 d\|F(\phi)z\|^2,$$

thus, for v defined above

$$\begin{aligned}
\|(U - I)v\|^2 &= ((U - I)v, (U - I)v)_H \\
&= ((U - I)^*(U - I)v, v)_H \\
&= \int_0^{2\pi} |e^{i\phi} - 1|^2 d\|F(\phi)v\|^2 \\
&= \int_0^{2\pi^-} |e^{i\phi} - 1|^2 d\|F(\phi)v\|^2 \\
&= 0.
\end{aligned} \tag{3.37}$$

The last two equalities result from $e^{2\pi i} - 1 = 0$ and $d\|F(\phi)v\| = \theta$ on $[0, 2\pi)$. Since $v \neq \theta$ the last equation implies that $1 \in P\sigma(U)$, which contradicts the existence of

$$(I - U)^{-1}.$$

Thus, F is continuous at $\phi = 2\pi$.

Now choose a sequence of real numbers $\{\phi_n\}$ such that $\phi_n \in (0, 2\pi)$, $n = 0, \pm 1, \pm 2, \pm 3, \ldots$ such that

$$-\cot\left(\frac{\phi_n}{2}\right) = n.$$

Now define $T_n = F(\phi_n) - F(\phi_{n-1})$. Since U commutes with $F(\phi)$, U commutes with T_n. Since

$$A = i(I + U)(I - U)^{-1},$$

this implies that the range of T_n is invariant under U and A. Observe that

$$\begin{aligned}\sum_n T_n &= \sum_n (F(\phi_n) - F(\phi_{n-1}))\\ &= \lim_{\phi\to 2\pi} F(\phi) - \lim_{\phi\to 0} F(\phi)\\ &= I - \theta = I. \end{aligned} \tag{3.38}$$

Hence

$$\sum_n R(T_n) = H.$$

Also, for $u \in H$, we have that

$$F(\phi)T_n u = \begin{cases} 0, \text{ if } \phi < \phi_{n-1}, \\ (F(\phi) - F(\phi_{n-1}))u, \text{ if } \phi_{n-1} \leq \phi \leq \phi_n, \\ F(\phi_n) - F(\phi_{n-1}))u, \text{ if } \phi > \phi_n, \end{cases} \tag{3.39}$$

so that

$$\begin{aligned}(I-U)T_n u &= \int_0^{2\pi} (1 - e^{i\phi})dF(\phi)T_n u\\ &= \int_{\phi_{n-1}}^{\phi_n} (1 - e^{i\phi})dF(\phi)u. \end{aligned} \tag{3.40}$$

Therefore

$$\begin{aligned}&\int_{\phi_{n-1}}^{\phi_n} (1 - e^{i\phi})^{-1} dF(\phi)(I-U)T_n u\\ &= \int_{\phi_{n-1}}^{\phi_n} (1 - e^{i\phi})^{-1} dF(\phi) \int_{\phi_{n-1}}^{\phi_n} (1 - e^{i\phi})dF(\phi)u\\ &= \int_{\phi_{n-1}}^{\phi_n} (1 - e^{i\phi})^{-1}(1 - e^{i\phi})dF(\phi)u\\ &= \int_{\phi_{n-1}}^{\phi_n} dF(\phi)u\\ &= \int_0^{2\pi} dF(\phi)T_n u = T_n u. \end{aligned} \tag{3.41}$$

Hence

$$\left[(I-U)|_{R(T_n)}\right]^{-1} = \int_{\phi_{n-1}}^{\phi_n} (1 - e^{i\phi})^{-1} dF(\phi).$$

From this, from above, and as

$$A = i(I+U)(I-U)^{-1}$$

we obtain

$$AT_nu = \int_{\phi_{n-1}}^{\phi_n} i(1+e^{i\phi})(1-e^{i\phi})^{-1}dF(\phi)u.$$

Therefore defining

$$\lambda = -\cot\left(\frac{\phi}{2}\right),$$

and

$$E(\lambda) = F(-2\cot^{-1}\lambda),$$

we get

$$i(1+e^{i\phi})(1-e^{i\phi})^{-1} = -\cot\left(\frac{\phi}{2}\right) = \lambda.$$

Hence,

$$AT_nu = \int_{n-1}^{n} \lambda dE(\lambda)u.$$

Finally, from

$$u = \sum_{n=-\infty}^{\infty} T_nu,$$

we can obtain

$$\begin{aligned} Au &= A(\sum_{n=-\infty}^{\infty} T_nu) \\ &= \sum_{n=-\infty}^{\infty} AT_nu \\ &= \sum_{n=-\infty}^{\infty} \int_{n-1}^{n} \lambda dE(\lambda)u. \end{aligned} \tag{3.42}$$

Being the convergence in question in norm, we may write

$$Au = \int_{-\infty}^{\infty} \lambda dE(\lambda)u.$$

Since $u \in H$ is arbitrary, we may denote

$$A = \int_{-\infty}^{\infty} \lambda dE(\lambda). \tag{3.43}$$

Chapter 4
Basic Results on Measure and Integration

The main references for this chapter are Rudin [57], Royden [59], and Stein and Shakarchi [62], where more details may be found. All these three books are excellent and we strongly recommend their reading.

4.1 Basic Concepts

In this chapter U denotes a topological space.

Definition 4.1.1 (σ-algebra). A collection $\mathcal{M}$ of subsets of U is said to be a σ-algebra if $\mathcal{M}$ has the following properties:

1. $U \in \mathcal{M}$,
2. if $A \in \mathcal{M}$, then $U \setminus A \in \mathcal{M}$,
3. if $A_n \in \mathcal{M}, \forall n \in \mathbb{N}$, then $\cup_{n=0}^{\infty} A_n \in \mathcal{M}$.

Definition 4.1.2 (Measurable Spaces). If $\mathcal{M}$ is a σ-algebra in U, we say that U is a measurable space. The elements of $\mathcal{M}$ are called the measurable sets of U.

Definition 4.1.3 (Measurable Function). If U is a measurable space and V is a topological space, we say that $f : U \to V$ is a measurable function if $f^{-1}(\mathcal{V})$ is measurable whenever $\mathcal{V} \subset V$ is an open set.

Remark 4.1.4.

1. Observe that $\emptyset = U \setminus U$ so that from 1 and 2 in Definition 4.1.1, we have that $\emptyset \in \mathcal{M}$.
2. From 1 and 3 from Definition 4.1.1, it is clear that $\cup_{i=1}^{n} A_i \in \mathcal{M}$ whenever $A_i \in \mathcal{M}, \forall i \in \{1, \ldots, n\}$.
3. Since $\cap_{i=1}^{\infty} A_i = (\cup_{i=1}^{\infty} A_i^c)^c$ also from Definition 4.1.1, it is clear that $\mathcal{M}$ is closed under countable intersections.
4. Since $A \setminus B = B^c \cap A$ we obtain : if $A, B \in \mathcal{M}$, then $A \setminus B \in \mathcal{M}$.

F. Botelho, *Functional Analysis and Applied Optimization in Banach Spaces: Applications to Non-Convex Variational Models*, DOI 10.1007/978-3-319-06074-3_4,

Theorem 4.1.5. *Let $\mathscr{F}$ be any collection of subsets of U. Then there exists a smallest σ-algebra $\mathscr{M}_0$ in U such that $\mathscr{F} \subset M_0$.*

Proof. Let Ω be the family of all σ-algebras that contain $\mathscr{F}$. Since the set of all subsets in U is a σ-algebra, Ω is nonempty.

Let $\mathscr{M}_0 = \cap_{\mathscr{M}_\lambda \subset \Omega} \mathscr{M}_\lambda$; it is clear that $\mathscr{M}_0 \supset \mathscr{F}$, and it remains to prove that in fact $\mathscr{M}_0$ is a σ-algebra. Observe that:

1. $U \in \mathscr{M}_\lambda$, $\forall \mathscr{M}_\lambda \in \Omega$, so that $U \in \mathscr{M}_0$,
2. $A \in \mathscr{M}_0$ implies $A \in \mathscr{M}_\lambda, \forall \mathscr{M}_\lambda \in \Omega$, so that $A^c \in \mathscr{M}_\lambda, \forall \mathscr{M}_\lambda \in \Omega$, which means $A^c \in \mathscr{M}_0$,
3. $\{A_n\} \subset \mathscr{M}_0$ implies $\{A_n\} \subset \mathscr{M}_\lambda$, $\forall \mathscr{M}_\lambda \in \Omega$, so that $\cup_{n=1}^\infty A_n \in \mathscr{M}_\lambda$, $\forall \mathscr{M}_\lambda \in \Omega$, which means $\cup_{n=1}^\infty A_n \in \mathscr{M}_0$.

From Definition 4.1.1 the proof is complete.

Definition 4.1.6 (Borel Sets). Let U be a topological space, considering the last theorem, there exists a smallest σ-algebra in U, denoted by $\mathscr{B}$, which contains the open sets of U. The elements of $\mathscr{B}$ are called the Borel sets.

Theorem 4.1.7. *Suppose $\mathscr{M}$ is a σ-algebra in U and V is a topological space. For $f : U \to V$, we have:*

1. *If $\Omega = \{E \subset V \mid f^{-1}(E) \in \mathscr{M}\}$, then Ω is a σ-algebra.*
2. *If $V = [-\infty, \infty]$, and $f^{-1}((\alpha, \infty]) \in \mathscr{M}$, for each $\alpha \in \mathbb{R}$, then f is measurable.*

Proof.

1. (a) $V \in \Omega$ since $f^{-1}(V) = U$ and $U \in \mathscr{M}$.
   (b) $E \in \Omega \Rightarrow f^{-1}(E) \in \mathscr{M} \Rightarrow U \setminus f^{-1}(E) \in \mathscr{M} \Rightarrow f^{-1}(V \setminus E) \in \mathscr{M} \Rightarrow V \setminus E \in \Omega$.
   (c) $\{E_i\} \subset \Omega \Rightarrow f^{-1}(E_i) \in \mathscr{M}, \forall i \in \mathbb{N} \Rightarrow \cup_{i=1}^\infty f^{-1}(E_i) \in \mathscr{M} \Rightarrow f^{-1}(\cup_{i=1}^\infty E_i) \in \mathscr{M} \Rightarrow \cup_{i=1}^\infty E_i \in \Omega$.
   Thus Ω is a σ-algebra.
2. Define $\Omega = \{E \subset [-\infty, \infty] \mid f^{-1}(E) \in \mathscr{M}\}$. From above Ω is a σ-algebra. Given $\alpha \in \mathbb{R}$, let $\{\alpha_n\}$ be a real sequence such that $\alpha_n \to \alpha$ as $n \to \infty$, $\alpha_n < \alpha, \forall n \in \mathbb{N}$. Since $(\alpha_n, \infty] \in \Omega$ for each n and
$$[-\infty, \alpha) = \cup_{n=1}^\infty [-\infty, \alpha_n] = \cup_{n=1}^\infty (\alpha_n, \infty]^C, \tag{4.1}$$
we obtain $[-\infty, \alpha) \in \Omega$. Furthermore, we have $(\alpha, \beta) = [-\infty, \beta) \cap (\alpha, \infty] \in \Omega$. Since every open set in $[-\infty, \infty]$ may be expressed as a countable union of intervals (α, β) we have that Ω contains all the open sets. Thus, $f^{-1}(E) \in \mathscr{M}$ whenever E is open, so that f is measurable.

Proposition 4.1.8. *If $\{f_n : U \to [-\infty, \infty]\}$ is a sequence of measurable functions and $g = \sup_{n \geq 1} f_n$ and $h = \limsup_{n \to \infty} f_n$, then g and h are measurable.*

Proof. Observe that $g^{-1}((\alpha, \infty]) = \cup_{n=1}^\infty f_n^{-1}((\alpha, \infty])$. From the last theorem g is measurable. By analogy $h = \inf_{k \geq 1}\{\sup_{i \geq k} f_i\}$ is measurable.

4.2 Simple Functions

Definition 4.2.1 (Simple Functions). A function $f : U \to \mathbb{C}$ is said to be a simple function if its range $(R(f))$ has only finitely many points. If $\{\alpha_1, \ldots, \alpha_n\} = R(f)$ and we set $A_i = \{u \in U \mid f(u) = \alpha_i\}$, clearly we have $f = \sum_{i=1}^n \alpha_i \chi_{A_i}$, where

$$\chi_{A_i}(u) = \begin{cases} 1, \text{ if } u \in A_i, \\ 0, \text{ otherwise.} \end{cases} \tag{4.2}$$

Theorem 4.2.2. *Let $f : U \to [0, \infty]$ be a measurable function. Thus there exists a sequence of simple functions $\{s_n : U \to [0, \infty]\}$ such that*

1. $0 \leq s_1 \leq s_2 \leq \ldots \leq f,$
2. $s_n(u) \to f(u)$ *as* $n \to \infty, \forall u \in U.$

Proof. Define $\delta_n = 2^{-n}$. To each $n \in \mathbb{N}$ and each $t \in \mathbb{R}^+$, there corresponds a unique integer $K = K_n(t)$ such that

$$K\delta_n \leq t \leq (K+1)\delta_n. \tag{4.3}$$

Defining

$$\varphi_n(t) = \begin{cases} K_n(t)\delta_n, & \text{if } 0 \leq t < n, \\ n, & \text{if } t \geq n, \end{cases} \tag{4.4}$$

we have that each φ_n is a Borel function on $[0, \infty]$, such that

1. $t - \delta_n < \varphi_n(t) \leq t$ if $0 \leq t \leq n,$
2. $0 \leq \varphi_1 \leq \ldots \leq t,$
3. $\varphi_n(t) \to t$ as $n \to \infty, \forall t \in [0, \infty].$

It follows that the sequence $\{s_n = \varphi_n \circ f\}$ corresponds to the results indicated above.

4.3 Measures

Definition 4.3.1 (Measure). Let $\mathcal{M}$ be a σ-algebra on a topological space U. A function $\mu : \mathcal{M} \to [0, \infty]$ is said to be a measure if $\mu(\emptyset) = 0$ and μ is countably additive, that is, given $\{A_i\} \subset U$, a sequence of pair wise disjoint sets then

$$\mu(\cup_{i=1}^\infty A_i) = \sum_{i=1}^\infty \mu(A_i). \tag{4.5}$$

In this case $(U, \mathcal{M}, \mu)$ is called a measure space.

Proposition 4.3.2. *Let $\mu : \mathcal{M} \to [0,\infty]$, where $\mathcal{M}$ is a σ-algebra of U. Then we have the following:*

1. *$\mu(A_1 \cup \ldots \cup A_n) = \mu(A_1) + \ldots + \mu(A_n)$ for any given $\{A_i\}$ of pairwise disjoint measurable sets of $\mathcal{M}$.*
2. *If $A, B \in \mathcal{M}$ and $A \subset B$, then $\mu(A) \leq \mu(B)$.*
3. *If $\{A_n\} \subset \mathcal{M}$, $A = \cup_{n=1}^{\infty} A_n$ and*

$$A_1 \subset A_2 \subset A_3 \subset \ldots \tag{4.6}$$

then $\lim_{n\to\infty} \mu(A_n) = \mu(A)$.

4. *If $\{A_n\} \subset \mathcal{M}$, $A = \cap_{n=1}^{\infty} A_n$, $A_1 \supset A_2 \supset A_3 \supset \ldots$., and $\mu(A_1)$ is finite, then*

$$\lim_{n\to\infty} \mu(A_n) = \mu(A). \tag{4.7}$$

Proof.

1. Take $A_{n+1} = A_{n+2} = \ldots = \emptyset$ in Definition 4.1.1 item 1.
2. Observe that $B = A \cup (B - A)$ and $A \cap (B - A) = \emptyset$ so that by the above, $\mu(A \cup (B - A)) = \mu(A) + \mu(B - A) \geq \mu(A)$.
3. Let $B_1 = A_1$ and let $B_n = A_n - A_{n-1}$; then $B_n \in \mathcal{M}$, $B_i \cap B_j = \emptyset$ if $i \neq j$, $A_n = B_1 \cup \ldots \cup B_n$, and $A = \cup_{i=1}^{\infty} B_i$. Thus

$$\mu(A) = \mu(\cup_{i=1}^{\infty} B_i) = \sum_{n=1}^{\infty} \mu(B_i) = \lim_{n\to\infty} \sum_{i=1}^{n} \mu(B_i) = \lim_{n\to\infty} \mu(A_n). \tag{4.8}$$

4. Let $C_n = A_1 \setminus A_n$. Then $C_1 \subset C_2 \subset \ldots$, $\mu(C_n) = \mu(A_1) - \mu(A_n)$, $A_1 \setminus A = \cup_{n=1}^{\infty} C_n$. Thus by 3 we have

$$\mu(A_1) - \mu(A) = \mu(A_1 \setminus A) = \lim_{n\to\infty} \mu(C_n) = \mu(A_1) - \lim_{n\to\infty} \mu(A_n). \tag{4.9}$$

4.4 Integration of Simple Functions

Definition 4.4.1 (Integral for Simple Functions). For $s : U \to [0,\infty]$, a measurable simple function, that is,

$$s = \sum_{i=1}^{n} \alpha_i \chi_{A_i}, \tag{4.10}$$

where

$$\chi_{A_i}(u) = \begin{cases} 1, & \text{if } u \in A_i, \\ 0, & \text{otherwise}, \end{cases} \tag{4.11}$$

we define the integral of s over $E \subset \mathcal{M}$, denoted by $\int_E s \, d\mu$ as

$$\int_E s\, d\mu = \sum_{i=1}^{n} \alpha_i \mu(A_i \cap E). \tag{4.12}$$

The convention $0.\infty = 0$ is used here.

Definition 4.4.2 (Integral for Nonnegative Measurable Functions). If $f : U \to [0,\infty]$ is measurable, for $E \in \mathscr{M}$, we define the integral of f on E, denoted by $\int_E f d\mu$, as

$$\int_E f d\mu = \sup_{s\in A} \left\{ \int_E s d\mu \right\}, \tag{4.13}$$

where

$$A = \{s \text{ simple and measurable} \mid 0 \leq s \leq f\}. \tag{4.14}$$

Definition 4.4.3 (Integrals for Measurable Functions). For a measurable $f : U \to [-\infty,\infty]$ and $E \in \mathscr{M}$, we define $f^+ = \max\{f,0\}$, $f^- = \max\{-f,0\}$ and the integral of f on E, denoted by $\int_E f\, d\mu$, as

$$\int_E f\, d\mu = \int_E f^+\, d\mu - \int_E f^-\, d\mu.$$

Theorem 4.4.4 (Lebesgue's Monotone Convergence Theorem). *Let $\{f_n\}$ be a sequence of real measurable functions on U and suppose that*

1. $0 \leq f_1(u) \leq f_2(u) \leq \ldots \leq \infty, \forall u \in U,$
2. $f_n(u) \to f(u)$ *as* $n \to \infty, \forall u \in U.$

Then,

(a) f *is measurable,*
(b) $\int_U f_n d\mu \to \int_U f d\mu$ *as* $n \to \infty$.

Proof. Since $\int_U f_n d\mu \leq \int_U f_{n+1} d\mu, \forall n \in \mathbb{N}$, there exists $\alpha \in [0,\infty]$ such that

$$\int_U f_n d\mu \to \alpha, \text{ as } n \to \infty, \tag{4.15}$$

By Proposition 4.1.8, f is measurable, and since $f_n \leq f$, we have

$$\int_U f_n d\mu \leq \int_U f\, d\mu. \tag{4.16}$$

From (4.15) and (4.16), we obtain

$$\alpha \leq \int_U f d\mu. \tag{4.17}$$

Let s be any simple function such that $0 \leq s \leq f$, and let $c \in \mathbb{R}$ such that $0 < c < 1$. For each $n \in \mathbb{N}$ we define

$$E_n = \{u \in U \mid f_n(u) \geq cs(u)\}. \tag{4.18}$$

Clearly E_n is measurable and $E_1 \subset E_2 \subset \ldots$ and $U = \cup_{n \in \mathbb{N}} E_n$. Observe that

$$\int_U f_n \, d\mu \geq \int_{E_n} f_n \, d\mu \geq c \int_{E_n} s d\mu. \tag{4.19}$$

Letting $n \to \infty$ and applying Proposition 4.3.2, we obtain

$$\alpha = \lim_{n \to \infty} \int_U f_n \, d\mu \geq c \int_U s d\mu, \tag{4.20}$$

so that

$$\alpha \geq \int_U s d\mu, \forall s \text{ simple and measurable such that } 0 \leq s \leq f. \tag{4.21}$$

This implies

$$\alpha \geq \int_U f d\mu. \tag{4.22}$$

From (4.17) and (4.22) the proof is complete.

We do not prove the next result (it is a direct consequence of the last theorem). For a proof see [57].

Corollary 4.4.5. *Let $\{f_n\}$ be a sequence of nonnegative measurable functions defined on U ($f_n : U \to [0, \infty], \forall n \in \mathbb{N}$). Defining $f(u) = \sum_{n=1}^{\infty} f_n(u), \forall u \in U$, we have*

$$\int_U f \, d\mu = \sum_{n=1}^{\infty} \int_U f_n \, d\mu.$$

Theorem 4.4.6 (Fatou's Lemma). *If $\{f_n : U \to [0, \infty]\}$ is a sequence of measurable functions, then*

$$\int_U \liminf_{n \to \infty} f_n \, d\mu \leq \liminf_{n \to \infty} \int_U f_n d\mu. \tag{4.23}$$

Proof. For each $k \in \mathbb{N}$ define $g_k : U \to [0, \infty]$ by

$$g_k(u) = \inf_{i \geq k} \{f_i(u)\}. \tag{4.24}$$

Then

$$g_k \leq f_k \tag{4.25}$$

so that

$$\int_U g_k \, d\mu \leq \int_U f_k \, d\mu, \forall k \in \mathbb{N}. \tag{4.26}$$

Also $0 \leq g_1 \leq g_2 \leq \ldots$, each g_k is measurable, and

$$\lim_{k\to\infty} g_k(u) = \liminf_{n\to\infty} f_n(u), \forall u \in U. \tag{4.27}$$

From the Lebesgue monotone convergence theorem

$$\liminf_{k\to\infty} \int_U g_k \, d\mu = \lim_{k\to\infty} \int_U g_k \, d\mu = \int_U \liminf_{n\to\infty} f_n \, d\mu. \tag{4.28}$$

From (4.26) we have

$$\liminf_{k\to\infty} \int_U g_k \, d\mu \leq \liminf_{k\to\infty} \left\{ \int_U f_k \, d\mu \right\}. \tag{4.29}$$

Thus, from (4.28) and (4.29), we obtain

$$\int_U \liminf_{n\to\infty} f_n \, d\mu \leq \liminf_{n\to\infty} \int_U f_n \, d\mu. \tag{4.30}$$

Theorem 4.4.7 (Lebesgue's Dominated Convergence Theorem). *Suppose $\{f_n\}$ is sequence of complex measurable functions on U such that*

$$\lim_{n\to\infty} f_n(u) = f(u), \forall u \in U. \tag{4.31}$$

If there exists a measurable function $g : U \to \mathbb{R}^+$ such that $\int_U g \, d\mu < \infty$ and $|f_n(u)| \leq g(u), \forall u \in U,\ n \in \mathbb{N}$, then

1. $\int_U |f| \, d\mu < \infty$,
2. $\lim_{n\to\infty} \int_U |f_n - f| \, d\mu = 0.$

Proof.

1. This inequality holds since f is measurable and $|f| \leq g$.
2. Since $2g - |f_n - f| \geq 0$, we may apply Fatou's lemma and obtain

$$\int_U 2g d\mu \leq \liminf_{n\to\infty} \int_U (2g - |f_n - f|) d\mu, \tag{4.32}$$

so that

$$\limsup_{n\to\infty} \int_U |f_n - f| \, d\mu \leq 0. \tag{4.33}$$

Hence

$$\lim_{n\to\infty} \int_U |f_n - f| \, d\mu = 0. \tag{4.34}$$

This completes the proof.

We finish this section with an important remark:

Remark 4.4.8. In a measurable space U we say that a property holds almost everywhere (a.e.) if it holds on U except for a set of measure zero. Finally, since integrals are not changed by the redefinition of the functions in question on sets of zero measure, the proprieties of items 1 and 2 of the Lebesgue monotone convergence may be considered a.e. in U, instead of in all U. Similar remarks are valid for Fatou's lemma and the Lebesgue dominated convergence theorem.

4.5 Signed Measures

In this section we study signed measures. We start with the following definition.

Definition 4.5.1. Let $(U,\mathcal{M})$ be a measurable space. We say that a measure μ is finite if $\mu(U)<\infty$. On the other hand, we say that μ is σ-finite if there exists a sequence $\{U_n\}\subset U$ such that $U=\cup_{n=1}^{\infty}U_n$ and $\mu(U_n)<\infty, \forall n\in\mathbb{N}$.

Definition 4.5.2 (Signed Measure). Let $(U,\mathcal{M})$ be a measurable space. We say that $\nu:\mathcal{M}\to[-\infty,+\infty]$ is a signed measure if

- ν may assume at most one the values $-\infty,+\infty$,
- $\nu(\emptyset)=0$,
- $\nu\left(\sum_{n=1}^{\infty}E_n\right)=\sum_{n=1}^{\infty}\nu(E_n)$ for all sequence of measurable disjoint sets $\{E_n\}$.

We say that $A\in\mathcal{M}$ is a positive set with respect to ν if A is measurable and $\nu(E)\geq 0$ for all E measurable such that $E\subset A$.

Similarly, We say that $B\in\mathcal{M}$ is a negative set with respect to ν if B is measurable and $\nu(E)\leq 0$ for all E measurable such that $E\subset B$.

Finally, if $A\in\mathcal{M}$ is both positive and negative with respect to ν, it is said to be a null set.

Lemma 4.5.3. *Considering the last definitions, we have that a countable union of positive measurable sets is positive.*

Proof. Let $A=\cup_{n=1}^{\infty}A_n$ where A_n is positive, $\forall n\in\mathbb{N}$. Choose a measurable set $E\subset A$. Set

$$E_n=(E\cap A_n)\setminus(\cup_{i=1}^{n-1}A_i).$$

Thus, E_n is a measurable subset of A_n so that $\nu(E_n)\geq 0$. Observe that

$$E=\cup_{n=1}^{\infty}E_n,$$

where $\{E_n\}$ is a sequence of measurable disjoint sets.

Therefore $\nu(E)=\sum_{n=1}^{\infty}\nu(E_n)\geq 0$.

Since $E\subset A$ is arbitrary, A is positive.

The proof is complete.

Lemma 4.5.4. *Considering the last definitions, let E be a measurable set such that*

$$0 < \nu(E) < \infty.$$

Then there exists a positive set $A \subset E$ such that $\nu(A) > 0$.

Proof. Observe that if E is not positive then it contains a set of negative measure. In such a case, let n_1 be the smallest positive integer such that there exists a measurable set $E_1 \subset E$ such that

$$\nu(E_1) < -1/n_1.$$

Reasoning inductively, if $E \setminus \left(\cup_{j=1}^{k-1} E_j\right)$ is not positive, let n_k be the smallest positive integer such that there exists a measurable set

$$E_k \subset E \setminus \left(\cup_{j=1}^{k-1} E_j\right)$$

such that

$$\nu(E_k) < -1/n_k.$$

Define

$$A = E \setminus (\cup_{k=1}^{\infty} E_k).$$

Then

$$E = A \cup (\cup_{k-1}^{\infty} E_k).$$

Since such a union is disjoint, we have

$$\nu(E) = \nu(A) + \sum_{k=1}^{\infty} \nu(E_k),$$

so that since $\nu(E) < \infty$, this last series is convergent.

Also, since

$$1/n_k < -\nu(E_k),$$

we have that

$$\sum_{k=1}^{\infty} 1/n_k$$

is convergent so that $n_k \to \infty$ as $k \to \infty$.

From $\nu(E) > 0$ we must have $\nu(A) > 0$.

Now, we will show that A is positive. Let $\varepsilon > 0$. Choose k sufficiently big such that $1/(n_k - 1) < \varepsilon$.

Since

$$A \subset E \setminus \left(\cup_{j=1}^{k} E_j\right),$$

A contains no measurable set with measure less than

$$-1/(n_k - 1) > -\varepsilon,$$

that is, A contains no measurable set with measure less than $-\varepsilon$.

Since $\varepsilon > 0$ is arbitrary, A contains no measurable negative set. Thus, A is positive.

This completes the proof.

Proposition 4.5.5 (Hahn Decomposition). *Let ν be a signed measure on a measurable space $(U, \mathcal{M})$. Then there exist a positive set A and a negative set B such that $U = A \cup B$ and $A \cap B = \emptyset$.*

Proof. Without losing generality, suppose ν does not assume the value $+\infty$ (the other case may be dealt similarly). Define

$$\lambda = \sup\{\nu(A) \mid A \text{ is positive }\}.$$

Since the empty set $\emptyset$ is positive, we obtain $\lambda \geq 0$.

Let $\{A_n\}$ be a sequence of positive sets such that

$$\lim_{n\to\infty} \nu(A_n) = \lambda.$$

Define

$$A = \cup_{i=1}^{\infty} A_i.$$

From Lemma 4.5.3, A is a positive set, so that

$$\lambda \geq \nu(A).$$

On the other hand,

$$A \setminus A_n \subset A$$

so that

$$\nu(A - A_n) \geq 0, \forall n \in \mathbb{N}.$$

Therefore

$$\nu(A) = \nu(A_n) + \nu(A \setminus A_n) \geq \nu(A_n), \forall n \in \mathbb{N}.$$

Hence

$$\nu(A) \geq \lambda,$$

so that $\lambda = \nu(A)$.

Let $B = U \setminus A$. Suppose $E \subset B$, so that E is positive. Hence,

$$\begin{aligned} \lambda &\geq \nu(E \cup A) \\ &= \nu(E) + \nu(A) \\ &= \nu(E) + \lambda, \end{aligned} \tag{4.35}$$

so that $\nu(E) = 0$.

Thus, B contains no positive set of positive measure, so that by Lemma 4.5.4, B contains no subsets of positive measure, that is, B is negative.

The proof is complete.

Remark 4.5.6. Denoting the Hahn decomposition of U relating ν by $\{A,B\}$, we may define the measures ν^+ and ν^- by

$$\nu^+(E) = \nu(E \cap A),$$

and

$$\nu^-(E) = -\nu(E \cap B),$$

so that

$$\nu = \nu^+ - \nu^-.$$

We recall that two measures ν_1 and ν_2 are mutually singular if there are disjoint measurable sets such that

$$U = A \cup B$$

and

$$\nu_1(A) = \nu_2(B) = 0.$$

Observe that the measures ν^+ and ν^- above defined are mutually singular. The decomposition

$$\nu = \nu^+ - \nu^-$$

is called the Jordan one of ν. The measures ν^+ and ν^- are called the positive and negative parts of ν, respectively.

Observe that either ν^+ or ν^- is finite since only one of the values $+\infty, -\infty$ may be assumed by ν. We may also define

$$|\nu|(E) = \nu^+(E) + \nu^-(E),$$

which is called the absolute value or total variation of ν.

4.6 The Radon–Nikodym Theorem

We start this section with the definition of absolutely continuous measures.

Definition 4.6.1 (Absolutely Continuous Measures). We say that a measure ν is absolutely continuous with respect to a measure μ and write $\nu \ll \mu$, if $\nu(A) = 0$ for all set such that $\mu(A) = 0$. In case of a signed measure we write $\nu \ll \mu$ if $|\nu| \ll |\mu|$.

Theorem 4.6.2 (The Radon–Nikodym Theorem). *Let $(U, \mathscr{M}, \mu)$ be a σ-finite measure space. Let ν be a measure defined on $\mathscr{M}$ which is absolutely continuous with respect to μ, that is, $\nu \ll \mu$.*

Then there exists a nonnegative measurable function f such that

$$\nu(E) = \int_E f\, d\mu, \forall E \in \mathscr{M}.$$

The function f is unique up to usual representatives.

Proof. First assume ν and μ are finite.

Define $\lambda = \nu + \mu$. Also define the functional F by

$$F(f) = \int_U f \, d\mu.$$

We recall that $f \in L^2(\mu)$ if f is measurable and

$$\int_U |f|^2 \, d\mu < \infty.$$

The space $L^2(\mu)$ is a Hilbert one with inner product

$$(f,g)_{L^2(\mu)} = \int_U fg \, d\mu.$$

Observe that from the Cauchy–Schwartz inequality, we may write

$$\begin{aligned} |F(f)| &= |(f,1)_{L^2(\mu)}| \\ &\leq \|f\|_{L^2(\mu)}[\mu(U)]^{1/2} \\ &\leq \|f\|_{L^2(\lambda)}[\mu(U)]^{1/2}, \end{aligned} \tag{4.36}$$

since

$$\|f\|^2_{L^2(\mu)} = \int_U |f|^2 \, d\mu \leq \int_U |f|^2 \, d\lambda = \|f\|^2_{L^2(\lambda)}.$$

Thus, F is a bounded linear functional on $L^2(\lambda)$, where $f \in L^2(\lambda)$, if f is measurable and

$$\int_U f^2 \, d\lambda < \infty.$$

Since $L^2(\lambda)$ is also a Hilbert space with the inner product

$$(f,g)_{L^2(\lambda)} = \int_U fg \, d\lambda,$$

from the Riesz representation theorem, there exists $g \in L^2(\lambda)$, such that

$$F(f) = \int_U fg \, d\lambda.$$

Thus,

$$\int_U f \, d\mu = \int_U fg \, d\lambda,$$

and in particular,

$$\int_U f \, d\mu = \int_U fg \, (d\mu + d\nu).$$

Hence

$$\int_U f(1-g)\,d\mu = \int_U fg\,d\nu. \tag{4.37}$$

Assume, to obtain contradiction, that $g<0$ in a set A such that $\mu(A)>0$.

Thus

$$\int_A (1-g)\,d\mu > 0,$$

so that from this and (4.37) with $f=\chi_A$ we get

$$\int_A g\,d\nu > 0.$$

Since $g<0$ on A we have a contradiction. Thus $g\geq 0$, a.e. $[\mu]$ on U.

Now, assume, also to obtain contradiction, that $g>1$ on set B such that $\mu(B)>0$.

Thus

$$\int_B (1-g)\,d\mu \leq 0,$$

so that from this and (4.37) with $f=\chi_B$ we obtain

$$\nu(B) \leq \int_B g\,d\nu \leq 0$$

and hence

$$\nu(B)=0.$$

Thus, $\int_B g\,d\nu = 0$ so that

$$\int_B (1-g)\,d\mu = 0,$$

which implies that $\mu(B)=0$, a contradiction.

From above we conclude that

$$0\leq g\leq 1, \text{ a.e. } [\mu] \text{ in } U.$$

On the other hand, for a fixed E μ-measurable again from (4.37) with $f=\chi_E$, we get

$$\int_E (1-g)\,d\mu = \int_E g\,d\nu,$$

so that

$$\int_E (1-g)\,d\mu = \int_E g\,d\nu - \int_E d\nu + \nu(E),$$

and therefore

$$\nu(E) = \int_E (1-g)\,(d\mu + d\nu),$$

that is,

$$\nu(E) = \int_E (1-g)\,d\lambda, \forall E \in \mathcal{M}.$$

Define

$$B = \{u \in U \, : \, g(u) = 0\}.$$

Hence, $\mu(B) = \int_B g \, d\lambda = 0$.

From this, since $\lambda \ll \mu$, we obtain

$$g^{-1} g = 1, \text{ a.e. } [\lambda].$$

Therefore, for a not relabeled $E \in \mathcal{M}$, we have

$$\lambda(E) = \int_E g^{-1} g \, d\lambda = \int_E g^{-1} \, d\mu.$$

Finally, observe that

$$\begin{aligned} \mu(E) + \nu(E) &= \lambda(E) \\ &= \int_E d\lambda \\ &= \int_E g^{-1} \, d\mu. \end{aligned} \tag{4.38}$$

Thus,

$$\begin{aligned} \nu(E) &= \int_E g^{-1} \, d\mu - \mu(E) \\ &= \int_E (g^{-1} - 1) \, d\mu \\ &= \int_E (1 - g) g^{-1} \, d\mu, \forall E \in \mathcal{M}. \end{aligned} \tag{4.39}$$

The proof for the finite case is complete. The proof for σ-finite is developed in the next lines.

Since U is σ-finite, there exists a sequence $\{U_n\}$ such that $U = \cup_{n=1}^{\infty} U_n$, and $\mu(U_n) < \infty$ and $\nu(U_n) < \infty, \forall n \in \mathbb{N}$.

Define

$$F_n = U_n \setminus \left(\cup_{j=1}^{n-1} U_j \right),$$

thus $U = \cup_{n=1}^{\infty} F_n$ and $\{F_n\}$ is a sequence of disjoint sets, such that $\mu(F_n) < \infty$ and $\nu(F_n) < \infty, \forall n \in \mathbb{N}$.

Let $E \in \mathcal{M}$. For each $n \in \mathbb{N}$ from above we may obtain f_n such that

$$\nu(E \cap F_n) = \int_{E \cap F_n} f_n \, d\mu, \forall E \in \mathcal{M}.$$

From this and the monotone convergence theorem corollary we may write

$$\nu(E) = \sum_{n=1}^{\infty} \nu(E \cap F_n)$$

$$
\begin{aligned}
&= \sum_{n=1}^{\infty} \int_{E\cap F_n} f_n \, d\mu \\
&= \sum_{n=1}^{\infty} \int_{E} f_n \chi_{F_n} \, d\mu \\
&= \int_{E} \sum_{n=1}^{\infty} f_n \chi_{F_n} \, d\mu \\
&= \int_{E} f \, d\mu, \qquad (4.40)
\end{aligned}
$$

where

$$f = \sum_{n=1}^{\infty} f_n \chi_{F_n}.$$

The proof is complete.

Theorem 4.6.3 (The Lebesgue Decomposition). *Let $(U, \mathcal{M}, \mu)$ be a σ-finite measure space and let ν be a σ-finite measure defined on $\mathcal{M}$.*

Then we may find a measure ν_0, singular with respect to μ, and a measure ν_1, absolutely continuous with respect to μ, such that

$$\nu = \nu_0 + \nu_1.$$

Furthermore, the measures ν_0 and ν_1 are unique.

Proof. Since μ and ν are σ-finite measures, so is

$$\lambda = \nu + \mu.$$

Observe that ν and μ are absolutely continuous with respect to λ. Hence, by the Radon–Nikodym theorem, there exist nonnegative measurable functions f and g such that

$$\mu(E) = \int_E f \, d\lambda, \ \forall E \in \mathcal{M}$$

and

$$\nu(E) = \int_E g \, d\lambda \ \forall E \in \mathcal{M}.$$

Define

$$A = \{u \in U \mid f(u) > 0\},$$

and

$$B = \{u \in U \mid f(u) = 0\}.$$

Thus,

$$U = A \cup B,$$

and

$$A \cap B = \emptyset.$$

Also define

$$\nu_0(E) = \nu(E \cap B), \forall E \in \mathcal{M}.$$

We have that $\nu_0(A) = 0$ so that

$$\nu_0 \perp \mu.$$

Define

$$\begin{aligned}\nu_1(E) &= \nu(E \cap A) \\ &= \int_{E \cap A} g \, d\lambda.\end{aligned} \quad (4.41)$$

Therefore,

$$\nu = \nu_0 + \nu_1.$$

To finish the proof, we have only to show that

$$\nu_1 \ll \mu.$$

Let $E \in \mathcal{M}$ such that $\mu(E) = 0$. Thus

$$0 = \mu(E) = \int_E f \, d\lambda,$$

and in particular

$$\int_{(E \cap A)} f d\lambda = 0.$$

Since $f > 0$ on $A \cap E$ we conclude that

$$\lambda(A \cap E) = 0.$$

Therefore, since $\nu \ll \lambda$, we obtain

$$\nu(E \cap A) = 0,$$

so that

$$\nu_1(E) = \nu(E \cap A) = 0.$$

From this we may infer that

$$\nu_1 \ll \mu.$$

The proof of uniqueness is left to the reader.

4.7 Outer Measure and Measurability

Let U be a set. Denote by $\mathcal{P}$ the set of all subsets of U. An outer measure $\mu^* : \mathcal{P} \to [0, +\infty]$ is a set function such that

1. $\mu^*(\emptyset) = 0$,
2. if $A \subset B$, then $\mu^*(A) \leq \mu^*(B)$, $\forall A, B \subset U$,
3. if $E \subset \cup_{n=1}^{\infty} E_n$, then

$$\mu^*(E) \leq \sum_{n=1}^{\infty} \mu^*(E_n).$$

The outer measure is called finite if $\mu^*(U) < \infty$.

Definition 4.7.1 (Measurable Set). A set $E \subset U$ is said to be measurable with respect to μ^* if

$$\mu^*(A) = \mu^*(A \cap E) + \mu^*(A \cap E^c), \forall A \subset U.$$

Theorem 4.7.2. *The set $\mathscr{B}$ of μ^*-measurable sets is a σ-algebra. If $\overline{\mu}$ is defined to be μ^* restricted to $\mathscr{B}$, then $\overline{\mu}$ is a complete measure on $\mathscr{B}$.*

Proof. Let $E = \emptyset$ and let $A \subset U$.

Thus,

$$\mu^*(A) = \mu^*(A \cap \emptyset) + \mu^*(A \cap \emptyset^c) = \mu^*(A \cap U) = \mu^*(A).$$

Therefore $\emptyset$ is μ^*-measurable.

Let $E_1, E_2 \in U$ be μ^*-measurable sets. Let $A \subset U$. Thus,

$$\mu^*(A) = \mu^*(A \cap E_2) + \mu^*(A \cap E_2^c),$$

so that from the measurability of E_1 we get

$$\mu^*(A) = \mu^*(A \cap E_2) + \mu^*(A \cap E_2^c \cap E_1) + \mu^*(A \cap E_2^c \cap E_1^c). \tag{4.42}$$

Since

$$A \cap (E_1 \cup E_2) = (A \cap E_2) \cup (A \cap E_1 \cap E_2^c),.$$

we obtain

$$\mu^*(A \cap (E_1 \cup E_2)) \leq \mu^*(A \cap E_2) + \mu^*(A \cap E_2^c \cap E_1). \tag{4.43}$$

From this and (4.42) we obtain

$$\begin{aligned} \mu^*(A) &\geq \mu^*(A \cap (E_1 \cup E_2)) + \mu^*(A \cap E_1^c \cap E_2^c) \\ &= \mu^*(A \cap (E_1 \cup E_2)) + \mu^*(A \cap (E_1 \cup E_2)^c). \end{aligned} \tag{4.44}$$

Hence $E_1 \cup E_2$ is μ^*-measurable.

By induction, the union of a finite number of μ^*-measurable sets is μ^*-measurable.

Assume $E = \cup_{i=1}^{\infty} E_i$ where $\{E_i\}$ is a sequence of disjoint μ^*-measurable sets.

Define $G_n = \cup_{i=1}^{n} E_i$. Then G_n is μ^*-measurable and for a given $A \subset U$ we have

$$\begin{aligned} \mu^*(A) &= \mu^*(A \cap G_n) + \mu^*(A \cap G_n^c) \\ &\geq \mu^*(A \cap G_n) + \mu^*(A \cap E^c), \end{aligned} \tag{4.45}$$

since $E^c \subset G_n^c$, $\forall n \in \mathbb{N}$.

Observe that

$$G_n \cap E_n = E_n$$

and

$$G_n \cap E_n^c = G_{n-1}.$$

Thus, from the measurability of E_n, we may get

$$\begin{aligned}\mu^*(A \cap G_n) &= \mu^*(A \cap G_n \cap E_n) + \mu^*(A \cap G_n \cap E_n^c) \\ &= \mu^*(A \cap E_n) + \mu^*(A \cap G_{n-1}). \end{aligned} \tag{4.46}$$

By induction we obtain

$$\mu(A \cap G_n) = \sum_{i=1}^{n} \mu^*(A \cap E_i),$$

so that

$$\mu^*(A) \geq \mu^*(A \cap E^c) + \sum_{i=1}^{n} \mu^*(A \cap E_i), \forall n \in \mathbb{N},$$

that is, considering that

$$A \cap E \subset \cup_{i=1}^{\infty}(A \cap E_i),$$

we get

$$\begin{aligned}\mu^*(A) &\geq \mu^*(A \cap E^c) + \sum_{i=1}^{\infty} \mu^*(A \cap E_i) \\ &\geq \mu^*(A \cap E^c) + \mu^*(A \cap E). \end{aligned} \tag{4.47}$$

Since $A \subset U$ is arbitrary we may conclude that $E = \cup_{i=1}^{\infty} E_i$ is μ^*-measurable. Therefore $\mathscr{B}$ is a σ-algebra.

Finally, we prove that $\overline{\mu}$ is a measure.

Let $E_1, E_2 \subset U$ be two disjoint μ^*-measurable sets.

Thus

$$\begin{aligned}\overline{\mu}(E_1 \cup E_2) &= \mu^*(E_1 \cup E_2) \\ &= \mu^*((E_1 \cup E_2) \cap E_2) + \mu^*((E_1 \cup E_2) \cap E_2^c) \\ &= \mu^*(E_2) + \mu^*(E_1). \end{aligned} \tag{4.48}$$

By induction we obtain the finite additivity.

Also, if

$$E = \cup_{i=1}^{\infty} E_i,$$

where $\{E_i\}$ is a sequence of disjoint measurable sets.

Thus,

$$\overline{\mu}(E) \geq \overline{\mu}(\cup_{i=1}^{n} E_i) = \sum_{i=1}^{n} \overline{\mu}(E_i), \ \forall n \in \mathbb{N}.$$

Therefore,

$$\overline{\mu}(E) \geq \sum_{i=1}^{\infty} \overline{\mu}(E_i).$$

Now observe that

$$\overline{\mu}(E) = \mu^*(\cup_{i=1}^{\infty} E_i) \leq \sum_{i=1}^{\infty} \mu^*(E_i) = \sum_{i=1}^{\infty} \overline{\mu}(E_i),$$

and thus

$$\overline{\mu}(E) = \sum_{i=1}^{\infty} \overline{\mu}(E_i).$$

The proof is complete.

Definition 4.7.3. A measure on an algebra $\mathscr{A} \subset U$ is a set function $\mu : \mathscr{A} \to [0, +\infty)$ such that

1. $\mu(\emptyset) = 0$,
2. if $\{E_i\}$ is a sequence of disjoint sets in $\mathscr{A}$ so that $E = \cup_{i=1}^{\infty} E_i \in \mathscr{A}$, then

$$\mu(E) = \sum_{i=1}^{\infty} \mu(E_i).$$

We may define an outer measure in U by

$$\mu^*(E) = \inf\left\{ \sum_{i=1}^{\infty} \mu(A_i) \mid E \subset \cup_{i=1}^{\infty} A_i \right\},$$

where $A_i \in \mathscr{A}$, $\forall i \in \mathbb{N}$.

Proposition 4.7.4. *Suppose $A \in \mathscr{A}$ and $\{A_i\} \subset \mathscr{A}$ is such that*

$$A \subset \cup_{i=1}^{\infty} A_i.$$

Under such hypotheses,

$$\mu(A) \leq \sum_{i=1}^{\infty} \mu(A_i).$$

Proof. Define

$$B_n = (A \cap A_n) \setminus (\cup_{i=1}^{n-1} A_i).$$

Thus

$$B_n \subset A_n, \ \forall n \in \mathbb{N},$$

$B_n \in \mathscr{A}$, $\forall n \in \mathbb{N}$, and

$$A = \cup_{i=1}^{\infty} B_i.$$

Moreover, $\{B_n\}$ is a sequence of disjoint sets, so that

$$\mu(A) = \sum_{i=1}^{\infty} \mu(B_i) \leq \sum_{i=1}^{\infty} \mu(A_i).$$

Corollary 4.7.5. *If $A \in \mathscr{A}$, then $\mu^*(A) = \mu(A)$.*

Theorem 4.7.6. *The set function μ^* is an outer measure.*

Proof. The only not immediate property to be proven is the countably sub-additivity.

Suppose $E \subset \cup_{i=1}^{\infty} E_i$. If $\mu^*(E_i) = +\infty$ for some $i \in \mathbb{N}$, the result holds.

Thus, assume $\mu^*(E_i) < +\infty$, $\forall i \in \mathbb{N}$.

Let $\varepsilon > 0$. Thus for each $i \in \mathbb{N}$ there exists $\{A_{ij}\} \subset \mathscr{A}$ such that $E_i \subset \cup_{j=1}^{\infty} A_{ij}$, and

$$\sum_{j=1}^{\infty} \mu(A_{ij}) \leq \mu^*(E_i) + \frac{\varepsilon}{2^i}.$$

Therefore,

$$\mu(E) \leq \sum_{i=1}^{\infty} \sum_{j=1}^{\infty} \mu(A_{ij}) \leq \sum_{i=1}^{\infty} \mu^*(E_i) + \varepsilon.$$

Since $\varepsilon > 0$ is arbitrary, we get

$$\mu^*(E) \leq \sum_{i=1}^{\infty} \mu^*(E_i).$$

Proposition 4.7.7. *Suppose $A \in \mathscr{A}$. Then A is μ^*-measurable.*

Proof. Let $E \in U$ such that $\mu^*(E) < +\infty$. Let $\varepsilon > 0$.

Thus, there exists $\{A_i\} \subset \mathscr{A}$ such that $E \subset \cup_{i=1}^{\infty} A_i$ and

$$\sum_{i=1}^{\infty} \mu(A_i) < \mu^*(E) + \varepsilon.$$

Observe that

$$\mu(A_i) = \mu(A_i \cap A) + \mu(A_i \cap A^c),$$

so that from the fact that

$$E \cap A \subset \cup_{i=1}^{\infty} (A_i \cap A),$$

and

$$(E \cap A^c) \subset \cup_{i=1}^{\infty} (A_i \cap A^c),$$

we obtain

$$\begin{aligned} \mu^*(E) + \varepsilon &> \sum_{i=1}^{\infty} (A_i \cap A) + \sum_{i=1}^{\infty} \mu(A_i \cap A^c) \\ &\geq \mu^*(E \cap A) + \mu^*(E \cap A^c). \end{aligned} \tag{4.49}$$

Since $\varepsilon > 0$ is arbitrary, we get

$$\mu^*(E) \geq \mu^*(E \cap A) + \mu^*(E \cap A^c).$$

The proof is complete.

Proposition 4.7.8. *Suppose μ is a measure on an algebra $\mathscr{A} \subset U$, μ^* is the outer measure induced by μ, and $E \subset U$ is a set. Then, for each $\varepsilon > 0$, there is a set $A \in \mathscr{A}_\sigma$ with $E \subset A$ and*

$$\mu^*(A) \leq \mu^*(E) + \varepsilon.$$

Also, there is a set $B \in \mathscr{A}_{\sigma\delta}$ such that $E \subset B$ and

$$\mu^*(E) = \mu^*(B).$$

Proof. Let $\varepsilon > 0$. Thus, there is a sequence $\{A_i\} \subset \mathscr{A}$ such that

$$E \subset \cup_{i=1}^{\infty} A_i$$

and

$$\sum_{i=1}^{\infty} \mu(A_i) \leq \mu^*(E) + \varepsilon.$$

Define $A = \cup_{i=1}^{\infty} A_i$, then

$$\begin{aligned} \mu^*(A) &\leq \sum_{i=1}^{\infty} \mu^*(A_i) \\ &= \sum_{i=1}^{\infty} \mu(A_i) \\ &\leq \mu^*(E) + \varepsilon. \end{aligned} \tag{4.50}$$

Now, observe that we write $A \in \mathscr{A}_\sigma$ if $A = \cup_{i=1}^{\infty} A_i$ where $A_i \in \mathscr{A}$, $\forall i \in \mathbb{N}$.

Also, we write $B \in \mathscr{A}_{\sigma\delta}$ if $B = \cap_{n=1}^{\infty} A_n$, where $A_n \in \mathscr{A}_\sigma$, $\forall n \in \mathbb{N}$.

From above, for each $n \in \mathbb{N}$, there exists $A_n \in \mathscr{A}_\sigma$ such that

$$E \subset A_n$$

and

$$\mu^*(A_n) \leq \mu(E) + 1/n.$$

Define $B = \cap_{n=1}^{\infty} A_n$. Thus, $B \subset \mathscr{A}_{\sigma\delta}$, $E \subset B$ and

$$\mu^*(B) \leq \mu^*(A_n) \leq \mu^*(E) + 1/n, \ \forall n \in \mathbb{N}.$$

Hence

$$\mu^*(B) = \mu^*(E).$$

The proof is complete.

Proposition 4.7.9. *Suppose μ is a σ-finite measure on a σ-algebra $\mathscr{A}$, and let μ^* be the outer measure induced by μ.*

Under such hypotheses, a set E is μ^* measurable if and only if $E = A \setminus B$ where $A \in \mathscr{A}_{\sigma\delta}$, $B \subset A$, $\mu^*(B) = 0$.

Finally, for each set B such that $\mu^*(B) = 0$, there exists $C \in \mathscr{A}_{\sigma\delta}$ such that $B \subset C$ and $\mu^*(C) = 0$.

Proof. The if part is obvious.

Now suppose E is μ^*-measurable. Let $\{U_i\}$ be a countable collection of disjoint sets of finite measure such that

$$U = \cup_{i=1}^{\infty} U_i.$$

Observe that

$$E = \cup_{i=1}^{\infty} E_i,$$

where

$$E_i = E \cap U_i,$$

is μ^*-measurable for each $i \in \mathbb{N}$.

Let $\varepsilon > 0$. From the last proposition for each $i, n \in \mathbb{N}$ there exists $A_{ni} \in \mathscr{A}_\sigma$ such that

$$\mu(A_{ni}) < \mu^*(E_i) + \frac{1}{n2^i}.$$

Define

$$A_n = \cup_{i=1}^{\infty} A_{ni}.$$

Thus,

$$E \subset A_n$$

and

$$A_n \setminus E \subset \cup_{i=1}^{\infty} (A_{ni} \setminus E_i),$$

and therefore,

$$\mu(A_n \setminus E) \leq \sum_{i=1}^{\infty} \mu(A_{ni} \setminus E_i) \leq \sum_{i=1}^{\infty} \frac{1}{n2^i} = \frac{1}{n}.$$

Since $A_n \in \mathscr{A}_\sigma$, defining

$$A = \cap_{n=1}^{\infty} A_n,$$

we have that $A \in \mathscr{A}_{\sigma\delta}$ and

$$A \setminus E \subset A_n \setminus E$$

so that

$$\mu^*(A \setminus E) \leq \mu^*(A_n \setminus E) \leq \frac{1}{n}, \forall n \in \mathbb{N}.$$

Hence $\mu^*(A \setminus E) = 0$.

The proof is complete.

Theorem 4.7.10 (Carathéodory). *Let μ be a measure on analgebra $\mathscr{A}$ and μ^* the respective induced outer measure.*

Then the restriction $\overline{\mu}$ of μ^* to the μ^*-measurable sets is an extension of μ to a σ-algebra containing $\mathscr{A}$. If μ is finite or σ-finite, so is $\overline{\mu}$. In particular, if μ is σ-finite, then $\overline{\mu}$ is the only measure on the smallest σ-algebra containing $\mathscr{A}$ which is an extension of μ.

Proof. From the Theorem 4.7.2, $\overline{\mu}$ is an extension of μ to a σ-algebra containing $\mathscr{A}$, that is, $\overline{\mu}$ is a measure on such a set.

Observe that from the last results, if μ is σ-finite, so is $\overline{\mu}$.

Now assume μ is σ-finite. We will prove the uniqueness of $\overline{\mu}$.

Let $\mathscr{B}$ be the smallest σ-algebra containing $\mathscr{A}$ and let $\tilde{\mu}$ be another measure on $\mathscr{B}$ which extends μ on $\mathscr{A}$.

Since each set $\mathscr{A}_\sigma$ may be expressed as a disjoint countable union of sets in $\mathscr{A}$, the measure $\tilde{\mu}$ equals $\overline{\mu}$ on $\mathscr{A}_\sigma$. Let B be a μ^*-measurable set such that $\mu^*(B) < \infty$.

Let $\varepsilon > 0$. By Proposition 4.7.9 there exists an $A \in \mathscr{A}_\sigma$ such that $B \subset A$ and

$$\mu^*(A) < \mu^*(B) + \varepsilon.$$

Since $B \subset A$, we obtain

$$\tilde{\mu}(B) \leq \tilde{\mu}(A) = \mu^*(A) \leq \mu^*(B) + \varepsilon.$$

Considering that $\varepsilon > 0$ is arbitrary, we get

$$\tilde{\mu}(B) \leq \mu^*(B).$$

Observe that the class of μ^*-measurable sets is a σ-algebra containing $\mathscr{A}$.

Therefore, as above indicated, we have obtained $A \in \mathscr{A}_\sigma$ such that $B \subset A$ and

$$\mu^*(A) \leq \mu^*(B) + \varepsilon$$

so that

$$\mu^*(A) = \mu^*(B) + \mu^*(A \setminus B),$$

from this and above

$$\tilde{\mu}(A \setminus B) \leq \mu^*(A \setminus B) \leq \varepsilon,$$

if $\mu^*(B) < \infty$.

Therefore,

$$\begin{aligned} \mu^*(B) &\leq \mu^*(A) - \tilde{\mu}(A) \\ &= \tilde{\mu}(B) + \tilde{\mu}(A \setminus B) \leq \tilde{\mu}(B) + \varepsilon. \end{aligned} \tag{4.51}$$

Since $\varepsilon > 0$ is arbitrary we have

$$\mu^*(B) \leq \tilde{\mu}(B),$$

so that $\mu^*(B) = \tilde{\mu}(B)$. Finally, since μ is σ-finite, there exists a sequence of countable disjoint sets $\{U_i\}$ such that $\mu(U_i) < \infty$, $\forall i \in \mathbb{N}$, and $U = \cup_{i=1}^\infty U_i$.

If $B \in \mathscr{B}$, then

$$B = \cup_{i=1}^{\infty}(U_i \cap B).$$

Thus, from above,

$$\begin{aligned}
\tilde{\mu}(B) &= \sum_{i=1}^{\infty} \tilde{\mu}(U_i \cap B) \\
&= \sum_{i=1}^{\infty} \overline{\mu}(U_i \cap B) \\
&= \overline{\mu}(B).
\end{aligned} \tag{4.52}$$

The proof is complete.

Remark 4.7.11. We may start the process of construction of a measure by the action of a set function on a semi-algebra. Here, a semi-algebra $\mathscr{C}$ is a collection of subsets of U such that the intersection of any two sets in $\mathscr{C}$ is in $\mathscr{C}$ and the complement of any set in $\mathscr{C}$ is a finite disjoint union of sets in $\mathscr{C}$.

If $\mathscr{C}$ is any semi-algebra of sets, then the collection consisting of the empty set and all finite disjoint unions of sets in $\mathscr{C}$ is an algebra, which is said to be generated by $\mathscr{C}$. We denote such algebra by $\mathscr{A}$.

If we have a set function acting on $\mathscr{C}$, we may extend it to $\mathscr{A}$ by defining

$$\mu(A) = \sum_{i=1}^{n} \mu(E_i),$$

where $A = \cup_{i=1}^{n} E_i$ and $E_i \in \mathscr{C}$, $\forall i \in \{1, \ldots, n\}$, so that this last union is disjoint. We recall that any $A \in \mathscr{A}$ admits such a representation.

4.8 The Fubini Theorem

We start this section with the definition of complete measure space.

Definition 4.8.1. We say that a measure space $(U, \mathscr{M}, \mu)$ is complete if $\mathscr{M}$ contains all subsets of sets of zero measure. That is, if $A \in \mathscr{M}$, $\mu(A) = 0$ and $B \subset A$, then $B \in \mathscr{M}$.

In the next lines we recall the formal definition of semi-algebra.

Definition 4.8.2. We say (in fact recall) that $\mathscr{C} \in U$ is a semi-algebra in U if the two conditions below are valid:

1. if $A, B \in \mathscr{C}$, then $A \cap B \in \mathscr{C}$,
2. for each $A \in \mathscr{C}$, A^c is a finite disjoint union of elements in $\mathscr{C}$.

4.8.1 Product Measures

Let $(U,\mathcal{M}_1,\mu_1)$ and $(V,\mathcal{M}_2,\mu_2)$ be two complete measure spaces. We recall that the Cartesian product between U and V, denoted by $U\times V$, is defined by

$$U\times V=\{(u,v)\mid u\in U \text{ and } v\in V\}.$$

If $A\subset U$ and $B\subset V$, we call $A\times B$ a rectangle. If $A\in\mathcal{M}_1$ and $B\in\mathcal{M}_2$, we say that $A\times B$ is a measurable rectangle. The collection $\mathcal{R}$ of measurable rectangles is a semi-algebra since

$$(A\times B)\cap(C\times D)=(A\cap C)\times(B\cap D),$$

and

$$(A\times B)^c=(A^c\times B)\cup(A\times B^c)\cup(A^c\times B^c).$$

We define $\lambda:\mathcal{M}_1\times\mathcal{M}_2\to\mathbb{R}^+$ by

$$\lambda(A\times B)=\mu_1(A)\mu_2(B).$$

Lemma 4.8.3. *Let $\{A_i\times B_i\}_{i\in\mathbb{N}}$ be a countable disjoint collection of measurable rectangles whose union is the rectangle $A\times B$. Then*

$$\lambda(A\times B)=\sum_{i=1}^{\infty}\mu_1(A_i)\mu_2(B_i).$$

Proof. Let $u\in A$. Thus each $v\in B$ is such that (u,v) is exactly in one $A_i\times B_i$. Therefore

$$\chi_{A\times B}(u,v)=\sum_{i=1}^{\infty}\chi_{A_i}(u)\chi_{B_i}(v).$$

Hence, for the fixed u in question, from the corollary of Lebesgue monotone convergence theorem, we may write

$$\begin{aligned}\int_V \chi_{A\times B}(u,v)d\mu_2(v) &= \int\sum_{i=1}^{\infty}\chi_{A_i}(u)\chi_{B_i}(v)d\mu_2(v)\\ &=\sum_{i=1}^{\infty}\chi_{A_i}(u)\mu_2(B_i)\end{aligned}\tag{4.53}$$

so that also from the mentioned corollary

$$\int_U d\mu_1(u)\int_V \chi_{A\times B}(u,v)d\mu_2(v)=\sum_{i=1}^{\infty}\mu_1(A_i)\mu_2(B_i).$$

Observe that

$$\int_U d\mu_1(u) \int_V \chi_{A\times B}(u,v)d\mu_2(v) = \int_U d\mu_1(u) \int_V \chi_A(u)\chi_B(v)d\mu_2(v)$$
$$= \mu_1(A)\mu_2(B).$$

From the last two equations we may write

$$\lambda(A\times B) = \mu_1(A)\mu_2(B) = \sum_{i=1}^{\infty} \mu_1(A_i)\mu_2(B_i).$$

Definition 4.8.4. Let $E \subset U \times V$. We define E_u and E_v by

$$E_u = \{v \mid (u,v) \in E\},$$

and

$$E_v = \{u \mid (u,v) \in E\}.$$

Observe that

$$\chi_{E_u}(v) = \chi_E(u,v),$$
$$(E^c)_u = (E_u)^c,$$

and

$$(\cup E_\alpha)_u = \cup (E_\alpha)_u,$$

for any collection $\{E_\alpha\}$.

We denote by $\mathscr{R}_\sigma$ as the collection of sets which are countable unions of measurable rectangles. Also, $\mathscr{R}_{\sigma\delta}$ will denote the collection of sets which are countable intersections of elements of $\mathscr{R}_\sigma$.

Lemma 4.8.5. *Let $u \in U$ and $E \in \mathscr{R}_{\sigma\delta}$. Then E_u is a measurable subset of V.*

Proof. If $E \in \mathscr{R}$, the result is trivial. Let $E \in \mathscr{R}_\sigma$. Then E may be expressed as a disjoint union

$$E = \cup_{i=1}^{\infty} E_i,$$

where $E_i \in \mathscr{R}, \forall i \in \mathbb{N}$. Thus,

$$\begin{aligned}\chi_{E_u}(v) &= \chi_E(u,v) \\ &= \sup_{i\in\mathbb{N}} \chi_{E_i}(u,v) \\ &= \sup_{i\in\mathbb{N}} \chi_{(E_i)_u}(v).\end{aligned} \tag{4.54}$$

Since each $(E_i)_u$ is measurable we have that

$$\chi_{(E_i)_u}(v)$$

is a measurable function of v, so that

$$\chi_{E_u}(v)$$

is measurable, which implies that E_u is measurable. Suppose now

$$E = \cap_{i=1}^{\infty} E_i,$$

where $E_{i+1} \subset E_i, \forall i \in \mathbb{N}$. Then

$$\begin{aligned}
\chi_{E_u}(v) &= \chi_E(u,v) \\
&= \inf_{i \in \mathbb{N}} \chi_{E_i}(u,v) \\
&= \inf_{i \in \mathbb{N}} \chi_{(E_i)_u}(v).
\end{aligned} \tag{4.55}$$

Thus as from above $\chi_{(E_i)_u}(v)$ is measurable for each $i \in \mathbb{N}$, we have that χ_{E_u} is also measurable so that E_u is measurable.

Lemma 4.8.6. *Let E be a set in $\mathscr{R}_{\sigma\delta}$ with $(\mu_1 \times \mu_2)(E) < \infty$. Then the function g defined by*

$$g(u) = \mu_2(E_u)$$

is a measurable function and

$$\int_U g \, d\mu_1(u) = (\mu_1 \times \mu_2)(E).$$

Proof. The lemma is true if E is a measurable rectangle. Let $\{E_i\}$ be a disjoint sequence of measurable rectangles and $E = \cup_{i=1}^{\infty} E_i$. Set

$$g_i(u) = \mu_2((E_i)_u).$$

Then each g_i is a nonnegative measurable function and

$$g = \sum_{i=1}^{\infty} g_i.$$

Thus, g is measurable, and by the corollary of the Lebesgue monotone convergence theorem, we have

$$\begin{aligned}
\int_U g(u) d\mu_1(u) &= \sum_{i=1}^{\infty} \int_U g_i(u) d\mu_1(u) \\
&= \sum_{i=1}^{\infty} (\mu_1 \times \mu_2)(E_i) \\
&= (\mu_1 \times \mu_2)(E).
\end{aligned} \tag{4.56}$$

Let E be a set of finite measure in $\mathscr{R}_{\sigma\delta}$. Then there is a sequence in $\mathscr{R}_\sigma$ such that

$$E_{i+1} \subset E_i$$

and

$$E = \cap_{i=1}^{\infty} E_i.$$

Let $g_i(u) = \mu_2((E_i)_u)$, since

$$\int_U g_1(u) = (\mu_1 \times \mu_2)(E_1) < \infty,$$

we have that

$$g_1(u) < \infty \text{ a.e. in } E_1.$$

For an $u \in E_1$ such that $g_1(u) < \infty$ we have that $\{(E_i)_u\}$ is a sequence of measurable sets of finite measure whose intersection is E_u. Thus

$$g(u) = \mu_2(E_u) = \lim_{i \to \infty} \mu_2((E_i)_u) = \lim_{i \to \infty} g_i(u), \tag{4.57}$$

that is,

$$g_i \to g, \text{ a.e. in } E.$$

We may conclude that g is also measurable. Since

$$0 \leq g_i \leq g, \forall i \in \mathbb{N}$$

the Lebesgue dominated convergence theorem implies that

$$\int_E g(u)\, d\mu_1(u) = \lim_{i \to \infty} \int g_i\, d\mu_1(u) = \lim_{i \to \infty} (\mu_1 \times \mu_2)(E_i) = (\mu_1 \times \mu_2)(E).$$

Lemma 4.8.7. *Let E be a set such that $(\mu_1 \times \mu_2)(E) = 0$. Then for almost all $u \in U$ we have*

$$\mu_2(E_u) = 0.$$

Proof. Observe that there is a set in $\mathscr{R}_{\sigma\delta}$ such that $E \subset F$ and

$$(\mu_1 \times \mu_2)(F) = 0.$$

From the last lemma

$$\mu_2(F_u) = 0$$

for almost all u. From $E_u \subset F_u$ we obtain

$$\mu_2(E_u) = 0$$

for almost all u, since μ_2 is complete.

Proposition 4.8.8. *Let E be a measurable subset of $U \times V$ such that $(\mu_1 \times \mu_2)(E)$ is finite. For almost all u the set E_u is a measurable subset of V. The function g defined by*

$$g(u) = \mu_2(E_u)$$

is measurable and

$$\int g \, d\mu_1(u) = (\mu_1 \times \mu_2)(E).$$

Proof. First observe that there is a set $F \in \mathscr{R}_{\sigma\delta}$ such that $E \subset F$ and

$$(\mu_1 \times \mu_2)(F) = (\mu_1 \times \mu_2)(E).$$

Let $G = F \setminus E$. Since F and E are measurable, G is measurable, and

$$(\mu_1 \times \mu_2)(G) = 0.$$

By the last lemma we obtain

$$\mu_2(G_u) = 0,$$

for almost all u so that

$$g(u) = \mu_2(E_u) = \mu_2(F_u) \text{ a.e. in } U.$$

By Lemma 4.8.6 we may conclude that g is measurable and

$$\int g \, d\mu_1(u) = (\mu_1 \times \mu_2)(F) = (\mu_1 \times \mu_2)(E).$$

Theorem 4.8.9 (Fubini). *Let $(U, \mathscr{M}_1, \mu_1)$ and $(V, \mathscr{M}_2, \mu_2)$ be two complete measure spaces and f an integrable function on $U \times V$. Then*

1. *$f_u(v) = f(u,v)$ is measurable and integrable for almost all u,*
2. *$f_v(u) = f(u,v)$ is measurable and integrable for almost all v,*
3. *$h_1(u) = \int_V f(u,v) \, d\mu_2(v)$ is integrable on U,*
4. *$h_2(v) = \int_U f(u,v) \, d\mu_1(u)$ is integrable on V,*
5.

$$\int_U \left[\int_V f \, d\mu_2(v) \right] d\mu_1(u) = \int_V \left[\int_U f \, d\mu_1(u) \right] d\mu_2(v) \\ = \int_{U \times V} f d(\mu_1 \times \mu_2). \tag{4.58}$$

Proof. It suffices to consider the case where f is nonnegative (we can then apply the result to $f^+ = \max(f,0)$ and $f^- = \max(-f,0)$). The last proposition asserts that the theorem is true if f is a simple function which vanishes outside a set of finite measure. Similarly as in Theorem 4.2.2, we may obtain a sequence of nonnegative simple functions $\{\phi_n\}$ such that

$$\phi_n \uparrow f.$$

Observe that given $u \in U$, f_u is such that

$$(\phi_n)_u \uparrow f_u, \text{ a.e.}\,.$$

By the Lebesgue monotone convergence theorem we get

$$\int_V f(u,v)\,d\mu_2(v) = \lim_{n\to\infty} \int_V \phi_n(u,v)\,d\mu_2(v),$$

so that this last resulting function is integrable in U. Again by the Lebesgue monotone convergence theorem, we obtain

$$\begin{aligned}
\int_U \left[\int_V f\,d\mu_2(v)\right] d\mu_1(u) &= \lim_{n\to\infty} \int_U \left[\int_V \phi_n\,d\mu_2(v)\right] d\mu_1(u) \\
&= \lim_{n\to\infty} \int_{U\times V} \phi_n\,d(\mu_1 \times \mu_2) \\
&= \int_{U\times V} f\,d(\mu_1 \times \mu_2).
\end{aligned} \tag{4.59}$$

Chapter 5
The Lebesgue Measure in $\mathbb{R}^n$

5.1 Introduction

In this chapter we will define the Lebesgue measure and the concept of Lebesgue measurable set. We show that the set of Lebesgue measurable sets is a σ-algebra so that the earlier results, proven for more general measure spaces, remain valid in the present context (such as the Lebesgue monotone and dominated convergence theorems). The main reference for this chapter is [62].

We start with the following theorems without proofs.

Theorem 5.1.1. *Every open set $A \subset \mathbb{R}$ may be expressed as a countable union of disjoint open intervals.*

Remark 5.1.2. In this text Q_j denotes a closed cube in $\mathbb{R}^n$ and $|Q_j|$ its volume, that is, $|Q_j| = \prod_{i=1}^n (b_i - a_i)$, where $Q_j = \prod_{i=1}^n [a_i, b_i]$. Also we assume that if two Q_1 and Q_2, closed or not, have the same interior, then $|Q_1| = |Q_2| = |\bar{Q}_1|$. We recall that two cubes $Q_1, Q_2 \subset \mathbb{R}^n$ are said to be quasi-disjoint if their interiors are disjoint.

Theorem 5.1.3. *Every open set $A \subset \mathbb{R}^n$, where $n \geq 1$ may be expressed as a countable union of quasi-disjoint closed cubes.*

Definition 5.1.4 (Outer Measure). Let $E \subset \mathbb{R}^n$. The outer measure of E, denoted by $m^*(E)$, is defined by

$$m^*(E) = \inf\left\{ \sum_{j=1}^{\infty} |Q_j| \ : \ E \subset \cup_{j=1}^{\infty} Q_j \right\},$$

where Q_j is a closed cube, $\forall j \in \mathbb{N}$.

F. Botelho, *Functional Analysis and Applied Optimization in Banach Spaces: Applications to Non-Convex Variational Models*, DOI 10.1007/978-3-319-06074-3_5,

5.2 Properties of the Outer Measure

First observe that given $\varepsilon > 0$, there exists a sequence $\{Q_j\}$ such that

$$E \subset \cup_{j=1}^{\infty} Q_j$$

and

$$\sum_{j=1}^{\infty} |Q_j| \leq m^*(E) + \varepsilon.$$

1. Monotonicity: If $E_1 \subset E_2$ then $m^*(E_1) \leq m^*(E_2)$. This follows from the fact that if $E_2 \subset \cup_{j=1}^{\infty} Q_j$ then $E_1 \subset \cup_{j=1}^{\infty} Q_j$.
2. Countable sub-additivity : If $E \subset \cup_{j=1}^{\infty} E_j$, then $m^*(E) \leq \sum_{j=1}^{\infty} m^*(E_j)$.

   Proof. First assume that $m^*(E_j) < \infty$, $\forall j \in \mathbb{N}$; otherwise, the result is obvious. Thus, given $\varepsilon > 0$ for each $j \in \mathbb{N}$, there exists a sequence $\{Q_{k,j}\}_{k \in \mathbb{N}}$ such that

$$E_j \subset \cup_{k=1}^{\infty} Q_{k,j}$$

   and

$$\sum_{k=1}^{\infty} |Q_{k,j}| < m^*(E_j) + \frac{\varepsilon}{2^j}.$$

   Hence

$$E \subset \cup_{j,k=1}^{\infty} Q_{k,j}$$

   and therefore

$$\begin{aligned} m^*(E) \leq \sum_{j,k=1}^{\infty} |Q_{k,j}| &= \sum_{j=1}^{\infty} \left(\sum_{k=1}^{\infty} |Q_{k,j}| \right) \\ &\leq \sum_{j=1}^{\infty} \left(m^*(E_j) + \frac{\varepsilon}{2^j} \right) \\ &= \sum_{j=1}^{\infty} m^*(E_j) + \varepsilon. \end{aligned} \tag{5.1}$$

   Being $\varepsilon > 0$ arbitrary, we obtain

$$m^*(E) \leq \sum_{j=1}^{\infty} m^*(E_j).$$

3. If

$$E \subset \mathbb{R}^n,$$

   and

$$\alpha = \inf\{m^*(A) \mid A \text{ is open and } E \subset A\},$$

then

$$m^*(E) = \alpha.$$

Proof. From the monotonicity, we have

$$m^*(E) \leq m^*(A), \forall A \supset E, A \text{ open}.$$

Thus

$$m^*(E) \leq \alpha.$$

Suppose given $\varepsilon > 0$. Choose a sequence $\{Q_j\}$ of closed cubes such that

$$E \subset \cup_{j=1}^{\infty} Q_j$$

and

$$\sum_{j=1}^{\infty} |Q_j| \leq m^*(E) + \varepsilon.$$

Let $\{\tilde{Q}_j\}$ be a sequence of open cubes such that $\tilde{Q}_j \supset Q_j$

$$|\tilde{Q}_j| \leq |Q_j| + \frac{\varepsilon}{2^j}, \forall j \in \mathbb{N}.$$

Define

$$A = \cup_{j=1}^{\infty} \tilde{Q}_j;$$

hence A is open, $A \supset E$, and

$$\begin{aligned} m^*(A) &\leq \sum_{j=1}^{\infty} |\tilde{Q}_j| \\ &\leq \sum_{j=1}^{\infty} \left(|Q_j| + \frac{\varepsilon}{2^j} \right) \\ &= \sum_{j=1}^{\infty} |Q_j| + \varepsilon \\ &\leq m^*(E) + 2\varepsilon. \end{aligned} \tag{5.2}$$

Therefore

$$\alpha \leq m^*(E) + 2\varepsilon.$$

Being $\varepsilon > 0$ arbitrary, we have

$$\alpha \leq m^*(E).$$

The proof is complete.

4. If $E = E_1 \cup E_2$ and $d(E_1, E_2) > 0$, then

$$m^*(E) = m^*(E_1) + m^*(E_2)$$

Proof. First observe that being $E = E_1 \cup E_2$ we have

$$m^*(E) \le m^*(E_1) + m^*(E_2).$$

Let $\varepsilon > 0$. Choose $\{Q_j\}$ a sequence of closed cubes such that

$$E \subset \cup_{j=1}^{\infty} Q_j,$$

and

$$\sum_{j=1}^{\infty} |Q_j| \le m^*(E) + \varepsilon.$$

Let $\delta > 0$ such that

$$d(E_1, E_2) > \delta > 0.$$

Dividing the cubes Q_j if necessary, we may assume that the diameter of each cube Q_j is smaller than δ. Thus each Q_j intersects just one of the sets E_1 and E_2. Denote by J_1 and J_2 the sets of indices j such that Q_j intersects E_1 and E_2, respectively. Thus,

$$E_1 \subset \cup_{j \in J_1} Q_j \text{ and } E_2 \subset \cup_{j \in J_2} Q_j.$$

Hence,

$$\begin{aligned} m^*(E_1) + m^*(E_2) &\le \sum_{j \in J_1} |Q_j| + \sum_{j \in J_2} |Q_j| \\ &\le \sum_{j=1}^{\infty} |Q_j| \le m^*(E) + \varepsilon. \end{aligned} \tag{5.3}$$

Being $\varepsilon > 0$ arbitrary,

$$m^*(E_1) + m^*(E_2) \le m^*(E).$$

This completes the proof.

5. If a set E is a countable union of cubes quasi disjoints, that is,

$$E = \cup_{j=1}^{\infty} Q_j,$$

then

$$m^*(E) = \sum_{j=1}^{\infty} |Q_j|.$$

Proof. Let $\varepsilon > 0$.
Let $\{\tilde{Q}_j\}$ be open cubes such that $\tilde{Q}_j \subset\subset Q_j^\circ$ (i.e., the closure of $\tilde{Q}_j$ is contained in the interior of Q_j) and

$$|Q_j| \le |\tilde{Q}_j| + \frac{\varepsilon}{2^j}.$$

Thus, for each $N \in \mathbb{N}$ the cubes $\tilde{Q}_1, \ldots, \tilde{Q}_N$ are disjoint and each pair have a finite distance. Hence,

$$m^*(\cup_{j=1}^N \tilde{Q}_j) = \sum_{j=1}^N |\tilde{Q}_j| \geq \sum_{j=1}^N \left(|Q_j| - \frac{\varepsilon}{2^j} \right).$$

Being

$$\cup_{j=1}^N \tilde{Q}_j \subset E,$$

we obtain

$$m^*(E) \geq \sum_{j=1}^N |\tilde{Q}_j| \geq \sum_{j=1}^N |Q_j| - \varepsilon.$$

Therefore

$$\sum_{j=1}^\infty |Q_j| \leq m^*(E) + \varepsilon.$$

Being $\varepsilon > 0$ arbitrary, we may conclude that

$$\sum_{j=1}^\infty |Q_j| \leq m^*(E).$$

The proof is complete.

5.3 The Lebesgue Measure

Definition 5.3.1. A set $E \subset \mathbb{R}^n$ is said to be Lebesgue measurable if for each $\varepsilon > 0$ there exists $A \subset \mathbb{R}^n$ open such that

$$E \subset A$$

and

$$m^*(A - E) \leq \varepsilon.$$

If E is measurable, we define its Lebesgue measure, denoted by $m(E)$, as

$$m(E) = m^*(E).$$

5.4 Properties of Measurable Sets

1. Each open set is measurable.
2. If $m^*(E) = 0$ then E is measurable. In particular if $E \subset A$ and $m^*(A) = 0$, then E is measurable.

Proof. Let $E \subset \mathbb{R}^n$ be such that $m^*(E) = 0$. Suppose given $\varepsilon > 0$, thus there exists $A \subset \mathbb{R}^n$ open such that $E \subset A$ and $m^*(A) < \varepsilon$. Therefore

$$m^*(A - E) < \varepsilon.$$

3. A countable union of measurable sets is measurable.

   Proof. Suppose

$$E = \cup_{j=1}^{\infty} E_j$$

where each E_j is measurable. Suppose given $\varepsilon > 0$. For each $j \in \mathbb{N}$, there exists $A_j \subset \mathbb{R}^n$ open such that

$$E_j \subset A_j$$

and

$$m^*(A_j - E_j) \leq \frac{\varepsilon}{2^j}.$$

Define $A = \cup_{j=1}^{\infty} A_j$. Thus $E \subset A$ and

$$(A - E) \subset \cup_{j=1}^{\infty} (A_j - E_j).$$

From the monotonicity and countable sub-additivity of the outer measure we have

$$m^*(A - E) \leq \sum_{j=1}^{\infty} m^*(A_j - E_j) < \varepsilon.$$

4. Closed sets are measurable.

   Proof. Observe that

$$F = \cup_{k=1}^{\infty} F \cap B_k,$$

where B_k denotes a closed ball of radius k with center at origin. Thus F may be expressed as a countable union of compact sets. Hence, we have only to show that if F is compact then it is measurable. Let F be a compact set. Observe that

$$m^*(F) < \infty.$$

Let $\varepsilon > 0$; thus, there exists an open $A \subset \mathbb{R}^n$ such that $F \subset A$ and

$$m^*(A) \leq m^*(F) + \varepsilon.$$

Being F closed, $A - F$ is open, and therefore, $A - F$ may be expressed as a countable union of quasi disjoint closed cubes. Hence

$$A - F = \cup_{j=1}^{\infty} Q_j.$$

For each $N \in \mathbb{N}$

$$K = \cup_{j=1}^{N} Q_j$$

is compact; therefore

$$d(K,F) > 0.$$

Being $K \cup F \subset A$, we have

$$m^*(A) \geq m^*(F \cup K) = m^*(F) + m^*(K) = m^*(F) + \sum_{j=1}^{N} |Q_j|.$$

Therefore

$$\sum_{j=1}^{N} |Q_j| \leq m^*(A) - m^*(F) \leq \varepsilon.$$

Finally,

$$m^*(A - F) \leq \sum_{j=1}^{\infty} |Q_j| < \varepsilon.$$

This completes the proof.

5. If $E \subset \mathbb{R}^n$ is measurable, then E^c is measurable.

Proof. A point $x \in \mathbb{R}^n$ is denoted by $x = (x_1, x_2, \ldots, x_n)$ where $x_i \in \mathbb{R}$ for each $i \in \{1, \ldots, n\}$. Let E be a measurable set. For each $k \in \mathbb{N}$ there exists an open $A_k \supset E$ such that

$$m^*(A_k - E) < \frac{1}{k}.$$

Observe that A_k^c is closed and therefore measurable, $\forall k \in \mathbb{N}$. Thus

$$S = \cup_{k=1}^{\infty} A_k^c$$

is also measurable. On the other hand

$$S \subset E^c$$

and if $x \in (E^c - S)$, then $x \in E^c$ and $x \notin S$, so that $x \notin E$ and $x \notin A_k^c$, $\forall k \in \mathbb{N}$. Hence $x \notin E$ and $x \in A_k$, $\forall k \in \mathbb{N}$ and finally $x \in (A_k - E), \forall k \in \mathbb{N}$, that is,

$$E^c - S \subset A_k - E, \forall k \in \mathbb{N}.$$

Therefore

$$m^*(E^c - S) \leq \frac{1}{k}, \forall k \in \mathbb{N}.$$

Thus

$$m^*(E^c - S) = 0.$$

This means that $E^c - S$ is measurable, so that

$$E^c = S \cup (E^c - S)$$

is measurable. The proof is complete.

6. A countable intersection of measurable sets is measurable.

   Proof. This follows from items 3 and 5 just observing that
$$\cap_{j=1}^{\infty} E_j = (\cup_{j=1}^{\infty} E_j^c)^c.$$

Theorem 5.4.1. *If $\{E_i\}$ is sequence of measurable pairwise disjoint sets and $E = \cup_{j=1}^{\infty} E_i$, then*
$$m(E) = \sum_{j=1}^{\infty} m(E_j).$$

Proof. First assume that E_j is bounded. Being E_j^c measurable, given $\varepsilon > 0$, there exists an open $H_j \supset E_j^c$ such that
$$m^*(H_j - E_j^c) < \frac{\varepsilon}{2^j}, \forall j \in \mathbb{N}.$$

Denoting $F_j = H_j^c$ we have that $F_j \subset E_j$ is closed and
$$m^*(E_j - F_j) < \frac{\varepsilon}{2^j}, \forall j \in \mathbb{N}.$$

For each $N \in \mathbb{N}$ the sets $F_1, \ldots, F_N$ are compact and disjoint, so that
$$m(\cup_{j=1}^{N} F_j) = \sum_{j=1}^{N} m(F_j).$$

As
$$\cup_{j=1}^{N} F_j \subset E$$

we have
$$m(E) \geq \sum_{j=1}^{N} m(F_j) \geq \sum_{j=1}^{N} m(E_j) - \varepsilon.$$

Hence
$$m(E) \geq \sum_{j=1}^{\infty} m(E_j) - \varepsilon.$$

Being $\varepsilon > 0$ arbitrary, we obtain
$$m(E) \geq \sum_{j=1}^{\infty} m(E_j).$$

As the reverse inequality is always valid, we have
$$m(E) = \sum_{j=1}^{\infty} m(E_j).$$

For the general case, select a sequence of cubes $\{Q_k\}$ such that
$$\mathbb{R}^n = \cup_{k=1}^{\infty} Q_k$$

and $Q_k \subset Q_{k+1} \forall k \in \mathbb{N}$. Define $S_1 = Q_1$ and $S_k = Q_k - Q_{k-1}, \forall k \geq 2$. Also define

$$E_{j,k} = E_j \cap S_k, \forall j,k \in \mathbb{N}.$$

Thus

$$E = \cup_{j=1}^{\infty} \left(\cup_{k=1}^{\infty} E_{j,k}\right) = \cup_{j,k=1}^{\infty} E_{j,k},$$

where such a union is disjoint and each $E_{j,k}$ is bounded. Through the last result, we get

$$\begin{aligned} m(E) &= \sum_{j,k=1}^{\infty} m(E_{j,k}) \\ &= \sum_{j=1}^{\infty} \sum_{k=1}^{\infty} m(E_{j,k}) \\ &= \sum_{j=1}^{\infty} m(E_j). \end{aligned} \tag{5.4}$$

The proof is complete.

Theorem 5.4.2. *Suppose $E \subset \mathbb{R}^n$ is a measurable set. Then for each $\varepsilon > 0$:*

1. *There exists an open set $A \subset \mathbb{R}^n$ such that $E \subset A$ and*

$$m(A - E) < \varepsilon.$$

2. *There exists a closed set $F \subset \mathbb{R}^n$ such that $F \subset E$ and*

$$m(E - F) < \varepsilon.$$

3. *If $m(E)$ is finite, there exists a compact set $K \subset E$ such that*

$$m(E \setminus K) < \varepsilon.$$

4. *If $m(E)$ is finite, there exist a finite union of closed cubes*

$$F = \cup_{j=1}^{N} Q_j$$

such that

$$m(E \triangle F) \leq \varepsilon,$$

where

$$E \triangle F = (E \setminus F) \cup (F \setminus E).$$

Proof.

1. This item follows from the definition of measurable set.
2. Being E^c measurable, there exists an open $B \subset \mathbb{R}^n$ such that $E^c \subset B$ and

$$m^*(B \setminus E^c) < \varepsilon.$$

Defining $F = B^c$, we have that F is closed, $F \subset E$, and $E \setminus F = B \setminus E^c$. Therefore

$$m(E - F) < \varepsilon.$$

3. Choose a closed set such that $F \subset E$ e

$$m(E \setminus F) < \frac{\varepsilon}{2}.$$

Let B_n be a closed ball with center at origin and radius n. Define $K_n = F \cap B_n$ and observe that K_n is compact, $\forall n \in \mathbb{N}$. Thus

$$E \setminus K_n \searrow E \setminus F.$$

Being $m(E) < \infty$ we have

$$m(E \setminus K_n) < \varepsilon,$$

for all n sufficiently big.

4. Choose a sequence of closed cubes $\{Q_j\}$ such that

$$E \subset \cup_{j=1}^{\infty} Q_j$$

and

$$\sum_{j=1}^{\infty} |Q_j| \leq m(E) + \frac{\varepsilon}{2}.$$

Being $m(E) < \infty$ the series converges and there exists $N_0 \in \mathbb{N}$ such that

$$\sum_{N_0+1}^{\infty} |Q_j| < \frac{\varepsilon}{2}.$$

Defining $F = \cup_{j=1}^{N_0} Q_j$, we have

$$\begin{aligned} m(E \triangle F) &= m(E - F) + m(F - E) \\ &\leq m\left(\cup_{j=N_0+1}^{\infty} Q_j\right) + m\left(\cup_{j=1}^{\infty} Q_j - E\right) \\ &\leq \sum_{j=N_0+1}^{\infty} |Q_j| + \sum_{j=1}^{\infty} |Q_j| - m(E) \\ &\leq \frac{\varepsilon}{2} + \frac{\varepsilon}{2} = \varepsilon. \end{aligned} \tag{5.5}$$

5.5 Lebesgue Measurable Functions

Definition 5.5.1. Let $E \subset \mathbb{R}^n$ be a measurable set. A function $f : E \to [-\infty, +\infty]$ is said to be Lebesgue measurable if for each $a \in \mathbb{R}$, the set

$$f^{-1}([-\infty, a)) = \{x \in E \mid f(x) < a\}$$

is measurable.

Observe that:

1. If
$$f^{-1}([-\infty,a))$$
is measurable for each $a \in \mathbb{R}$, then
$$f^{-1}([-\infty,a]) = \cap_{k=1}^{\infty} f^{-1}([-\infty,a+1/k))$$
is measurable for each $a \in \mathbb{R}$.
2. If
$$f^{-1}([-\infty,a])$$
is measurable for each $a \in \mathbb{R}$, then
$$f^{-1}([-\infty,a)) = \cup_{k=1}^{\infty} f^{-1}([-\infty,a-1/k])$$
is also measurable for each $a \in \mathbb{R}$.
3. Given $a \in \mathbb{R}$, observe that
$$\begin{aligned} & f^{-1}([-\infty,a)) \text{ is measurable } \Leftrightarrow E - f^{-1}([-\infty,a)) \text{ is measurable} \\ & \Leftrightarrow f^{-1}(\mathbb{R}) - f^{-1}([-\infty,a)) \Leftrightarrow f^{-1}(\mathbb{R} - [-\infty,a)) \text{ is measurable} \\ & \Leftrightarrow f^{-1}([a,+\infty]) \text{ is measurable .} \end{aligned} \tag{5.6}$$
4. From above, we can prove that
$$f^{-1}([-\infty,a))$$
is measurable $\forall a \in \mathbb{R}$ if and only if
$$f^{-1}((a,b))$$
is measurable for each $a,b \in \mathbb{R}$ such that $a < b$. Therefore f is measurable if and only if $f^{-1}(\mathscr{O})$ is measurable whenever $\mathscr{O} \subset \mathbb{R}$ is open.
5. Thus f is measurable if $f^{-1}(\mathscr{F})$ is measurable whenever $\mathscr{F} \subset \mathbb{R}$ is closed.

Proposition 5.5.2. *If f is continuous in $\mathbb{R}^n$, then f is measurable. If f is measurable and real and ϕ is continuous, then $\phi \circ f$ is measurable.*

Proof. The first implication is obvious. For the second, being ϕ continuous
$$\phi^{-1}([-\infty,a))$$
is open, and therefore
$$(\phi \circ f)^{-1}(([-\infty,a)) = f^{-1}(\phi^{-1}([-\infty,a)))$$
is measurable, $\forall a \in \mathbb{R}$.

Proposition 5.5.3. *Suppose $\{f_k\}$ is a sequence of measurable functions. Then*

$$\sup_{k\in\mathbb{N}} f_k(x), \quad \inf_{k\in\mathbb{N}} f_k(x)$$

and

$$\limsup_{k\to\infty} f_k(x), \quad \liminf_{k\to\infty} f_k(x)$$

are measurable.

Proof. We will prove only that $\sup_{n\in\mathbb{N}} f_n(x)$ is measurable. The remaining proofs are analogous. Let

$$f(x) = \sup_{n\in\mathbb{N}} f_n(x).$$

Thus

$$f^{-1}((a,+\infty]) = \cup_{n=1}^{\infty} f_n^{-1}((a,+\infty]).$$

Being each f_n measurable, such a set is measurable, $\forall a \in \mathbb{R}$. By analogy

$$\inf_{k\in\mathbb{N}} f_k(x)$$

is measurable and

$$\limsup_{k\to\infty} f_k(x) = \inf_{k\geq 1} \sup_{j\geq k} f_j(x),$$

and

$$\liminf_{k\to\infty} f_k(x) = \sup_{k\geq 1} \inf_{j\geq k} f_j(x)$$

are measurable.

Proposition 5.5.4. *Let $\{f_k\}$ be a sequence of measurable functions such that*

$$\lim_{k\to\infty} f_k(x) = f(x).$$

Then f is measurable.

Proof. Just observe that

$$f(x) = \lim_{k\to\infty} f_k(x) = \limsup_{k\to\infty} f_k(x).$$

The next result we do not prove it. For a proof see [62].

Proposition 5.5.5. *If f and g are measurable functions, then*

1. *f^2 is measurable,*
2. *$f+g$ and $f\cdot g$ are measurable if both assume finite values.*

Proposition 5.5.6. *Let $E \subset \mathbb{R}^n$ a measurable set. Suppose $f : E \to \mathbb{R}$ is measurable. Thus if $g : E \to \mathbb{R}$ is such that*

$$g(x) = f(x), \text{ a.e. in } E,$$

then g is measurable.

Proof. Define

$$A = \{x \in E \mid f(x) \neq g(x)\}$$

and

$$B = \{x \in E \mid f(x) = g(x)\}.$$

A is measurable since $m^*(A) = m(A) = 0$ and therefore $B = E - A$ is also measurable. Let $a \in \mathbb{R}$. Hence

$$g^{-1}((a,+\infty]) = \left(g^{-1}((a,+\infty]) \cap A\right) \cup \left(g^{-1}((a,+\infty]) \cap B\right).$$

Observe that

$$\begin{aligned} x \in g^{-1}((a,+\infty]) \cap B &\Leftrightarrow x \in B \text{ and } g(x) \in (a,+\infty] \\ &\Leftrightarrow x \in B \text{ and } f(x) \in (a,+\infty] \\ &\Leftrightarrow x \in B \cap f^{-1}((a,+\infty]). \end{aligned} \tag{5.7}$$

Thus $g^{-1}((a,+\infty]) \cap B$ is measurable. As $g^{-1}((a,+\infty]) \cap A \subset A$ we have $m^*(g^{-1}((a,+\infty]) \cap A) = 0$, that is, such a set is measurable. Hence being $g^{-1}((a,+\infty])$ the union of two measurable sets is also measurable. Being $a \in \mathbb{R}$ arbitrary, g is measurable.

Theorem 5.5.7. *Suppose f is a nonnegative measurable function on $\mathbb{R}^n$. Then there exists an increasing sequence of nonnegative simple functions $\{\varphi_k\}$ such that*

$$\lim_{k \to +\infty} \varphi_k(x) = f(x), \forall x \in \mathbb{R}^n.$$

Proof. Let $N \in \mathbb{N}$. Let Q_N be the cube with center at origin and side of measure N. Define

$$F_N(x) = \begin{cases} f(x), & \text{if } x \in Q_N \text{ and } f(x) \leq N, \\ N, & \text{if } x \in Q_N \text{ and } f(x) > N, \\ 0, & \text{otherwise.} \end{cases}$$

Thus $F_N(x) \to f(x)$ as $N \to \infty, \forall x \in \mathbb{R}^n$. Fixing $M, N \in \mathbb{N}$ define

$$E_{l,M} = \left\{x \in Q_N : \frac{l}{M} \leq F_N(x) \leq \frac{l+1}{M}\right\},$$

for $0 \leq l \leq N \cdot M$. Defining

$$F_{N,M} = \sum_{l=0}^{NM} \frac{l}{M} \chi_{E_{l,M}},$$

we have that $F_{N,M}$ is a simple function and

$$0 \leq F_N(x) - F_{N,M}(x) \leq \frac{1}{M}.$$

If $\varphi_K(x) = F_{K,K}(x)$, we obtain

$$0 \leq |F_K(x) - \varphi_K(x)| \leq \frac{1}{K}.$$

Hence

$$|f(x) - \varphi_K(x)| \leq |f(x) - F_K(x)| + |F_K(x) - \varphi_K(x)|.$$

Therefore

$$\lim_{K \to \infty} |f(x) - \varphi_K(x)| = 0, \forall x \in \mathbb{R}^n.$$

The proof is complete.

Theorem 5.5.8. *Suppose that f is a measurable function defined on $\mathbb{R}^n$. Then there exists a sequence of simple functions $\{\varphi_k\}$ such that*

$$|\varphi_k(x)| \leq |\varphi_{k+1}(x)|, \forall x \in \mathbb{R}^n, k \in \mathbb{N}$$

and

$$\lim_{k \to \infty} \varphi_k(x) = f(x), \forall x \in \mathbb{R}^n.$$

Proof. Write

$$f(x) = f^+(x) - f^-(x),$$

where

$$f^+(x) = \max\{f(x), 0\}$$

and

$$f^-(x) = \max\{-f(x), 0\}.$$

Thus f^+ and f^- are nonnegative measurable functions so that from the last theorem there exist increasing sequences of nonnegative simple functions such that

$$\varphi_k^{(1)}(x) \to f^+(x), \forall x \in \mathbb{R}^n,$$

and

$$\varphi_k^{(2)}(x) \to f^-(x), \forall x \in \mathbb{R}^n,$$

as $k \to \infty$. Defining

$$\varphi_k(x) = \varphi_k^{(1)}(x) - \varphi_k^{(2)}(x),$$

we obtain

$$\varphi_k(x) \to f(x), \forall x \in \mathbb{R}^n$$

as $k \to \infty$ and

$$|\varphi_k(x)| = \varphi_k^{(1)}(x) + \varphi_k^{(2)}(x) \nearrow |f(x)|, \forall x \in \mathbb{R}^n,$$

as $k \to \infty$.

Theorem 5.5.9. *Suppose f is a measurable function in $\mathbb{R}^n$. Then there exists a sequence of step functions $\{\varphi_k\}$ which converges to f a.e. in $\mathbb{R}^n$.*

Proof. From the last theorem, it suffices to prove that if E is measurable and $m(E) < \infty$, then χ_E may be approximated almost everywhere in E by step functions. Suppose given $\varepsilon > 0$. Observe that from Proposition 5.4.2, there exist cubes $Q_1, \ldots, Q_N$ such that

$$m(E \triangle \cup_{j=1}^N Q_j) < \varepsilon.$$

We may obtain almost disjoints rectangles $\tilde{R}_j$ such that $\cup_{j=1}^M \tilde{R}_j = \cup_{j=1}^N Q_j$ and disjoints rectangles $R_j \subset \tilde{R}_j$ such that

$$m(E \triangle \cup_{j=1}^M R_j) < 2\varepsilon.$$

Thus

$$f(x) = \sum_{j=1}^M \chi_{R_j},$$

possibly except in a set of measure $< 2\varepsilon$. Hence, for each $k > 0$, there exists a step function φ_k such that $m(E_k) < 2^{-k}$ where

$$E_k = \{x \in \mathbb{R}^n \mid f(x) \neq \varphi_k(x)\}.$$

Defining

$$F_k = \cup_{j=k+1}^\infty E_j$$

we have

$$\begin{aligned} m(F_k) &\leq \sum_{j=k+1}^\infty m(E_j) \\ &\leq \sum_{j=k+1}^\infty 2^{-j} \\ &= \frac{2^{-(k+1)}}{1-1/2} \\ &= 2^{-k}. \end{aligned} \tag{5.8}$$

Therefore also defining

$$F = \cap_{k=1}^\infty F_k$$

we have $m(F) = 0$ considering that

$$m(F) \leq 2^{-k}, \forall k \in \mathbb{N}.$$

Finally, observe that

$$\varphi_k(x) \to f(x), \forall x \in F^c.$$

The proof is complete.

Theorem 5.5.10 (Egorov). *Suppose that $\{f_k\}$ is a sequence of measurable functions defined in a measurable set E such that $m(E) < \infty$. Assume that $f_k \to f$, a.e in E. Thus given $\varepsilon > 0$ we may find a closed set $A_\varepsilon \subset E$ such that $f_k \to f$ uniformly in A_ε and $m(E - A_\varepsilon) < \varepsilon$.*

Proof. Without losing generality we may assume that

$$f_k \to f, \forall x \in E.$$

For each $N, k \in \mathbb{N}$ define

$$E_k^N = \{x \in E \mid |f_j(x) - f(x)| < 1/N, \forall j \geq k\}.$$

Fixing $N \in \mathbb{N}$, we may observe that

$$E_k^N \subset E_{k+1}^N$$

and that $\cup_{k=1}^{\infty} E_k^N = E$. Thus we may obtain k_N such that

$$m(E - E_{k_N}^N) < \frac{1}{2^N}.$$

Observe that

$$|f_j(x) - f(x)| < \frac{1}{N}, \forall j \geq k_N,\ x \in E_{k_N}^N.$$

Choose $M \in \mathbb{N}$ such that

$$\sum_{k=M}^{\infty} 2^{-k} \leq \frac{\varepsilon}{2}.$$

Define

$$\tilde{A}_\varepsilon = \cap_{N \geq M}^{\infty} E_{k_N}^N.$$

Thus

$$m(E - \tilde{A}_\varepsilon) \leq \sum_{N=M}^{\infty} m(E - E_{k_N}^N) < \frac{\varepsilon}{2}.$$

Suppose given $\delta > 0$. Let $N \in \mathbb{N}$ be such that $N > M$ and $1/N < \delta$. Thus if $x \in \tilde{A}_\varepsilon$ then $x \in E_{k_N}^N$ so that

$$|f_j(x) - f(x)| < \delta, \forall j > k_N.$$

Hence $f_k \to f$ uniformly in $\tilde{A}_\varepsilon$. Observe that $\tilde{A}_\varepsilon$ is measurable and thus there exists a closed set $A_\varepsilon \subset \tilde{A}_\varepsilon$ such that

$$m(\tilde{A}_\varepsilon - A_\varepsilon) < \frac{\varepsilon}{2}.$$

That is

$$m(E - A_\varepsilon) \leq m(E - \tilde{A}_\varepsilon) + m(\tilde{A}_\varepsilon - A_\varepsilon) < \frac{\varepsilon}{2} + \frac{\varepsilon}{2} = \varepsilon,$$

and

$$f_k \to f$$

uniformly in A_ε. The proof is complete.

Definition 5.5.11. We say that $f : \mathbb{R}^n \to [-\infty, +\infty] \in L^1(\mathbb{R}^n$ if f is measurable and

$$\int_{\mathbb{R}^n} |f| \, dx < \infty.$$

Definition 5.5.12. We say that a set $A \subset L^1(\mathbb{R}^n)$ is dense in $L^1(\mathbb{R}^n)$, if for each $f \in L^1(\mathbb{R}^n)$ and each $\varepsilon > 0$ there exists $g \in A$ such that

$$\|f - g\|_{L^1(\mathbb{R}^n)} = \int_{\mathbb{R}^n} |f - g| \, dx < \varepsilon.$$

Theorem 5.5.13. *About dense sets in $L^1(\mathbb{R}^n)$ we have:*

1. *The set of simple functions is dense in $L^1(\mathbb{R}^n)$.*
2. *The set of step functions is dense in $L^1(\mathbb{R}^n)$.*
3. *The set of continuous functions with compact support is dense in $L^1(\mathbb{R}^n)$.*

Proof.

1. From the last theorems given $f \in L^1(\mathbb{R}^n)$ there exists a sequence of simple functions such that

$$\varphi_k(x) \to f(x) \text{ a.e. in } \mathbb{R}^n.$$

Since $\{\varphi_k\}$ may be also such that

$$|\varphi_k| \leq |f|, \forall k \in \mathbb{N}$$

from the Lebesgue dominated converge theorem, we have

$$\|\varphi_k - f\|_{L^1(\mathbb{R}^n)} \to 0,$$

as $k \to \infty$.

2. From the last item, it suffices to show that simple functions may be approximated by step functions. As a simple function is a linear combination of characteristic functions of sets of finite measure, it suffices to prove that given $\varepsilon > 0$ and a set of finite measure, there exists φ a step function such that

$$\|\chi_E - \varphi\|_{L^1(\mathbb{R}^n)} < \varepsilon.$$

This may be made similar as in Theorem 5.5.9.

3. From the last item, it suffices to establish the result as f is a characteristic function of a rectangle in $\mathbb{R}^n$. First consider the case of a interval $[a,b]$. We may approximate $f = \chi_{[a,b]}$ by $g(x)$, where g is continuous, and be linear on $(a-\varepsilon, a)$ and $(b, b+\varepsilon)$ and

$$g(x) = \begin{cases} 1, & \text{if } a \leq x \leq b, \\ 0, & \text{if } x \leq a-\varepsilon \text{ or } x \geq b+\varepsilon. \end{cases}$$

Thus

$$\|f-g\|_{L^1(\mathbb{R}^n)} < 2\varepsilon.$$

for the general case of a rectangle in $\mathbb{R}^n$, we just recall that in this case f is the product of the characteristic functions of n intervals. Therefore we may approximate f by the product of n functions similar to g defined above.

Chapter 6
Other Topics in Measure and Integration

In this chapter we present some important results which may be found in similar form at Chapters 2, 6, and 7 in the excellent book Real and Complex Analysis, [57] by Rudin, where more details may be found.

6.1 Some Preliminary Results

In the next results μ is a measure on U. We start with the following theorem.

Theorem 6.1.1. *Let $f : U \to [0,\infty]$ be a measurable function. If $E \in \mathcal{M}$ and*

$$\int_E f\, d\mu = 0,$$

then

$$f = 0, \text{ a.e. in } E.$$

Proof. Define

$$A_n = \{u \in E \mid f(u) > 1/n\}, \forall n \in \mathbb{N}.$$

Thus

$$\mu(A_n)/n \leq \int_{A_n} f\, d\mu \leq \int_E f\, d\mu = 0.$$

Therefore $\mu(A_n) = 0, \forall n \in \mathbb{N}$.

Define

$$A = \{u \in E \mid f(u) > 0\}.$$

Hence,

$$A = \cup_{n=1}^{\infty} A_n,$$

so that $\mu(A) = 0$.

F. Botelho, *Functional Analysis and Applied Optimization in Banach Spaces: Applications to Non-Convex Variational Models*, DOI 10.1007/978-3-319-06074-3_6,

Thus,

$$f = 0, \text{ a.e. in } E.$$

Theorem 6.1.2. *Assume $f \in L^1(\mu)$ and $\int_E f\,d\mu = 0, \forall E \in \mathcal{M}$. Under such hypotheses, $f = 0$, a.e. in U.*

Proof. Consider first the case $f : U \to [-\infty, +\infty]$. Define

$$A_n = \{u \in U \mid f(u) > 1/n\}, \forall n \in \mathbb{N}.$$

Thus,

$$\mu(A_n)/n \leq \int_{A_n} f\,d\mu = 0.$$

Hence, $\mu(A_n) = 0, \forall n \in \mathbb{N}$.

Define

$$A = \{u \in E \mid f(u) > 0\}.$$

Therefore,

$$A = \cup_{n=1}^{\infty} A_n,$$

so that $\mu(A) = 0$.

Thus,

$$f \leq 0, \text{ a.e. in } U.$$

By analogy we get

$$f \geq 0, \text{ a.e. in } U,$$

so that

$$f = 0, \text{ a.e. in } U.$$

To complete the proof, just apply this last result to the real and imaginary parts of a complex f.

Theorem 6.1.3. *Suppose $\mu(U) < \infty$ and $f \in L^1(\mu)$. Moreover, assume*

$$\frac{\int_E |f|\,d\mu}{\mu(E)} \leq \alpha \in [0, \infty), \ \forall E \in \mathcal{M}.$$

Under such hypothesis we have

$$|f| \leq \alpha, \text{ a.e. in } U.$$

Proof. Define

$$A_n = \{u \in U \mid |f(u)| > \alpha + 1/n\}, \forall n \in \mathbb{N}.$$

Thus, if $\mu(A_n) > 0$, we get

$$1/n \leq \frac{\int_{A_n} (|f| - \alpha)\,d\mu}{\mu(A_n)} = \frac{\int_{A_n} |f|\,d\mu}{\mu(A_n)} - \alpha \leq 0,$$

a contradiction. Hence, $\mu(A_n) = 0, \forall n \in \mathbb{N}$.

Define

$$A = \{u \in U \mid |f(u)| > \alpha\}.$$

Therefore,

$$A = \cup_{n=1}^{\infty} A_n,$$

so that $\mu(A) = 0$.

Thus,

$$|f(u)| \leq \alpha, \text{ a.e. in } U.$$

The proof is complete.

At this point we present some preliminary results to the development of the well-known Urysohn's lemma.

Theorem 6.1.4. *Let U be a Hausdorff space and $K \subset U$ compact. Let $v \in K^c$. Then there exist open sets V and $W \subset U$ such that $v \in V$, $K \subset W$ and $V \cap W = \emptyset$.*

Proof. For each $u \in K$ there exist open sets $W_u, V_v^u \subset U$ such that $u \in W_u$, $v \in W_v^u$, and $W_u \cap V_v^u = \emptyset$.

Observe that $K \subset \cup_{u \in K} W_u$ so that, since K is compact, there exist $u_1, u_2, \ldots, u_n \in K$ such that

$$K \subset \cup_{i=1}^{n} W_{u_i}.$$

Finally, defining the open sets

$$V = \cap_{i=1}^{n} V_v^{u_i}$$

and

$$W = \cup_{i=1}^{n} W_{u_i},$$

we get

$$V \cap W = \emptyset,$$

$v \in V$, and $K \subset W$.

The proof is complete.

Theorem 6.1.5. *Let $\{K_\alpha,\ \alpha \in L\}$ be a collection of compact subsets of a Hausdorff space U.*

Assume $\cap_{\alpha \in L} K_\alpha = \emptyset$. Under such hypotheses some finite subcollection of $\{K_\alpha,\ \alpha \in L\}$ has empty intersection.

Proof. Define $V_\alpha = K_\alpha^c$, $\forall \alpha \in L$. Fix $\alpha_0 \in L$. From the hypotheses

$$K_{\alpha_0} \cap [\cap_{\alpha \in L \setminus \{\alpha_0\}} K_\alpha] = \emptyset.$$

Hence

$$K_{\alpha_0} \subset [\cap_{\alpha \in L \setminus \{\alpha_0\}} K_\alpha]^c,$$

that is,

$$K_{\alpha_0} \subset \cup_{\alpha \in L \setminus \{\alpha_0\}} K_\alpha^c = \cup_{\alpha \in L \setminus \{\alpha_0\}} V_\alpha.$$

Since K_{α_0} is compact, there exists $\alpha_1,\ldots,\alpha_n \in L$ such that

$$K_{\alpha_0} \subset V_{\alpha_1} \cup \ldots \cup V_{\alpha_n} = (K_{\alpha_1} \cap \ldots \cap K_{\alpha_n})^c,$$

so that

$$K_{\alpha_0} \cap K_{\alpha_1} \cap \ldots \cap K_{\alpha_n} = \emptyset.$$

The proof is complete.

Definition 6.1.6. We say that a space U is locally compact if each $u \in U$ has a neighborhood whose closure is compact.

Theorem 6.1.7. *Let U be a locally compact Hausdorff space. Suppose $W \subset U$ is open and $K \subset W$, where K is compact. Then there exists an open set $V \subset U$ whose closure is compact and such that*

$$K \subset V \subset \overline{V} \subset W.$$

Proof. Let $u \in K$. Since U is locally compact there exists an open $V_u \subset U$ such that $u \in V_u$ and $\overline{V}_u$ is compact.

Observe that

$$K \subset \cup_{u \in K} V_u$$

and since K is compact there exist $u_1, u_2, \ldots, u_n \in K$ such that

$$K \subset \cup_{j=1}^n V_{u_j}.$$

Hence, defining $G = \cup_{j=1}^n V_{u_j}$, we get

$$K \subset G,$$

where $\overline{G}$ is compact.

If $W = U$ define $V = G$ and the proof would be complete.

Otherwise, if $W \neq U$ define $C = U \setminus W$. From Theorem 6.1.4, for each $v \in C$, there exists an open W_v such that $K \subset W_v$ and $v \notin \overline{W}_v$.

Hence $\{C \cap \overline{G} \cap \overline{W}_v \ : \ v \in C\}$ is a collection of compact sets with empty intersection.

From Theorem 6.1.5 there are points $v_1, \ldots, v_n \in C$ such that

$$C \cap \overline{G} \cap \overline{W}_{v_1} \cap \ldots \cap \overline{W}_{v_n} = \emptyset.$$

Defining

$$V = G \cap W_{v_1} \cap \ldots \cap W_{v_n}$$

we obtain

$$\overline{V} \subset \overline{G} \cap \overline{W}_{v_1} \cap \ldots \cap \overline{W}_{v_n}.$$

Also,

$$K \subset V \subset \overline{V} \subset W.$$

This completes the proof.

Definition 6.1.8. Let $f : U \to [-\infty, +\infty]$ be a function on a topological space U.

We say that f is lower semicontinuous if $A_\alpha = \{u \in U \,:\, f(u) > \alpha\}$ is open for all $\alpha \in \mathbb{R}$. Similarly, we say that f is upper semicontinuous if $B_\alpha = \{u \in U \,:\, f(u) < \alpha\}$ is open for all $\alpha \in \mathbb{R}$.

Observe that from this last definition f is continuous if and only if it is both lower and upper semicontinuous.

Here we state and prove a very important result, namely, the Uryshon's lemma.

Lemma 6.1.9 (Urysohn's Lemma). *Assume U is a locally compact Hausdorff space and $V \subset U$ is an open set which contains a compact set K. Under such assumptions, there exists a function $f \in C_c(V)$ such that*

- $0 \leq f(u) \leq 1, \forall u \in V,$
- $f(u) = 1, \forall u \in K.$

Proof. Set $r_1 = 0$ and $r_2 = 1$, and let $r_3, r_4, r_5, \ldots$ be an enumeration of the rational numbers in $(0,1)$. Observe that we may find open sets V_0 and V_1 such that $\overline{V}_0$ is compact and

$$K \subset V_1 \subset \overline{V}_1 \subset V_0 \subset \overline{V}_0 \subset V.$$

Reasoning by induction, suppose $n \geq 2$ and that $V_{r_1}, \ldots, V_{r_n}$ have been chosen so that if $r_i < r_j$ then $\overline{V}_{r_j} \subset V_{r_i}$. Denote

$$r_i = \max\{r_k \mid k \in \{1, \ldots, n\} \text{ and } r_k < r_{n+1}\}$$

and

$$r_j = \min\{r_k, \mid k \in \{1, \ldots, n\} \text{ and } r_k > r_{n+1}\}.$$

We may find again an open set $V_{r_{n+1}}$ such that

$$\overline{V}_{r_j} \subset V_{r_{n+1}} \subset \overline{V}_{r_{n+1}} \subset V_{r_i}.$$

Thus, we have obtained a sequence V_r of open sets such that for every r rational in $(0,1)$, $\overline{V}_r$ is compact and if $s > r$ then $\overline{V}_s \subset V_r$. Define

$$f_r(u) = \begin{cases} r, & \text{if } u \in V_r, \\ 0, & \text{otherwise,} \end{cases}$$

and

$$g_s(u) = \begin{cases} 1, & \text{if } u \in \overline{V}_s, \\ s, & \text{otherwise.} \end{cases}$$

Also define

$$f(u) = \sup_{r \in \mathbb{Q} \cap (0,1)} f_r(u), \forall u \in V$$

and

$$g(u) = \inf_{s \in \mathbb{Q} \cap (0,1)} g_s(u), \forall u \in V.$$

Observe that f is lower semicontinuous and g is upper semicontinuous. Moreover,

$$0 \leq f \leq 1$$

and

$$f = 1, \text{ if } u \in K.$$

Observe also that the support of f is contained in $\overline{V}_0$.

To complete the proof, it suffices to show that

$$f = g.$$

The inequality

$$f_r(u) > g_s(u)$$

is possible only if $r > s$, $u \in V_r$, and $u \notin \overline{V}_s$.

But if $r > s$, then $V_r \subset V_s$, and hence $f_r \leq g_s, \forall r, s \in \mathbb{Q} \cap (0,1)$, so that $f \leq g$. Suppose there exists $u \in V$ such that

$$f(u) < g(u).$$

Thus there exist rational numbers r, s such that

$$f(u) < r < s < g(u).$$

Since $f(u) < r$, $u \notin V_r$. Since $g(u) > s$, $u \in \overline{V}_s$.

As $\overline{V}_s \subset V_r$, we have a contradiction. Hence $f = g$, and such a function is continuous.

The proof is complete.

Theorem 6.1.10 (Partition of Unity). *Let U be a locally compact Hausdorff space. Assume $K \subset U$ is compact so that*

$$K \subset \cup_{i=1}^n V_i,$$

where V_i is open $\forall i \in \{1, \ldots, n\}$. Under such hypotheses, there exists functions $h_1, \ldots, h_n$ such that

$$\sum_{i=1}^n h_i = 1, \text{ on } K,$$

$$h_i \in C_c(V_i) \text{ and } 0 \leq h_i \leq 1, \ \forall i \in \{1, \ldots, n\}.$$

Proof. Let $u \in K \subset \cup_{i=1}^n V_i$. Thus there exists $j \in \{1, \ldots, n\}$ such that $u \in V_j$. We may select an open set W_u such that $\overline{W}_u$ is compact and $\overline{W}_u \subset V_j$.

Observe that

$$K \subset \cup_{u \in K} W_u.$$

From this, since K is compact, there exist $u_1, \ldots, u_N$ such that

$$K \subset \cup_{j=1}^N W_{u_j}.$$

For each $i \in \{1, \ldots, n\}$ define by $\tilde{W}_i$ the union of those $\overline{W_{u_j}}$, contained in V_i.

By the Uryshon's lemma we may find continuous functions g_i such that

$$g_i = 1, \text{ on } \tilde{W}_i,$$

$$g_i \in C_c(V_i),$$

$$0 \leq g_i \leq 1, \forall i \in \{1,\ldots,n\}.$$

Define

$$\begin{aligned} h_1 &= g_1 \\ h_2 &= (1-g_1)g_2 \\ h_3 &= (1-g_1)(1-g_2)g_3 \\ &\ldots\ldots\ldots\ldots\ldots\ldots \\ h_n &= (1-g_1)(1-g_2)\ldots(1-g_{n-1})g_n. \end{aligned} \tag{6.1}$$

Thus,

$$0 \leq h_i \leq 1 \text{ and } h_i \in C_c(V_i), \ \forall i \in \{1,\ldots,n\}.$$

Furthermore, by induction, we may obtain

$$h_1 + h_2 + \ldots + h_n = 1 - (1-g_1)(1-g_2)\ldots(1-g_n).$$

Finally, if $u \in K$ then $u \in \tilde{W}_i$ for some $i \in \{1,\ldots,n\}$, so that $g_i(u) = 1$ and hence

$$(h_1 + \ldots + h_n)(u) = 1, \forall u \in K.$$

The set $\{h_1,\ldots,h_n\}$ is said to be a partition of unity on K subordinate to the open cover $\{V_1,\ldots,V_n\}$.

The proof is complete.

6.2 The Riesz Representation Theorem

In the next lines we introduce the main result in this section, namely, the Riesz representation theorem.

Theorem 6.2.1 (Riesz Representation Theorem). *Let U be a locally compact Hausdorff space and let F be a positive linear functional on $C_c(U)$. Then there exists a σ-algebra $\mathcal{M}$ in U which contains all the Borel sets and there exists a unique positive measure μ on $\mathcal{M}$ such that*

1. $F(f) = \int_U f \, d\mu, \forall f \in C_c(U)$,
2. $\mu(K) < \infty$, *for every compact* $K \subset U$,
3. $\mu(E) = \inf\{\mu(V) \mid E \subset V, \ V \text{ open}\}, \ \forall E \in \mathcal{M}$,
4. $\mu(E) = \sup\{\mu(K) \mid K \subset E, \ K \text{ compact}\}$ *holds for all open E and all $E \in \mathcal{M}$ such that* $\mu(E) < \infty$,
5. *If $E \in \mathcal{M}$, $A \subset E$ and $\mu(E) = 0$ then $A \in \mathcal{M}$.*

Proof. We start by proving the uniqueness of μ. If μ satisfies 3 and 4, then μ is determined by its values on compact sets. Then, if μ_1 and μ_2 are two measures for which the theorem holds, to prove uniqueness, it suffices to show that

$$\mu_1(K) = \mu_2(K)$$

for every compact $K \subset U$. Let $\varepsilon > 0$. Fix a compact $K \subset U$. By 2 and 3, there exists an open $V \supset K$ such that

$$\mu_2(V) < \mu_2(K) + \varepsilon.$$

By the Urysohn's lemma, there exists a $f \in C_c(V)$ such that

$$0 \leq f(u) \leq 1, \forall u \in V$$

and

$$f(u) = 1, \; \forall u \in K.$$

Thus,

$$\begin{aligned} \mu_1(K) &= \int_U \chi_K \, d\mu_1 \\ &\leq \int_U f \, d\mu_1 \\ &= F(f) \\ &= \int_U f \, d\mu_2 \\ &\leq \int_U \chi_V \, d\mu_2 \\ &= \mu_2(V) \\ &< \mu_2(K) + \varepsilon. \end{aligned} \tag{6.2}$$

Since $\varepsilon > 0$ is arbitrary, we get

$$\mu_1(K) \leq \mu_2(K).$$

Interchanging the roles of μ_1 and μ_2 we similarly obtain

$$\mu_2(K) \leq \mu_1(K),$$

so that

$$\mu_1(K) = \mu_2(K).$$

The proof of uniqueness is complete.

Now for every open $V \subset U$, define

$$\mu(V) = \sup\{F(f) \mid f \in C_c(V) \text{ and } 0 \leq f \leq 1\}.$$

If V_1, V_2 are open and $V_1 \subset V_2$, then

$$\mu(V_1) \leq \mu(V_2).$$

Hence,

$$\mu(E) = \inf\{\mu(V) \mid E \subset V, \ V \text{ open}\},$$

if E is an open set. Define

$$\mu(E) = \inf\{\mu(V) \mid E \subset V, \ V \text{ open}\},$$

$\forall E \subset U$. Define by $\mathcal{M}_F$ the collection of all $E \subset U$ such that $\mu(E) < \infty$ and

$$\mu(E) = \sup\{\mu(K) \mid K \subset E, \ K \text{ compact}\}.$$

Finally, define by $\mathcal{M}$ the collection of all sets such that $E \subset U$ and $E \cap K \in \mathcal{M}_F$ for all compact $K \subset U$. Since

$$\mu(A) \leq \mu(B),$$

if $A \subset B$ we have that $\mu(E) = 0$ implies $E \cap K \in \mathcal{M}_F$ for all K compact, so that $E \in \mathcal{M}$. Thus, 5 holds and so does 3 by definition.

Observe that if $f \geq 0$, then $F(f) \geq 0$, that is, if $f \leq g$ then $F(f) \leq F(g)$.

Now we prove that if $\{E_n\} \subset U$ is a sequence, then

$$\mu\left(\cup_{n=1}^{\infty} E_n\right) \leq \sum_{n=1}^{\infty} \mu(E_n). \tag{6.3}$$

First we show that

$$\mu(V_1 \cup V_2) \leq \mu(V_1) + \mu(V_2),$$

if V_1, V_2 are open sets.

Choose $g \in C_c(V_1 \cup V_2)$ such that

$$0 \leq g \leq 1.$$

By Theorem 6.1.10 there exist functions h_1 and h_2 such that $h_i \in C_c(V_i)$ and

$$0 \leq h_i \leq 1$$

and so that $h_1 + h_2 = 1$ on the support of g. Hence, $h_i \in C_c(V_i)$, $0 \leq h_i g \leq 1$, and $g = (h_1 + h_2)g$ and thus

$$F(g) = F(h_1 g) + F(h_2 g) \leq \mu(V_1) + \mu(V_2).$$

Since g is arbitrary, from the definition of μ, we obtain

$$\mu(V_1 \cup V_2) \leq \mu(V_1) + \mu(V_2).$$

Furthermore, if $\mu(E_n) = \infty$, for some $n \in \mathbb{N}$, then (6.3) is obviously valid. Assume then $\mu(E_n) < \infty, \forall n \in \mathbb{N}$.

Let a not relabeled $\varepsilon > 0$. Therefore for each $n \in \mathbb{N}$ there exists an open $V_n \supset E_n$ such that

$$\mu(V_n) < \mu(E_n) + \frac{\varepsilon}{2^n}.$$

Define

$$V = \cup_{n=1}^{\infty} V_n,$$

and choose $f \in C_c(V)$ such that $0 \leq f \leq 1$. Since the support of f is compact, there exists $N \in \mathbb{N}$ such that

$$spt(f) \subset \cup_{n=1}^{N} V_n.$$

Therefore

$$\begin{aligned} F(f) &\leq \mu\left(\cup_{n=1}^{N} V_n\right) \\ &\leq \sum_{n=1}^{N} \mu(V_n) \\ &\leq \sum_{n=1}^{\infty} \mu(E_n) + \varepsilon. \end{aligned} \tag{6.4}$$

Since this holds for any $f \in C_c(V)$ with $0 \leq f \leq 1$ and $\cup_{n=1}^{\infty} E_n \subset V$, we get

$$\mu\left(\cup_{n=1}^{\infty} E_n\right) \leq \mu(V) \leq \sum_{i=1}^{\infty} \mu(E_n) + \varepsilon.$$

Since $\varepsilon > 0$ is arbitrary, we have proven (6.3).

In the next lines we prove that if K is compact, then $K \in \mathcal{M}_F$ and

$$\mu(K) = \inf\{F(f) \mid f \in C_c(U),\ f = 1 \text{ on } K\}. \tag{6.5}$$

For if $f \in C_c(U)$, $f = 1$ on K, and $0 < \alpha < 1$, define

$$V_\alpha = \{u \in U \mid f(u) > \alpha\}.$$

Thus, $K \subset V_\alpha$ and if $g \in C_c(V_\alpha)$ and $0 \leq g \leq 1$ we get

$$\alpha g \leq f.$$

Hence,

$$\begin{aligned} \mu(K) &\leq \mu(V_\alpha) \\ &= \sup\{F(g) \mid g \in C_c(V_\alpha),\ 0 \leq g \leq 1\} \\ &\leq \alpha^{-1} F(f). \end{aligned} \tag{6.6}$$

Letting $\alpha \to 1$ we obtain

$$\mu(K) \leq F(f).$$

Thus $\mu(K) < \infty$, and obviously $K \in \mathcal{M}_F$.

Also there exists an open $V \supset K$ such that

$$\mu(V) < \mu(K) + \varepsilon.$$

By the Urysohn's lemma, we may find $f \in C_c(V)$ such that $f = 1$ on K and $0 \le f \le 1$. Thus

$$F(f) \le \mu(V) < \mu(K) + \varepsilon.$$

Since $\varepsilon > 0$ is arbitrary, (6.5) holds.

At this point we prove that for every open V we have

$$\mu(V) = \sup\{\mu(K) \mid K \subset V,\ K \text{ compact}\} \tag{6.7}$$

and hence $\mathcal{M}_F$ contains every open set such that $\mu(V) < \infty$.

Let $V \subset U$ be an open set such that $\mu(V) < \infty$.

Let $\alpha \in \mathbb{R}$ be such that $\alpha < \mu(V)$. Therefore there exists $f \in C_c(V)$ such that $0 \le f \le 1$ and such that $\alpha < F(f)$.

If $W \subset U$ is an open set such that $K = spt(f) \subset W$, we have that $f \in C_c(W)$ and $0 \le f \le 1$ so that

$$F(f) \le \mu(W).$$

Thus, since $W \supset K$ is arbitrary, we obtain

$$F(f) \le \mu(K),$$

so that

$$\alpha < \mu(K),$$

where $K \subset U$ is a compact set.

Hence (6.7) holds.

Suppose that

$$E = \cup_{n=1}^{\infty} E_n,$$

where $\{E_n\}$ is a sequence of disjoint sets in $\mathcal{M}_F$.

We are going to show that

$$\mu(E) = \sum_{n=1}^{\infty} \mu(E_n). \tag{6.8}$$

In addition if $\mu(E) < \infty$, then also $E \subset \mathcal{M}_F$.

First we show that if $K_1, K_2 \subset U$ are compact disjoint sets, then

$$\mu(K_1 \cup K_2) = \mu(K_1) + \mu(K_2). \tag{6.9}$$

From the Urysohn's lemma there exists $f \in C_c(U)$ such that $f = 1$ on K_1, $f = 0$ on K_2, and

$$0 \le f \le 1.$$

From (6.5) there exists $g \in C_c(U)$ such that $g = 1$ on $K_1 \cup K_2$ and

$$F(g) < \mu(K_1 \cup K_2) + \varepsilon.$$

Observe that $fg = 1$ on K_1 and $(1-f)g = 1$ on K_2 and also $fg,\ (1-f)g \in C_c(U)$ and $0 \leq fg \leq 1$ and $0 \leq (1-f)g \leq 1$ so that

$$\begin{aligned}\mu(K_1)+\mu(K_2) &\leq F(fg)+F((1-f)g)\\ &= F(g)\\ &\leq \mu(K_1 \cup K_2)+\varepsilon. \end{aligned} \tag{6.10}$$

Since $\varepsilon > 0$ is arbitrary we obtain

$$\mu(K_1)+\mu(K_2) \leq \mu(K_1 \cup K_2).$$

From this (6.9) holds.

Also if $\mu(E) = \infty$, (6.8) follows from (6.3).

Thus assume $\mu(E) < \infty$.

Since $E_n \in \mathscr{M}_F, \forall n \in \mathbb{N}$ we may obtain compact sets $H_n \subset E_n$ such that

$$\mu(H_n) > \mu(E_n) - \frac{\varepsilon}{2^n}, \forall n \in \mathbb{N}.$$

Defining $K_N = \cup_{n=1}^N H_n$, by 3 we get

$$\begin{aligned}\mu(E) &\geq \mu(K_N)\\ &= \sum_{n=1}^{N} \mu(H_n)\\ &\geq \sum_{n=1}^{N} \mu(E_n) - \varepsilon, \forall N \in \mathbb{N}. \end{aligned} \tag{6.11}$$

Since $N \in \mathbb{N}$ and $\varepsilon > 0$ are arbitrary we get

$$\mu(E) \geq \sum_{n=1}^{\infty} \mu(E_n).$$

From this and (6.3) we obtain

$$\mu(E) = \sum_{n=1}^{\infty} \mu(E_n). \tag{6.12}$$

Let $\varepsilon_0 > 2\varepsilon$. If $\mu(E) < \infty$, there exists $N_0 \in \mathbb{N}$ such that $\mu(K_{N_0}) > \sum_{n=1}^{\infty} \mu(E_n) - \varepsilon_0$.

From this and (6.12) we obtain

$$\mu(E) \leq \mu(K_{N_0}) + \varepsilon_0.$$

Therefore, since $\varepsilon > 0$ and $\varepsilon_0 > 2\varepsilon$ are arbitrary, we may conclude that E satisfies 4 so that $E \in \mathscr{M}_F$.

Now we prove the following.

If $E \in \mathcal{M}_F$ there is a compact $K \subset U$ and an open $V \subset U$ such that $K \subset E \subset V$ and

$$\mu(V \setminus K) < \varepsilon.$$

From above, there exists a compact K and an open V such that

$$K \subset E \subset V$$

and

$$\mu(V) - \frac{\varepsilon}{2} < \mu(E) < \mu(K) + \frac{\varepsilon}{2}.$$

Since $V \setminus K$ is open and of finite measure, it is in $\mathcal{M}_F$. From the last chain of inequalities we obtain

$$\mu(K) + \mu(V \setminus K) = \mu(V) < \mu(K) + \varepsilon,$$

so that

$$\mu(V \setminus K) < \varepsilon.$$

In the next lines we prove that if $A, B \in \mathcal{M}_F$ then

$$A \setminus B, A \cup B \text{ and } A \cap B \in \mathcal{M}_F.$$

By above there exist compact sets K_1, K_2 and open sets V_1, V_2 such that

$$K_1 \subset A \subset V_1, \; K_2 \subset B \subset V_2$$

and

$$\mu(V_i \setminus K_i) < \varepsilon, \forall i \in \{1,2\}.$$

Since

$$(A \setminus B) \subset (V_1 \setminus K_2) \subset (V_1 \setminus K_1) \cup (K_1 \setminus V_2) \cup (V_2 \setminus K_2),$$

we get

$$\mu(A \setminus B) < \varepsilon + \mu(K_1 \setminus V_2) + \varepsilon,$$

Since $K_1 \setminus V_2 \subset A \setminus B$ is compact and $\varepsilon > 0$ is arbitrary, we get

$$A \setminus B \in \mathcal{M}_F.$$

Since

$$A \cup B = (A \setminus B) \cup B,$$

we obtain

$$A \cup B \in \mathcal{M}_F.$$

Since

$$A \cap B = A \setminus (A \setminus B)$$

we get

$$A \cap B \in \mathcal{M}_F.$$

At this point we prove that $\mathcal{M}$ is a σ-algebra in U which contains all the Borel sets.

Let $K \subset U$ be a compact subset. If $A \in \mathcal{M}$ then

$$A^c \cap K = K \setminus (A \cap K),$$

so that $A^c \cap K \in \mathcal{M}_F$ considering that $K \in \mathcal{M}_F$ and $A \cap K \in \mathcal{M}_F$.

Thus if $A \in \mathcal{M}$ then $A^c \in \mathcal{M}$.

Next suppose

$$A = \cup_{n=1}^{\infty} A_n,$$

where $A_n \in \mathcal{M}, \forall n \in \mathbb{N}$.

Define $B_1 = A_1 \cap K$ and

$$B_n = (A_n \cap K) \setminus (B_1 \cup B_2 \cup \ldots \cup B_{n-1}),$$

$\forall n \geq 2, n \in \mathbb{N}$.

Then $\{B_n\}$ is disjoint sequence of sets in $\mathcal{M}_F$.

Thus

$$A \cap K = \cup_{n=1}^{\infty} B_n \in \mathcal{M}_F.$$

Hence $A \in \mathcal{M}$. Finally, if $C \subset U$ is a closed subset, then $C \cap K$ is compact, so that $C \cap K \in \mathcal{M}_F$. Hence $C \in \mathcal{M}$.

Therefore $\mathcal{M}$ is a σ-algebra which contains the closed sets, so that it contains the Borel sets.

Finally, we will prove that

$$\mathcal{M}_F = \{E \in \mathcal{M} \mid \mu(E) < \infty\}.$$

For, if $E \in \mathcal{M}_F$ then $E \cap K \in \mathcal{M}_F$ for all compact $K \subset U$, hence $E \in \mathcal{M}$.

Conversely, assume $E \in \mathcal{M}$ and $\mu(E) < \infty$. There is an open $V \supset E$ such that $\mu(V) < \infty$. Pick a compact $K \subset V$ such that

$$\mu(V \setminus K) < \varepsilon.$$

Since $E \cap K \in \mathcal{M}_F$ there is a compact $K_1 \subset (E \cap K)$ such that

$$\mu(E \cap K) < \mu(K_1) + \varepsilon.$$

Since

$$E \subset (E \cap K) \cup (V \setminus K),$$

it follows that

$$\mu(E) \leq \mu(E \cap K) + \mu(V \setminus K) < \mu(K_1) + 2\varepsilon.$$

This implies $E \in \mathcal{M}_F$.

To finish the proof, we show that

$$F(f) = \int_U f\, d\mu, \forall f \in C_c(U).$$

From linearity it suffices to prove the result for the case where f is real.

Let $f \in C_c(U)$. Let K be the support of f and let $[a,b] \subset \mathbb{R}$ be such that

$$R(f) \subset (a,b),$$

where $R(f)$ denotes the range of f.

Suppose given a not relabeled $\varepsilon > 0$. Choose a partition of $[a,b]$ denoted by

$$\{y_i\} = \{a = y_0 < y_1 < y_2 < \ldots < y_n = b\},$$

such that $y_i - y_{i-1} < \varepsilon, \forall i \in \{1,\ldots,n\}$.

Denote

$$E_i = \{u \in K \mid y_{i-1} < f(u) \leq y_i\},$$

$\forall i \in \{1,\ldots,n\}$.

Since f is continuous, it is Borel measurable, and the sets E_i are disjoint Borel ones such that

$$\cup_{i=1}^n E_i = K.$$

Select open sets $V_i \supset E_i$ such that

$$\mu(V_i) < \mu(E_i) + \frac{\varepsilon}{n}, \forall i \in \{1,\ldots,n\},$$

and such that

$$f(u) < y_i + \varepsilon, \forall u \in V_i.$$

From Theorem 6.1.10 there exists a partition of unity subordinate to $\{V_i\}_{i=1}^n$ such that $h_i \in C_c(V_i)$, $0 \leq h_i \leq 1$ and

$$\sum_{i=1}^n h_i = 1, \text{ on } K.$$

Hence

$$f = \sum_{i=1}^n h_i f$$

and

$$\mu(K) \leq F\left(\sum_{i=1}^n h_i f\right) = \sum_{i=1}^n F(h_i f).$$

Observe that

$$\begin{aligned}
\mu(E_i) + \frac{\varepsilon}{n} &> \mu(V_i) \\
&= \sup\{F(f) \mid f \in C_c(V_i),\ 0 \leq f \leq 1\} \\
&> F(h_i), \forall i \in \{1,\ldots,n\}.
\end{aligned} \tag{6.13}$$

Thus

$$\begin{aligned}
F(f) &= \sum_{i=1}^{n} F(h_i f) \\
&\leq \sum_{i=1}^{n} F(h_i(y_{i-1}+2\varepsilon)) \\
&= \sum_{i=1}^{n} (y_{i-1}+2\varepsilon)F(h_i) \\
&< \sum_{i=1}^{n} (y_{i-1}+2\varepsilon)\left(\mu(E_i)+\frac{\varepsilon}{n}\right) \\
&< \sum_{i=1}^{n} y_{i-1}\mu(E_i)+\sum_{i=1}^{n}(y_{i-1})\frac{\varepsilon}{n}+2\varepsilon\sum_{i=1}^{n}\mu(E_i)+2\varepsilon^2 \\
&< \int_U f\,d\mu + b\varepsilon + 2\varepsilon\mu(K)+2\varepsilon^2. \qquad (6.14)
\end{aligned}$$

Since $\varepsilon > 0$ is arbitrary, we obtain

$$F(f) \leq \int_U f\,d\mu, \forall f \in C_c(U).$$

From this

$$F(-f) \leq \int_U (-f)\,d\mu, \forall f \in C_c(U),$$

that is,

$$F(f) \geq \int_U f\,d\mu, \forall f \in C_c(U).$$

Hence

$$F(f) = \int_U f\,d\mu, \forall f \in C_c(U).$$

The proof is complete.

6.3 The Lebesgue Points

In this section we introduce a very important concept in analysis, namely, the definition of Lebesgue points.

We recall that in $\mathbb{R}^n$ the open ball with center u and radius r is defined by

$$B_r(u) = \{v \in \mathbb{R}^n \mid |v-u|_2 < r\}.$$

Consider a Borel measure μ on $\mathbb{R}^n$. We may associate to μ, the function $F_{r\mu}(u)$, denoted by

$$F_{r\mu}(u) = \frac{\mu(B_r(u))}{m(B_r(u))},$$

where m denotes the Lebesgue measure.

We define the symmetric derivative of μ at u, by $(D\mu)(u)$, by

$$(D\mu)(u) = \lim_{r \to 0} F_{r\mu}(u),$$

whenever such a limit exists.

We also define the function G_μ for a positive measure μ by

$$G_\mu(u) = \sup_{0<r<\infty} F_{r\mu}(u).$$

The function $G_\mu : \mathbb{R}^n \to [0, +\infty]$ is lower semicontinuous and hence measurable.

Lemma 6.3.1. *Let $W = \cup_{i=1}^N B_{r_i}(u_i)$ be a finite union of open balls. Then there is a set $S \subset \{1, 2, \ldots, N\}$ such that*

1. *The balls $B_{r_i}(u_i), i \in S$ are disjoint.*
2. *$W \subset \cup_{i \in S} B_{3r_i}(u_i)$.*

Proof. Let us first order the balls $B_{r_i}(u_i)$ so that

$$r_1 \geq r_2 \geq \ldots \geq r_N.$$

Set $i_1 = 1$, and discard all balls such that

$$B_{i_1} \cap B_j \neq \emptyset.$$

Let B_{i_2} be the first of the remaining balls, if any. Discard all B_j such that $j > i_2$ and $B_{i_2} \cap B_j \neq \emptyset$.

Let B_{i_3} the first of the remaining balls as long as possible. Such a process stops after a finite number of steps. Define $S = \{i_1, i_2, \ldots\}$. It is clear that 1 holds. Now we prove that each discarded B_j is contained in

$$\{B_{3r_i}, \ i \in S\}.$$

Just observe that if $r' < r$ and $B_{r'}(u')$ intersects $B_r(u)$, then $B_{r'}(u') \subset B_{3r}(u)$.

The proof is complete.

Theorem 6.3.2. *Suppose μ is a finite Borel measure on $\mathbb{R}^n$ and $\lambda > 0$. Then*

$$m(A_\lambda) \leq 3^n \lambda^{-1} \|\mu\|,$$

where

$$A_\lambda = \{u \in U \mid G_\mu(u) > \lambda\}$$

and

$$\|\mu\| = |\mu|(\mathbb{R}^n).$$

Proof. Let K be a compact subset of the open set A_λ.

As $G_\mu(u) = \sup_{0<r<\infty}\{F_{r\mu}(u)\}$, each $u \in K$ is the center of an open ball B_u such that

$$\mu(B_u) > \lambda m(B_u).$$

Since K is compact, there exists a finite number of such balls which covers K. By Lemma 6.3.1, there exists a disjoint subcollection here denoted by $\{B_{r_1},\ldots,B_{r_N}\}$ such that $K \subset \cup_{k=1}^N B_{3r_k}$, so that

$$\begin{aligned} m(K) &\leq 3^n \sum_{k=1}^N m(B_{r_k}) \\ &\leq 3^n \lambda^{-1} \sum_{k=1}^N |\mu|(B_{r_k}) \\ &\leq 3^n \lambda^{-1} \|\mu\|. \end{aligned} \tag{6.15}$$

The result follows taking the supremum relating all compact $K \subset A_\lambda$.

Remark 6.3.3. Observe that, if $f \in L^1(\mathbb{R}^n)$ and $\lambda > 0$, for $A_\lambda = \{u \in \mathbb{R}^n \mid |f| > \lambda\}$, we have

$$m(A_\lambda) \leq \lambda^{-1}\|f\|_1.$$

This follows from the fact that

$$\lambda m(A_\lambda) \leq \int_{A_\lambda} |f|\, dm \leq \int_{\mathbb{R}^n} |f|\, dm = \|f\|_1.$$

Observe also that defining $d\eta = |f|\, dm$, for every $\lambda > 0$, defining

$$G_f(u) = \sup_{0<r<\infty} \frac{\eta(B_r(u))}{m(B_r(u))}$$

and

$$A_\lambda = \{u \in U \mid G_f(u) > \lambda\},$$

we have

$$m(A_\lambda) \leq 3^n \lambda^{-1} \|f\|_1.$$

6.3.1 Lebesgue Points

Finally in this section we present the main definition of Lebesgue points and some relating results.

Definition 6.3.4. Let $f \in L^1(\mathbb{R}^n)$. A point $u \in L^1(\mathbb{R}^n)$ such that

$$\lim_{r\to 0} \frac{1}{m(B_r(u))} \int_{B_r(u)} |f(v) - f(u)|\, dm(v) = 0$$

is called a Lebesgue point of f.

Theorem 6.3.5. *If $f \in L^1(\mathbb{R}^n)$, then almost all $u \in \mathbb{R}^n$ is a Lebesgue point of f.*

Proof. Define

$$H_{r_f}(u) = \frac{1}{m(B_r(u))} \int_{B_r(u)} |f - f(u)|\, dm, \forall u \in \mathbb{R}^n,\ r > 0,$$

and also define

$$H_f(u) = \limsup_{r \to 0} H_{r_f}(u).$$

We have to show that $H_f = 0$, a.e. $[m]$.

Select $y > 0$ and fix $k \in \mathbb{N}$. Observe that there exists $g \in C(\mathbb{R}^n)$ such that

$$\|f - g\|_1 < 1/k.$$

Define $h = f - g$. Since g is continuous, $H_g = 0$ in $\mathbb{R}^n$. Observe that

$$\begin{aligned} H_{r_h}(u) &= \frac{1}{m(B_r(u))} \int_{B_r(u)} |h - h(u)|\, dm \\ &\leq \frac{1}{m(B_r(u))} \int_{B_r(u)} |h|\, dm + |h(u)|, \end{aligned} \tag{6.16}$$

so that

$$H_h < G_h + |h|.$$

Since

$$H_{r_f} \leq H_{r_g} + H_{r_h},$$

we obtain

$$H_f \leq G_h + |h|.$$

Define

$$\begin{aligned} A_y &= \{u \in \mathbb{R}^n \mid H_f(u) > 2y\}, \\ B_{y,k} &= \{u \in \mathbb{R}^n \mid G_h(u) > y\}, \end{aligned}$$

and

$$C_{y,k} = \{u \in \mathbb{R}^n \mid |h| > y\}.$$

Observe that $\|h\|_1 < 1/k$, so that from Remark 6.3.3 we obtain

$$m(B_{y,k}) \leq \frac{3^n}{yk}$$

and

$$m(C_{y,k}) \leq \frac{1}{yk}$$

and hence

$$m(B_{y,k} \cup C_{y,k}) \leq \frac{3^n + 1}{yk}.$$

Therefore

$$m(A_y) \le m(B_{y,k} \cup C_{y,k}) \le \frac{3^n+1}{yk}.$$

Since k is arbitrary, we get $m(A_y) = 0, \forall y > 0$ so that $m\{u \in \mathbb{R}^n \mid H_f(u) > 0\} = 0$.
The proof is complete.

We finish this section with the following result.

Theorem 6.3.6. *Suppose μ is a complex Borel measure on $\mathbb{R}^n$ such that $\mu \ll m$. Suppose f is the Radon–Nikodym derivative of μ with respect to m. Under such assumptions,*

$$D\mu = f, \text{ a.e. } [m]$$

and

$$\mu(E) = \int_E D\mu \, dm,$$

for all Borel set $E \subset \mathbb{R}^n$.

Proof. From the Radon–Nikodym theorem we have

$$\mu(E) = \int_E f \, dm,$$

for all measurable set $E \subset \mathbb{R}^n$.

Observe that at any Lebesgue point u of f we have

$$\begin{aligned} f(u) &= \lim_{r \to 0} \frac{1}{m(B_r(u))} \int_{B_r(u)} f \, dm \\ &= \lim_{r \to 0} \frac{\mu(B_r(u))}{m(B_r(u))} \\ &= D\mu(u). \end{aligned} \tag{6.17}$$

The proof is complete.

Chapter 7
Distributions

The main reference for this chapter is Rudin [58].

7.1 Basic Definitions and Results

Definition 7.1.1 (Test Functions, the Space $\mathscr{D}(\Omega)$). Let $\Omega \subset \mathbb{R}^n$ be a nonempty open set. For each $K \subset \Omega$ compact, consider the space $\mathscr{D}_K$, the set of all $C^\infty(\Omega)$ functions with support in K. We define the space of test functions, denoted by $\mathscr{D}(\Omega)$ as

$$\mathscr{D}(\Omega) = \cup_{K\subset\Omega}\mathscr{D}_K,\ K\ compact. \tag{7.1}$$

Thus $\phi \in \mathscr{D}(\Omega)$ if and only if $\phi \in C^\infty(\Omega)$ and the support of ϕ is a compact subset of Ω.

Definition 7.1.2 (Topology for $\mathscr{D}(\Omega)$). Let $\Omega \subset \mathbb{R}^n$ be an open set.

1. For every $K \subset \Omega$ compact, σ_K denotes the topology which a local base is defined by $\{\mathscr{V}_{N,k}\}$, where $N, k \in \mathbb{N}$,

$$\mathscr{V}_{N,k} = \{\phi \in \mathscr{D}_K \mid \|\phi\|_N < 1/k\} \tag{7.2}$$

   and

$$\|\phi\|_N = \max\{|D^\alpha \phi(x)| \mid x \in \Omega, |\alpha| \leq N\}. \tag{7.3}$$

2. $\hat{\sigma}$ denotes the collection of all convex balanced sets $\mathscr{W} \in \mathscr{D}(\Omega)$ such that $\mathscr{W} \cap \mathscr{D}_K \in \sigma_K$ for every compact $K \subset \Omega$.
3. We define σ in $\mathscr{D}(\Omega)$ as the collection of all unions of sets of the form $\phi + \mathscr{W}$, for $\phi \in \mathscr{D}(\Omega)$ and $\mathscr{W} \in \hat{\sigma}$.

F. Botelho, *Functional Analysis and Applied Optimization in Banach Spaces: Applications to Non-Convex Variational Models*, DOI 10.1007/978-3-319-06074-3_7,

Theorem 7.1.3. *Concerning the last definition we have the following:*

1. *σ is a topology in $\mathscr{D}(\Omega)$.*
2. *Through σ, $\mathscr{D}(\Omega)$ is made into a locally convex topological vector space.*

Proof.

1. From item 3 of Definition 7.1.2, it is clear that arbitrary unions of elements of σ are elements of σ. Let us now show that finite intersections of elements of σ also belong to σ. Suppose $\mathscr{V}_1 \in \sigma$ and $\mathscr{V}_2 \in \sigma$; if $\mathscr{V}_1 \cap \mathscr{V}_2 = \emptyset$, we are done. Thus, suppose $\phi \in \mathscr{V}_1 \cap \mathscr{V}_2$. By the definition of σ there exist two sets of indices L_1 and L_2, such that

$$\mathscr{V}_i = \cup_{\lambda \in L_i}(\phi_{i\lambda} + \mathscr{W}_{i\lambda}), \text{ for } i = 1,2, \tag{7.4}$$

and as $\phi \in \mathscr{V}_1 \cap \mathscr{V}_2$ there exist $\phi_i \in \mathscr{D}(\Omega)$ and $\mathscr{W}_i \in \hat{\sigma}$ such that

$$\phi \in \phi_i + \mathscr{W}_i, \text{ for } i = 1,2. \tag{7.5}$$

Thus there exists $K \subset \Omega$ such that $\phi_i \in \mathscr{D}_K$ for $i \in \{1,2\}$. Since $\mathscr{D}_K \cap \mathscr{W}_i \in \sigma_K$, $\mathscr{D}_K \cap \mathscr{W}_i$ is open in $\mathscr{D}_K$ so that from (7.5) there exists $0 < \delta_i < 1$ such that

$$\phi - \phi_i \in (1 - \delta_i)\mathscr{W}_i, \text{ for } i \in \{1,2\}. \tag{7.6}$$

From (7.6) and from the convexity of $\mathscr{W}_i$ we have

$$\phi - \phi_i + \delta_i \mathscr{W}_i \subset (1 - \delta_i)\mathscr{W}_i + \delta_i \mathscr{W}_i = \mathscr{W}_i \tag{7.7}$$

so that

$$\phi + \delta_i \mathscr{W}_i \subset \phi_i + \mathscr{W}_i \subset \mathscr{V}_i, \text{ for } i \in \{1,2\}. \tag{7.8}$$

Define $\mathscr{W}_\phi = (\delta_1 \mathscr{W}_1) \cap (\delta_2 \mathscr{W}_2)$ so that

$$\phi + \mathscr{W}_\phi \subset \mathscr{V}_i, \tag{7.9}$$

and therefore we may write

$$\mathscr{V}_1 \cap \mathscr{V}_2 = \cup_{\phi \in \mathscr{V}_1 \cap \mathscr{V}_2}(\phi + \mathscr{W}_\phi) \in \sigma. \tag{7.10}$$

This completes the proof.

2. It suffices to show that single points are closed sets in $\mathscr{D}(\Omega)$ and the vector space operations are continuous.

   (a) Pick $\phi_1,\ \phi_2 \in \mathscr{D}(\Omega)$ such that $\phi_1 \neq \phi_2$ and define

$$\mathscr{V} = \{\phi \in \mathscr{D}(\Omega) \mid \|\phi\|_0 < \|\phi_1 - \phi_2\|_0\}. \tag{7.11}$$

Thus $\mathscr{V} \in \hat{\sigma}$ and $\phi_1 \notin \phi_2 + \mathscr{V}$. As $\phi_2 + \mathscr{V}$ is open and also is contained in $\mathscr{D}(\Omega) \setminus \{\phi_1\}$ and $\phi_2 \neq \phi_1$ is arbitrary, it follows that $\mathscr{D}(\Omega) \setminus \{\phi_1\}$ is open, so that $\{\phi_1\}$ is closed.

(b) The proof that addition is σ-continuous follows from the convexity of any element of $\hat{\sigma}$. Thus given $\phi_1, \phi_2 \in \mathscr{D}(\Omega)$ and $\mathscr{V} \in \hat{\sigma}$ we have

$$\phi_1 + \frac{1}{2}\mathscr{V} + \phi_2 + \frac{1}{2}\mathscr{V} = \phi_1 + \phi_2 + \mathscr{V}. \tag{7.12}$$

(c) To prove the continuity of scalar multiplication, first consider $\phi_0 \in \mathscr{D}(\Omega)$ and $\alpha_0 \in \mathbb{R}$. Then,

$$\alpha\phi - \alpha_0\phi_0 = \alpha(\phi - \phi_0) + (\alpha - \alpha_0)\phi_0. \tag{7.13}$$

For $\mathscr{V} \in \hat{\sigma}$ there exists $\delta > 0$ such that $\delta\phi_0 \in \frac{1}{2}\mathscr{V}$. Let us define $c = \frac{1}{2}(|\alpha_0| + \delta)$. Thus if $|\alpha - \alpha_0| < \delta$ then $(\alpha - \alpha_0)\phi_0 \in \frac{1}{2}\mathscr{V}$. Let $\phi \in \mathscr{D}(\Omega)$ such that

$$\phi - \phi_0 \in c\mathscr{V} = \frac{1}{2(|\alpha_0| + \delta)}\mathscr{V}, \tag{7.14}$$

so that

$$(|\alpha_0| + \delta)(\phi - \phi_0) \in \frac{1}{2}\mathscr{V}. \tag{7.15}$$

This means

$$\alpha(\phi - \phi_0) + (\alpha - \alpha_0)\phi_0 \in \frac{1}{2}\mathscr{V} + \frac{1}{2}\mathscr{V} = \mathscr{V}. \tag{7.16}$$

Therefore $\alpha\phi - \alpha_0\phi_0 \in \mathscr{V}$ whenever $|\alpha - \alpha_0| < \delta$ and $\phi - \phi_0 \in c\mathscr{V}$.

For the next result the proof may be found in Rudin [58].

Proposition 7.1.4. *A convex balanced set $\mathscr{V} \subset \mathscr{D}(\Omega)$ is open if and only if $\mathscr{V} \in \sigma$.*

Proposition 7.1.5. *The topology σ_K of $\mathscr{D}_K \subset \mathscr{D}(\Omega)$ coincides with the topology that $\mathscr{D}_K$ inherits from $\mathscr{D}(\Omega)$.*

Proof. From Proposition 7.1.4 we have

$$\mathscr{V} \in \sigma \text{ implies } \mathscr{D}_K \cap \mathscr{V} \in \sigma_K. \tag{7.17}$$

Now suppose $\mathscr{V} \in \sigma_K$, we must show that there exists $A \in \sigma$ such that $\mathscr{V} = A \cap \mathscr{D}_K$. The definition of σ_K implies that for every $\phi \in \mathscr{V}$, there exist $N \in \mathbb{N}$ and $\delta_\phi > 0$ such that

$$\{\varphi \in \mathscr{D}_K \mid \|\varphi - \phi\|_N < \delta_\phi\} \subset \mathscr{V}. \tag{7.18}$$

Define

$$\mathscr{U}_\phi = \{\varphi \in \mathscr{D}(\Omega) \mid \|\varphi\|_N < \delta_\phi\}. \quad (7.19)$$

Then $\mathscr{U}_\phi \in \hat{\sigma}$ and

$$\mathscr{D}_K \cap (\phi + \mathscr{U}_\phi) = \phi + (\mathscr{D}_K \cap \mathscr{U}_\phi) \subset \mathscr{V}. \quad (7.20)$$

Defining $A = \cup_{\phi \in \mathscr{V}}(\phi + \mathscr{U}_\phi)$, we have completed the proof.

The proof for the next two results may also be found in Rudin [58].

Proposition 7.1.6. *If A is a bounded set of $\mathscr{D}(\Omega)$, then $A \subset \mathscr{D}_K$ for some $K \subset \Omega$, and there are $M_N < \infty$ such that $\|\phi\|_N \leq M_N, \forall \phi \in A,\ N \in \mathbb{N}$.*

Proposition 7.1.7. *If $\{\phi_n\}$ is a Cauchy sequence in $\mathscr{D}(\Omega)$, then $\{\phi_n\} \subset \mathscr{D}_K$ for some $K \subset \Omega$ compact, and*

$$\lim_{i,j\to\infty} \|\phi_i - \phi_j\|_N = 0, \forall N \in \mathbb{N}. \quad (7.21)$$

Proposition 7.1.8. *If $\phi_n \to 0$ in $\mathscr{D}(\Omega)$, then there exists a compact $K \subset \Omega$ which contains the support of $\phi_n, \forall n \in \mathbb{N}$ and $D^\alpha \phi_n \to 0$ uniformly, for each multi-index α.*

The proof follows directly from the last proposition.

Theorem 7.1.9. *Suppose $T : \mathscr{D}(\Omega) \to V$ is linear, where V is a locally convex space. Then the following statements are equivalent:*

1. *T is continuous.*
2. *T is bounded.*
3. *If $\phi_n \to \theta$ in $\mathscr{D}(\Omega)$, then $T(\phi_n) \to \theta$ as $n \to \infty$.*
4. *The restrictions of T to each $\mathscr{D}_K$ are continuous.*

Proof.

- $1 \Rightarrow 2$. This follows from Proposition 1.9.3.
- $2 \Rightarrow 3$. Suppose T is bounded and $\phi_n \to 0$ in $\mathscr{D}(\Omega)$, by the last proposition $\phi_n \to 0$ in some $\mathscr{D}_K$ so that $\{\phi_n\}$ is bounded and $\{T(\phi_n)\}$ is also bounded. Hence, by Proposition 1.9.3, $T(\phi_n) \to 0$ in V.
- $3 \Rightarrow 4$. Assume 3 holds and consider $\{\phi_n\} \subset \mathscr{D}_K$. If $\phi_n \to \theta$, then by Proposition 7.1.5, $\phi_n \to \theta$ in $\mathscr{D}(\Omega)$, so that by above, $T(\phi_n) \to \theta$ in V. Since $\mathscr{D}_K$ is metrizable, also by Proposition 1.9.3, we have that 4 follows.
- $4 \Rightarrow 1$. Assume 4 holds and let $\mathscr{V}$ be a convex balanced neighborhood of zero in V. Define $\mathscr{U} = T^{-1}(\mathscr{V})$. Thus $\mathscr{U}$ is balanced and convex. By Proposition 7.1.5, $\mathscr{U}$ is open in $\mathscr{D}(\Omega)$ if and only if $\mathscr{D}_K \cap \mathscr{U}$ is open in $\mathscr{D}_K$ for each compact $K \subset \Omega$; thus, if the restrictions of T to each $\mathscr{D}_K$ are continuous at θ, then T is continuous at θ; hence, 4 implies 1.

Definition 7.1.10 (Distribution). A linear functional in $\mathscr{D}(\Omega)$ which is continuous with respect to σ is said to be a distribution.

Proposition 7.1.11. *Every differential operator is a continuous mapping from $\mathscr{D}(\Omega)$ into $\mathscr{D}(\Omega)$.*

Proof. Since $\|D^\alpha \phi\|_N \leq \|\phi\|_{|\alpha|+N}, \forall N \in \mathbb{N}$, D^α is continuous on each $\mathscr{D}_K$, so that by Theorem 7.1.9, D^α is continuous on $\mathscr{D}(\Omega)$.

Theorem 7.1.12. *Denoting by $\mathscr{D}'(\Omega)$ the dual space of $\mathscr{D}(\Omega)$ we have that $T : \mathscr{D}(\Omega) \to \mathbb{R} \in \mathscr{D}'(\Omega)$ if and only if for each compact set $K \subset \Omega$ there exists an $N \in \mathbb{N}$ and $c \in \mathbb{R}^+$ such that*

$$|T(\phi)| \leq c\|\phi\|_N, \forall \phi \in \mathscr{D}_K. \tag{7.22}$$

Proof. The proof follows from the equivalence of 1 and 4 in Theorem 7.1.9.

7.2 Differentiation of Distributions

Definition 7.2.1 (Derivatives for Distributions). Given $T \in \mathscr{D}'(\Omega)$ and a multi-index α, we define the D^α derivative of T as

$$D^\alpha T(\phi) = (-1)^{|\alpha|} T(D^\alpha \phi), \forall \phi \in \mathscr{D}(\Omega). \tag{7.23}$$

Remark 7.2.2. Observe that if $|T(\phi)| \leq c\|\phi\|_N, \forall \phi \in \mathscr{D}(\Omega)$ for some $c \in \mathbb{R}^+$, then

$$|D^\alpha T(\phi)| \leq c\|D^\alpha \phi\|_N \leq c\|\phi\|_{N+|\alpha|}, \forall \phi \in \mathscr{D}(\Omega), \tag{7.24}$$

thus $D^\alpha T \in \mathscr{D}'(\Omega)$. Therefore, derivatives of distributions are also distributions.

Theorem 7.2.3. *Suppose $\{T_n\} \subset \mathscr{D}'(\Omega)$. Let $T : \mathscr{D}(\Omega) \to \mathbb{R}$ be defined by*

$$T(\phi) = \lim_{n \to \infty} T_n(\phi), \forall \phi \in \mathscr{D}(\Omega). \tag{7.25}$$

Then $T \in \mathscr{D}'(\Omega)$, and

$$D^\alpha T_n \to D^\alpha T \text{ in } \mathscr{D}'(\Omega). \tag{7.26}$$

Proof. Let K be an arbitrary compact subset of Ω. Since (7.25) holds for every $\phi \in \mathscr{D}_K$, the principle of uniform boundedness implies that the restriction of T to $\mathscr{D}_K$ is continuous. It follows from Theorem 7.1.9 that T is continuous in $\mathscr{D}(\Omega)$, that is, $T \in \mathscr{D}'(\Omega)$. On the other hand

$$\begin{aligned} (D^\alpha T)(\phi) &= (-1)^{|\alpha|} T(D^\alpha \phi) = (-1)^{|\alpha|} \lim_{n \to \infty} T_n(D^\alpha \phi) \\ &= \lim_{n \to \infty} (D^\alpha T_n(\phi)), \forall \phi \in \mathscr{D}(\Omega). \end{aligned} \tag{7.27}$$

7.3 Examples of Distributions

7.3.1 First Example

Let $\Omega \subset \mathbb{R}^n$ be an open bounded set. As a first example of distribution consider the functional

$$T : \mathscr{D}(\Omega) \to \mathbb{R}$$

given by

$$T(\phi) = \int_\Omega f\phi \, dx,$$

where $f \in L^1(\Omega)$. Observe that

$$\begin{aligned} |T(\phi)| &\leq \int_\Omega |f\phi| \, dx \\ &\leq \int_\Omega |f| \, dx \|\phi\|_\infty, \end{aligned} \tag{7.28}$$

so that T is a bounded linear functional on $\mathscr{D}(\Omega)$, that is, T is a distribution.

7.3.2 Second Example

For the second example, define $\Omega = (0,1)$ and $T : \mathscr{D}(\Omega) \to \mathbb{R}$ by

$$T(\phi) = \phi(1/2) + \phi'(1/3).$$

Thus,

$$|T(\phi)| = |\phi(1/2) + \phi'(1/3)| \leq \|\phi\|_\infty + \|\phi'\|_\infty \leq 2\|\phi\|_1,$$

so that T is also a distribution (bounded and linear).

7.3.3 Third Example

For the third example, consider an open bounded $\Omega \subset \mathbb{R}^n$ and $T : \mathscr{D}(\Omega) \to \mathbb{R}$ by

$$T(\phi) = \int_\Omega f\phi \, dx,$$

where $f \in L^1(\Omega)$.

Observe that the derivative of T for the multi-index $\alpha = (\alpha_1, \ldots, \alpha_n)$ is defined by

$$D^\alpha T(\phi) = (-1)^{|\alpha|} T(D^\alpha \phi) = (-1)^{|\alpha|} \int_\Omega f D^\alpha \phi \, dx.$$

If there exists $g \in L^1(\Omega)$, such that

$$(-1)^{|\alpha|}\int_\Omega fD^\alpha\phi\,dx = \int_\Omega g\phi\,dx, \forall\phi \in \mathscr{D}(\Omega),$$

we say that g is the derivative D^α of f in the distributional sense.

For example, for $\Omega = (0,1)$ and $f : \overline{\Omega} \to \mathbb{R}$ given by

$$f(x) = \begin{cases} 0, & \text{if } x \in [0,1/2], \\ 1, & \text{if } x \in (1/2,1], \end{cases}$$

and

$$T(\phi) = \int_\Omega f\phi\,dx,$$

where $\phi \in C_c^\infty(\Omega)$,we have

$$\begin{aligned} D_xT(\phi) &= -\int_\Omega f\frac{d\phi}{dx}\,dx \\ &= -\int_{1/2}^1 (1)\frac{d\phi}{dx}\,dx \\ &= -\phi(1) + \phi(1/2) = \phi(1/2), \end{aligned} \tag{7.29}$$

that is,

$$D_xT(\phi) = \phi(1/2), \forall\phi \in C_c^\infty(\Omega).$$

Finally, defining $f : \overline{\Omega} \to \mathbb{R}$ by

$$f(x) = \begin{cases} x, & \text{if } x \in [0,1/2], \\ -x+1, & \text{if } x \in (1/2,1], \end{cases}$$

and

$$T(\phi) = \int_\Omega f\phi\,dx,$$

where $\phi \in C_c^\infty(\Omega)$ we have

$$\begin{aligned} D_xT(\phi) &= -\int_\Omega f\frac{d\phi}{dx}\,dx \\ &= -\int_0^1 f\frac{d\phi}{dx}\,dx \\ &= \int_0^1 g\phi\,dx, \end{aligned} \tag{7.30}$$

where

$$g(x) = \begin{cases} 1, & \text{if } x \in [0,1/2], \\ -1, & \text{if } x \in (1/2,1]. \end{cases}$$

In such a case we denote $g = D_x f$ and say that g is the derivative of f in the distributional sense.

We emphasize that in this last example the classical derivative of f is not defined, since f is not differentiable ant $x = 1/2$.

Chapter 8
The Lebesgue and Sobolev Spaces

Here, we emphasize that the two main references for this chapter are Adams [2] and Evans [26]. We start with the definition of Lebesgue spaces, denoted by $L^p(\Omega)$, where $1 \leq p \leq \infty$ and $\Omega \subset \mathbb{R}^n$ is an open set. In this chapter, integrals always refer to the Lebesgue measure.

8.1 Definition and Properties of L^p Spaces

Definition 8.1.1 (L^p Spaces). For $1 \leq p < \infty$, we say that $u \in L^p(\Omega)$ if $u : \Omega \to \mathbb{R}$ is measurable and

$$\int_\Omega |u|^p dx < \infty. \tag{8.1}$$

We also denote $\|u\|_p = [\int_\Omega |u|^p dx]^{1/p}$ and will show that $\|\cdot\|_p$ is a norm.

Definition 8.1.2 (L^∞ Spaces). We say that $u \in L^\infty(\Omega)$ if u is measurable and there exists $M \in \mathbb{R}^+$, such that $|u(x)| \leq M$, a.e. in Ω. We define

$$\|u\|_\infty = \inf\{M > 0 \mid |u(x)| \leq M, \text{ a.e. in } \Omega\}. \tag{8.2}$$

We will show that $\|\cdot\|_\infty$ is a norm. For $1 \leq p \leq \infty$, we define q by the relations

$$q = \begin{cases} +\infty, & \text{if } p = 1, \\ \frac{p}{p-1}, & \text{if } 1 < p < +\infty, \\ 1, & \text{if } p = +\infty, \end{cases}$$

so that symbolically we have

$$\frac{1}{p} + \frac{1}{q} = 1.$$

The next result is fundamental in the proof of the Sobolev imbedding theorem.

F. Botelho, *Functional Analysis and Applied Optimization in Banach Spaces: Applications to Non-Convex Variational Models*, DOI 10.1007/978-3-319-06074-3_8,

Theorem 8.1.3 (Hölder Inequality). *Consider $u \in L^p(\Omega)$ and $v \in L^q(\Omega)$, with $1 \leq p \leq \infty$. Then $uv \in L^1(\Omega)$ and*

$$\int_\Omega |uv|dx \leq \|u\|_p \|v\|_q. \tag{8.3}$$

Proof. The result is clear if $p = 1$ or $p = \infty$. You may assume $\|u\|_p, \|v\|_q > 0$; otherwise the result is also obvious. Thus suppose $1 < p < \infty$. From the concavity of log function on $(0, \infty)$ we obtain

$$\log\left(\frac{1}{p}a^p + \frac{1}{q}b^q\right) \geq \frac{1}{p}\log a^p + \frac{1}{q}\log b^q = \log(ab). \tag{8.4}$$

Thus,

$$ab \leq \frac{1}{p}(a^p) + \frac{1}{q}(b^q), \ \forall a \geq 0, b \geq 0. \tag{8.5}$$

Therefore

$$|u(x)||v(x)| \leq \frac{1}{p}|u(x)|^p + \frac{1}{q}|v(x)|^q, \text{ a.e. in } \Omega. \tag{8.6}$$

Hence $|uv| \in L^1(\Omega)$ and

$$\int_\Omega |uv|dx \leq \frac{1}{p}\|u\|_p^p + \frac{1}{q}\|v\|_q^q. \tag{8.7}$$

Replacing u by λu in (8.7) $\lambda > 0$, we obtain

$$\int_\Omega |uv|dx \leq \frac{\lambda^{p-1}}{p}\|u\|_p^p + \frac{1}{\lambda q}\|v\|_q^q. \tag{8.8}$$

For $\lambda = \|u\|_p^{-1}\|v\|_q^{q/p}$ we obtain the Hölder inequality.

The next step is to prove that $\|\cdot\|_p$ is a norm.

Theorem 8.1.4. *$L^p(\Omega)$ is a vector space and $\|\cdot\|_p$ is norm $\forall p$ such that $1 \leq p \leq \infty$.*

Proof. The only nontrivial property to be proved concerning the norm definition is the triangle inequality. If $p = 1$ or $p = \infty$, the result is clear. Thus, suppose $1 < p < \infty$. For $u, v \in L^p(\Omega)$ we have

$$|u(x) + v(x)|^p \leq (|u(x)| + |v(x)|)^p \leq 2^p(|u(x)|^p + |v(x)|^p), \tag{8.9}$$

so that $u + v \in L^p(\Omega)$. On the other hand

$$\begin{aligned}\|u+v\|_p^p &= \int_\Omega |u+v|^{p-1}|u+v|dx \\ &\leq \int_\Omega |u+v|^{p-1}|u|dx + \int_\Omega |u+v|^{p-1}|v|dx,\end{aligned} \tag{8.10}$$

and hence, from the Hölder inequality,

$$\|u+v\|_p^p \leq \|u+v\|_p^{p-1}\|u\|_p + \|u+v\|_p^{p-1}\|v\|_p, \tag{8.11}$$

that is,

$$\|u+v\|_p \leq \|u\|_p + \|v\|_p, \forall u,v \in L^p(\Omega). \tag{8.12}$$

Theorem 8.1.5. *$L^p(\Omega)$ is a Banach space for any p such that $1 \leq p \leq \infty$.*

Proof. Suppose $p = \infty$. Suppose $\{u_n\}$ is Cauchy sequence in $L^\infty(\Omega)$. Thus, given $k \in \mathbb{N}$, there exists $N_k \in \mathbb{N}$ such that if $m,n \geq N_k$, then

$$\|u_m - u_n\|_\infty < \frac{1}{k}. \tag{8.13}$$

Therefore, for each k, there exist a set E_k such that $m(E_k) = 0$, and

$$|u_m(x) - u_n(x)| < \frac{1}{k}, \ \forall x \in \Omega \setminus E_k, \ \forall m,n \geq N_k. \tag{8.14}$$

Observe that $E = \cup_{k=1}^\infty E_k$ is such that $m(E) = 0$. Thus $\{u_n(x)\}$ is a real Cauchy sequence at each $x \in \Omega \setminus E$. Define $u(x) = \lim_{n\to\infty} u_n(x)$ on $\Omega \setminus E$. Letting $m \to \infty$ in (8.14) we obtain

$$|u(x) - u_n(x)| < \frac{1}{k}, \ \forall x \in \Omega \setminus E, \ \forall n \geq N_k. \tag{8.15}$$

Thus $u \in L^\infty(\Omega)$ and $\|u_n - u\|_\infty \to 0$ as $n \to \infty$.

Now suppose $1 \leq p < \infty$. Let $\{u_n\}$ be a Cauchy sequence in $L^p(\Omega)$. We can extract a subsequence $\{u_{n_k}\}$ such that

$$\|u_{n_{k+1}} - u_{n_k}\|_p \leq \frac{1}{2^k}, \forall k \in \mathbb{N}. \tag{8.16}$$

To simplify the notation we write u_k in place of u_{n_k}, so that

$$\|u_{k+1} - u_k\|_p \leq \frac{1}{2^k}, \forall k \in \mathbb{N}. \tag{8.17}$$

Defining

$$g_n(x) = \sum_{k=1}^n |u_{k+1}(x) - u_k(x)|, \tag{8.18}$$

we obtain

$$\|g_n\|_p \leq 1, \ \forall n \in \mathbb{N}. \tag{8.19}$$

From the monotone convergence theorem and (8.19), $g_n(x)$ converges to a limit $g(x)$ with $g \in L^p(\Omega)$. On the other hand, for $m \geq n \geq 2$, we have

$$\begin{aligned} |u_m(x) - u_n(x)| &\leq |u_m(x) - u_{m-1}(x)| + \ldots + |u_{n+1}(x) - u_n(x)| \\ &\leq g(x) - g_{n-1}(x), \text{ a.e. in } \Omega. \end{aligned} \tag{8.20}$$

Hence $\{u_n(x)\}$ is Cauchy a.e. in Ω and converges to a limit $u(x)$ so that

$$|u(x) - u_n(x)| \leq g(x), \text{ a.e. in } \Omega, \text{ for } n \geq 2, \tag{8.21}$$

which means $u \in L^p(\Omega)$. Finally from $|u_n(x) - u(x)| \to 0$, a.e. in Ω, $|u_n(x) - u(x)|^p \leq |g(x)|^p$, and the Lebesgue dominated convergence theorem we get

$$\|u_n - u\|_p \to 0 \text{ as } n \to \infty. \tag{8.22}$$

Theorem 8.1.6. *Let $\{u_n\} \subset L^p(\Omega)$ and $u \in L^p(\Omega)$ such that $\|u_n - u\|_p \to 0$. Then there exists a subsequence $\{u_{n_k}\}$ such that*

1. *$u_{n_k}(x) \to u(x)$, a.e. in Ω,*
2. *$|u_{n_k}(x)| \leq h(x)$, a.e. in $\Omega, \forall k \in \mathbb{N}$, for some $h \in L^p(\Omega)$.*

Proof. The result is clear for $p = \infty$. Suppose $1 \leq p < \infty$. From the last theorem we can easily obtain that $|u_{n_k}(x) - u(x)| \to 0$ as $k \to \infty$, a.e. in Ω. To complete the proof, just take $h = u + g$, where g is defined in the proof of last theorem.

Theorem 8.1.7. *$L^p(\Omega)$ is reflexive for all p such that $1 < p < \infty$.*

Proof. We divide the proof into 3 parts.

1. For $2 \leq p < \infty$ we have that

$$\left\|\frac{u+v}{2}\right\|^p_{L^p(\Omega)} + \left\|\frac{u-v}{2}\right\|^p_{L^p(\Omega)} \leq \frac{1}{2}(\|u\|^p_{L^p(\Omega)} + \|v\|^p_{L^p(\Omega)}), \forall u, v \in L^p(\Omega). \tag{8.23}$$

Proof. Observe that

$$\alpha^p + \beta^p \leq (\alpha^2 + \beta^2)^{p/2}, \forall \alpha, \beta \geq 0. \tag{8.24}$$

Now taking $\alpha = \left|\frac{a+b}{2}\right|$ and $\beta = \left|\frac{a-b}{2}\right|$ in (8.24), we obtain (using the convexity of $t^{p/2}$)

$$\begin{aligned} \left|\frac{a+b}{2}\right|^p + \left|\frac{a-b}{2}\right|^p &\leq \left(\left|\frac{a+b}{2}\right|^2 + \left|\frac{a-b}{2}\right|^2\right)^{p/2} = \left(\frac{a^2}{2} + \frac{b^2}{2}\right)^{p/2} \\ &\leq \frac{1}{2}|a|^p + \frac{1}{2}|b|^p. \end{aligned} \tag{8.25}$$

The inequality (8.23) follows immediately.

2. $L^p(\Omega)$ is uniformly convex and therefore reflexive for $2 \leq p < \infty$.
Proof. Suppose given $\varepsilon > 0$ and suppose that

$$\|u\|_p \leq 1, \ \|v\|_p \leq 1 \text{ and } \|u - v\|_p > \varepsilon. \tag{8.26}$$

From part 1, we obtain

$$\left\|\frac{u+v}{2}\right\|_p^p < 1 - \left(\frac{\varepsilon}{2}\right)^p, \tag{8.27}$$

and therefore

$$\left\|\frac{u+v}{2}\right\|_p < 1 - \delta, \tag{8.28}$$

for $\delta = 1 - (1 - (\varepsilon/2)^p)^{1/p} > 0$. Thus $L^p(\Omega)$ is uniformly convex and from Theorem 2.7.2 it is reflexive.
3. $L^p(\Omega)$ is reflexive for $1 < p \leq 2$. Let $1 < p \leq 2$; from 2 we can conclude that L^q is reflexive. We will define $T : L^p(\Omega) \to (L^q)^*$ by

$$\langle Tu, f\rangle_{L^q(\Omega)} = \int_\Omega uf dx, \forall u \in L^p(\Omega), \ \ f \in L^q(\Omega). \tag{8.29}$$

From the Hölder inequality, we obtain

$$|\langle Tu, f\rangle_{L^q(\Omega)}| \leq \|u\|_p \|f\|_q, \tag{8.30}$$

so that

$$\|Tu\|_{(L^q(\Omega))^*} \leq \|u\|_p. \tag{8.31}$$

Pick $u \in L^p(\Omega)$ and define $f_0(x) = |u(x)|^{p-2}u(x)$ ($f_0(x) = 0$ if $u(x) = 0$). Thus, we have that $f_0 \in L^q(\Omega)$, $\|f_0\|_q = \|u\|_p^{p-1}$, and $\langle Tu, f_0\rangle_{L^q(\Omega)} = \|u\|_p^p$. Therefore,

$$\|Tu\|_{(L^q(\Omega))^*} \geq \frac{\langle Tu, f_0\rangle_{L^q(\Omega)}}{\|f_0\|_q} = \|u\|_p. \tag{8.32}$$

Hence from (8.31) and (8.32) we have

$$\|Tu\|_{(L^q(\Omega))^*} = \|u\|_p, \forall u \in L^p(\Omega). \tag{8.33}$$

Thus T is an isometry from $L^p(\Omega)$ to a closed subspace of $(L^q(\Omega))^*$. Since from the first part $L^q(\Omega)$ is reflexive, we have that $(L^q(\Omega))^*$ is reflexive. Hence $T(L^p(\Omega))$ and $L^p(\Omega)$ are reflexive.

Theorem 8.1.8 (Riesz Representation Theorem). *Let $1 < p < \infty$ and let f be a continuous linear functional on $L^p(\Omega)$. Then there exists a unique $u_0 \in L^q$ such that*

$$f(v) = \int_\Omega vu_0 \, dx, \ \forall v \in L^p(\Omega). \tag{8.34}$$

Furthermore

$$\|f\|_{(L^p)^*} = \|u_0\|_q. \tag{8.35}$$

Proof. First we define the operator $T : L^q(\Omega) \to (L^p(\Omega))^*$ by

$$\langle Tu, v\rangle_{L^p(\Omega)} = \int_\Omega uv\,dx, \forall v \in L^p(\Omega). \tag{8.36}$$

Similarly to last theorem, we obtain

$$\|Tu\|_{(L^p(\Omega))^*} = \|u\|_q. \tag{8.37}$$

We have to show that T is onto. Define $E = T(L^q(\Omega))$. As E is a closed subspace, it suffices to show that E is dense in $(L^p(\Omega))^*$. Suppose $h \in (L^p)^{**} = L^p$ is such that

$$\langle Tu, h\rangle_{L^p(\Omega)} = 0, \forall u \in L^q(\Omega). \tag{8.38}$$

Choosing $u = |h|^{p-2}h$ we may conclude that $h = 0$ which, by Corollary 2.2.13, completes the first part of the proof. The proof of uniqueness is left to the reader.

Definition 8.1.9. Let $1 \leq p \leq \infty$. We say that $u \in L^p_{loc}(\Omega)$ if $u\chi_K \in L^p(\Omega)$ for all compact $K \subset \Omega$.

8.1.1 Spaces of Continuous Functions

We introduce some definitions and properties concerning spaces of continuous functions. First, we recall that by a domain we mean an open set in $\mathbb{R}^n$. Thus for a domain $\Omega \subset \mathbb{R}^n$ and for any nonnegative integer m we define by $C^m(\Omega)$ the set of all functions u which the partial derivatives $D^\alpha u$ are continuous on Ω for any α such that $|\alpha| \leq m$, where if $D^\alpha = D_1^{\alpha_1} D_2^{\alpha_2} \ldots D_n^{\alpha_n}$, we have $|\alpha| = \alpha_1 + \ldots + \alpha_n$. We define $C^\infty(\Omega) = \cap_{m=0}^\infty C^m(\Omega)$ and denote $C^0(\Omega) = C(\Omega)$. Given a function $\phi : \Omega \to \mathbb{R}$, its support, denoted by $spt(\phi)$, is given by

$$spt(\phi) = \overline{\{x \in \Omega \mid \phi(x) \neq 0\}}.$$

$C_c^\infty(\Omega)$ denotes the set of functions in $C^\infty(\Omega)$ with compact support contained in Ω.

The sets $C_0(\Omega)$ and $C_0^\infty(\Omega)$ consist of the closure of $C_c(\Omega)$ (which is the set of functions in $C(\Omega)$ with compact support in Ω) and $C_c^\infty(\Omega)$, respectively, relating the uniform convergence norm. On the other hand, $C_B^m(\Omega)$ denotes the set of functions $u \in C^m(\Omega)$ for which $D^\alpha u$ is bounded on Ω for $0 \leq |\alpha| \leq m$. Observe that $C_B^m(\Omega)$ is a Banach space with the norm denoted by $\|\cdot\|_{B,m}$ given by

$$\|u\|_{B,m} = \max_{0 \leq |\alpha| \leq m} \sup_{x \in \Omega} \{|D^\alpha u(x)|\}.$$

Also, we define $C^m(\bar{\Omega})$ as the set of functions $u \in C^m(\Omega)$ for which $D^\alpha u$ is bounded and uniformly continuous on Ω for $0 \leq |\alpha| \leq m$. Observe that $C^m(\bar{\Omega})$ is a closed subspace of $C_B^m(\Omega)$ and is also a Banach space with the norm inherited from $C_B^m(\Omega)$. An important space is one of the Hölder continuous functions.

Definition 8.1.10 (Spaces of the Hölder Continuous Functions). If $0 < \lambda < 1$, for a nonnegative integer m, we define the space of the Hölder continuous functions denoted by $C^{m,\lambda}(\bar{\Omega})$, as the subspace of $C^m(\bar{\Omega})$ consisting of those functions u for which, for $0 \leq |\alpha| \leq m$, there exists a constant K such that

$$|D^\alpha u(x) - D^\alpha u(y)| \leq K|x-y|^\lambda, \forall x, y \subset \Omega.$$

$C^{m,\lambda}(\bar{\Omega})$ is a Banach space with the norm denoted by $\|\cdot\|_{m,\lambda}$ given by

$$\|u\|_{m,\lambda} = \|u\|_{B,m} + \max_{0\leq|\alpha|\leq m} \sup_{x,y\in\Omega} \left\{ \frac{|D^\alpha u(x) - D^\alpha u(y)|}{|x-y|^\lambda}, \ x \neq y \right\}.$$

From now on we say that $f : \Omega \to \mathbb{R}$ is locally integrable, if it is Lebesgue integrable on any compact $K \subset \Omega$. Furthermore, we say that $f \in L^p_{loc}(\Omega)$ if $f \in L^p(K)$ for any compact $K \subset \Omega$. Finally, given an open $\Omega \subset \mathbb{R}^n$, we denote $W \subset\subset \Omega$ whenever $\overline{W}$ is compact and $\overline{W} \subset \Omega$.

Theorem 8.1.11. *The space $C_0(\Omega)$ is dense in $L^p(\Omega)$, for $1 \leq p < \infty$.*

Proof. For the proof we need the following lemma:

Lemma 8.1.12. *Let $f \in L^1_{loc}(\Omega)$ such that*

$$\int_\Omega fu \, dx = 0, \forall u \in C_0(\Omega). \tag{8.39}$$

Then $f = 0$ a.e. in Ω.

First suppose $f \in L^1(\Omega)$ and Ω bounded, so that $m(\Omega) < \infty$. Given $\varepsilon > 0$, since $C_0(\Omega)$ is dense in $L^1(\Omega)$, there exists $f_1 \in C_0(\Omega)$ such that $\|f - f_1\|_1 < \varepsilon$ and thus, from (8.39), we obtain

$$\left| \int_\Omega f_1 u \, dx \right| \leq \varepsilon \|u\|_\infty, \forall u \in C_0(\Omega). \tag{8.40}$$

Defining

$$K_1 = \{x \in \Omega \mid f_1(x) \geq \varepsilon\} \tag{8.41}$$

and

$$K_2 = \{x \in \Omega \mid f_1(x) \leq -\varepsilon\}, \tag{8.42}$$

as K_1 and K_2 are disjoint compact sets, by the Urysohn theorem, there exists $u_0 \in C_0(\Omega)$ such that

$$u_0(x) = \begin{cases} +1, & \text{if } x \in K_1, \\ -1, & \text{if } x \in K_2 \end{cases} \tag{8.43}$$

and

$$|u_0(x)| \leq 1, \forall x \in \Omega. \tag{8.44}$$

Also defining $K = K_1 \cup K_2$, we may write

$$\int_\Omega f_1 u_0 \, dx = \int_{\Omega - K} f_1 u_0 \, dx + \int_K f_1 u_0 \, dx. \tag{8.45}$$

Observe that, from (8.40),

$$\int_K |f_1| \, dx \leq \int_\Omega |f_1 u_0| \, dx \leq \varepsilon \tag{8.46}$$

so that

$$\int_\Omega |f_1| \, dx = \int_K |f_1| \, dx + \int_{\Omega - K} |f_1| \, dx \leq \varepsilon + \varepsilon m(\Omega). \tag{8.47}$$

Hence

$$\|f\|_1 \leq \|f - f_1\|_1 + \|f_1\|_1 \leq 2\varepsilon + \varepsilon m(\Omega). \tag{8.48}$$

Since $\varepsilon > 0$ is arbitrary, we have that $f = 0$ a.e. in Ω. Finally, if $m(\Omega) = \infty$, define

$$\Omega_n = \{x \in \Omega \mid \text{dist}(x, \Omega^c) > 1/n \text{ and } |x| < n\}. \tag{8.49}$$

It is clear that $\Omega = \cup_{n=1}^\infty \Omega_n$ and from above $f = 0$ a.e. on $\Omega_n, \forall n \in \mathbb{N}$, so that $f = 0$ a.e. in Ω.

Finally, to finish the proof of Theorem 8.1.11, suppose $h \in L^q(\Omega)$ is such that

$$\int_\Omega hu \, dx = 0, \forall u \in C_0(\Omega). \tag{8.50}$$

Observe that $h \in L^1_{loc}(\Omega)$ since $\int_K |h| \, dx \leq \|h\|_q m(K)^{1/p} < \infty$. From last lemma $h = 0$ a.e. in Ω, which by Corollary 2.2.13 completes the proof.

Theorem 8.1.13. *$L^p(\Omega)$ is separable for any $1 \leq p < \infty$.*

Proof. The result follows from last theorem and from the fact that $C_0(K)$ is separable for each $K \subset \Omega$ compact [from the Weierstrass theorem, polynomials with rational coefficients are dense $C_0(K)$]. Observe that $\Omega = \cup_{n=1}^\infty \Omega_n$, Ω_n defined as in (8.49), where $\bar{\Omega}_n$ is compact, $\forall n \in \mathbb{N}$.

8.2 The Sobolev Spaces

Now we define the Sobolev spaces, denoted by $W^{m,p}(\Omega)$.

Definition 8.2.1 (Sobolev Spaces). We say that $u \in W^{m,p}(\Omega)$ if $u \in L^p(\Omega)$ and $D^\alpha u \in L^p(\Omega)$, for all α such that $0 \leq |\alpha| \leq m$, where the derivatives are understood in the distributional sense.

Definition 8.2.2. We define the norm $\|\cdot\|_{m,p}$ for $W^{m,p}(\Omega)$, where $m \in \mathbb{N}$ and $1 \leq p \leq \infty$, as

$$\|u\|_{m,p} = \left\{ \sum_{0 \leq |\alpha| \leq m} \|D^\alpha u\|_p^p \right\}^{1/p}, \text{ if } 1 \leq p < \infty, \tag{8.51}$$

and

$$\|u\|_{m,\infty} = \max_{0 \leq |\alpha| \leq m} \|D^\alpha u\|_\infty. \tag{8.52}$$

Theorem 8.2.3. *$W^{m,p}(\Omega)$ is a Banach space.*

Proof. Consider $\{u_n\}$ a Cauchy sequence in $W^{m,p}(\Omega)$. Then $\{D^\alpha u_n\}$ is a Cauchy sequence for each $0 \leq |\alpha| \leq m$. Since $L^p(\Omega)$ is complete there exist functions u and u_α, for $0 \leq |\alpha| \leq m$, in $L^p(\Omega)$ such that $u_n \to u$ and $D^\alpha u_n \to u_\alpha$ in $L^p(\Omega)$ as $n \to \infty$. From above $L^p(\Omega) \subset L^1_{loc}(\Omega)$ and so u_n determines a distribution $T_{u_n} \in \mathscr{D}'(\Omega)$. For any $\phi \in \mathscr{D}(\Omega)$ we have, by the Hölder inequality,

$$|T_{u_n}(\phi) - T_u(\phi)| \leq \int_\Omega |u_n(x) - u(x)||\phi(x)|dx \leq \|\phi\|_q \|u_n - u\|_p. \tag{8.53}$$

Hence $T_{u_n}(\phi) \to T_u(\phi)$ for every $\phi \in \mathscr{D}(\Omega)$ as $n \to \infty$. Similarly $T_{D^\alpha u_n}(\phi) \to T_{u_\alpha}(\phi)$ for every $\phi \in \mathscr{D}(\Omega)$. We have that

$$\begin{aligned} T_{u_\alpha}(\phi) &= \lim_{n \to \infty} T_{D^\alpha u_n}(\phi) = \lim_{n \to \infty} (-1)^{|\alpha|} T_{u_n}(D^\alpha \phi) \\ &= (-1)^{|\alpha|} T_u(D^\alpha \phi) = T_{D^\alpha u}(\phi), \end{aligned} \tag{8.54}$$

for every $\phi \in \mathscr{D}(\Omega)$. Thus $u_\alpha = D^\alpha u$ in the sense of distributions, for $0 \leq |\alpha| \leq m$, and $u \in W^{m,p}(\Omega)$. As $\lim_{n \to \infty} \|u - u_n\|_{m,p} = 0$, $W^{m,p}(\Omega)$ is complete.

Remark 8.2.4. Observe that distributional and classical derivatives coincide when the latter exist and are continuous. We define $S \subset W^{m,p}(\Omega)$ by

$$S = \{\phi \in C^m(\Omega) \mid \|\phi\|_{m,p} < \infty\}. \tag{8.55}$$

Thus, the completion of S concerning the norm $\|\cdot\|_{m,p}$ is denoted by $H^{m,p}(\Omega)$.

Corollary 8.2.5. $H^{m,p}(\Omega) \subset W^{m,p}(\Omega)$.

Proof. Since $W^{m,p}(\Omega)$ is complete we have that $H^{m,p}(\Omega) \subset W^{m,p}(\Omega)$.

Theorem 8.2.6. *$W^{m,p}(\Omega)$ is separable if $1 \leq p < \infty$ and is reflexive and uniformly convex if $1 < p < \infty$. Particularly, $W^{m,2}(\Omega)$ is a separable Hilbert space with the inner product*

$$(u,v)_m = \sum_{0 \leq |\alpha| \leq m} \langle D^\alpha u, D^\alpha v \rangle_{L^2(\Omega)}. \tag{8.56}$$

Proof. We can see $W^{m,p}(\Omega)$ as a subspace of $L^p(\Omega, \mathbb{R}^N)$, where $N = \sum_{0 \leq |\alpha| \leq m} 1$. From the relevant properties for $L^p(\Omega)$, we have that $L^p(\Omega; \mathbb{R}^N)$ is a reflexive and uniformly convex for $1 < p < \infty$ and separable for $1 \leq p < \infty$. Given $u \in W^{m,p}(\Omega)$, we may associate the vector $Pu \in L^p(\Omega; \mathbb{R}^N)$ defined by

$$Pu = \{D^\alpha u\}_{0 \leq |\alpha| \leq m}. \tag{8.57}$$

Since $\|Pu\|_{p^N} = \|u\|_{m,p}$, we have that $W^{m,p}$ is closed subspace of $L^p(\Omega; \mathbb{R}^N)$. Thus, from Theorem 1.21 in Adams [1], we have that $W^{m,p}(\Omega)$ is separable if $1 \leq p < \infty$ and reflexive and uniformly convex if $1 < p < \infty$.

Lemma 8.2.7. *Let $1 \leq p < \infty$ and define $U = L^p(\Omega; \mathbb{R}^N)$. For every continuous linear functional f on U, there exists a unique $v \in L^q(\Omega; \mathbb{R}^N) = U^*$ such that*

$$f(u) = \sum_{i=1}^N \langle u_i, v_i \rangle, \forall u \in U. \tag{8.58}$$

Moreover,

$$\|f\|_{U^*} = \|v\|_{q^N}, \tag{8.59}$$

where $\| \cdot \|_{q^N} = \| \cdot \|_{L^q(\Omega, \mathbb{R}^N)}$.

Proof. For $u = (u_1, \ldots, u_n) \in L^p(\Omega; \mathbb{R}^N)$, we may write

$$\begin{aligned} f(u) &= f((u_1, 0, \ldots, 0)) + \ldots + f((0, \ldots, 0, u_j, 0, \ldots, 0)) \\ &\quad + \ldots + f((0, \ldots, 0, u_n)), \end{aligned} \tag{8.60}$$

and since $f((0, \ldots, 0, u_j, 0, \ldots, 0))$ is continuous linear functional on $u_j \in L^p(\Omega)$, there exists a unique $v_j \in L^q(\Omega)$ such that $f(0, \ldots, 0, u_j, 0, \ldots, 0) = \langle u_j, v_j \rangle_{L^2(\Omega)}, \forall u_j \in L^p(\Omega), \forall\, 1 \leq j \leq N$, so that

$$f(u) = \sum_{i=1}^N \langle u_i, v_i \rangle, \forall u \in U. \tag{8.61}$$

From the Hölder inequality we obtain

$$|f(u)| \leq \sum_{j=1}^N \|u_j\|_p \|v_j\|_q \leq \|u\|_{p^N} \|v\|_{q^N}, \tag{8.62}$$

and hence $\|f\|_{U^*} \leq \|v\|_{q^N}$. The equality in (8.62) is achieved for $u \in L^p(\Omega, \mathbb{R}^N)$, $1 < p < \infty$ such that

$$u_j(x) = \begin{cases} |v_j|^{q-2}\bar{v}_j, & \text{if } v_j \neq 0 \\ 0, & \text{if } v_j = 0. \end{cases} \tag{8.63}$$

If $p = 1$, choose k such that $\|v_k\|_\infty = \max_{1 \leq j \leq N} \|v_j\|_\infty$. Given $\varepsilon > 0$, there is a measurable set A such that $m(A) > 0$ and $|v_k(x)| \geq \|v_k\|_\infty - \varepsilon, \forall x \in A$. Defining $u(x)$ as

$$u_i(x) = \begin{cases} \bar{v}_k/v_k, & \text{if } i = k,\ x \in A \text{ and } v_k(x) \neq 0 \\ 0, & \text{otherwise,} \end{cases} \tag{8.64}$$

we have

$$\begin{aligned} f(u_k) = \langle u, v_k \rangle_{L^2(\Omega)} &= \int_A |v_k| dx \geq (\|(v_k\|_\infty - \varepsilon)\|u_k\|_1 \\ &= (\|v\|_{\infty^N} - \varepsilon)\|u\|_{1^N}. \end{aligned} \tag{8.65}$$

Since ε is arbitrary, the proof is complete.

Theorem 8.2.8. *Let $1 \leq p < \infty$. Given a continuous linear functional f on $W^{m,p}(\Omega)$, there exists $v \in L^q(\Omega, \mathbb{R}^N)$ such that*

$$f(u) = \sum_{0 \leq |\alpha| \leq m} \langle D^\alpha u, v_\alpha \rangle_{L^2(\Omega)}. \tag{8.66}$$

Proof. Consider f a continuous linear operator on $U = W^{m,p}(\Omega)$. By the Hahn–Banach theorem, we can extend f to $\tilde{f}$, on $L^p(\Omega; \mathbb{R}^N)$, so that $\|\tilde{f}\|_{q^N} = \|f\|_{U^*}$ and by the last theorem there exists $\{v_\alpha\} \in L^q(\Omega; \mathbb{R}^N)$ such that

$$\tilde{f}(\hat{u}) = \sum_{0 \leq |\alpha| \leq m} \langle \hat{u}_\alpha, v_\alpha \rangle_{L^2(\Omega)}, \forall v \in L^p(\Omega; \mathbb{R}^N). \tag{8.67}$$

In particular for $u \in W^{m,p}(\Omega)$, defining $\hat{u} = \{D^\alpha u\} \in L^p(\Omega; \mathbb{R}^N)$, we obtain

$$f(u) = \tilde{f}(\hat{u}) = \sum_{1 \leq |\alpha| \leq m} \langle D^\alpha u, v_\alpha \rangle_{L^2(\Omega)}. \tag{8.68}$$

Finally, observe that, also from the Hahn–Banach theorem. $\|f\|_{U^*} = \|\tilde{f}\|_{q^N} = \|v\|_{q^N}$.

Definition 8.2.9. Let $\Omega \subset \mathbb{R}^n$ be a domain. For m a positive integer and $1 \leq p < \infty$ we define $W_0^{m,p}(\Omega)$ as the closure in $\|\cdot\|_{m,p}$ of $C_c^\infty(\Omega)$, where we recall that $C_c^\infty(\Omega)$ denotes the set of $C^\infty(\Omega)$ functions with compact support contained in Ω. Finally, we also recall that the support of $\phi : \Omega \to \mathbb{R}$, denoted by $spt(\phi)$, is given by

$$spt(\phi) = \overline{\{x \in \Omega \mid \phi(x) \neq 0\}}$$

8.3 The Sobolev Imbedding Theorem

8.3.1 The Statement of the Sobolev Imbedding Theorem

Now we present the Sobolev imbedding theorem. We recall that for normed spaces X, Y the notation

$$X \hookrightarrow Y$$

means that $X \subset Y$ and there exists a constant $K > 0$ such that

$$\|u\|_Y \leq K\|u\|_X, \forall u \in X.$$

If in addition the imbedding is compact, then for any bounded sequence $\{u_n\} \subset X$ there exists a convergent subsequence $\{u_{n_k}\}$, which converges to some u in the norm $\|\cdot\|_Y$. At this point, we first introduce the following definition.

Definition 8.3.1. Let $\Omega \subset \mathbb{R}^n$ be an open bounded set. We say that $\partial\Omega$ is $\hat{C}^1$ if for each $x_0 \in \partial\Omega$, denoting $\hat{x} = (x_1, \ldots, x_{n-1})$ for a local coordinate system, there exist $r > 0$ and a function $f(x_1, \ldots, x_{n-1}) = f(\hat{x})$ such that

$$W = \overline{\Omega} \cap B_r(x_0) = \{x \in B_r(x_0) \mid x_n \geq f(x_1, \ldots, x_{n-1})\}.$$

Moreover, $f(\hat{x})$ is a Lipschitz continuous function, so that

$$|f(\hat{x}) - f(\hat{y})| \leq C_1|\hat{x} - \hat{y}|_2, \text{ on its domain,}$$

for some $C_1 > 0$. Finally, we assume

$$\left\{\frac{\partial f(\hat{x})}{\partial x_k}\right\}_{k=1}^{n-1}$$

is classically defined, almost everywhere also on its concerning domain, so that $f \in W^{1,2}$.

Theorem 8.3.2 (The Sobolev Imbedding Theorem). *Let Ω be an open bounded set in $\mathbb{R}^n$ such that $\partial\Omega$ is $\hat{C}^1$. Let $j \geq 0$ and $m \geq 1$ be integers and let $1 \leq p < \infty$.*

1. ***Part I***

(a) ***Case A*** *If either $mp > n$ or $m = n$ and $p = 1$, then*

$$W^{j+m,p}(\Omega) \hookrightarrow C_B^j(\Omega). \tag{8.69}$$

Moreover,

$$W^{j+m,p}(\Omega) \hookrightarrow W^{j,q}(\Omega), \text{ for } p \leq q \leq \infty, \tag{8.70}$$

and, in particular,

$$W^{m,p}(\Omega) \hookrightarrow L^q(\Omega), \text{ for } p \leq q \leq \infty. \tag{8.71}$$

(b) ***Case B*** *If $mp = n$, then*

$$W^{j+m,p}(\Omega) \hookrightarrow W^{j,q}(\Omega), \text{ for } p \leq q < \infty, \tag{8.72}$$

and, in particular,

$$W^{m,p}(\Omega) \hookrightarrow L^q(\Omega), \text{ for } p \leq q < \infty. \tag{8.73}$$

(c) ***Case C*** *If $mp < n$ and $p = 1$, then*

$$W^{j+m,p}(\Omega) \hookrightarrow W^{j,q}(\Omega), \text{ for } p \leq q \leq p^* = \frac{np}{n-mp}, \tag{8.74}$$

and, in particular,

$$W^{m,p}(\Omega) \hookrightarrow L^q(\Omega), \text{ for } p \leq q \leq p^* = \frac{np}{n-mp}. \tag{8.75}$$

2. ***Part II*** *If $mp > n > (m-1)p$, then*

$$W^{j+m,p} \hookrightarrow C^{j,\lambda}(\overline{\Omega}), \text{ for } 0 < \lambda \leq m - (n/p), \tag{8.76}$$

and if $n = (m-1)p$, then

$$W^{j+m,p} \hookrightarrow C^{j,\lambda}(\overline{\Omega}), \text{ for } 0 < \lambda < 1. \tag{8.77}$$

Also, if $n = m-1$ and $p = 1$, then (8.77) holds for $\lambda = 1$ as well.

3. ***Part III*** *All imbeddings in Parts A and B are valid for arbitrary domains Ω if the W-space undergoing the imbedding is replaced with the corresponding W_0-space.*

8.4 The Proof of the Sobolev Imbedding Theorem

Now we present a collection of results which imply the proof of the Sobolev imbedding theorem. We start with the approximation by smooth functions.

Definition 8.4.1. Let $\Omega \subset \mathbb{R}^n$ be an open bounded set. For each $\varepsilon > 0$ define

$$\Omega_\varepsilon = \{x \in \Omega \mid dist(x, \partial\Omega) > \varepsilon\}.$$

Definition 8.4.2. Define $\eta \in C_c^\infty(\mathbb{R}^n)$ by

$$\eta(x) = \begin{cases} C\exp\left(\frac{1}{|x|_2^2 - 1}\right), & \text{if } |x|_2 < 1, \\ 0, & \text{if } |x|_2 \geq 1, \end{cases}$$

where $|\cdot|_2$ refers to the Euclidean norm in $\mathbb{R}^n$, that is, for $x = (x_1, \ldots, x_n) \in \mathbb{R}^n$, we have

$$|x|_2 = \sqrt{x_1^2 + \ldots + x_n^2}.$$

Moreover, $C>0$ is chosen so that

$$\int_{\mathbb{R}^n} \eta \, dx = 1.$$

For each $\varepsilon > 0$, set

$$\eta_\varepsilon(x) = \frac{1}{\varepsilon^n}\eta\left(\frac{x}{\varepsilon}\right).$$

The function η is said to be the fundamental mollifier. The functions $\eta_\varepsilon \in C_c^\infty(\mathbb{R}^n)$ and satisfy

$$\int_{\mathbb{R}^n} \eta_\varepsilon \, dx = 1,$$

and $spt(\eta_\varepsilon) \subset B(0,\varepsilon)$.

Definition 8.4.3. If $f : \Omega \to \mathbb{R}^n$ is locally integrable, we define its mollification, denoted by $f_\varepsilon : \Omega_\varepsilon \to \mathbb{R}$ as

$$f_\varepsilon = \eta_\varepsilon * f,$$

that is,

$$\begin{aligned} f_\varepsilon(x) &= \int_\Omega \eta_\varepsilon(x-y) f(y)\, dy \\ &= \int_{B(0,\varepsilon)} \eta_\varepsilon(y) f(x-y)\, dy. \end{aligned} \tag{8.78}$$

Theorem 8.4.4 (Properties of Mollifiers). *The mollifiers have the following properties:*

1. $f_\varepsilon \in C^\infty(\Omega_\varepsilon)$,
2. $f_\varepsilon \to f$ *a.e. as* $\varepsilon \to 0$,
3. *If* $f \in C(\Omega)$, *then* $f_\varepsilon \to f$ *uniformly on compact subsets of* Ω.

Proof.

1. Fix $x \in \Omega_\varepsilon$, $i \in \{1,\ldots,n\}$ and a h small enough such that

$$x + he_i \in \Omega_\varepsilon.$$

Thus

$$\begin{aligned} \frac{f_\varepsilon(x+he_i) - f_\varepsilon(x)}{h} =& \frac{1}{\varepsilon^n}\int_\Omega \frac{1}{h}\left[\eta\left(\frac{x+he_i-y}{\varepsilon}\right) - \eta\left(\frac{x-y}{\varepsilon}\right)\right] \\ & \times f(y)\, dy \\ =& \frac{1}{\varepsilon^n}\int_V \frac{1}{h}\left[\eta\left(\frac{x+he_i-y}{\varepsilon}\right) - \eta\left(\frac{x-y}{\varepsilon}\right)\right] \\ & \times f(y)\, dy, \end{aligned} \tag{8.79}$$

for an appropriate compact $V \subset\subset \Omega$. As

$$\frac{1}{h}\left[\eta\left(\frac{x+he_i-y}{\varepsilon}\right)-\eta\left(\frac{x-y}{\varepsilon}\right)\right] \to \frac{1}{\varepsilon}\frac{\partial \eta}{\partial x_i}\left(\frac{x-y}{\varepsilon}\right),$$

as $h \to 0$, uniformly on V, we obtain

$$\frac{\partial f_\varepsilon(x)}{\partial x_i} = \int_\Omega \frac{\partial \eta_\varepsilon(x-y)}{\partial x_i} f(y)\, dy.$$

By analogy, we may show that

$$D^\alpha f_\varepsilon(x) = \int_\Omega D^\alpha \eta_\varepsilon(x-y) f(y)\, dy, \forall x \in \Omega_\varepsilon.$$

2. From the Lebesgue differentiation theorem we have

$$\lim_{r\to 0} \frac{1}{|B(x,r)|}\int_{B(x,r)} |f(y)-f(x)|\, dy = 0, \tag{8.80}$$

for almost all $x \in \Omega$. Fix $x \in \Omega$ such that (8.80) holds. Hence,

$$\begin{aligned}
|f_\varepsilon(x)-f(x)| &= \int_{B(x,\varepsilon)} \eta_\varepsilon(x-y)[f(x)-f(y)]\, dy \\
&\leq \frac{1}{\varepsilon^n}\int_{B(x,\varepsilon)} \eta\left(\frac{x-y}{\varepsilon}\right)[f(x)-f(y)]\, dy \\
&\leq \frac{C}{|B(x,\varepsilon)|}\int_{B(x,\varepsilon)} |f(y)-f(x)|\, dy
\end{aligned} \tag{8.81}$$

for an appropriate constant $C > 0$. From (8.80), we obtain $f_\varepsilon \to f$ as $\varepsilon \to 0$.

3. Assume $f \in C(\Omega)$. Given $V \subset\subset \Omega$ choose W such that

$$V \subset\subset W \subset\subset \Omega,$$

and note that f is uniformly continuous on $\overline{W}$. Thus the limit indicated in (8.80) holds uniformly on V, and therefore $f_\varepsilon \to f$ uniformly on V.

Theorem 8.4.5. *Let $u \in L^p(\Omega)$, where $1 \leq p < \infty$. Then*

$$\eta_\varepsilon * u \in L^p(\Omega),$$

$$\|\eta_\varepsilon * u\|_p \leq \|u\|_p, \forall \varepsilon > 0$$

and

$$\lim_{\varepsilon\to 0^+} \|\eta_\varepsilon * u - u\|_p = 0.$$

Proof. Suppose $u \in L^p(\Omega)$ and $1 < p < \infty$. Defining $q = p/(p-1)$, from the Hölder inequality, we have

$$\begin{aligned}
|\eta_\varepsilon * u(x)| &= \left|\int_{\mathbb{R}^n} \eta_\varepsilon(x-y)u(y)\,dy\right| \\
&= \left|\int_{\mathbb{R}^n} [\eta_\varepsilon(x-y)]^{(1-1/p)}[\eta_\varepsilon(x-y)]^{1/p}u(y)\,dy\right| \\
&\leq \left[\int_{\mathbb{R}^n} \eta_\varepsilon(x-y)\,dy\right]^{1/q}\left[\int_{\mathbb{R}^n} \eta_\varepsilon(x-y)|u(y)|^p\,dy\right]^{1/p} \\
&= \left[\int_{\mathbb{R}^n} \eta_\varepsilon(x-y)|u(y)|^p\,dy\right]^{1/p}. \qquad (8.82)
\end{aligned}$$

From this and the Fubini theorem, we obtain

$$\begin{aligned}
\int_\Omega |\eta_\varepsilon * u(x)|^p\,dx &\leq \int_{\mathbb{R}^n}\int_{\mathbb{R}^n} \eta_\varepsilon(x-y)|u(y)|^p\,dy\,dx \\
&= \int_{\mathbb{R}^n} |u(y)|^p\left(\int_{\mathbb{R}^n} \eta_\varepsilon(x-y)\,dx\right)dy \\
&= \|u\|_p^p. \qquad (8.83)
\end{aligned}$$

Suppose given $\rho > 0$. As $C_0(\Omega)$ is dense in $L^p(\Omega)$, there exists $\phi \in C_0(\Omega)$ such that

$$\|u - \phi\|_p < \rho/3.$$

From the fact that

$$\eta_\varepsilon * \phi \to \phi$$

as $\varepsilon \to 0$, uniformly in Ω we have that there exists $\delta > 0$ such that

$$\|\eta_\varepsilon * \phi - \phi\|_p < \rho/3$$

if $\varepsilon < \delta$. Thus, for any $\varepsilon < \delta(\rho)$, we get

$$\begin{aligned}
\|\eta_\varepsilon * u - u\|_p &= \|\eta_\varepsilon * u - \eta_\varepsilon * \phi + \eta_\varepsilon * \phi - \phi + \phi - u\|_p \\
&\leq \|\eta_\varepsilon * u - \eta_\varepsilon * \phi\|_p + \|\eta_\varepsilon * \phi - \phi\|_p + \|\phi - u\|_p \\
&\leq \rho/3 + \rho/3 + \rho/3 = \rho. \qquad (8.84)
\end{aligned}$$

Since $\rho > 0$ is arbitrary, the proof is complete.

For the next theorem we denote

$$\tilde{u}(x) = \begin{cases} u(x), & \text{if } x \in \Omega, \\ 0, & \text{if } x \in \mathbb{R}^n \setminus \Omega. \end{cases}$$

8.4.1 Relatively Compact Sets in $L^p(\Omega)$

Theorem 8.4.6. *Consider $1 \leq p < \infty$. A bounded set $K \subset L^p(\Omega)$ is relatively compact if and only if for each $\varepsilon > 0$, there exist $\delta > 0$ and $G \subset\subset \Omega$ (we recall that $G \subset\subset \Omega$ means that $\overline{G}$ is compact and $\overline{G} \subset \Omega$) such that for each $u \in K$ and $h \in \mathbb{R}^n$ such that $|h| < \delta$ we have*

1.

$$\int_{\Omega} |\tilde{u}(x+h) - \tilde{u}(x)|^p \, dx < \varepsilon^p, \tag{8.85}$$

2.

$$\int_{\Omega-\overline{G}} |u(x)|^p \, dx < \varepsilon^p. \tag{8.86}$$

Proof. Suppose K is relatively compact in $L^p(\Omega)$. Suppose given $\varepsilon > 0$. As $\overline{K}$ is compact we may find a finite $\varepsilon/6$-net for K. Denote such a $\varepsilon/6$-net by N where

$$N = \{v_1, \ldots, v_m\} \subset L^p(\Omega).$$

Since $C_c(\Omega)$ is dense in $L^p(\Omega)$, for each $k \in \{1, \ldots, m\}$, there exists $\phi_k \in C_c(\Omega)$ such that

$$\|\phi_k - v_k\|_p < \frac{\varepsilon}{6}.$$

Thus defining

$$S = \{\phi_1, \ldots, \phi_m\},$$

given $u \in K$, we may select $v_k \in N$ such that

$$\|u - v_k\|_p < \frac{\varepsilon}{6},$$

so that

$$\begin{aligned} \|\phi_k - u\|_p &\leq \|\phi_k - v_k\|_p + \|v_k - u\|_p \\ &\leq \frac{\varepsilon}{6} + \frac{\varepsilon}{6} = \frac{\varepsilon}{3}. \end{aligned} \tag{8.87}$$

Define

$$G = \cup_{k-1}^{m} spt(\phi_k),$$

where

$$spt(\phi_k) = \overline{\{x \in \mathbb{R}^n \mid \phi_k(x) \neq 0\}}.$$

We have that

$$G \subset\subset \Omega,$$

where as abovementioned this means $\overline{G} \subset \Omega$. Observe that

$$\varepsilon^p > \|u - \phi_k\|_p^p \geq \int_{\Omega - \overline{G}} |u(x)|^p \, dx.$$

Since $u \in K$ is arbitrary, (8.86) is proven. Since ϕ_k is continuous and $spt(\phi_k)$ is compact we have that ϕ_k is uniformly continuous, that is, for the ε given above, there exists $\tilde{\delta} > 0$ such that if $|h| < \min\{\tilde{\delta}, 1\}$, then

$$|\phi_k(x+h) - \phi_k(x)| < \frac{\varepsilon}{3(|\overline{G}|+1)}, \forall x \in \overline{G}.$$

Thus,

$$\int_\Omega |\phi_k(x+h) - \phi_k(x)|^p \, dx < \left(\frac{\varepsilon}{3}\right)^p.$$

Also observe that since

$$\|u - \phi_k\|_p < \frac{\varepsilon}{3},$$

we have that

$$\|T_h u - T_h \phi_k\|_p < \frac{\varepsilon}{3},$$

where $T_h u = u(x+h)$. Thus, if $|h| < \delta = \min\{\tilde{\delta}, 1\}$, we obtain

$$\begin{aligned} \|T_h \tilde{u} - \tilde{u}\|_p &\leq \|T_h \tilde{u} - T_h \phi_k\|_p + \|T_h \phi_k - \phi_k\|_p \\ &\quad + \|\phi_k - u\|_p \\ &< \frac{\varepsilon}{3} + \frac{\varepsilon}{3} + \frac{\varepsilon}{3} = \varepsilon. \end{aligned} \tag{8.88}$$

For the converse, it suffices to consider the special case $\Omega = \mathbb{R}^n$, because for the general Ω we can define $\tilde{K} = \{\tilde{u} \mid u \in K\}$. Suppose given $\varepsilon > 0$ and choose $G \subset\subset \mathbb{R}^n$ such that for all $u \in K$ we have

$$\int_{\mathbb{R}^n - \overline{G}} |u(x)|^p \, dx < \frac{\varepsilon}{3}.$$

For each $\rho > 0$ the function $\eta_\rho * u \in C^\infty(\mathbb{R}^n)$, and in particular $\eta_\rho * u \in C(\overline{G})$. Suppose $\phi \in C_0(\mathbb{R}^n)$. Fix $\rho > 0$. By the Hölder inequality we have

$$\begin{aligned} |\eta_\rho * \phi(x) - \phi(x)|^p &= \left| \int_{\mathbb{R}^n} \eta_\rho(y)(\phi(x-y) - \phi(x)) \, dy \right|^p \\ &= \left| \int_{\mathbb{R}^n} (\eta_\rho(y))^{1-1/p} (\eta_\rho(y))^{1/p} (T_{-y}\phi(x) - \phi(x)) \, dy \right|^p \\ &\leq \int_{B_\rho(\theta)} (\eta_\rho(y)) |T_{-y}\phi(x) - \phi(x)|^p \, dy. \end{aligned} \tag{8.89}$$

Hence, from the Fubini theorem, we may write

$$\int_{\mathbb{R}^n} |\eta_\rho * \phi(x) - \phi(x)|^p \, dx \leq \int_{B_\rho(\theta)} (\eta_\rho(y)) \int_{\mathbb{R}^n} |T_{-y}\phi(x) - \phi(x)|^p \, dx \, dy, \tag{8.90}$$

so that we may write

$$\|\eta_\rho * \phi - \phi\|_p \leq \sup_{h \in B_\rho(\theta)} \{\|T_h\phi - \phi\|_p\}. \tag{8.91}$$

Fix $u \in L^p(\mathbb{R}^n)$. We may obtain a sequence $\{\phi_k\} \subset C_c(\mathbb{R}^n)$ such that

$$\phi_k \to u, \text{ in } L^p(\mathbb{R}^n).$$

Observe that

$$\eta_\rho * \phi_k \to \eta_\rho * u, \text{ in } L^p(\mathbb{R}^n),$$

as $k \to \infty$. Also

$$T_h\phi_k \to T_h u, \text{ in } L^p(\mathbb{R}^n),$$

as $k \to \infty$. Thus

$$\|T_h\phi_k - \phi_k\|_p \to \|T_h u - u\|_p,$$

in particular

$$\limsup_{k \to \infty} \left\{ \sup_{h \in B_\rho(\theta)} \{\|T_h\phi_k - \phi_k\| \right\} \leq \sup_{h \in B_\rho(\theta)} \{\|T_h u - u\|_p\}.$$

Therefore as

$$\|\eta_\rho * \phi_k - \phi_k\|_p \to \|\eta_\rho * u - u\|_p,$$

as $k \to \infty$, from (8.91) we get

$$\|\eta_\rho * u - u\|_p \leq \sup_{h \in B_{\rho(\theta)}} \{\|T_h u - u\|_p\}.$$

From this and (8.85) we obtain

$$\|\eta_\rho * u - u\|_p \to 0, \text{ uniformly in } K \text{ as } \rho \to 0.$$

Fix $\rho_0 > 0$ such that

$$\int_{\overline{G}} |\eta_{\rho_0} * u - u|^p \, dx < \frac{\varepsilon}{3 \cdot 2^{p-1}}, \forall u \in K.$$

Observe that

$$\begin{aligned}|\eta_{\rho_0} * u(x)| &= \left|\int_{\mathbb{R}^n} \eta_{\rho_0}(x-y)u(y)\,dy\right| \\ &= \left|\int_{\mathbb{R}^n} [\eta_{\rho_0}(x-y)]^{(1-1/p)}[\eta_{\rho_0}(x-y)]^{1/p}u(y)\,dy\right| \\ &\leq \left[\int_{\mathbb{R}^n} \eta_{\rho_0}(x-y)\,dy\right]^{1/q}\left[\int_{\mathbb{R}^n} \eta_{\rho_0}(x-y)|u(y)|^p\,dy\right]^{1/p} \\ &= \left[\int_{\mathbb{R}^n} \eta_{\rho_0}(x-y)|u(y)|^p\,dy\right]^{1/p}. \end{aligned} \tag{8.92}$$

From this, we may write

$$|\eta_{\rho_0} * u(x)| \leq \left(\sup_{y\in\mathbb{R}^n} \eta_{\rho_0}(y)\right)^{1/p} \|u\|_p \leq K_1, \forall x \in \mathbb{R}^n,\ u \in K,$$

where $K_1 = K_2K_3$,

$$K_2 = \left(\sup_{y\in\mathbb{R}^n} \eta_{\rho_0}(y)\right)^{1/p},$$

and K_3 is any constant such that

$$\|u\|_p < K_3, \forall u \in K.$$

Similarly

$$|\eta_{\rho_0} * u(x+h) - \eta_{\rho_0}u(x)| \leq \left(\sup_{y\in\mathbb{R}^n} \eta_{\rho_0}(y)\right)^{1/p} \|T_hu - u\|_p,$$

and thus from (8.85) we obtain

$$\eta_{\rho_0} * u(x+h) \to \eta_{\rho_0} * u(x), \text{ as } h \to 0$$

uniformly in $\mathbb{R}^n$ and for $u \in K$.

By the Arzela–Ascoli theorem

$$\{\eta_{\rho_0} * u \mid u \in K\}$$

is relatively compact in $C(\overline{G})$, and it is totally bounded so that there exists a ε_0-net $N = \{v_1, \ldots, v_m\}$ where

$$\varepsilon_0 = \left(\frac{\varepsilon}{3 \cdot 2^{p-1}|\overline{G}|}\right)^{1/p}.$$

Thus for some $k \in \{1, \ldots, m\}$ we have

$$\|v_k - \eta_{\rho_0} * u\|_\infty < \varepsilon_0.$$

Hence,

$$\begin{aligned}\int_{\mathbb{R}^n} |u(x)-\tilde{v}_k(x)|^p\,dx &= \int_{\mathbb{R}^n-\overline{G}} |u(x)|^p\,dx + \int_{\overline{G}} |u(x)-v_k(x)|^p\,dx \\ &\leq \frac{\varepsilon}{3} + 2^{p-1}\int_{\overline{G}} (|u(x)-(\eta_{\rho_0}*u)(x)|^p \\ &\quad + |\eta_{\rho_0}*u(x)-v_k(x)|^p)\,dx \\ &\leq \frac{\varepsilon}{3} + 2^{p-1}\left(\frac{\varepsilon}{3\cdot 2^{p-1}} + \frac{\varepsilon|\overline{G}|}{3\cdot 2^{p-1}|\overline{G}|}\right) \\ &= \varepsilon. \end{aligned} \tag{8.93}$$

Thus K is totally bounded and therefore it is relatively compact.
The proof is complete.

8.4.2 Some Approximation Results

Theorem 8.4.7. *Let $\Omega \subset \mathbb{R}^n$ be an open set. Assume $u \in W^{m,p}(\Omega)$ for some $1 \leq p < \infty$, and set*

$$u_\varepsilon = \eta_\varepsilon * u \text{ in } \Omega_\varepsilon.$$

Then,

1. *$u_\varepsilon \in C^\infty(\Omega_\varepsilon), \forall \varepsilon > 0$,*
2. *$u_\varepsilon \to u$ in $W^{m,p}_{loc}(\Omega)$, as $\varepsilon \to 0$.*

Proof. Assertion 1 has been already proved. Let us prove 2. We will show that if $|\alpha| \leq m$, then

$$D^\alpha u_\varepsilon = \eta_\varepsilon * D^\alpha u, \text{ in } \Omega_\varepsilon.$$

For that, let $x \in \Omega_\varepsilon$. Thus,

$$\begin{aligned} D^\alpha u_\varepsilon(x) &= D^\alpha\left(\int_\Omega \eta_\varepsilon(x-y)u(y)\,dy\right) \\ &= \int_\Omega D^\alpha_x \eta_\varepsilon(x-y)u(y)\,dy \\ &= (-1)^{|\alpha|}\int_\Omega D^\alpha_y(\eta_\varepsilon(x-y))u(y)\,dy. \end{aligned} \tag{8.94}$$

Observe that for fixed $x \in \Omega_\varepsilon$ the function

$$\phi(y) = \eta_\varepsilon(x-y) \in C^\infty_c(\Omega).$$

Therefore,

$$\int_\Omega D^\alpha_y(\eta_\varepsilon(x-y))u(y)\,dy = (-1)^{|\alpha|}\int_\Omega \eta_\varepsilon(x-y)D^\alpha_y u(y)\,dy,$$

and hence,

$$\begin{aligned} D^\alpha u_\varepsilon(x) &= (-1)^{|\alpha|+|\alpha|} \int_\Omega \eta_\varepsilon(x-y) D^\alpha u(y)\,dy \\ &= (\eta_\varepsilon * D^\alpha u)(x). \end{aligned} \tag{8.95}$$

Now choose any open bounded set such that $V \subset\subset \Omega$. We have that

$$D^\alpha u_\varepsilon \to D^\alpha u, \text{ in } L^p(V) \text{ as } \varepsilon \to 0,$$

for each $|\alpha| \leq m$.

Thus,

$$\|u_\varepsilon - u\|_{m,p,V}^p = \sum_{|\alpha|\leq m} \|D^\alpha u_\varepsilon - D^\alpha u\|_{p,V} \to 0,$$

as $\varepsilon \to 0$.

Theorem 8.4.8. *Let $\Omega \subset \mathbb{R}^n$ be a bounded open set and suppose $u \in W^{m,p}(\Omega)$ for some $1 \leq p < \infty$. Then there exists a sequence $\{u_k\} \subset C^\infty(\Omega)$ such that*

$$u_k \to u \text{ in } W^{m,p}(\Omega).$$

Proof. Observe that

$$\Omega = \cup_{i=1}^\infty \Omega_i,$$

where

$$\Omega_i = \{x \in \Omega \mid dist(x, \partial\Omega) > 1/i\}.$$

Define

$$V_i = \Omega_{i+3} - \bar{\Omega}_{i+1},$$

and choose any open set V_0 such that $V_0 \subset\subset \Omega$, so that

$$\Omega = \cup_{i=0}^\infty V_i.$$

Let $\{\zeta_i\}_{i=0}^\infty$ be a smooth partition of unit subordinate to the open sets $\{V_i\}_{i=0}^\infty$. That is,

$$\begin{cases} 0 \leq \zeta_i \leq 1, \quad \zeta_i \in C_c^\infty(V_i) \\ \sum_{i=0}^\infty \zeta_i = 1, \quad \text{on } \Omega. \end{cases}$$

Now suppose $u \in W^{m,p}(\Omega)$. Thus $\zeta_i u \in W^{m,p}(\Omega)$ and $\text{spt}(\zeta_i u) \subset V_i \subset \Omega$. Choose $\delta > 0$. For each $i \in \mathbb{N}$ choose $\varepsilon_i > 0$ small enough so that

$$u_i = \eta_{\varepsilon_i} * (\zeta_i u)$$

satisfies

$$\|u_i - \zeta_i u\|_{m,p,\Omega} \leq \frac{\delta}{2^{i+1}},$$

and $\text{spt}(u_i) \subset W_i$ where $W_i = \Omega_{i+4} - \bar{\Omega}_i \supset V_i$. Define

$$v = \sum_{i=0}^{\infty} u_i.$$

Thus such a function belongs to $C^\infty(\Omega)$, since for each open $V \subset\subset \Omega$ there are at most finitely many nonzero terms in the sum. Since

$$u = \sum_{i=0}^{\infty} \zeta_i u,$$

we have that for a fixed $V \subset\subset \Omega$,

$$\begin{aligned} \|v - u\|_{m,p,V} &\leq \sum_{i=0}^{\infty} \|u_i - \zeta_i u\|_{m,p,V} \\ &\leq \delta \sum_{i=0}^{\infty} \frac{1}{2^{i+1}} = \delta. \end{aligned} \tag{8.96}$$

Taking the supremum over sets $V \subset\subset \Omega$ we obtain

$$\|v - u\|_{m,p,\Omega} < \delta.$$

Since $\delta > 0$ is arbitrary, the proof is complete.

The next result is also relevant. For a proof see Evans [26], p. 232.

Theorem 8.4.9. *Let $\Omega \subset \mathbb{R}^n$ be a bounded set such that $\partial\Omega$ is C^1. Suppose $u \in W^{m,p}(\Omega)$ where $1 \leq p < \infty$. Thus there exists a sequence $\{u_n\} \subset C^\infty(\overline{\Omega})$ such that*

$$u_n \to u \text{ in } W^{m,p}(\Omega), \text{ as } n \to \infty.$$

Anyway, now we prove a more general result.

Theorem 8.4.10. *Let $\Omega \subset \mathbb{R}^n$ be an open bounded set such that $\partial\Omega$ is $\hat{C}^1$. Let $u \in W^{m,p}(\Omega)$ where m is a nonnegative integer and $1 \leq p < \infty$.*

Under such assumptions, there exists $\{u_k\} \subset C^\infty(\overline{\Omega})$ such that

$$\|u_k - u\|_{m,p,\Omega} \to 0, \text{ as } k \to \infty.$$

Proof. Fix $x_0 \in \partial\Omega$. Since $\partial\Omega$ is $\hat{C}^1$, denoting $\hat{x} = (x_1, \ldots, x_{n-1})$ for a local coordinate system, there exists $r > 0$ and a function $f(x_1, \ldots, x_{n-1}) = f(\hat{x})$ such that

$$W = \overline{\Omega} \cap B_r(x_0) = \{x \in B_r(x_0) \mid x_n \geq f(x_1, \ldots, x_{n-1})\}.$$

We emphasize $f(\hat{x})$ is a Lipschitz continuous function, so that

$$|f(\hat{x}) - f(\hat{y})| \leq C_1 |\hat{x} - \hat{y}|_2, \text{ on its domain,}$$

for some $C_1 > 0$. Furthermore

$$\left\{\frac{\partial f(\hat{x})}{\partial x_k}\right\}_{k=1}^{n-1}$$

is classically defined, almost everywhere also on its concerning domain.

Let $\varepsilon > 0$. For each $\delta > 0$ define $x_\delta = x + C\delta\mathbf{e}_n$, where $C > 1$ is a fixed constant. Define $u_\delta = u(x_\delta)$. Now choose $\delta > 0$ sufficiently small such that

$$\|u_\delta - u\|_{m,p,W} < \varepsilon/2.$$

For each $n \in \mathbb{N}$, $x \in W$ define

$$v_n(x) = (\eta_{1/n} * u_\delta)(x).$$

Observe that

$$\|v_n - u\|_{m,p,W} \leq \|v_n - u_\delta\|_{m,p,W} + \|u_\delta - u\|_{m,p,W}.$$

For the fixed $\delta > 0$, there exists $N_\varepsilon \in \mathbb{N}$ such that if $n > N_\varepsilon$ we have

$$\|v_n - u_\delta\|_{m,p,W} < \varepsilon/2,$$

and

$$v_n \in C^\infty(\overline{W}).$$

Hence

$$\|v_n - u\|_{m,p,W} \leq \|v_n - u_\delta\|_{m,p,W} + \|u_\delta - u\|_{m,p,W} < \varepsilon/2 + \varepsilon/2 = \varepsilon.$$

Clarifying the dependence of r on $x_0 \in \partial\Omega$ we denote $r = r_{x_0}$. Observe that

$$\partial\Omega \subset \cup_{x_0 \in \partial\Omega} B_{r_{x_0}}(x_0)$$

so that since $\partial\Omega$ is compact, there exists $x_1, \ldots, x_M \in \partial\Omega$ such that

$$\partial\Omega \subset \cup_{i=1}^M B_{r_i}(x_i).$$

We denote $B_{r_i}(x_i) = B_i$ and $W_i = \Omega \cap B_i$, $\forall i \in \{1, \ldots, M\}$. We also choose an appropriate open set $B_0 \subset\subset \Omega$ such that

$$\Omega \subset \cup_{i=0}^M B_i.$$

Let $\{\zeta_i\}_{i=0}^M$ be a concerned partition of unity relating $\{B_i\}_{i=0}^M$.

Thus $\zeta_i \in C_c^\infty(B_i)$ and $0 \leq \zeta_i \leq 1$, $\forall i \in \{0, \ldots, M\}$ and also

$$\sum_{i=0}^M \zeta_i = 1 \text{ on } \Omega.$$

From above, we may find $v_i \in C^\infty(\overline{W}_i)$ such that $\|v_i - u\|_{m,p,W_i} < \varepsilon$, $\forall i \in \{1,\ldots,M\}$. Define $u_0 = v_0 = u$ on $B_0 \equiv W_0$,

$$u_i = \zeta_i u, \ \forall i \in \{0,\ldots,M\}$$

and

$$v = \sum_{i=0}^{M} \zeta_i v_i.$$

We emphasize

$$v \in C^\infty(\overline{\Omega}).$$

Therefore

$$\begin{aligned}
\|v-u\|_{m,p,\Omega} &= \left\| \sum_{i=0}^{M} (\zeta_i u - \zeta_i v_i) \right\|_{m,p,\Omega} \\
&\leq C_2 \sum_{i=0}^{M} \|u - v_i\|_{m,p,(\Omega \cap B_i)} \\
&= C_2 \sum_{i=0}^{M} \|u - v_i\|_{m,p,W_i} \\
&< C_2 M \varepsilon. \qquad (8.97)
\end{aligned}$$

Since neither C_2 nor M depends on $\varepsilon > 0$, the proof is complete.

8.4.3 Extensions

In this section we study extensions of the Sobolev spaces from a domain $\Omega \subset \mathbb{R}^n$ to $\mathbb{R}^n$. First we enunciate a result found in Evans [26].

Theorem 8.4.11. *Assume $\Omega \subset \mathbb{R}^n$ is an open bounded set and that $\partial\Omega$ is C^1. Let V be a bounded open set such that $\Omega \subset\subset V$. Then there exists a bounded linear operator*

$$E : W^{1,p}(\Omega) \to W^{1,p}(\mathbb{R}^n),$$

such that for each $u \in W^{1,p}(\Omega)$ we have:

1. *$Eu = u$, a.e. in Ω,*
2. *Eu has support in V,*
3. *$\|Eu\|_{1,p,\mathbb{R}^n} \leq C\|u\|_{1,p,\Omega}$, where the constant depends only on p, Ω, and V.*

The next result, which we prove, is a more general one.

Theorem 8.4.12. *Assume $\Omega \subset \mathbb{R}^n$ is an open bounded set and that $\partial\Omega$ is $\hat{C}^1$. Let V be a bounded open set such that $\Omega \subset\subset V$. Then there exists a bounded linear operator*

$$E : W^{1,p}(\Omega) \to W^{1,p}(\mathbb{R}^n),$$

such that for each $u \in W^{1,p}(\Omega)$ *we have:*

1. $Eu = u$, *a.e. in* Ω,
2. Eu *has support in* V,
3. $\|Eu\|_{1,p,\mathbb{R}^n} \leq C\|u\|_{1,p,\Omega}$, *where the constant depends only on* p, Ω, *and* V.

Proof. Let $u \in W^{1,p}(\Omega)$. Fix $N \in \mathbb{N}$ and select $\phi_N \in C^\infty(\overline{\Omega})$ such that

$$\|\phi_N - u\|_{1,p,\Omega} < 1/N.$$

Choose $x_0 \in \partial\Omega$. From the hypothesis we may write

$$\overline{\Omega} \cap B_r(x_0) = \{x \in B_r(x_0) \mid x_n \geq f(x_1, \ldots, x_{n-1})\},$$

for some $r > 0$ and so that denoting $\hat{x} = (x_1, \ldots, x_{n-1})$, $f(x_1, \ldots, x_{n-1}) = f(\hat{x})$ is a Lipschitz continuous function such that

$$\left\{\frac{\partial f(\hat{x})}{\partial x_k}\right\}_{k=1}^{n-1}$$

is classically defined almost everywhere on its domain and

$$|f(\hat{x}) - f(\hat{y})| \leq C_1|\hat{x} - \hat{y}|_2, \forall \hat{x}, \hat{y} \text{ on its domain,}$$

for some $C_1 > 0$.

Define the variable $y \in \mathbb{R}^n$ by $y_i = x_i, \forall i \in \{1, \ldots, n-1\}$, and $y_n = x_n - f(x_1, \ldots, x_{n-1})$.

Thus

$$\phi_N(x_1, \ldots, x_n) = \phi_N(y_1, \ldots, y_{n-1}, y_n + f(y_1, .., y_{n-1})) = \overline{\phi}_N(y_1, \ldots, y_n).$$

Observe that defining $\psi(x) = y$ from the continuity of ψ^{-1}, there exists $r_1 > 0$ such that

$$\psi^{-1}(B^+_{r_1}(y_0)) \subset \Omega \cap B_r(x_0),$$

where $y_0 = (x_{0^1}, \ldots, x_{0^{n-1}}, 0)$. We define $W^+ = \psi^{-1}(\overline{B}^+_{r_1}(y_0))$ and $W^- = \psi^{-1}$ $(\overline{B}^-_{r_1}(y_0))$ where we denote

$$B^+ = B^+_{r_1}(y_0) = \{y \in B_{r_1}(y_0) \mid y_n \geq 0\},$$

and

$$B^- = B^-_{r_1}(y_0) = \{y \in B_{r_1}(y_0) \mid y_n < 0\}.$$

We emphasize that locally about x_0 we have that $\partial\Omega$ and $\psi(\partial\Omega)$ correspond to the equations $x_n - f(x_1, \ldots, x_{n-1}) = 0$ and $y_n = 0$, respectively.

Moreover, $\overline{\phi}_N$ is Lipschitz continuous on B^+ so that $\overline{\phi} \in W^{1,p}(B^+)$, and therefore there exists $\tilde{\phi}_N \in C^\infty(\overline{B}^+)$ such that

$$\|\tilde{\phi}_N - \overline{\phi}_N\|_{1,p,B^+} < 1/N.$$

Define $\hat{\phi}_N : B \to \mathbb{R}$ by

$$\hat{\phi}_N(y) = \begin{cases} \tilde{\phi}_N(y) & \text{if } y \in \overline{B}^+ \\ -3\tilde{\phi}_N(y_1,\ldots,y_{n-1},-y_n) + 4\tilde{\phi}_N(y_1,\ldots,y_{n-1},-y_n/2) & \text{if } y \in B^-. \end{cases}$$

It may be easily verified that $\hat{\phi}_N \in C^1(B)$. Also, there exists $C_2 > 0$ such that

$$\begin{aligned} \|\hat{\phi}_N\|_{1,p,B} &\leq C_2\|\hat{\phi}_N\|_{1,p,B^+} \\ &= C_2\|\tilde{\phi}_N\|_{1,p,B^+} \\ &\leq C_2\|\overline{\phi}_N\|_{1,p,B^+} + C_2/N, \end{aligned} \tag{8.98}$$

where C_2 depends only on Ω and p.

We claim that $\{\hat{\phi}_N\}$ is a Cauchy sequence in $W^{1,p}(B)$.

For $N_1, N_2 \in \mathbb{N}$ we have

$$\begin{aligned} \|\hat{\phi}_{N_1} - \hat{\phi}_{N_2}\|_{1,p,B} &\leq C_1\|\hat{\phi}_{N_1} - \hat{\phi}_{N_2}\|_{1,p,B^+} \\ &\leq C_1\|\tilde{\phi}_{N_1} - \overline{\phi}_{N_1} + \overline{\phi}_{N_1} - \overline{\phi}_{N_2} + \overline{\phi}_{N_2} - \hat{\phi}_{N_2}\|_{1,p,B^+} \\ &< C_1\|\tilde{\phi}_{N_1} - \overline{\phi}_{N_1}\|_{1,p,B^+} + C_1\|\overline{\phi}_{N_1} - \overline{\phi}_{N_2}\|_{1,p,B^+} \\ &\quad + C_1\|\overline{\phi}_{N_2} - \hat{\phi}_{N_2}\|_{1,p,B^+} \\ &\leq C_1/N_1 + C_1\|\overline{\phi}_{N_1} - \overline{\phi}_{N_2}\|_{1,p,B^+} + C_1/N_2 \\ &\to 0, \text{ as } N_1, N_2 \to \infty. \end{aligned} \tag{8.99}$$

Also, since $\hat{\phi}_N \to u(x(y))$, in $W^{1,p}(B^+)$, up to a subsequence not relabeled,

$$\hat{\phi}_N(y) \to u(x(y)), \text{ a.e. in } B^+.$$

Define $\hat{u} = \lim_{N\to\infty} \hat{\phi}_N$ in $W^{1,p}(B)$. Therefore

$$\hat{u}(y(x)) = u(x), \text{ a.e. in } W^+.$$

Now denoting simply

$$\hat{u}(y(x)) = \overline{u}(x)$$

we obtain

$$\overline{u} = u, \text{ a.e. in } W^+.$$

Now choose $\varepsilon > 0$. Thus there exists $N_0 \in \mathbb{N}$ such that if $N > N_0$ we have

$$\begin{aligned} \|\overline{u}\|_{1,p,W} &< \|\hat{\phi}_N(y(x))\|_{1,p,W} + \varepsilon \\ &\leq C_3\|\hat{\phi}_N(y)\|_{1,p,B} + \varepsilon \\ &\leq C_4\|\hat{\phi}_N\|_{1,p,B^+} + \varepsilon \\ &\leq C_5\|\hat{\phi}_N\|_{1,p,W^+} + \varepsilon \end{aligned} \tag{8.100}$$

so that letting $N \to \infty$, since $\varepsilon > 0$ is arbitrary, we get

$$\|\overline{u}\|_{1,p,W} \leq C_5 \|u\|_{1,p,W^+}.$$

Now denoting $W = W_{x_0}$ we have that $\partial\Omega \subset \cup_{x_0 \in \partial\Omega} W_{x_0}$ and since $\partial\Omega$ is compact, there exist $x_1, \ldots, x_M \in \partial\Omega$, such that

$$\partial\Omega \subset \cup_{i=1}^M W_{x_i}.$$

Hence for an appropriate open $W_0 \subset\subset \Omega$, we get

$$\Omega \subset \cup_{i=0}^M W_i.$$

where we have denoted $W_i = W_{x_i}, \forall i \in \{0, \ldots, M\}$.

Let $\{\zeta_i\}_{i=0}^M$ be a concerned partition of unity relating $\{W_i\}_{i=0}^M$, so that

$$\sum_{i=0}^M \zeta_i = 1, \text{ in } \Omega,$$

and $\zeta_i \in C_c^\infty(W_i)$, $0 \leq \zeta_i \leq 1$, $\forall i \in \{0, \ldots, M\}$.

Define

$$u_i = \zeta_i u, \forall i \in \{0, \ldots, M\}.$$

For each i we denote the extension of u from W_i^+ to W_i by $\overline{u}_i$. Also define $\overline{u}_0 = u, \in W_0$, and $\overline{u} = \sum_{i=0}^M \zeta_i \overline{u}_i$.

Recalling that $u = \overline{u}_i, a.e.$ on W_i^+ and that $\overline{\Omega} = \cup_{i=1}^M W_i^+ \cup W_0$, we obtain $\overline{u} = \sum_{i=0}^M \zeta_i \overline{u}_i = \sum_{i=0}^M \zeta_i u = u$, a.e. in Ω. Furthermore

$$\begin{aligned}
\|\overline{u}\|_{1,p,\mathbb{R}^n} &\leq \sum_{i=0}^M \|\zeta_i \overline{u}_i\|_{1,p,\mathbb{R}^n} \\
&\leq C_5 \sum_{i=0}^M \|\overline{u}_i\|_{1,p,W_i} \\
&\leq C_5 \|u\|_{1,p,W_0} + C_5 \sum_{i=1}^M \|u_i\|_{1,p,W_i^+} \\
&\leq (M+1) C_5 \|u\|_{1,p,\Omega} \\
&= C \|u\|_{1,p,\Omega}, \qquad (8.101)
\end{aligned}$$

where $C = (M+1)C_5$.

We recall that the partition of unity may be chosen so that its support is on V.

Finally, we denote $Eu = \overline{u}$.

The proof is complete.

8.4.4 The Main Results

Definition 8.4.13. For $1 \leq p < n$ we define $r = \frac{np}{n-p}$.

Theorem 8.4.14 (Gagliardo–Nirenberg–Sobolev Inequality). *Let $1 \leq p < n$. Thus there exists a constant $K > 0$ depending only p and n such that*

$$\|u\|_{r,\mathbb{R}^n} \leq K\|Du\|_{p,\mathbb{R}^n}, \forall u \in C_c^1(\mathbb{R}^n).$$

Proof. Suppose $p = 1$. Let $u \in C_c^1(\mathbb{R}^n)$. From the fundamental theorem of calculus we have

$$u(x) = \int_{-\infty}^{x_i} \frac{\partial u(x_1,\ldots,x_{i-1},y_i,x_{i+1},\ldots,x_n)}{\partial x_i}\, dy_i,$$

so that

$$|u(x)| \leq \int_{-\infty}^{\infty} |Du(x_1,\ldots,x_{i-1},y_i,x_{i+1},\ldots,x_n)|\, dy_i.$$

Therefore,

$$|u(x)|^{n/(n-1)} \leq \prod_{i=1}^{n}\left(\int_{-\infty}^{\infty} |Du(x_1,\ldots,x_{i-1},y_i,x_{i+1},\ldots,x_n)|\, dy_i\right)^{1/(n-1)}.$$

From this, we get

$$\begin{aligned}
&\int_{-\infty}^{\infty} |u(x)|^{n/(n-1)}\, dx_1 \\
&\leq \int_{-\infty}^{\infty} \prod_{i=1}^{n}\left(\int_{-\infty}^{\infty} |Du|\, dy_i\right)^{1/(n-1)} dx_1 \\
&\leq \left(\int_{-\infty}^{\infty} |Du|\, dy_1\right)^{1/(n-1)} \\
&\times \int_{-\infty}^{\infty}\left(\prod_{i=2}^{n}\left(\int_{-\infty}^{\infty} |Du|\, dy_i\right)^{1/(n-1)}\right) dx_1. \qquad (8.102)
\end{aligned}$$

From this and the generalized Hölder inequality, we obtain

$$\begin{aligned}
&\int_{-\infty}^{\infty} |u(x)|^{n/(n-1)}\, dx_1 \\
&\leq \left(\int_{-\infty}^{\infty} |Du|\, dy_1\right)^{1/(n-1)} \\
&\times \prod_{i=2}^{n}\left(\int_{-\infty}^{\infty}\int_{-\infty}^{\infty} |Du|\, dx_1 dy_i\right)^{1/(n-1)}. \qquad (8.103)
\end{aligned}$$

Integrating in x_2 we obtain

$$\begin{aligned}\int_{-\infty}^{\infty}\int_{-\infty}^{\infty} |u(x)|^{n/(n-1)}\,dx_1dx_2 \\ \leq \int_{-\infty}^{\infty}\left(\left(\int_{-\infty}^{\infty}|Du|\,dy_1\right)^{1/(n-1)}\right. \\ \left.\times\prod_{i=2}^{n}\left(\int_{-\infty}^{\infty}\int_{-\infty}^{\infty}|Du|\,dx_1dy_i\right)^{1/(n-1)}\right)\,dx_2,\end{aligned}$$

so that

$$\begin{aligned}\int_{-\infty}^{\infty}\int_{-\infty}^{\infty} |u(x)|^{n/(n-1)}\,dx_1dx_2 \\ \leq \left(\int_{-\infty}^{\infty}\int_{-\infty}^{\infty}|Du|\,dy_2dx_1\right)^{1/(n-1)} \\ \times\int_{-\infty}^{\infty}\left(\left(\int_{-\infty}^{\infty}|Du|\,dy_1\right)^{1/(n-1)}\right. \\ \left.\times\prod_{i=3}^{n}\left(\int_{-\infty}^{\infty}\int_{-\infty}^{\infty}|Du|\,dx_1dy_i\right)^{1/(n-1)}\right)\,dx_2.\end{aligned} \tag{8.104}$$

By applying the generalized Hölder inequality we get

$$\begin{aligned}\int_{-\infty}^{\infty}\int_{-\infty}^{\infty} |u(x)|^{n/(n-1)}\,dx_1dx_2 \\ \leq \left(\int_{-\infty}^{\infty}\int_{-\infty}^{\infty}|Du|\,dy_2dx_1\right)^{1/(n-1)} \\ \times\left(\int_{-\infty}^{\infty}\int_{-\infty}^{\infty}|Du|\,dy_1dx_2\right)^{1/(n-1)} \\ \times\prod_{i=3}^{n}\left(\int_{-\infty}^{\infty}\int_{-\infty}^{\infty}\int_{-\infty}^{\infty}|Du|\,dx_1dx_2dy_i\right)^{1/(n-1)}.\end{aligned} \tag{8.105}$$

Therefore, reasoning inductively, after n steps, we get

$$\begin{aligned}\int_{\mathbb{R}^n} |u(x)|^{n/(n-1)}\,dx \\ \leq \prod_{i=1}^{n}\left(\int_{-\infty}^{\infty}\cdots\int_{-\infty}^{\infty}|Du|\,dx\right)^{1/(n-1)} \\ = \left(\int_{\mathbb{R}^n}|Du|\,dx\right)^{n/(n-1)}.\end{aligned} \tag{8.106}$$

This is the result for $p = 1$. Now suppose $1 < p < n$.

For $\gamma > 1$ apply the above result for

$$v = |u|^\gamma,$$

to obtain

$$\begin{aligned}\left(\int_{\mathbb{R}^n} |u(x)|^{\gamma n/(n-1)}\,dx\right)^{(n-1)/n} &\leq \int_{\mathbb{R}^n} |D|u|^\gamma|;dx \\ &\leq \gamma \int_{\mathbb{R}^n} |u|^{\gamma-1}|Du|;dx \\ &\leq \gamma \left(\int_{\mathbb{R}^n} |u|^{(\gamma-1)p/(p-1)}\,dx\right)^{(p-1)/p} \\ &\times \left(\int_{\mathbb{R}^n} |Du|^p\,dx\right)^{1/p}. \qquad (8.107)\end{aligned}$$

In particular for γ such that

$$\frac{\gamma n}{n-1} = \frac{(\gamma-1)p}{p-1},$$

that is, $\gamma = \frac{p(n-1)}{n-p}$, so that

$$\frac{\gamma n}{n-1} = \frac{(\gamma-1)p}{p-1} = \frac{np}{n-p},$$

we get

$$\left(\int_{\mathbb{R}^n} |u|^r\,dx\right)^{((n-1)/n-(p-1)/p)} \leq C\left(\int_{\mathbb{R}^n} |Du|^p\,dx\right)^{1/p}.$$

From this and considering that

$$\frac{n-1}{n} - \frac{p-1}{p} = \frac{n-p}{np} = \frac{1}{r},$$

we finally obtain

$$\left(\int_{\mathbb{R}^n} |u|^r\,dx\right)^{1/r} \leq C\left(\int_{\mathbb{R}^n} |Du|^p\,dx\right)^{1/p}.$$

The proof is complete.

Theorem 8.4.15. *Let $\Omega \subset \mathbb{R}^n$ be a bounded open set. Suppose $\partial\Omega$ is $\hat{C}^1$, $1 \leq p < n$, and $u \in W^{1,p}(\Omega)$.*

Then $u \in L^r(\Omega)$ and

$$\|u\|_{r,\Omega} \leq K\|u\|_{1,p,\Omega},$$

where the constant depends only on p, n, and Ω.

Proof. Since $\partial\Omega$ is $\hat{C}^1$, from Theorem 8.4.12, there exists an extension $Eu = \bar{u} \in W^{1,p}(\mathbb{R}^n)$ such that $\bar{u} = u$ in Ω the support of $\bar{u}$ is compact and

$$\|\bar{u}\|_{1,p,\mathbb{R}^n} \leq C\|u\|_{1,p,\Omega},$$

where C does not depend on u. As $\bar{u}$ has compact support, from Theorem 8.4.10, there exists a sequence $\{u_k\} \in C_c^\infty(\mathbb{R}^n)$ such that

$$u_k \to \bar{u} \text{ in } W^{1,p}(\mathbb{R}^n),$$

from the last theorem

$$\|u_k - u_l\|_{r,\mathbb{R}^n} \leq K\|Du_k - Du_l\|_{p,\mathbb{R}^n}.$$

Hence,

$$u_k \to \bar{u} \text{ in } L^r(\mathbb{R}^n),$$

also from the last theorem

$$\|u_k\|_{r,\mathbb{R}^n} \leq K\|Du_k\|_{p,\mathbb{R}^n}, \forall k \in \mathbb{N},$$

so that

$$\|\bar{u}\|_{r,\mathbb{R}^n} \leq K\|D\bar{u}\|_{p,\mathbb{R}^n}.$$

Therefore, we may get

$$\begin{aligned} \|u\|_{r,\Omega} &\leq \|\bar{u}\|_{r,\mathbb{R}^n} \\ &\leq K\|D\bar{u}\|_{p,\mathbb{R}^n} \\ &\leq K_1\|\bar{u}\|_{1,p,\mathbb{R}^n} \\ &\leq K_2\|u\|_{1,p,\Omega}. \end{aligned} \tag{8.108}$$

The proof is complete.

Theorem 8.4.16. *Let $\Omega \subset \mathbb{R}^n$ be a bounded open set such that $\partial\Omega \in \hat{C}^1$. If $mp < n$, then $W^{m,p}(\Omega) \hookrightarrow L^q(\Omega)$ for $p \leq q \leq (np)/(n-mp)$.*

Proof. Define $q_0 = np/(n-mp)$. We first prove by induction on m that

$$W^{m,p} \hookrightarrow L^{q_0}(\Omega).$$

The last result is exactly the case for $m = 1$. Assume

$$W^{m-1,p} \hookrightarrow L^{r_1}(\Omega), \tag{8.109}$$

where

$$r_1 = np/(n-(m-1)p) = np/(n-np+p),$$

whenever $n > (m-1)p$. If $u \in W^{m,p}(\Omega)$ where $n > mp$, then u and $D_j u$ are in $W^{m-1,p}(\Omega)$, so that from (8.109) we have $u \in W^{1,r_1}(\Omega)$ and

$$\|u\|_{1,r_1,\Omega} \le K\|u\|_{m,p,\Omega}. \tag{8.110}$$

Since $n > mp$ we have that $r_1 = np/((n-mp)+p) < n$, from $q_0 = nr_1/(n-r_1) = np/(n-mp)$ by the last theorem, we have

$$\|u\|_{q_0,\Omega} \le K_2\|u\|_{1,r_1,\Omega},$$

where the constant K_2 does not depend on u, and therefore from this and (8.110) we obtain

$$\|u\|_{q_0,\Omega} \le K_2\|u\|_{1,r_1,\Omega} \le K_3\|u\|_{m,p,\Omega}. \tag{8.111}$$

The induction is complete. Now suppose $p \le q \le q_0$. Define

$$s = (q_0-q)p/(q_0-p) \text{ and } t = p/s = (q_0-p)/(q_0-q).$$

Through the Hölder inequality, we get

$$\begin{aligned}
\|u\|^q_{q,\Omega} &= \int_\Omega |u(x)|^s |u(x)|^{q-s}\,dx \\
&\le \left(\int_\Omega |u(x)|^{st}\,dx\right)^{1/t} \left(\int_\Omega |u(x)|^{(q-s)t'}\,dx\right)^{1/t'} \\
&= \|u\|^{p/t}_{p,\Omega}\|u\|^{q_0/t'}_{q_0,\Omega} \\
&\le \|u\|^{p/t}_{p,\Omega}(K_3)^{q_0/t'}\|u\|^{q_0/t'}_{m,p,\Omega} \\
&\le (K_3)^{q_0/t'}\|u\|^{p/t}_{m,p,\Omega}\|u\|^{q_0/t'}_{m,p,\Omega} \\
&= (K_3)^{q_0/t'}\|u\|^q_{m,p,\Omega},
\end{aligned} \tag{8.112}$$

since

$$p/t + q_0/t' = q.$$

This completes the proof.

Corollary 8.4.17. *If $mp = n$, then $W^{m,p}(\Omega) \hookrightarrow L^q$ for $p \le q < \infty$.*

Proof. If $q > p' = p/(p-1)$, then $q = ns/(n-ms)$ where $s = pq/(p+q)$ is such that $1 \le s \le p$. Observe that

$$W^{m,p}(\Omega) \hookrightarrow W^{m,s}(\Omega)$$

with the imbedding constant depending only on $|\Omega|$. Since $ms < n$, by the last theorem, we obtain

$$W^{m,p}(\Omega) \hookrightarrow W^{m,s}(\Omega) \hookrightarrow L^q(\Omega).$$

Now if $p \leq q \leq p'$, from above we have $W^{m,p}(\Omega) \hookrightarrow L^{p'}(\Omega)$ and the obvious imbedding $W^{m,p}(\Omega) \hookrightarrow L^p(\Omega)$. Define $s = (p'-q)p/(p'-p)$, and the result follows from a reasoning analogous to the final chain of inequalities of last theorem, indicated in (8.112).

About the next theorem, note that its hypotheses are satisfied if $\partial\Omega$ is $\hat{C}^1$ (here we do not give the details).

Theorem 8.4.18. *Let $\Omega \subset \mathbb{R}^n$ be an open bounded set, such that for each $x \in \overline{\Omega}$ there exists a convex set $C_x \subset \overline{\Omega}$ whose shape depends on x but such that $|C_x| > \alpha$, for some $\alpha > 0$ that does not depend on x. Thus, if $mp > n$, then*

$$W^{m,p}(\Omega) \hookrightarrow C_B^0(\Omega).$$

Proof. Suppose first $m = 1$ so that $p > n$. Fix $x \in \overline{\Omega}$ and pick $y \in C_x$. For $\phi \in C^\infty(\overline{\Omega})$, from the fundamental theorem of calculus, we have

$$\phi(y) - \phi(x) = \int_0^1 \frac{d(\phi(x+t(y-x))}{dt}\, dt.$$

Thus,

$$|\phi(x)| \leq |\phi(y)| + \int_0^1 \left|\frac{d(\phi(x+t(y-x))}{dt}\right| dt,$$

and hence

$$\int_{C_x} |\phi(x)|\, dy \leq \int_{C_x} |\phi(y)|\, dy + \int_{C_x}\int_0^1 \left|\frac{d(\phi(x+t(y-x))}{dt}\right| dt\, dy,$$

so that, from the Hölder inequality and the Fubini theorem, we get

$$\begin{aligned} |\phi(x)|\alpha &\leq |\phi(x)| \cdot |C_x| \\ &\leq \|\phi\|_{p,\Omega}|C_x|^{1/p'} + \int_0^1 \int_{C_x} \left|\frac{d(\phi(x+t(y-x))}{dt}\right| dy\, dt. \end{aligned}$$

Therefore

$$|\phi(x)|\alpha \leq \|\phi\|_{p,\Omega}|\Omega|^{1/p'} + \int_0^1 \int_V |\nabla\phi(z)|\delta t^{-n}\, dz\, dt,$$

where $|V| = t^n|C_x|$ and δ denote the diameter of Ω. From the Hölder inequality again, we obtain

$$|\phi(x)|\alpha \leq \|\phi\|_{p,\Omega}|\Omega|^{1/p'} + \delta \int_0^1 \left(\int_V |\nabla\phi(z)|^p\, dy\right)^{1/p} t^{-n}(t^n|C_x|)^{1/p'}\, dt,$$

and thus

$$|\phi(x)|\alpha \leq \|\phi\|_{p,\Omega}|\Omega|^{1/p'} + \delta|C_x|^{1/p'}\|\nabla\phi\|_{p,\Omega} \int_0^1 t^{-n(1-1/p')}\, dt.$$

Since $p > n$ we obtain

$$\int_0^1 t^{-n(1-1/p')}\,dt = \int_0^1 t^{-n/p}\,dt = \frac{1}{1-n/p}.$$

From this, the last inequality and from the fact that $|C_x| \leq |\Omega|$, we have that there exists $K > 0$ such that

$$|\phi(x)| \leq K\|\phi\|_{1,p,\Omega}, \forall x \in \overline{\Omega},\ \phi \in C^\infty(\overline{\Omega}). \tag{8.113}$$

Here the constant K depends only on p, n, and Ω. Consider now $u \in W^{1,p}(\Omega)$.

Thus there exists a sequence $\{\phi_k\} \subset C^\infty(\overline{\Omega})$ such that

$$\phi_k \to u, \text{ in } W^{1,p}(\Omega).$$

Up to a not relabeled subsequence, we have

$$\phi_k \to u, \text{ a.e. in } \overline{\Omega}. \tag{8.114}$$

Fix $x \in \overline{\Omega}$ such that the limit indicated in (8.114) holds. Suppose given $\varepsilon > 0$. Therefore, there exists $k_0 \in \mathbb{N}$ such that

$$|\phi_{k_0}(x) - u(x)| \leq \varepsilon/2$$

and

$$\|\phi_{k_0} - u\|_{1,p,\Omega} < \varepsilon/(2K).$$

Thus,

$$\begin{aligned} |u(x)| &\leq |\phi_{k_0}(x)| + \varepsilon/2 \\ &\leq K\|\phi_{k_0}\|_{1,p,\Omega} + \varepsilon/2 \\ &\leq K\|u\|_{1,p,\Omega} + \varepsilon. \end{aligned} \tag{8.115}$$

Since $\varepsilon > 0$ is arbitrary, the proof for $m = 1$ is complete, because for $\{\phi_k\} \in C^\infty(\overline{\Omega})$ such that $\phi_k \to u$ in $W^{1,p}(\Omega)$, from (8.113), we have that $\{\phi_k\}$ is a uniformly Cauchy sequence, so that it converges to a continuous u^*, where $u^* = u$, a.e. in $\overline{\Omega}$.

For $m > 1$ but $p > n$ we still have

$$|u(x)| \leq K\|u\|_{1,p,\Omega} \leq K_1\|u\|_{m,p,\Omega}, \text{ a.e. in } \Omega, \forall u \in W^{m,p}(\Omega)$$

If $p \leq n \leq mp$, there exists j satisfying $1 \leq j \leq m-1$ such that $jp \leq n \leq (j+1)p$. If $jp < n$, set

$$\hat{r} = np/(n-jp).$$

Let $1 \leq p_1 \leq n$ such that

$$\hat{r} = np_1/(n-p_1).$$

Thus we have that

$$np/(n-jp) = np_1/(n-p_1),$$

so that

$$p_1 = np/(n-(j-1)p),$$

so that by above and the last theorem:

$$\|u\|_\infty \leq K_1\|u\|_{1,\hat{r},\Omega} \leq K_1\|u\|_{m-j,\hat{r},\Omega} \leq K_2\|u\|_{m-(j-1),p_1,\Omega}.$$

Now define

$$\hat{r}_1 = p_1 = np/(n-(j-1)p)$$

and $1 \leq p_2 \leq n$ such that

$$\hat{r}_1 = np_2/(n-p_2),$$

so that

$$np/(n-(j-1)p) = np_2/(n-p_2).$$

Hence $p_2 = np/(n-(j-2)p)$ so that by the last theorem

$$\|u\|_{m-(j-1),p_1,\Omega} = \|u\|_{m-(j-1),\hat{r}_1,\Omega} \leq K_3\|u\|_{m-(j-2),p_2,\Omega}.$$

Proceeding inductively in this fashion, after j steps, observing that $p_j = p$, we get

$$\|u\|_\infty \leq K_1\|u\|_{1,\hat{r},\Omega} \leq K_1\|u\|_{m-j,\hat{r},\Omega} \leq K_j\|u\|_{m,p,\Omega},$$

for some appropriate K_j. Finally, if $jp = n$, choosing $\hat{r} = \max\{n,p\}$ also by the last theorem we obtain the same last chain of inequalities. For that, assume $\hat{r} = \max\{n,p\} = n > p$. Let p_1 be such that

$$r_1 = \frac{np_1}{n-p_1} = n,$$

that is,

$$p_1 = \frac{n}{2}.$$

Since $n > p$, we have that $n \geq 2$ so that $1 \leq p_1 < n$. From the last theorem we obtain

$$\|u\|_\infty \leq C\|u\|_{m-j,r_1,\Omega} \leq C_1\|u\|_{m-(j-1),p_1,\Omega}.$$

Let $r_2 = p_1 = n/2$, and define p_2 such that

$$r_2 = n/2 = \frac{np_2}{n-p_2},$$

that is, $p_2 = n/3$.

Hence, again by the last theorem, we get

$$\|u\|_\infty \leq C_1\|u\|_{m-(j-1),r_2,\Omega} \leq C_2\|u\|_{m-(j-2),p_2,\Omega}.$$

Reasoning inductively, after $j-1$ steps, we get $p_{j-1} = n/j = p$, so that

$$\|u\|_\infty \leq C\|u\|_{m-(j-1),r_1,\Omega} \leq C_3\|u\|_{m-(j-(j-1)),p_{j-1},\Omega} \leq C_4\|u\|_{m,p,\Omega}.$$

Finally, if $r_1 = \max\{n,p\} = p \geq n$, define p_1 such that

$$r_1 = p = \frac{np_1}{n-p_1},$$

that is,

$$p_1 = \frac{np}{n+p} \leq p,$$

so that by last theorem

$$\|u\|_\infty \leq \|u\|_{m-j,r_1,\Omega} \leq C_5\|u\|_{m-(j-1),p_1,\Omega} \leq C_6\|u\|_{m,p,\Omega}.$$

This completes the proof.

Theorem 8.4.19. *Let $\Omega \subset \mathbb{R}^n$ be a set with a boundary $\hat{C}^1$. If $mp > n$, then $W^{m,p}(\Omega) \hookrightarrow L^q(\Omega)$ for $p \leq q \leq \infty$.*

Proof. From the proof of the last theorem, we may obtain

$$\|u\|_{\infty,\Omega} < K\|u\|_{m,p,\Omega}, \forall u \in W^{m,p}(\Omega).$$

If $p \leq q < \infty$, we have

$$\begin{aligned}
\|u\|_{q,\Omega}^q &= \int_\Omega |u(x)|^p |u(x)|^{q-p}\,dx \\
&\leq \int_\Omega |u(x)|^p \left(K\|u\|_{m,p,\Omega}\right)^{q-p}\,dx \\
&\leq K^{q-p}\|u\|_{p,\Omega}^p \|u\|_{m,p,\Omega}^{q-p} \\
&\leq K^{q-p}\|u\|_{m,p,\Omega}^p \|u\|_{m,p,\Omega}^{q-p} \\
&= K^{q-p}\|u\|_{m,p,\Omega}^q.
\end{aligned} \tag{8.116}$$

The proof is complete.

Theorem 8.4.20. *Let $S \subset \mathbb{R}^n$ be an n-dimensional ball of radius bigger than 3. If $n < p$, then there exists a constant C, depending only on p and n, such that*

$$\|u\|_{C^{0,\lambda}(S)} \leq C\|u\|_{1,p,S}, \forall u \in C^1(S),$$

where $0 < \lambda \leq 1 - n/p$.

Proof. First consider $\lambda = 1 - n/p$ and $u \in C^1(S)$. Let $x, y \in S$ such that $|x-y| < 1$ and define $\sigma = |x-y|$. Consider a fixed cube denoted by $R_\sigma \subset S$ such that $|R_\sigma| = \sigma^n$ and $x, y \in \bar{R}_\sigma$. For $z \in R_\sigma$, we may write

$$u(x) - u(z) = -\int_0^1 \frac{du(x+t(z-x))}{dt}\, dt,$$

that is,

$$u(x)\sigma^n = \int_{R_\sigma} u(z)\, dz - \int_{R_\sigma}\int_0^1 \nabla u(x+t(z-x)) \cdot (z-x)\, dt\, dz.$$

Thus, denoting in the next lines V by an appropriate set such that $|V| = t^n|R_\sigma|$, we obtain

$$\begin{aligned}
|u(x) - \int_{R_\sigma} u(z)\, dz/\sigma^n| \leq & \sqrt{n}\sigma^{1-n}\int_{R_\sigma}\int_0^1 |\nabla u(x+t(z-x))|\, dt\, dz \\
\leq & \sqrt{n}\sigma^{1-n}\int_0^1 t^{-n}\int_V |\nabla u(z)|\, dz\, dt \\
\leq & \sqrt{n}\sigma^{1-n}\int_0^1 t^{-n}\|\nabla u\|_{p,S}|V|^{1/p'}\, dt \\
\leq & \sqrt{n}\sigma^{1-n}\sigma^{n/p'}\|\nabla u\|_{p,S}\int_0^1 t^{-n}t^{n/p'}\, dt \\
\leq & \sqrt{n}\sigma^{1-n/p}\|\nabla u\|_{p,S}\int_0^1 t^{-n/p}\, dt \\
\leq & \sigma^{1-n/p}\|u\|_{1,p,S}K,
\end{aligned} \tag{8.117}$$

where

$$K = \sqrt{n}\int_0^1 t^{-n/p}\, dt = \sqrt{n}/(1-n/p).$$

A similar inequality holds with y in place of x, so that

$$|u(x) - u(y)| \leq 2K|x-y|^{1-n/p}\|u\|_{1,p,S}, \forall x, y \in R_\sigma.$$

Now consider $0 < \lambda < 1 - n/p$. Observe that, as $|x-y|^\lambda \geq |x-y|^{1-n/p}$, if $|x-y| < 1$, we have

$$\begin{aligned}
&\sup_{x,y\in S}\left\{\frac{|u(x)-u(y)|}{|x-y|^\lambda} \mid x \neq y, |x-y| < 1\right\} \\
&\leq \sup_{x,y\in S}\left\{\frac{|u(x)-u(y)|}{|x-y|^{1-n/p}} \mid x \neq y, |x-y| < 1\right\} \leq K\|u\|_{1,p,S}.
\end{aligned} \tag{8.118}$$

Also,

$$\sup_{x,y\in S}\left\{\frac{|u(x)-u(y)|}{|x-y|^\lambda} \mid |x-y| \geq 1\right\} \leq 2\|u\|_{\infty,S} \leq 2K_1\|u\|_{1,p,S}$$

so that

$$\sup_{x,y\in S}\left\{\frac{|u(x)-u(y)|}{|x-y|^{\lambda}} \mid x\neq y\right\}\leq (K+2K_1)\|u\|_{1,p,S}, \forall u\in C^1(S).$$

The proof is complete.

Theorem 8.4.21. *Let $\Omega\subset\mathbb{R}^n$ be an open bounded set such that $\partial\Omega$ is $\hat{C}^1$. Assume $n<p\leq\infty$.*

Then

$$W^{1,p}(\Omega)\hookrightarrow C^{0,\lambda}(\overline{\Omega}),$$

for all $0<\lambda\leq 1-n/p$.

Proof. Fix $0<\lambda\leq 1-n/p$ and let $u\in W^{1,p}(\Omega)$. Since $\partial\Omega$ is $\hat{C}^1$, from Theorem 8.4.12, there exists an extension $Eu=\bar{u}$ such that $\bar{u}=u$, a.e. in Ω, and

$$\|\bar{u}\|_{1,p,\mathbb{R}^n}\leq K\|u\|_{1,p,\Omega},$$

where the constant K does not depend on u. From the proof of this same theorem, we may assume that $\text{spt}(\bar{u})$ is on an n-dimensional sphere $S\supset\Omega$ with sufficiently big radius and such sphere does not depend on u. Thus, in fact, we have

$$\|\bar{u}\|_{1,p,S}\leq K\|u\|_{1,p,\Omega}.$$

Since $C^\infty(S)$ is dense in $W^{1,p}(S)$, there exists a sequence $\{\phi_k\}\subset C^\infty(S)$ such that

$$u_k\to\bar{u}, \text{ in } W^{1,p}(S). \tag{8.119}$$

Up to a not relabeled subsequence, we have

$$u_k\to\bar{u}, \text{ a.e. in } \Omega.$$

From last theorem we have

$$\|u_k-u_l\|_{C^{0,\lambda}(S)}\leq C\|u_k-u_l\|_{1,p,S},$$

so that $\{u_k\}$ is a Cauchy sequence in $C^{0,\lambda}(\overline{S})$, and thus $u_k\to u^*$ for some $u^*\in C^{0,\lambda}(S)$. Hence, from this and (8.119), we have

$$u^*=\bar{u}, \text{ a.e. in } S.$$

Finally, from above and last theorem, we may write

$$\|u^*\|_{C^{0,\lambda}(\overline{\Omega})}\leq\|u^*\|_{C^{0,\lambda}(S)}\leq K_1\|\bar{u}\|_{1,p,S}\leq K_2\|u\|_{1,p,\Omega}.$$

The proof is complete.

8.5 The Trace Theorem

In this section we state and prove the trace theorem.

Theorem 8.5.1. *Let $1 < p < \infty$ and let $\Omega \subset \mathbb{R}^n$ be an open bounded set such that $\partial\Omega$ is $\hat{C}^1$. Then there exists a bounded linear operator*

$$T : W^{1,p}(\Omega) \to L^p(\partial\Omega),$$

such that

- *$Tu = u|_{\partial\Omega}$ if $u \in W^{1,p}(\Omega) \cap C(\overline{\Omega})$,*
- $$\|Tu\|_{p,\partial\Omega} \leq C\|u\|_{1,p,\Omega}, \forall u \in W^{1,p}(\Omega),$$

  where the constant C depends only on p and Ω.

Proof. Let $u \in W^{1,p}(\Omega) \cap C(\overline{\Omega})$. Choose $x_0 \in \partial\Omega$.

Since $\partial\Omega$ is $\hat{C}^1$, there exists $r > 0$ such that for a local coordinate system we may write

$$\overline{\Omega} \cap B_r(x_0) = \{x \in B_r(x_0) \mid x_n \geq f(x_1, \ldots, x_{n-1})\},$$

where denoting $\hat{x} = (x_1, \ldots, x_{n-1})$, $f(\hat{x})$ is continuous and such that its partial derivatives are classically defined a.e. and bounded on its domain. Furthermore

$$|f(\hat{x}) - f(\hat{y})| \leq K|\hat{x} - \hat{y}|_2, \forall \hat{x}, \hat{y}$$

for some $K > 0$ also on its domain.

Define the coordinates y by

$$y_i = x_i, \forall i \in \{1, \ldots, n-1\},$$

and

$$y_n = x_n - f(x_1, \ldots, x_{n-1}).$$

Define $\hat{u}(y)$ by

$$u(x_1, \ldots, x_n) = u(y_1, \ldots, y_{n-1}, y_n + f(y_1, \ldots, y_{n-1})) = \hat{u}(y).$$

Also define $y_0 = (x_{0^1}, \ldots, x_{0^{n-1}}, x_{0^n} - f(x_{0^1}, \ldots, x_{0^{n-1}})) = (y_{0^1}, \ldots, y_{0^{n-1}}, 0)$ and choose $r_1 > 0$ such that

$$\Psi^{-1}(B^+_{r_1}(y_0)) \subset \Omega \cap B_r(x_0).$$

Observe that this is possible since Ψ and Ψ^{-1} are continuous, where $y = \Psi(x)$. Here

$$B^+_{r_1}(y_0) = \{y \in B_{r_1}(y_0) \mid y_n > 0\}.$$

For each $N \in \mathbb{N}$, choose, by mollification, for example, $\phi_N \in C^\infty(\overline{B}^+_{r_1}(y_0))$ such that

$$\|\phi_N - \hat{u}\|_{\infty,\overline{B}^+_{r_1}(y_0)} < \frac{1}{N}.$$

Denote $B = B_{r_1/2}(y_0)$, and $B^+ = B^+_{r_1/2}(y_0)$. Now choose $\eta \in C^\infty_c(B_{r_1}(y_0))$, such that $\eta > 0$ and $\eta \equiv 1$ on B. Also denote

$$\tilde{\Gamma} = \{y \in B \mid y_n = 0\},$$

and

$$\tilde{\Gamma}_1 = \{y \in B_{r_1}(y_0) \mid y_n = 0\}.$$

Observe that

$$\begin{aligned}
\int_{\tilde{\Gamma}} |\phi_N|^p \, d\Gamma &\leq \int_{\tilde{\Gamma}_1} \eta |\phi_N|^p \, d\Gamma \\
&= -\int_{B^+_{r_1}} (\eta |\phi_N|^p)_{y_n} \, dy \\
&\leq -\int_{B^+_{r_1}} (\eta_{y_n} |\phi_N|^p) \, dy \\
&\quad + \int_{B^+_{r_1}} (p|\phi_N|^{p-1} |(\phi_N)_{y_n}| \eta) \, dy.
\end{aligned} \tag{8.120}$$

Here we recall the Young inequality

$$ab \leq \frac{a^p}{p} + \frac{b^q}{q}, \forall a,b \geq 0, \text{ where } \frac{1}{p} + \frac{1}{q} = 1.$$

Thus,

$$(|\phi_N|^{p-1})(|(\phi_N)_{y_n}|\eta) \leq \frac{|(\phi_N)_{y_n}|^p \eta^p}{p} + \frac{|\phi_N|^{(p-1)q}}{q},$$

so that replacing such an inequality in (8.120), since $(p-1)q = p$, we get

$$\int_{\tilde{\Gamma}} |\phi_N|^p \, d\Gamma \leq C_1 \left(\int_{B^+_{r_1}} |\phi_N|^p \, dy + \int_{B^+_{r_1}} |D\phi_N|^p \, dy \right). \tag{8.121}$$

Letting $N \to +\infty$ we obtain

$$\begin{aligned}
\int_{\Gamma} |u(x)|^p \, d\Gamma &< C_2 \int_{\tilde{\Gamma}} |\hat{u}(y)|^p \, d\Gamma \\
&\leq C_3 \left(\int_{B^+_{r_1}} |\hat{u}|^p \, dy + \int_{B^+_{r_1}} |D\hat{u}|^p \, dy \right) \\
&\leq C_4 \left(\int_{W^+} |u|^p \, dx + \int_{W^+} |Du|^p \, dx \right),
\end{aligned} \tag{8.122}$$

where $\Gamma = \psi^{-1}(\tilde{\Gamma})$ and $W^+ = \Psi^{-1}(B^+_{r_1})$.

Observe that denoting $W = W_{x_0}$ we have that $\partial\Omega \subset \cup_{x\in\partial\Omega} W_x$, and thus, since $\partial\Omega$ is compact, we may select $x_1, \ldots, x_M$ such that $\partial\Omega \subset \cup_{i=1}^M W_i$. We emphasize to have denoted $W_{x_i} = W_i, \forall i \in \{1, \ldots, M\}$. Denoting $W_i^+ = W_i \cap \Omega$ we may obtain

$$\begin{aligned}\int_{\partial\Omega} |u(x)|^p \, d\Gamma &\leq \sum_{i=1}^M \int_{\Gamma_i} |u(x)|^p \, d\Gamma \\ &\leq \sum_{i=1}^M C_{4^i} \left(\int_{W_i^+} |u|^p \, dx + \int_{W_i^+} |Du|^p \, dx \right) \\ &\leq C_5 M \left(\int_\Omega |u|^p \, dx + \int_\Omega |Du|^p \, dx \right) \\ &= C \left(\int_\Omega |u|^p \, dx + \int_\Omega |Du|^p \, dx \right). \end{aligned} \tag{8.123}$$

At this point we denote $Tu = u|_{\partial\Omega}$.

Finally, for the case $u \in W^{1,p}(\Omega)$, select $\{u_k\} \subset C^\infty(\overline{\Omega})$ such that

$$\|u_k - u\|_{1,p,\Omega} \to 0, \text{ as } k \to \infty.$$

From above

$$\|Tu_k - Tu_l\|_{p,\partial\Omega} \leq C\|u_k - u_l\|_{1,p,\Omega},$$

so that

$$\{Tu_k\}$$

is a Cauchy sequence. Hence we may define

$$Tu = \lim_{k\to\infty} Tu_k, \text{ in } L^p(\partial\Omega).$$

The proof is complete.

Remark 8.5.2. Similar results are valid for $W_0^{m,p}$; however, in this case the traces relative to derivatives of order up to $m-1$ are involved.

8.6 Compact Imbeddings

Theorem 8.6.1. *Let m be a nonnegative integer and let $0 < \nu < \lambda \leq 1$. Then the following imbeddings exist:*

$$C^{m+1}(\overline{\Omega}) \hookrightarrow C^m(\overline{\Omega}), \tag{8.124}$$

$$C^{m,\lambda}(\overline{\Omega}) \hookrightarrow C^m(\overline{\Omega}), \tag{8.125}$$

$$C^{m,\lambda}(\overline{\Omega}) \hookrightarrow C^{m,\nu}(\overline{\Omega}). \tag{8.126}$$

If Ω is bounded, then Imbeddings (8.125) and (8.126) are compact.

Proof. Imbeddings (8.124) and (8.125) follow from the inequalities

$$\|\phi\|_{C^m(\overline{\Omega})} \leq \|\phi\|_{C^{m+1}(\overline{\Omega})},$$

$$\|\phi\|_{C^m(\overline{\Omega})} \leq \|\phi\|_{C^{m,\lambda}(\overline{\Omega})}.$$

To establish (8.126) note that for $|\alpha| \leq m$

$$\sup_{x,y\in\Omega}\left\{\frac{|D^\alpha\phi(x)-D^\alpha\phi(y)|}{|x-y|^\nu} \mid x\neq y,\ |x-y|<1\right\}$$
$$\leq \sup_{x,y\in\Omega}\left\{\frac{|D^\alpha\phi(x)-D^\alpha\phi(y)|}{|x-y|^\lambda} \mid x\neq y,\ |x-y|<1\right\}, \tag{8.127}$$

and also,

$$\sup_{x,y\in\Omega}\left\{\frac{|D^\alpha\phi(x)-D^\alpha\phi(y)|}{|x-y|^\nu} \mid |x-y|\geq 1\right\} \leq 2\sup_{x\in\Omega}\{|D^\alpha\psi|\}. \tag{8.128}$$

Therefore, we may conclude that

$$\|\phi\|_{C^{m,\nu}(\bar{\Omega})} \leq 3\|\phi\|_{C^{m,\lambda}(\overline{\Omega})}, \forall \phi \in C^{m,\nu}(\overline{\Omega}).$$

Now suppose Ω is bounded. If A is a bounded set in $C^{0,\lambda}(\overline{\Omega})$, then there exists $M>0$ such that

$$\|\phi\|_{C^{0,\lambda}(\overline{\Omega})} \leq M, \forall \phi \in A.$$

But then

$$|\phi(x)-\phi(y)| \leq M|x-y|^\lambda, \forall x,y \in \overline{\Omega}, \phi \in A,$$

so that by the Ascoli–Arzela theorem, A is pre-compact in $C(\overline{\Omega})$. This proves the compactness of (8.125) for $m=0$.

If $m \geq 1$ and A is bounded in $C^{m,\lambda}(\overline{\Omega})$, then A is bounded in $C^{0,\lambda}(\overline{\Omega})$. Thus, by above there is a sequence $\{\phi_k\} \subset A$ and $\phi \in C^{0,\lambda}(\overline{\Omega})$ such that

$$\phi_k \to \phi \text{ in } C(\overline{\Omega}).$$

However, $\{D_l\phi_k\}$ is also bounded in $C^{0,\lambda}(\overline{\Omega})$, so that there exists a not relabeled subsequence, also denoted by $\{\phi_k\}$ and ψ_i such that

$$D_i\phi_k \to \psi_i, \text{ in } C(\overline{\Omega}).$$

The convergence in $C(\bar{\Omega})$ being the uniform one, we have $\psi_i = D_i\phi$. We can proceed extracting (not relabeled) subsequences until obtaining

$$D^\alpha \phi_k \to D^\alpha \phi, \text{ in } C(\overline{\Omega}), \forall\, 0 \le |\alpha| \le m.$$

This completes the proof of compactness of (8.125). For (8.126), let S be a bounded set in $C^{m,\lambda}(\overline{\Omega})$. Observe that

$$\begin{aligned} \frac{|D^\alpha \phi(x) - D^\alpha \phi(y)|}{|x-y|^\nu} &= \left(\frac{|D^\alpha \phi(x) - D^\alpha \phi(y)|}{|x-y|^\lambda}\right)^{\nu/\lambda} \\ &\quad \cdot |D^\alpha \phi(x) - D^\alpha \phi(y)|^{1-\nu/\lambda} \\ &\le K|D^\alpha \phi(x) - D^\alpha \phi(y)|^{1-\nu/\lambda}, \end{aligned} \tag{8.129}$$

for all $\phi \in S$. From (8.125), S has a converging subsequence in $C^m(\overline{\Omega})$. From (8.129) such a subsequence is also converging in $C^{m,\nu}(\overline{\Omega})$. The proof is complete.

Theorem 8.6.2 (Rellich–Kondrachov). *Let $\Omega \subset \mathbb{R}^n$ be an open bounded set such that $\partial\Omega$ is $\hat{C}^1$. Let j,m be integers, $j \ge 0, m \ge 1$, and let $1 \le p < \infty$.*

1. ***Part I** If $mp \le n$, then the following imbeddings are compact:*

$$W^{j+m,p}(\Omega) \hookrightarrow W^{j,q}(\Omega),$$
$$\text{if } 0 < n - mp < n \text{ and } 1 \le np/(n-mp), \tag{8.130}$$

$$W^{j+m,p}(\Omega) \hookrightarrow W^{j,q}(\Omega), \text{ if } n = mp,\ 1 \le q < \infty. \tag{8.131}$$

2. ***Part II** If $mp > n$, then the following imbeddings are compact:*

$$W^{j+m,p} \hookrightarrow C^j_B(\Omega), \tag{8.132}$$

$$W^{j+m,p}(\Omega) \hookrightarrow W^{j,q}(\Omega), \text{ if } 1 \le q \le \infty. \tag{8.133}$$

3. ***Part III** The following imbeddings are compact:*

$$W^{j+m,p}(\Omega) \hookrightarrow C^j(\overline{\Omega}), \text{ if } mp > n, \tag{8.134}$$

$$W^{j+m,p}(\Omega) \hookrightarrow C^{j,\lambda}(\overline{\Omega}),$$
$$\text{if } mp > n \ge (m-1)p \text{ and } 0 < \lambda < m - n/p. \tag{8.135}$$

4. ***Part IV** All the above imbeddings are compact if we replace $W^{j+m,p}(\Omega)$ by $W_0^{j+m,p}(\Omega)$.*

Remark 8.6.3. Given X,Y,Z spaces, for which we have the imbeddings $X \hookrightarrow Y$ and $Y \hookrightarrow Z$ and if one of these imbeddings is compact then the composite imbedding $X \hookrightarrow Z$ is compact. Since the extension operator $u \to \tilde{u}$ where $\tilde{u}(x) = u(x)$ if $x \in \Omega$ and $\tilde{u}(x) = 0$ if $x \in \mathbb{R}^n - \Omega$ defines an imbedding $W_0^{j+m,p}(\Omega) \hookrightarrow W^{j+m,p}(\mathbb{R}^n)$ we have that Part IV of above theorem follows from the application of Parts I–III to $\mathbb{R}^n$

(despite the fact we are assuming Ω bounded, the general results may be found in Adams [1]).

Remark 8.6.4. To prove the compactness of any of above imbeddings it is sufficient to consider the case $j=0$. Suppose, for example, that the first imbedding has been proved for $j=0$. For $j\geq 1$ and $\{u_i\}$ bounded sequence in $W^{j+m,p}(\Omega)$ we have that $\{D^\alpha u_i\}$ is bounded in $W^{m,p}(\Omega)$ for each α such that $|\alpha|\leq j$. From the case $j=0$ it is possible to extract a subsequence (similarly to a diagonal process) $\{u_{i_k}\}$ for which $\{D^\alpha u_{i_k}\}$ converges in $L^q(\Omega)$ for each α such that $|\alpha|\leq j$, so that $\{u_{i_k}\}$ converges in $W^{j,q}(\Omega)$.

Remark 8.6.5. Since Ω is bounded, $C_B^0(\Omega)\hookrightarrow L^q(\Omega)$ for $1\leq q\leq\infty$. In fact

$$\|u\|_{0,q,\Omega}\leq\|u\|_{C_B^0}[vol(\Omega)]^{1/q}. \tag{8.136}$$

Thus the compactness of (8.133) (for $j=0$) follows from that of (8.132).

Proof of Parts II and III. If $mp>n>(m-1)p$ and $0<\lambda<(m-n)/p$, then there exists μ such that $\lambda<\mu<m-(n/p)$. Since Ω is bounded, the imbedding $C^{0,\mu}(\overline{\Omega})\hookrightarrow C^{0,\lambda}(\overline{\Omega})$ is compact by Theorem 8.6.1. Since by the Sobolev imbedding theorem we have $W^{m,p}(\Omega)\hookrightarrow C^{0,\mu}(\overline{\Omega})$, we have that Imbedding (8.135) is compact.

If $mp>n$, let j^* be the nonnegative integer satisfying $(m-j^*)p>n\geq(m-j^*-1)p$. Thus we have the chain of imbeddings

$$W^{m,p}(\Omega)\hookrightarrow W^{m-j^*,p}(\Omega)\hookrightarrow C^{0,\mu}(\overline{\Omega})\hookrightarrow C(\overline{\Omega}), \tag{8.137}$$

where $0<\mu<m-j^*-(n/p)$. The last imbedding in (8.137) is compact by Theorem 8.6.1, so that (8.134) is compact for $j=0$. By analogy (8.132) is compact for $j=0$. Therefore from the above remarks (8.133) is also compact. For the proof of Part I, we need the following lemma:

Lemma 8.6.6. *Let Ω be a bounded domain in $\mathbb{R}^n$. Let $1\leq q_1\leq q_0$ and suppose*

$$W^{m,p}(\Omega)\hookrightarrow L^{q_0}(\Omega), \tag{8.138}$$

$$W^{m,p}(\Omega)\hookrightarrow L^{q_1}. \tag{8.139}$$

Suppose also that (8.139) is compact. If $q_1\leq q<q_0$, then the imbedding

$$W^{m,p}\hookrightarrow L^q(\Omega) \tag{8.140}$$

is compact.

Proof. Define $\lambda=q_1(q_0-q)/(q(q_0-q_1))$ and $\mu=q_0(q-q_1)/(q(q_0-q_1))$. We have that $\lambda>0$ and $\mu\geq 0$. From the Hölder inequality and (8.138) there exists $K\in\mathbb{R}^+$ such that

$$\|u\|_{0,q,\Omega} \leq \|u\|_{0,q_1,\Omega}^{\lambda} \|u\|_{0,q_0,\Omega}^{\mu} \leq K \|u\|_{0,q_1,\Omega}^{\lambda} \|u\|_{m,p,\Omega}^{\mu}, \forall u \in W^{m,p}(\Omega). \quad (8.141)$$

Thus considering a sequence $\{u_i\}$ bounded in $W^{m,p}(\Omega)$, since (8.139) is compact there exists a subsequence $\{u_{nk}\}$ that converges and is therefore a Cauchy sequence in $L^{q_1}(\Omega)$. From (8.141), $\{u_{nk}\}$ is also a Cauchy sequence in $L^q(\Omega)$, so that (8.140) is compact.

Proof of Part I. Consider $j = 0$. Define $q_0 = np/(n - mp)$. To prove the imbedding

$$W^{m,p}(\Omega) \hookrightarrow L^q(\Omega), \; 1 \leq q < q_0, \quad (8.142)$$

is compact, by last lemma it suffices to do so only for $q = 1$. For $k \in \mathbb{N}$, define

$$\Omega_k = \{x \in \Omega \mid \text{dist}(x, \partial\Omega) > 2/k\}. \quad (8.143)$$

Suppose A is a bounded set of functions in $W^{m,p}(\Omega)$, that is, suppose there exists $K_1 > 0$ such that

$$\|u\|_{W^{m,p}(\Omega)} < K_1, \forall u \in A.$$

Also, suppose given $\varepsilon > 0$, and define, for $u \in W^{m,p}(\Omega)$, $\tilde{u}(x) = u(x)$ if $x \in \Omega$, $\tilde{u}(x) = 0$, if $x \in \mathbb{R}^n \setminus \Omega$. Fix $u \in A$. From the Hölder inequality and considering that $W^{m,p}(\Omega) \to L^{q_0}(\Omega)$, we have

$$\begin{aligned} \int_{\Omega - \Omega_k} |u(x)| dx &\leq \left\{ \int_{\Omega - \Omega_k} |u(x)|^{q_0} dx \right\}^{1/q_0} \left\{ \int_{\Omega - \Omega_k} 1 dx \right\}^{1 - 1/q_0} \\ &\leq K_1 \|u\|_{m,p,\Omega} [vol(\Omega - \Omega_k)]^{1 - 1/q_0}. \end{aligned} \quad (8.144)$$

Thus, since A is bounded in $W^{m,p}(\Omega)$, there exists $K_0 \in \mathbb{N}$ such that if $k \geq K_0$, then

$$\int_{\Omega - \Omega_k} |u(x)| dx < \varepsilon, \forall u \in A, \quad (8.145)$$

and, now fixing a not relabeled $k > K_0$, we get

$$\int_{\Omega - \Omega_k} |\tilde{u}(x + h) - \tilde{u}(x)| dx < 2\varepsilon, \forall u \in A, \forall h \in \mathbb{R}^n. \quad (8.146)$$

Observe that if $|h| < 1/k$, then $x + th \in \Omega_{2k}$ provided $x \in \Omega_k$ and $0 \leq t \leq 1$. If $u \in C^\infty(\Omega)$, we have that

$$\begin{aligned} \int_{\Omega_k} |u(x + h) - u(x)| &\leq \int_{\Omega_k} dx \int_0^1 \left|\frac{du(x + th)}{dt}\right| dt \\ &\leq |h| \int_0^1 dt \int_{\Omega_{2k}} |\nabla u(y)| dy \leq |h| \|u\|_{1,1,\Omega} \\ &\leq K_2 |h| \|u\|_{m,p,\Omega}. \end{aligned} \quad (8.147)$$

Since $C^\infty(\Omega)$ is dense in $W^{m,p}(\Omega)$, from above for $|h|$ sufficiently small,

$$\int_\Omega |\tilde{u}(x+h) - \tilde{u}(x)|dx < 3\varepsilon, \forall u \in A. \tag{8.148}$$

From Theorem 8.4.6, A is relatively compact in $L^1(\Omega)$, and therefore the imbedding indicated (8.142) is compact for $q = 1$. This completes the proof.

Part II
Variational Convex Analysis

Chapter 9
Basic Concepts on the Calculus of Variations

9.1 Introduction to the Calculus of Variations

We emphasize the main references for this chapter are [37, 38, 68].

Here we recall that a functional is a function whose co-domain is the real set. We denote such functionals by $F : U \to \mathbb{R}$, where U is a Banach space. In our work format, we consider the special cases:

1. $F(u) = \int_\Omega f(x, u, \nabla u)\, dx$, where $\Omega \subset \mathbb{R}^n$ is an open, bounded, and connected set.
2. $F(u) = \int_\Omega f(x, u, \nabla u, D^2 u)\, dx$, here

$$Du = \nabla u = \left\{ \frac{\partial u_i}{\partial x_j} \right\}$$

and

$$D^2 u = \{D^2 u_i\} = \left\{ \frac{\partial^2 u_i}{\partial x_k \partial x_l} \right\},$$

for $i \in \{1, \ldots, N\}$ and $j, k, l \in \{1, \ldots, n\}$.

Also, $f : \overline{\Omega} \times \mathbb{R}^N \times \mathbb{R}^{N \times n} \to \mathbb{R}$ is denoted by $f(x, s, \xi)$ and we assume

1.
$$\frac{\partial f(x, s, \xi)}{\partial s}$$
and
2.
$$\frac{\partial f(x, s, \xi)}{\partial \xi}$$
are continuous $\forall (x, s, \xi) \in \overline{\Omega} \times \mathbb{R}^N \times \mathbb{R}^{N \times n}$.

Remark 9.1.1. We also recall that the notation $\nabla u = Du$ may be used.

Now we define our general problem, namely problem $\mathscr{P}$ where

F. Botelho, *Functional Analysis and Applied Optimization in Banach Spaces: Applications to Non-Convex Variational Models*, DOI 10.1007/978-3-319-06074-3_9,

$$\text{Problem } \mathscr{P}: \text{ minimize } F(u) \text{ on } U,$$

that is, to find $u_0 \in U$ such that

$$F(u_0) = \min_{u \in U}\{F(u)\}.$$

At this point, we introduce some essential definitions.

Definition 9.1.2 (Space of Admissible Variations). Given $F : U \to \mathbb{R}$ we define the space of admissible variations for F, denoted by $\mathscr{V}$ as

$$\mathscr{V} = \{\varphi \mid u + \varphi \in U, \forall u \in U\}.$$

For example, for $F : U \to \mathbb{R}$ given by

$$F(u) = \frac{1}{2}\int_\Omega \nabla u \cdot \nabla u \, dx - \langle u, f\rangle_U,$$

where $\Omega \subset \mathbb{R}^3$ and

$$U = \{u \in W^{1,2}(\Omega) \mid u = \hat{u} \text{ on } \partial\Omega\}$$

we have

$$\mathscr{V} = W_0^{1,2}(\Omega).$$

Observe that in this example U is a subset of a Banach space.

Definition 9.1.3 (Local Minimum). Given $F : U \to \mathbb{R}$, we say that $u_0 \in U$ is a local minimum for F if there exists $\delta > 0$ such that

$$F(u) \geq F(u_0), \forall u \in U, \text{ such that } \|u - u_0\|_U < \delta,$$

or equivalently

$$F(u_0 + \varphi) \geq F(u_0), \forall \varphi \in \mathscr{V}, \text{ such that } \|\varphi\|_U < \delta.$$

Definition 9.1.4 (Gâteaux Variation). Given $F : U \to \mathbb{R}$ we define the Gâteaux variation of F at $u \in U$ on the direction $\varphi \in \mathscr{V}$, denoted by $\delta F(u, \varphi)$ as

$$\delta F(u, \varphi) = \lim_{\varepsilon \to 0} \frac{F(u + \varepsilon\varphi) - F(u)}{\varepsilon},$$

if such a limit is well defined. Furthermore, if there exists $u^* \in U^*$ such that

$$\delta F(u, \varphi) = \langle \varphi, u^*\rangle_U, \forall \varphi \in U,$$

we say that F is Gâteaux differentiable at $u \in U$, and $u^* \in U^*$ is said to be the Gâteaux derivative of F at u. Finally we denote

$$u^* = \delta F(u) \; or \; u^* = \frac{\partial F(u)}{\partial u}.$$

9.2 Evaluating the Gâteaux Variations

Consider $F : U \to \mathbb{R}$ such that

$$F(u) = \int_\Omega f(x, u, \nabla u)\, dx$$

where the hypothesis indicated in the last section is assumed. Consider $u \in C^1(\bar{\Omega}; \mathbb{R}^N)$ and $\varphi \in C_c^1(\bar{\Omega}; \mathbb{R}^N)$ and let us evaluate $\delta F(u, \varphi)$:

From Definition 9.1.4,

$$\delta F(u, \varphi) = \lim_{\varepsilon \to 0} \frac{F(u + \varepsilon\varphi) - F(u)}{\varepsilon}.$$

Observe that

$$\lim_{\varepsilon \to 0} \frac{f(x, u + \varepsilon\varphi, \nabla u + \varepsilon\nabla\varphi) - f(x, u, \nabla u)}{\varepsilon} = \frac{\partial f(x, u, \nabla u)}{\partial s} \cdot \varphi + \frac{\partial f(x, u, \nabla u)}{\partial \xi} \cdot \nabla\varphi.$$

Define

$$G(x, u, \varphi, \varepsilon) = \frac{f(x, u + \varepsilon\varphi, \nabla u + \varepsilon\nabla\varphi) - f(x, u, \nabla u)}{\varepsilon},$$

and

$$\tilde{G}(x, u, \varphi) = \frac{\partial f(x, u, \nabla u)}{\partial s} \cdot \varphi + \frac{\partial f(x, u, \nabla u)}{\partial \xi} \cdot \nabla\varphi.$$

Thus we have

$$\lim_{\varepsilon \to 0} G(x, u, \varphi, \varepsilon) = \tilde{G}(x, u, \varphi).$$

Now we will show that

$$\lim_{\varepsilon \to 0} \int_\Omega G(x, u, \varphi, \varepsilon)\, dx = \int_\Omega \tilde{G}(x, u, \varphi)\, dx.$$

Suppose to obtain contradiction that we do not have

$$\lim_{\varepsilon \to 0} \int_\Omega G(x, u, \varphi, \varepsilon)\, dx = \int_\Omega \tilde{G}(x, u, \varphi)\, dx.$$

Hence, there exists $\varepsilon_0 > 0$ such that for each $n \in \mathbb{N}$ there exists $0 < \varepsilon_n < 1/n$ such that

$$\left| \int_\Omega G(x, u, \varphi, \varepsilon_n)\, dx - \int_\Omega \tilde{G}(x, u, \varphi)\, dx \right| > \varepsilon_0. \tag{9.1}$$

Define

$$c_n = \max_{x \in \overline{\Omega}} \{|G(x,u(x),\varphi(x),\varepsilon_n) - \tilde{G}(x,u(x),\varphi(x))|\}.$$

Since the function in question is continuous on the compact set $\overline{\Omega}$, $\{x_n\}$ is well defined. Also from the fact that $\overline{\Omega}$ is compact, there exists a subsequence $\{x_{n_j}\}$ and $x_0 \in \bar{\Omega}$ such that

$$\lim_{j \to +\infty} x_{n_j} = x_0.$$

Thus

$$\begin{aligned} \lim_{j \to +\infty} c_{n_j} &= c_0 \\ &= \lim_{j \to +\infty} \{|G(x_{n_j}, u(x_{n_j}), \varphi(x_{n_j}), \varepsilon_{n_j}) - \tilde{G}(x_0, u(x_0), \varphi(x_0))|\} = 0. \end{aligned}$$

Therefore there exists $j_0 \in \mathbb{N}$ such that if $j > j_0$, then

$$c_{n_j} < \varepsilon_0 / |\Omega|.$$

Thus, if $j > j_0$, we have

$$\begin{aligned} &\left| \int_\Omega G(x,u,\varphi,\varepsilon_{n_j})\,dx - \int_\Omega \tilde{G}(x,u,\varphi)\,dx \right| \\ &\leq \int_\Omega |G(x,u,\varphi,\varepsilon_{n_j}) - \tilde{G}(x,u,\varphi)|\,dx \leq c_{n_j}|\Omega| < \varepsilon_0, \end{aligned} \tag{9.2}$$

which contradicts (9.1). Hence, we may write

$$\lim_{\varepsilon \to 0} \int_\Omega G(x,u,\varphi,\varepsilon)\,dx = \int_\Omega \tilde{G}(x,u,\varphi)\,dx,$$

that is,

$$\delta F(u,\varphi) = \int_\Omega \left\{ \frac{\partial f(x,u,\nabla u)}{\partial s} \cdot \varphi + \frac{\partial f(x,u,\nabla u)}{\partial \xi} \cdot \nabla \varphi \right\} dx.$$

Theorem 9.2.1 (Fundamental Lemma of Calculus of Variations). *Consider an open set $\Omega \subset \mathbb{R}^n$ and $u \in L^1_{loc}(\Omega)$ such that*

$$\int_\Omega u\varphi\,dx = 0, \forall \varphi \in C_c^\infty(\Omega).$$

Then $u = 0$, a.e. in Ω.

Remark 9.2.2. Of course a similar result is valid for the vectorial case. A proof of such a result was given in Chap. 8.

Theorem 9.2.3 (Necessary Conditions for a Local Minimum). *Suppose $u \in U$ is a local minimum for a Gâteaux differentiable $F : U \to \mathbb{R}$. Then*

$$\delta F(u,\varphi) = 0, \forall \varphi \in \mathscr{V}.$$

Proof. Fix $\varphi \in \mathscr{V}$. Define $\phi(\varepsilon) = F(u+\varepsilon\varphi)$. Since by hypothesis ϕ is differentiable and attains a minimum at $\varepsilon = 0$, from the standard necessary condition $\phi'(0) = 0$, we obtain $\phi'(0) = \delta F(u,\varphi) = 0$.

Theorem 9.2.4. *Consider the hypotheses stated in Section 9.1 on $F : U \to \mathbb{R}$. Suppose F attains a local minimum at $u \in C^2(\bar{\Omega};\mathbb{R}^N)$ and additionally assume that $f \in C^2(\overline{\Omega}, \mathbb{R}^N, \mathbb{R}^{N\times n})$. Then the necessary conditions for a local minimum for F are given by the Euler–Lagrange equations:*

$$\frac{\partial f(x,u,\nabla u)}{\partial s} - div\left(\frac{\partial f(x,u,\nabla u)}{\partial \xi}\right) = \theta, \text{ in } \Omega.$$

Proof. From Theorem 9.2.3, the necessary condition stands for $\delta F(u,\varphi) = 0, \forall \varphi \in \mathscr{V}$. From the above this implies, after integration by parts

$$\int_\Omega \left(\frac{\partial f(x,u,\nabla u)}{\partial s} - div\left(\frac{\partial f(x,u,\nabla u)}{\partial \xi}\right)\right)\cdot \varphi \, dx = 0,$$

$$\forall \varphi \in C_c^\infty(\Omega, \mathbb{R}^N).$$

The result then follows from the fundamental lemma of calculus of variations.

9.3 The Gâteaux Variation: A More General Case

Theorem 9.3.1. *Consider the functional $F : U \to \mathbb{R}$, where*

$$U = \{u \in W^{1,2}(\Omega, \mathbb{R}^N) \mid u = u_0 \text{ in } \partial\Omega\}.$$

Suppose

$$F(u) = \int_\Omega f(x,u,\nabla u)\, dx,$$

where $f : \Omega \times \mathbb{R}^N \times \mathbb{R}^{N\times n}$ is such that for each $K > 0$ there exists $K_1 > 0$ which does not depend on x such that

$$|f(x,s_1,\xi_1) - f(x,s_2,\xi_2)| < K_1(|s_1 - s_2| + |\xi_1 - \xi_2|)$$

$$\forall s_1, s_2 \in \mathbb{R}^N, \xi_1, \xi_2 \in \mathbb{R}^{N\times n}, \text{ such that } |s_1| < K, |s_2| < K,$$

$$|\xi_1| < K, |\xi_2| < K.$$

Also assume the hypotheses of Section 9.1 except for the continuity of derivatives of f. Under such assumptions, for each $u \in C^1(\overline{\Omega};\mathbb{R}^N)$ and $\varphi \in C_c^\infty(\Omega;\mathbb{R}^N)$, we have

$$\delta F(u,\varphi) = \int_\Omega \left\{\frac{\partial f(x,u,\nabla u)}{\partial s}\cdot \varphi + \frac{\partial f(x,u,\nabla u)}{\partial \xi}\cdot \nabla\varphi\right\} dx.$$

Proof. From Definition 9.1.4,

$$\delta F(u,\varphi) = \lim_{\varepsilon \to 0} \frac{F(u+\varepsilon\varphi) - F(u)}{\varepsilon}.$$

Observe that

$$\lim_{\varepsilon \to 0} \frac{f(x,u+\varepsilon\varphi,\nabla u+\varepsilon\nabla\varphi) - f(x,u,\nabla u)}{\varepsilon}$$
$$= \frac{\partial f(x,u,\nabla u)}{\partial s}\cdot\varphi + \frac{\partial f(x,u,\nabla u)}{\partial \xi}\cdot\nabla\varphi, \text{ a.e in } \Omega.$$

Define

$$G(x,u,\varphi,\varepsilon) = \frac{f(x,u+\varepsilon\varphi,\nabla u+\varepsilon\nabla\varphi) - f(x,u,\nabla u)}{\varepsilon},$$

and

$$\tilde{G}(x,u,\varphi) = \frac{\partial f(x,u,\nabla u)}{\partial s}\cdot\varphi + \frac{\partial f(x,u,\nabla u)}{\partial \xi}\cdot\nabla\varphi.$$

Thus we have

$$\lim_{\varepsilon \to 0} G(x,u,\varphi,\varepsilon) = \tilde{G}(x,u,\varphi), \text{ a.e in } \Omega.$$

Now we will show that

$$\lim_{\varepsilon \to 0} \int_\Omega G(x,u,\varphi,\varepsilon)\,dx = \int_\Omega \tilde{G}(x,u,\varphi)\,dx.$$

It suffices to show that (we do not provide details here)

$$\lim_{n \to \infty} \int_\Omega G(x,u,\varphi,1/n)\,dx = \int_\Omega \tilde{G}(x,u,\varphi)\,dx.$$

Observe that for an appropriate $K > 0$, we have

$$|G(x,u,\varphi,1/n)| \leq K(|\varphi| + |\nabla\varphi|), \text{ a.e. in } \Omega. \tag{9.3}$$

By the Lebesgue dominated convergence theorem, we obtain

$$\lim_{n \to +\infty} \int_\Omega G(x,u,\varphi,1/(n))\,dx = \int_\Omega \tilde{G}(x,u,\varphi)\,dx,$$

that is,

$$\delta F(u,\varphi) = \int_\Omega \left\{ \frac{\partial f(x,u,\nabla u)}{\partial s}\cdot\varphi + \frac{\partial f(x,u,\nabla u)}{\partial \xi}\cdot\nabla\varphi \right\} dx.$$

9.4 Fréchet Differentiability

In this section we introduce a very important definition, namely, Fréchet differentiability.

Definition 9.4.1. Let U, Y be Banach spaces and consider a transformation $T : U \to Y$. We say that T is Fréchet differentiable at $u \in U$ if there exists a bounded linear transformation $T'(u) : U \to Y$ such that

$$\lim_{v \to \theta} \frac{\|T(u+v) - T(u) - T'(u)(v)\|_Y}{\|v\|_U} = 0, \; v \neq \theta.$$

In such a case $T'(u)$ is called the Fréchet derivative of T at $u \in U$.

9.5 Elementary Convexity

In this section we develop some proprieties concerning elementary convexity.

Definition 9.5.1. A function $f : \mathbb{R}^n \to \mathbb{R}$ is said to be convex if

$$f(\lambda x + (1-\lambda)y) \leq \lambda f(x) + (1-\lambda) f(y), \forall x, y \in \mathbb{R}^n, \lambda \in [0,1].$$

Proposition 9.5.2. *If $f : \mathbb{R}^n \to \mathbb{R}$ is convex and differentiable, then*

$$f(y) - f(x) \geq \langle f'(x), y - x \rangle_{\mathbb{R}^n}, \forall x, y \in \mathbb{R}^n.$$

Proof. Pick $x, y \in \mathbb{R}^n$. By hypothesis

$$f((1-\lambda)x + \lambda y) \leq (1-\lambda) f(x) + \lambda f(y), \forall \lambda \in [0,1].$$

Thus

$$\frac{f(x + \lambda(y-x)) - f(x)}{\lambda} \leq f(y) - f(x), \forall \lambda \in (0,1].$$

Letting $\lambda \to 0^+$ we obtain

$$f(y) - f(x) \geq \langle f'(x), y - x \rangle_{\mathbb{R}^n}.$$

Since $x, y \in \mathbb{R}^n$ are arbitrary, the proof is complete.

Proposition 9.5.3. *Let $f : \mathbb{R}^n \to \mathbb{R}$ be a differentiable function. If*

$$f(y) - f(x) \geq \langle f'(x), y - x \rangle_{\mathbb{R}^n}, \forall x, y \in \mathbb{R}^n,$$

then f is convex.

Proof. Define $f^*(x^*)$ by

$$f(x^*) = \sup_{x\in\mathbb{R}^n}\{\langle x,x^*\rangle_{\mathbb{R}^n} - f(x)\}.$$

Such a function f^* is called the Fenchel conjugate of f. Observe that by hypothesis,

$$f^*(f'(x)) = \sup_{y\in\mathbb{R}^n}\{\langle y,f'(x)\rangle_{\mathbb{R}^n} - f(y)\} = \langle x,f'(x)\rangle_{\mathbb{R}^n} - f(x). \tag{9.4}$$

On the other hand

$$f^*(x^*) \geq \langle x,x^*\rangle_{\mathbb{R}^n} - f(x), \forall x,x^* \in \mathbb{R}^n,$$

that is,

$$f(x) \geq \langle x,x^*\rangle_{\mathbb{R}^n} - f^*(x^*), \forall x,x^* \in \mathbb{R}^n.$$

Observe that from (9.4)

$$f(x) = \langle x,f'(x)\rangle_{\mathbb{R}^n} - f^*(f'(x))$$

and thus

$$f(x) = \sup_{x^*\in\mathbb{R}^n}\{\langle x,x^*\rangle_{\mathbb{R}^n} - f(x^*)\}, \forall x \in \mathbb{R}^n.$$

Pick $x,y \in \mathbb{R}^n$ and $\lambda \in [0,1]$. Thus, we may write

$$\begin{aligned}
f(\lambda x+(1-\lambda)y) &= \sup_{x^*\in\mathbb{R}^n}\{\langle \lambda x+(1-\lambda)y,x^*\rangle_{\mathbb{R}^n} - f^*(x^*)\} \\
&= \sup_{x^*\in\mathbb{R}^n}\{\lambda\langle x,x^*\rangle_{\mathbb{R}^n} + (1-\lambda)\langle y,x^*\rangle_{\mathbb{R}^n} - \lambda f^*(x^*) \\
&\quad -(1-\lambda)f^*(x^*)\} \\
&\leq \lambda\{\sup_{x^*\in\mathbb{R}^n}\{\langle x,x^*\rangle_{\mathbb{R}^n} - f^*(x^*)\}\} \\
&\quad +(1-\lambda)\{\sup_{x^*\in\mathbb{R}^n}\{\langle y,x^*\rangle_{\mathbb{R}^n} - f^*(x^*)\}\} \\
&= \lambda f(x) + (1-\lambda)f(y).
\end{aligned} \tag{9.5}$$

Since $x,y \in \mathbb{R}^n$ and $\lambda \in [0,1]$ are arbitrary, we have that f is convex.

Corollary 9.5.4. *Let $f : \mathbb{R}^n \to \mathbb{R}$ be twice differentiable and*

$$\left\{\frac{\partial^2 f(x)}{\partial x_i \partial x_j}\right\},$$

positive definite, for all $x \in \mathbb{R}^n$. Then f is convex.

Proof. Pick $x,y \in \mathbb{R}^n$. Using Taylor's expansion we obtain

$$f(y) = f(x) + \langle f'(x), y-x\rangle_{\mathbb{R}^n} + \sum_{i=1}^n\sum_{j=1}^n \frac{\partial^2 f(\bar{x})}{\partial x_i \partial x_j}(y_i - x_i)(y_j - x_j),$$

for $\bar{x} = \lambda x + (1-\lambda)y$ (for some $\lambda \in [0,1]$). From the hypothesis we obtain

$$f(y) - f(x) - \langle f'(x), y - x \rangle_{\mathbb{R}^n} \geq 0.$$

Since $x, y \in \mathbb{R}^n$ are arbitrary, the proof is complete.

Similarly we may obtain the following result.

Corollary 9.5.5. *Let U be a Banach space. Consider $F : U \to \mathbb{R}$ Gâteaux differentiable. Then F is convex if and only if*

$$F(v) - F(u) \geq \langle F'(u), v - u \rangle_U, \forall u, v \in U.$$

Definition 9.5.6 (The Second Variation). Let U be a Banach space. Suppose $F : U \to \mathbb{R}$ is a Gâteaux differentiable functional. Given φ, $\eta \in \mathscr{V}$, we define the second variation of F at u, relating the directions φ, η, denoted by

$$\delta^2 F(u, \varphi, \eta),$$

by

$$\delta^2 F(u, \varphi, \eta) = \lim_{\varepsilon \to 0} \frac{\delta F(u + \varepsilon\eta, \varphi) - \delta F(u, \varphi)}{\varepsilon}.$$

If such a limit exists $\forall \varphi$, $\eta \in \mathscr{V}$, we say that F is twice Gâteaux differentiable at u. Finally, if $\eta = \varphi$, we denote $\delta^2 F(u, \varphi, \eta) = \delta^2 F(u, \varphi)$.

Corollary 9.5.7. *Let U be a Banach space. Suppose $F : U \to \mathbb{R}$ is a twice Gâteaux differentiable functional and that*

$$\delta^2 F(u, \varphi) \geq 0, \forall u \in U, \varphi \in \mathscr{V}.$$

Then, F is convex.

Proof. Pick $u, v \in U$. Define $\phi(\varepsilon) = F(u + \varepsilon(v - u))$. By hypothesis, ϕ is twice differentiable, so that

$$\phi(1) = \phi(0) + \phi'(0) + \phi''(\tilde{\varepsilon})/2,$$

where $|\tilde{\varepsilon}| \leq 1$. Thus

$$F(v) = F(u) + \delta F(u, v - u) + \delta^2 F(u + \tilde{\varepsilon}(v - u), v - u)/2.$$

Therefore, by hypothesis,

$$F(v) \geq F(u) + \delta F(u, v - u).$$

Since F is Gâteaux differentiable, we obtain

$$F(v) > F(u) + \langle F'(u), v - u \rangle_U.$$

Being $u, v \in U$ arbitrary, the proof is complete.

Corollary 9.5.8. *Let U be a Banach space. Let $F : U \to \mathbb{R}$ be a convex Gâteaux differentiable functional. If $F'(u) = \theta$, then*

$$F(v) \geq F(u), \forall v \in U,$$

that is, $u \in U$ is a global minimizer for F.

Proof. Just observe that

$$F(v) \geq F(u) + \langle F'(u), v-u \rangle_U, \forall u, v \in U.$$

Therefore, from $F'(u) = \theta$, we obtain

$$F(v) \geq F(u), \forall v \in U.$$

Theorem 9.5.9 (Sufficient Condition for a Local Minimum). *Let U be a Banach space. Suppose $F : U \to \mathbb{R}$ is a twice Gâteaux differentiable functional at a neighborhood of u_0, so that*

$$\delta F(u_0) = \theta$$

and

$$\delta^2 F(u, \varphi) \geq 0, \forall u \in B_r(u_0),\ \varphi \in \mathscr{V},$$

for some $r > 0$. Under such hypotheses, we have

$$F(u_0) \leq F(u_0 + \varepsilon\varphi), \forall\ \varepsilon,\ \varphi \text{ such that } |\varepsilon| < \min\{r, 1\},\ \|\varphi\|_U < 1.$$

Proof. Fix $\varphi \in \mathscr{V}$ such that $\|\varphi\|_U < 1$. Define

$$\phi(\varepsilon) = F(u_0 + \varepsilon\varphi).$$

Observe that for $|\varepsilon| < \min\{r, 1\}$, for some $\tilde{\varepsilon}$ such that $|\tilde{\varepsilon}| \leq |\varepsilon|$, we have

$$\begin{aligned} \phi(\varepsilon) &= \phi(0) + \phi'(0)\varepsilon + \phi''(\tilde{\varepsilon})\varepsilon^2/2 \\ &= F(u_0) + \varepsilon\langle \varphi, \delta F(u_0) \rangle_U + (\varepsilon^2/2)\delta^2 F(u_0 + \tilde{\varepsilon}\varphi, \varphi) \\ &= F(u_0) + (\varepsilon^2/2)\delta^2 F(u_0 + \tilde{\varepsilon}\varphi, \varphi) \geq F(u_0). \end{aligned}$$

Hence,

$$F(u_0) \leq F(u_0 + \varepsilon\varphi), \forall\ \varepsilon,\ \varphi \text{ such that } |\varepsilon| < r,\ \|\varphi\|_U < 1.$$

The proof is complete.

9.6 The Legendre–Hadamard Condition

Theorem 9.6.1. *If $u \in C^1(\bar{\Omega}; \mathbb{R}^N)$ is such that*

$$\delta^2 F(u, \varphi) \geq 0, \forall \varphi \in C_c^\infty(\Omega, \mathbb{R}^N),$$

then

$$f_{\xi^i_\alpha \xi^k_\beta}(x, u(x), \nabla u(x))\rho^i \rho^k \eta_\alpha \eta_\beta \geq 0, \forall x \in \Omega, \rho \in \mathbb{R}^N, \eta \in \mathbb{R}^n.$$

Such a condition is known as the Legendre-Hadamard condition.

Proof. Suppose

$$\delta^2 F(u, \varphi) \geq 0, \forall \varphi \in C_c^\infty(\Omega; \mathbb{R}^N).$$

We denote $\delta^2 F(u, \varphi)$ by

$$\begin{aligned} \delta^2 F(u, \varphi) = &\int_\Omega a(x) D\varphi(x) \cdot D\varphi(x)\, dx \\ &+ \int_\Omega b(x)\varphi(x) \cdot D\varphi(x)\, dx + \int_\Omega c(x)\varphi(x) \cdot \varphi(x)\, dx, \end{aligned} \tag{9.6}$$

where

$$a(x) = f_{\xi\xi}(x, u(x), Du(x)),$$
$$b(x) = 2f_{s\xi}(x, u(x), Du(x)),$$

and

$$c(x) - f_{ss}(x, u(x), Du(x)).$$

Now consider $v \in C_c^\infty(B_1(0), \mathbb{R}^N)$. Thus given $x_0 \in \Omega$ for λ sufficiently small we have that $\varphi(x) = \lambda v\left(\frac{x-x_0}{\lambda}\right)$ is an admissible direction. Now we introduce the new coordinates $y = (y^1, \ldots, y^n)$ by setting $y = \lambda^{-1}(x - x_0)$ and multiply (9.6) by λ^{-n} to obtain

$$\begin{aligned} &\int_{B_1(0)} \{a(x_0 + \lambda y)Dv(y) \cdot Dv(y) + 2\lambda b(x_0 + \lambda y)v(y) \cdot Dv(y) \\ &\quad + \lambda^2 c(x_0 + \lambda y)v(y) \cdot v(y)\}\, dy > 0, \end{aligned}$$

where $a = \{a_{ij}^{\alpha\beta}\}, b = \{b_{jk}^\beta\}$ and $c = \{c_{jk}\}$. Since a, b and c are continuous, we have

$$a(x_0 + \lambda y)Dv(y) \cdot Dv(y) \to a(x_0)Dv(y) \cdot Dv(y),$$

$$\lambda b(x_0 + \lambda y)v(y) \cdot Dv(y) \to 0,$$

and

$$\lambda^2 c(x_0 + \lambda y)v(y) \cdot v(y) \to 0,$$

uniformly on Ω as $\lambda \to 0$. Thus this limit gives us

$$\int_{B_1(0)} \tilde{f}_{jk}^{\alpha\beta} D_\alpha v^j D_\beta v^k\, dx \geq 0, \forall v \in C_c^\infty(B_1(0); \mathbb{R}^N), \tag{9.7}$$

where

$$\tilde{f}_{jk}^{\alpha\beta} = a_{jk}^{\alpha\beta}(x_0) = f_{\xi^i_\alpha \xi^k_\beta}(x_0, u(x_0), \nabla u(x_0)).$$

Now define $v = (v^1, \ldots, v^N)$, where

$$v^j = \rho^j cos((\eta \cdot y)t)\zeta(y)$$

$$\rho = (\rho^1, \ldots, \rho^N) \in \mathbb{R}^N$$

and

$$\eta = (\eta_1, \ldots, \eta_n) \in \mathbb{R}^n$$

and $\zeta \in C_c^\infty(B_1(0))$. From (9.7) we obtain

$$0 \leq \tilde{f}_{jk}^{\alpha\beta} \rho^j \rho^k \left\{ \int_{B_1(0)} (\eta_\alpha t(-sin((\eta \cdot y)t)\zeta + cos((\eta \cdot y)t)D_\alpha \zeta) \right.$$
$$\left. \cdot (\eta_\beta t(-sin((\eta \cdot y)t)\zeta + cos((\eta \cdot y)t)D_\beta \zeta) \; dy \right\} \tag{9.8}$$

By analogy for

$$v^j = \rho^j sin((\eta \cdot y)t)\zeta(y)$$

we obtain

$$0 \leq \tilde{f}_{jk}^{\alpha\beta} \rho^j \rho^k \left\{ \int_{B_1(0)} (\eta_\alpha t(cos((\eta \cdot y)t)\zeta + sin((\eta \cdot y)t)D_\alpha \zeta) \right.$$
$$\left. \cdot (\eta_\beta t(cos((\eta \cdot y)t)\zeta + sin((\eta \cdot y)t)D_\beta \zeta) \; dy \right\} \tag{9.9}$$

Summing up these last two equations, dividing the result by t^2, and letting $t \to +\infty$ we obtain

$$0 \leq \tilde{f}_{jk}^{\alpha\beta} \rho^j \rho^k \eta_\alpha \eta_\beta \int_{B_1(0)} \zeta^2 \; dy,$$

for all $\zeta \in C_c^\infty(B_1(0))$, which implies

$$0 \leq \tilde{f}_{jk}^{\alpha\beta} \rho^j \rho^k \eta_\alpha \eta_\beta.$$

The proof is complete.

9.7 The Weierstrass Condition for $n = 1$

Here we present the Weierstrass condition for the special case $N \geq 1$ and $n = 1$. We start with a definition.

Definition 9.7.1. We say that $u \in \hat{C}^1([a,b];\mathbb{R}^N)$ if $u : [a,b] \to \mathbb{R}^N$ is continuous in $[a,b]$ and Du is continuous except on a finite set of points in $[a,b]$.

Theorem 9.7.2 (Weierstrass). *Let $\Omega = (a,b)$ and $f : \bar{\Omega} \times \mathbb{R}^N \times \mathbb{R}^N \to \mathbb{R}$ be such that $f_s(x,s,\xi)$ and $f_\xi(x,s,\xi)$ are continuous on $\bar{\Omega} \times \mathbb{R}^N \times \mathbb{R}^N$.*

Define $F : U \to \mathbb{R}$ by

$$F(u) = \int_a^b f(x, u(x), u'(x)) \; dx,$$

where

$$U = \{u \in \hat{C}^1([a,b];\mathbb{R}^N) \mid u(a) = \alpha,\ u(b) = \beta\}.$$

Suppose $u \in U$ minimizes locally F on U, that is, suppose that there exists $\varepsilon_0 > 0$ such that

$$F(u) \leq F(v), \forall v \in U, \text{ such that } \|u - v\|_\infty < \varepsilon_0.$$

Under such hypotheses, we have

$$E(x,u(x),u'(x+),w) \geq 0, \forall x \in [a,b],\ w \in \mathbb{R}^N,$$

and

$$E(x,u(x),u'(x-),w) \geq 0, \forall x \in [a,b],\ w \in \mathbb{R}^N,$$

where

$$u'(x+) = \lim_{h \to 0^+} u'(x+h),$$

$$u'(x-) = \lim_{h \to 0^-} u'(x+h),$$

and

$$E(x,s,\xi,w) = f(x,s,w) - f(x,s,\xi) - f_\xi(x,s,\xi)(w - \xi).$$

Remark 9.7.3. The function E is known as the Weierstrass excess function.

Proof. Fix $x_0 \in (a,b)$ and $w \in \mathbb{R}^N$. Choose $0 < \varepsilon < 1$ and $h > 0$ such that $u + v \in U$ and

$$\|v\|_\infty < \varepsilon_0$$

where $v(x)$ is given by

$$v(x) = \begin{cases} (x - x_0)w, & \text{if } 0 \leq x - x_0 \leq \varepsilon h, \\ \tilde{\varepsilon}(h - x + x_0)w, & \text{if } \varepsilon h \leq x - x_0 \leq h, \\ 0, & \text{otherwise,} \end{cases}$$

where

$$\tilde{\varepsilon} = \frac{\varepsilon}{1 - \varepsilon}.$$

From

$$F(u + v) - F(u) \geq 0$$

we obtain

$$\int_{x_0}^{x_0+h} f(x,u(x) + v(x),u'(x) + v'(x))\,dx - \int_{x_0}^{x_0+h} f(x,u(x),u'(x))\,dx \geq 0. \tag{9.10}$$

Define

$$\tilde{x} = \frac{x - x_0}{h},$$

so that

$$d\tilde{x} = \frac{dx}{h}.$$

From (9.10) we obtain

$$\begin{aligned} h\int_0^1 & f(x_0+\tilde{x}h, u(x_0+\tilde{x}h)+v(x_0+\tilde{x}h), u'(x_0+\tilde{x}h)+v'(x_0+\tilde{x}h)\, d\tilde{x} \\ & -h\int_0^1 f(x_0+\tilde{x}h, u(x_0+\tilde{x}h), u'(x_0+\tilde{x}h))\, d\tilde{x} \geq 0. \end{aligned} \tag{9.11}$$

where the derivatives are related to x.

Therefore

$$\begin{aligned} & \int_0^{\varepsilon} f(x_0+\tilde{x}h, u(x_0+\tilde{x}h)+v(x_0+\tilde{x}h), u'(x_0+\tilde{x}h)+w)\, d\tilde{x} \\ & \quad -\int_0^{\varepsilon} f(x_0+\tilde{x}h, u(x_0+\tilde{x}h), u'(x_0+\tilde{x}h))\, d\tilde{x} \\ & \quad +\int_{\varepsilon}^1 f(x_0+\tilde{x}h, u(x_0+\tilde{x}h)+v(x_0+\tilde{x}h), u'(x_0+\tilde{x}h)-\tilde{\varepsilon}w)\, d\tilde{x} \\ & \quad -\int_{\varepsilon}^1 f(x_0+\tilde{x}h, u(x_0+\tilde{x}h), u'(x_0+\tilde{x}h))\, d\tilde{x} \\ & \geq 0. \end{aligned} \tag{9.12}$$

Letting $h \to 0$ we obtain

$$\begin{aligned} & \varepsilon(f(x_0, u(x_0), u'(x_0+)+w) - f(x_0, u(x_0), u'(x_0+)) \\ & \quad +(1-\varepsilon)(f(x_0, u(x_0), u'(x_0+)-\tilde{\varepsilon}w) - f(x_0, u(x_0), u'(x_0+))) \geq 0. \end{aligned}$$

Hence, by the mean value theorem, we get

$$\begin{aligned} & \varepsilon(f(x_0, u(x_0), u'(x_0+)+w) - f(x_0, u(x_0), u'(x_0+)) \\ & \quad -(1-\varepsilon)\tilde{\varepsilon}(f_{\xi}(x_0, u(x_0), u'(x_0+)+\rho(\tilde{\varepsilon})w)) \cdot w \geq 0. \end{aligned} \tag{9.13}$$

Dividing by ε and letting $\varepsilon \to 0$, so that $\tilde{\varepsilon} \to 0$ and $\rho(\tilde{\varepsilon}) \to 0$, we finally obtain

$$\begin{aligned} & f(x_0, u(x_0), u'(x_0+)+w) - f(x_0, u(x_0), u'(x_0+)) \\ & \quad -f_{\xi}(x_0, u(x_0), u'(x_0+)) \cdot w \geq 0. \end{aligned}$$

Similarly we may get

$$\begin{aligned} & f(x_0, u(x_0), u'(x_0-)+w) - f(x_0, u(x_0), u'(x_0-)) \\ & \quad - f_{\xi}(x_0, u(x_0), u'(x_0-)) \cdot w \geq 0. \end{aligned}$$

Since $x_0 \in [a,b]$ and $w \in \mathbb{R}^N$ are arbitrary, the proof is complete.

9.8 The Weierstrass Condition: The General Case

In this section we present a proof for the Weierstrass necessary condition for $N \geq 1, n \geq 1$. Such a result may be found in similar form in [37].

Theorem 9.1. *Assume $u \in C^1(\overline{\Omega};\mathbb{R}^N)$ is a point of strong minimum for a Fréchet differentiable functional $F : U \to \mathbb{R}$ that is, in particular, there exists $\varepsilon > 0$ such that*

$$F(u+\varphi) \geq F(u),$$

for all $\varphi \in C_c^\infty(\Omega;\mathbb{R}^n)$ such that

$$\|\varphi\|_\infty < \varepsilon.$$

Here

$$F(u) = \int_\Omega f(x,u,Du)\,dx,$$

where we recall to have denoted

$$Du = \nabla u = \left\{\frac{\partial u_i}{\partial x_j}\right\}.$$

Under such hypotheses, for all $x \in \Omega$ and each rank-one matrix $\eta = \{\rho_i\beta^\alpha\} = \{\rho \otimes \beta\}$, we have that

$$E(x,u(x),Du(x),Du(x)+\rho\otimes\beta) \geq 0,$$

where

$$\begin{aligned} &E(x,u(x),Du(x),Du(x)+\rho\otimes\beta)\\ &= f(x,u(x),Du(x)+\rho\otimes\beta) - f(x,u(x),Du(x))\\ &\quad -\rho^i\beta_\alpha f_{\xi_\alpha^i}(x,u(x),Du(x)). \end{aligned} \tag{9.14}$$

Proof. Since u is a point of local minimum for F, we have that

$$\delta F(u;\varphi) = 0, \forall \varphi \in C_c^\infty(\Omega;\mathbb{R}^N),$$

that is,

$$\int_\Omega (\varphi \cdot f_s(x,u(x),Du(x)) + D\varphi \cdot f_\xi(x,u(x),Du(x))\,dx = 0,$$

and hence,

$$\begin{aligned} &\int_\Omega (f(x,u(x),Du(x)+D\varphi(x)) - f(x,u(x),Du(x))\,dx\\ &\quad -\int_\Omega (\varphi(x)\cdot f_s(x,u(x),Du(x)) - D\varphi(x)\cdot f_\xi(x,u(x),Du(x))\,dx\\ &\geq 0, \end{aligned} \tag{9.15}$$

$\forall \varphi \in \mathscr{V}$, where

$$\mathscr{V} = \{\varphi \in C_c^\infty(\Omega;\mathbb{R}^N) \ : \ \|\varphi\|_\infty < \varepsilon\}.$$

Choose a unit vector $e \in \mathbb{R}^n$ and write

$$x = (x \cdot e)e + \overline{x},$$

where

$$\overline{x} \cdot e = 0.$$

Denote $D_e v = Dv \cdot e$ and let $\rho = (\rho_1, \ldots, \rho_N) \in \mathbb{R}^N$.
Also, let x_0 be any point of Ω. Without loss of generality assume $x_0 = 0$.
Choose $\lambda_0 \in (0,1)$ such that $C_{\lambda_0} \subset \Omega$, where

$$C_{\lambda_0} = \{x \in \mathbb{R}^n \ : \ |x \cdot e| \leq \lambda_0 \text{ and } \|\overline{x}\| \leq \lambda_0\}.$$

Let $\lambda \in (0, \lambda_0)$ and

$$\phi \in C_c((-1,1);\mathbb{R})$$

and choose a sequence

$$\phi_k \in C_c^\infty((-\lambda^2, \lambda);\mathbb{R})$$

which converges uniformly to the Lipschitz function ϕ_λ given by

$$\phi_\lambda = \begin{cases} t + \lambda^2, & \text{if } -\lambda^2 \leq t \leq 0, \\ \lambda(\lambda - t), & \text{if } 0 < t < \lambda \\ 0, & \text{otherwise} \end{cases} \tag{9.16}$$

and such that ϕ_k' converges uniformly to ϕ_λ' on each compact subset of

$$A_\lambda = \{t : -\lambda^2 < t < \lambda, \, t \neq 0\}.$$

We emphasize the choice of $\{\phi_k\}$ may be such that for some $K > 0$ we have $\|\phi\|_\infty < K$, $\|\phi_k\|_\infty < K$ and $\|\phi_k'\|_\infty < K, \forall k \in \mathbb{N}$.

Observe that for any sufficiently small $\lambda > 0$ we have that φ_k defined by

$$\varphi_k(x) = \rho \phi_k(x \cdot e)\phi(|\overline{x}|^2/\lambda^2) \in \mathscr{V}, \forall k \in \mathbb{N}$$

so that letting $k \to \infty$ we obtain that

$$\varphi(x) = \rho \phi_\lambda(x \cdot e)\phi(|\overline{x}|^2/\lambda^2),$$

is such that (9.15) is satisfied.

Moreover,

$$D_e\varphi(x) = \rho \phi_\lambda'(x \cdot e)\phi(|\overline{x}|^2/\lambda^2),$$

and

$$\overline{D}\varphi(x) = \rho \phi_\lambda(x \cdot e)\phi'(|\overline{x}|^2/\lambda^2)2\lambda^{-2}\overline{x},$$

where $\overline{D}$ denotes the gradient relating the variable $\overline{x}$.

Note that for such a $\varphi(x)$, the integrand of (9.15) vanishes if $x \notin C_\lambda$, where

$$C_\lambda = \{x \in \mathbb{R}^n \,:\, |x \cdot e| \leq \lambda \text{ and } \|\overline{x}\| \leq \lambda\}.$$

Define C_λ^+ and C_λ^- by

$$C_\lambda^- = \{x \in C_\lambda \,:\, x \cdot e \leq 0\},$$

and

$$C_\lambda^+ = \{x \in C_\lambda \,:\, x \cdot e > 0\}.$$

Hence, denoting

$$\begin{aligned} g_k(x) = & \,(f(x,u(x),Du(x)+D\varphi_k(x)) - f(x,u(x),Du(x)) \\ & -(\varphi_k(x) \cdot f_s(x,u(x),Du(x)+D\varphi_k(x) \cdot f_\xi(x,u(x),Du(x)) \end{aligned} \quad (9.17)$$

and

$$\begin{aligned} g(x) = & \,(f(x,u(x),Du(x)+D\varphi(x)) - f(x,u(x),Du(x)) \\ & -(\varphi(x) \cdot f_s(x,u(x),Du(x)+D\varphi(x) \cdot f_\xi(x,u(x),Du(x)) \end{aligned} \quad (9.18)$$

letting $k \to \infty$, using the Lebesgue dominated converge theorem, we obtain

$$\begin{aligned} & \int_{C_\lambda^-} g_k(x)\,dx + \int_{C_\lambda^+} g_k(x)\,dx \\ & \to \int_{C_\lambda^-} g(x)\,dx + \int_{C_\lambda^+} g(x)\,dx \geq 0, \end{aligned} \quad (9.19)$$

Now define

$$y = y^e e + \overline{y},$$

where

$$y^e = \frac{x \cdot e}{\lambda^2},$$

and

$$\overline{y} = \frac{\overline{x}}{\lambda}.$$

The sets C_λ^- and C_λ^+ correspond, concerning the new variables, to the sets B_λ^- and B_λ^+, where

$$B_\lambda^- = \{y \,:\, \|\overline{y}\| \leq 1, \text{ and } -\lambda^{-1} \leq y^e \leq 0\},$$

$$B_\lambda^+ = \{y \,:\, \|\overline{y}\| \leq 1, \text{ and } 0 < y^e \leq \lambda^{-1}\}.$$

Therefore, since $dx = \lambda^{n+1} dy$, multiplying (9.19) by λ^{-n-1}, we obtain

$$\int_{B_1^-} g(x(y))\,dy + \int_{B_\lambda^- \setminus B_1^-} g(x(y))\,dy + \int_{B_\lambda^+} g(x(y))\,dy \geq 0, \quad (9.20)$$

where

$$x = (x \cdot e)e + \bar{x} = \lambda^2 y^e + \lambda \bar{y} \equiv x(y).$$

Observe that

$$D_e\varphi(x) = \begin{cases} \rho\phi(\|\bar{y}\|^2) & \text{if } -1 \leq y^e \leq 0, \\ \rho\phi(\|\bar{y}\|^2)(-\lambda) & \text{if } 0 \leq y^e \leq \lambda^{-1}, \\ 0, & \text{otherwise.} \end{cases} \tag{9.21}$$

Observe also that

$$|g(x(y))| \leq o(\sqrt{|\varphi(x)|^2 + |D\varphi(x)|^2}),$$

so that from the expression of $\varphi(x)$ and $D\varphi(x)$ we obtain, for

$$y \in B_\lambda^+, \text{ or } y \in B_\lambda^- \setminus B_1^-,$$

that

$$|g(x(y))| \leq o(\lambda), \text{ as } \lambda \to 0.$$

Since the Lebesgue measures of B_λ^- and B_λ^+ are bounded by

$$2^{n-1}/\lambda$$

the second and third terms in (9.20) are of $o(1)$ where

$$\lim_{\lambda \to 0^+} o(1)/\lambda = 0,$$

so that letting $\lambda \to 0^+$, considering that

$$x(y) \to 0,$$

and on B_1^- (up to the limit set B)

$$\begin{aligned} g(x(y)) \to\ & f(0,u(0),Du(0) + \rho\phi(\|\bar{y}\|^2)e) \\ & -f(0,u(0),Du(0)) - \\ & \rho\phi(\|\bar{y}\|^2)ef_\xi(0,u(0),Du(0)) \end{aligned} \tag{9.22}$$

we get

$$\begin{aligned} & \int_B [f(0,u(0),Du(0) + \rho\phi(\|\bar{y}\|^2)e) - f(0,u(0),Du(0)) \\ & -\rho\phi(\|\bar{y}\|^2)ef_\xi(0,u(0),Du(0))]\, d\bar{y}_2 \dots d\bar{y}_n \\ \geq\ & 0, \end{aligned} \tag{9.23}$$

where B is an appropriate limit set (we do not provide more details here) such that

$$B = \{y \in \mathbb{R}^n \ : \ y^e = 0 \text{ and } \|\bar{y}\| \leq 1\}.$$

Here we have used the fact that on the set in question,

$$D\varphi(x) \to \rho\phi(\|\bar{y}\|^2)e, \text{ as } \lambda \to 0^+.$$

Finally, inequality (9.23) is valid for a sequence $\{\phi_n\}$ (in place of ϕ) such that

$$0 \leq \phi_n \leq 1 \text{ and } \phi_n(t) = 1, \text{ if } |t| < 1 - 1/n,$$

$\forall n \in \mathbb{N}$.

Letting $n \to \infty$, from (9.23), we obtain

$$\begin{aligned} &f(0,u(0),Du(0)+\rho \otimes e) - f(0,u(0),Du(0)) \\ &-\rho \cdot e f_\xi(0,u(0),Du(0)) \geq 0. \end{aligned} \tag{9.24}$$

9.9 The du Bois–Reymond Lemma

We present now a simpler version of the fundamental lemma of calculus of variations. The result is specific for $n = 1$ and is known as the du Bois–Reymond lemma.

Lemma 9.9.1 (du Bois–Reymond). *If $u \in C([a,b])$ and*

$$\int_a^b u\varphi' \, dx = 0, \forall \varphi \in \mathcal{V},$$

where

$$\mathcal{V} = \{\varphi \in C^1[a,b] \mid \varphi(a) = \varphi(b) = 0\},$$

then there exists $c \in \mathbb{R}$ such that

$$u(x) = c, \forall x \in [a,b].$$

Proof. Define

$$c = \frac{1}{b-a}\int_a^b u(t)\, dt,$$

and

$$\varphi(x) = \int_a^x (u(t) - c)\, dt.$$

Thus we have $\varphi(a) = 0$ and

$$\varphi(b) = \int_a^b u(t)\, dt - c(b-a) = 0.$$

Moreover $\varphi \in C^1([a,b])$ so that

$$\varphi \in \mathcal{V}.$$

Therefore

$$\begin{aligned} 0 &\leq \int_a^b (u(x)-c)^2\,dx \\ &= \int_a^b (u(x)-c)\varphi'(x)\,dx \\ &= \int_a^b u(x)\varphi'(x)\,dx - c[\varphi(x)]_a^b = 0. \end{aligned} \tag{9.25}$$

Thus

$$\int_a^b (u(x)-c)^2\,dx = 0,$$

and being $u(x)-c$ continuous, we finally obtain

$$u(x)-c=0, \forall x \in [a,b].$$

This completes the proof.

Proposition 9.9.2. *If $u, v \in C([a,b])$ and*

$$\int_a^b (u(x)\varphi(x)+v(x)\varphi'(x))\,dx = 0,$$

$\forall \varphi \in \mathscr{V}$, where

$$\mathscr{V} = \{\varphi \in C^1[a,b] \mid \varphi(a)=\varphi(b)=0\},$$

then

$$v \in C^1([a,b])$$

and

$$v'(x) = u(x), \forall x \in [a,b].$$

Proof. Define

$$u_1(x) = \int_a^x u(t)\,dt, \forall x \in [a,b].$$

Thus $u_1 \in C^1([a,b])$ and

$$u_1'(x) = u(x), \forall x \in [a,b].$$

Hence, for $\varphi \in \mathscr{V}$, we have

$$\begin{aligned} 0 &= \int_a^b (u(x)\varphi(x)+v(x)\varphi'(x)\,dx \\ &= \int_a^b (-u_1(x)\varphi'(x)+v\varphi'(x))\,dx + [u_1(x)\varphi(x)]_a^b \\ &= \int_a^b (v(x)-u_1(x))\varphi'(x)\,dx. \end{aligned} \tag{9.26}$$

That is,

$$\int_a^b (v(x) - u_1(x))\varphi'(x)\,dx, \forall \varphi \in \mathscr{V}.$$

By the du Bois–Reymond lemma, there exists $c \in \mathbb{R}$ such that

$$v(x) - u_1(x) = c, \forall x \in [a,b].$$

Hence

$$v = u_1 + c \in C^1([a,b]),$$

so that

$$v'(x) = u_1'(x) = u(x), \forall x \in [a,b].$$

The proof is complete.

9.10 The Weierstrass–Erdmann Conditions

We start with a definition.

Definition 9.10.1. Define $I = [a,b]$. A function $u \in \hat{C}^1([a,b];\mathbb{R}^N)$ is said to be a weak extremal of

$$F(u) = \int_a^b f(x,u(x),u'(x))\,dx,$$

if

$$\int_a^b (f_s(x,u(x),u'(x)) \cdot \varphi + f_\xi(x,u(x),u'(x)) \cdot \varphi'(x))\,dx = 0,$$

$\forall \varphi \in C_c^\infty([a,b];\mathbb{R}^N)$.

Proposition 9.10.2. *For any weak extremal of*

$$F(u) = \int_a^b f(x,u(x),u'(x))\,dx$$

there exists a constant $c \in \mathbb{R}^N$ *such that*

$$f_\xi(x,u(x),u'(x)) = c + \int_a^x f_s(t,u(t),u'(t))\,dt, \forall x \in [a,b]. \tag{9.27}$$

Proof. Fix $\varphi \in C_c^\infty([a,b];\mathbb{R}^N)$. Integration by parts of the extremal condition

$$\delta F(u,\varphi) = 0,$$

implies that

$$\int_a^b f_\xi(x,u(x),u'(x))\cdot\varphi'(x)\,dx \\ -\int_a^b\int_a^x f_s(t,u(t),u'(t))\,dt\cdot\varphi'(x)\,dx=0.$$

Since φ is arbitrary, considering the du Bois-Reymond lemma is valid also for $u \in L^1([a,b])$ and the respective N-dimensional version (see [37], page 32 for details), there exists, $c\in\mathbb{R}^N$ such that

$$f_\xi(x,u(x),u'(x))-\int_a^x f_s(t,u(t),u'(t))\,dt=c,\forall x\in[a,b].$$

The proof is complete.

Theorem 9.10.3 (Weierstrass–Erdmann Corner Conditions). *Let $I=[a,b]$. Suppose $u\in\hat{C}^1([a,b];\mathbb{R}^N)$ is such that*

$$F(u)\leq F(v),\forall v\in\mathscr{C}_r,$$

for some $r>0$ where

$$\mathscr{C}_r=\{v\in\hat{C}^1([a,b];\mathbb{R}^N)\mid v(a)=u(a),\ v(b)=u(b), \\ \text{and } \|u-v\|_\infty<r\}.$$

Let $x_0\in(a,b)$ be a corner point of u. Denoting $u_0=u(x_0)$, $\xi_0^+=u'(x_0+0)$, and $\xi_0^-=u'(x_0-0)$, then the following relations are valid:

1. $f_\xi(x_0,u_0,\xi_0^-)=f_\xi(x_0,u_0,\xi_0^+)$,
2. $$f(x_0,u_0,\xi_0^-)-\xi_0^- f_\xi(x_0,u_0,\xi_0^-) \\ = f(x_0,u_0,\xi_0^+)-\xi_0^+ f_\xi(x_0,u_0,\xi_0^+).$$

Remark 9.10.4. The conditions above are known as the Weierstrass–Erdmann corner conditions.

Proof. Condition (1) is just a consequence of (9.27). For (2), define

$$\tau_\varepsilon(x)=x+\varepsilon\lambda(x),$$

where $\lambda\in C_c^\infty(I)$. Observe that $\tau_\varepsilon(a)=a$ and $\tau_\varepsilon(b)=b$, $\forall\varepsilon>0$. Also $\tau_0(x)=x$. Choose $\varepsilon_0>0$ sufficiently small such that for each ε satisfying $|\varepsilon|<\varepsilon_0$, we have $\tau_\varepsilon'(x)>0$ and

$$\tilde{u}_\varepsilon(x)=(u\circ\tau_\varepsilon^{-1})(x)\in\mathscr{C}_r.$$

Define

$$\phi(\varepsilon) = F(x, \tilde{u}_\varepsilon, \tilde{u}'_\varepsilon(x)).$$

Thus ϕ has a local minimum at 0, so that $\phi'(0) = 0$, that is,

$$\frac{d(F(x, \tilde{u}_\varepsilon, \tilde{u}'_\varepsilon(x)))}{d\varepsilon}|_{\varepsilon=0} = 0.$$

Observe that

$$\frac{d\tilde{u}_\varepsilon}{dx} = u'(\tau_\varepsilon^{-1}(x))\frac{d\tau_\varepsilon^{-1}(x)}{dx},$$

and

$$\frac{d\tau_\varepsilon^{-1}(x)}{dx} = \frac{1}{1+\varepsilon\lambda'(\tau_\varepsilon^{-1}(x))}.$$

Thus,

$$F(\tilde{u}_\varepsilon) = \int_a^b f\left(x, u(\tau_\varepsilon^{-1}(x)), u'(\tau_\varepsilon^{-1}(x))\left(\frac{1}{1+\varepsilon\lambda'(\tau_\varepsilon^{-1}(x))}\right)\right) dx.$$

Defining

$$\bar{x} = \tau_\varepsilon^{-1}(x),$$

we obtain

$$d\bar{x} = \frac{1}{1+\varepsilon\lambda'(\bar{x})}\, dx,$$

that is,

$$dx = (1+\varepsilon\lambda'(\bar{x}))\, d\bar{x}.$$

Dropping the bar for the new variable, we may write

$$F(\tilde{u}_\varepsilon) = \int_a^b f\left(x+\varepsilon\lambda(x), u(x), \frac{u'(x)}{1+\varepsilon\lambda'(x)}\right)\left(1+\varepsilon\lambda'(x)\right) dx.$$

From

$$\frac{dF(\tilde{u}_\varepsilon)}{d\varepsilon}|_{\varepsilon=0},$$

we obtain

$$\int_a^b (\lambda f_x(x,u(x),u'(x)) + \lambda'(x)(f(x,u(x),u'(x)) \\ - u'(x) f_\xi(x,u(x),u'(x)))) \, dx = 0. \tag{9.28}$$

Since λ is arbitrary, from Proposition 9.9.2, (in fact from its version for $u \in L^1([a,b])$ and respective extension for the N dimensional case, please see [37] for details), we obtain

$$f(x,u(x),u'(x)) - u'(x) f_\xi(x,u(x),u'(x)) - \int_a^x f_x(t,u(t),u'(t))\, dt = c_1$$

for some $c_1 \in \mathbb{R}^N$.

Since $\int_a^x f_x(t,u(t),u'(t))\,dt + c_1$ is a continuous function (in fact absolutely continuous), the proof is complete.

9.11 Natural Boundary Conditions

Consider the functional $f : U \to \mathbb{R}$, where

$$F(u)\int_\Omega f(x,u(x),\nabla u(x))\,dx,$$

$$f(x,s,\xi) \in C^1(\bar{\Omega},\mathbb{R}^N,\mathbb{R}^{N\times n}),$$

and $\Omega \subset \mathbb{R}^n$ is an open bounded connected set.

Proposition 9.11.1. *Assume*

$$U = \{u \in W^{1,2}(\Omega;\mathbb{R}^N); u = u_0 \text{ on } \Gamma_0\},$$

where $\Gamma_0 \subset \partial\Omega$ is closed and $\partial\Omega = \Gamma = \Gamma_0 \cup \Gamma_1$ being Γ_1 open in Γ and $\Gamma_0 \cap \Gamma_1 = \emptyset$. Thus if $\partial\Omega \in C^1$, $f \in C^2(\bar{\Omega},\mathbb{R}^N,\mathbb{R}^{N\times n})$ and $u \in C^2(\bar{\Omega};\mathbb{R}^N)$, and also

$$\delta F(u,\varphi) = 0, \forall \varphi \in C^1(\bar{\Omega};\mathbb{R}^N), \text{ such that } \varphi = 0 \text{ on } \Gamma_0,$$

then u is a extremal of F which satisfies the following natural boundary conditions:

$$n_\alpha f_{\xi^i_\alpha}(x,u(x)\nabla u(x)) = 0, \text{ a.e. on } \Gamma_1, \forall i \in \{1,\ldots,N\}.$$

Proof. Observe that $\delta F(u,\varphi) = 0, \forall \varphi \in C_c^\infty(\Omega;\mathbb{R}^N)$; thus u is an extremal of F and through integration by parts and the fundamental lemma of calculus of variations, we obtain

$$L_f(u) = 0, \text{ in } \Omega,$$

where

$$L_f(u) = f_s(x,u(x),\nabla u(x)) - div(f_\xi(x,u(x),\nabla u(x)).$$

Defining

$$\mathscr{V} = \{\varphi \in C^1(\Omega;\mathbb{R}^N) \mid \varphi = 0 \text{ on } \Gamma_0\},$$

for an arbitrary $\varphi \in \mathscr{V}$, we obtain

$$\begin{aligned}
\delta F(u,\varphi) &= \int_\Omega L_f(u)\cdot\varphi\,dx \\
&\quad + \int_{\Gamma_1} n_\alpha f_{\xi^i_\alpha}(x,u(x),\nabla u(x))\varphi^i(x)\,d\Gamma \\
&= \int_{\Gamma_1} n_\alpha f_{\xi^i_\alpha}(x,u(x),\nabla u(x))\varphi^i(x)\,d\Gamma \\
&= 0, \forall \varphi \in \mathscr{V}.
\end{aligned} \tag{9.29}$$

Suppose, to obtain contradiction, that

$$n_\alpha f_{\xi_\alpha^i}(x_0, u(x_0), \nabla u(x_0)) = \beta > 0,$$

for some $x_0 \in \Gamma_1$ and some $i \in \{1, \ldots, N\}$. Defining

$$G(x) = n_\alpha f_{\xi_\alpha^i}(x, u(x), \nabla u(x)),$$

by the continuity of G, there exists $r > 0$ such that

$$G(x) > \beta/2, \text{ in } B_r(x_0),$$

and in particular

$$G(x) > \beta/2, \text{ in } B_r(x_0) \cap \Gamma_1.$$

Choose $0 < r_1 < r$ such that $B_{r_1}(x_0) \cap \Gamma_0 = \emptyset$. This is possible since Γ_0 is closed and $x_0 \in \Gamma_1$.

Choose $\varphi^i \in C_c^\infty(B_{r_1}(x_0))$ such that $\varphi^i \geq 0$ in $B_{r_1}(x_0)$ and $\varphi^i > 0$ in $B_{r_1/2}(x_0)$. Therefore

$$\int_{\Gamma_1} G(x)\varphi^i(x)\, dx > \frac{\beta}{2}\int_{\Gamma_1} \varphi^i\, dx > 0,$$

and this contradicts (9.29). Thus

$$G(x) \leq 0, \forall x \in \Gamma_1,$$

and by analogy

$$G(x) \geq 0, \forall x \in \Gamma_1,$$

so that

$$G(x) = 0, \forall x \in \Gamma_1.$$

The proof is complete.

Chapter 10
Basic Concepts on Convex Analysis

For this chapter the most relevant reference is Ekeland and Temam, [25].

10.1 Convex Sets and Convex Functions

Let S be a subset of a vector space U. We recall that S is convex if given $u, v \in S$ then

$$\lambda u + (1-\lambda)v \in S, \forall \lambda \in [0,1]. \tag{10.1}$$

Definition 10.1.1 (Convex Hull). Let S be a subset of a vector space U. We define the convex hull of S, denoted by $\text{Co}(S)$ as

$$\text{Co}(S) = \left\{ \sum_{i=1}^{n} \lambda_i u_i \mid n \in \mathbb{N}, \right.$$
$$\left. \sum_{i=1}^{n} \lambda_i = 1, \ \ \lambda_i \geq 0, \ u_i \in S, \forall i \in \{1, \ldots, n\} \right\}. \tag{10.2}$$

Definition 10.1.2 (Convex Functional). Let S be convex subset of the vector space U. A functional $F : S \to \bar{\mathbb{R}} = \mathbb{R} \cup \{+\infty, -\infty\}$ is said to be convex if

$$F(\lambda u + (1-\lambda)v) \leq \lambda F(u) + (1-\lambda)F(v), \forall u, v \in S, \lambda \in [0,1]. \tag{10.3}$$

10.1.1 Lower Semicontinuity

We start with the definition of epigraph.

F. Botelho, *Functional Analysis and Applied Optimization in Banach Spaces: Applications to Non-Convex Variational Models*, DOI 10.1007/978-3-319-06074-3_10,

Definition 10.1.3 (Epigraph). Given $F : U \to \bar{\mathbb{R}}$ we define its epigraph, denoted by $Epi(F)$ as

$$Epi(F) = \{(u,a) \in U \times \mathbb{R} \mid a \geq F(u)\}.$$

Definition 10.1.4. Let U be a Banach space. Consider the weak topology $\sigma(U,U^*)$ and let $F : U \to \mathbb{R} \cup \{+\infty\}$. Such a function is said to be weakly lower semicontinuous if $\forall \lambda$ such that $\lambda < F(u)$, there exists a weak neighborhood $V_\lambda(u) \in \sigma(U,U^*)$ such that

$$F(v) > \lambda, \ \forall v \in V_\lambda(u).$$

Theorem 10.1.5. *Let U be a Banach space and let $F : U \to \mathbb{R} \cup \{+\infty\}$. The following statements are equivalent:*

1. *F is weakly lower semicontinuous (w-l.s.c.).*
2. *$Epi(F)$ is closed in $U \times \mathbb{R}$ with the product topology between $\sigma(U,U^*)$ and the usual topology in $\mathbb{R}$.*
3. *$H_\gamma^F = \{u \in U \mid F(u) \leq \gamma\}$ is closed in $\sigma(U,U^*)$, $\forall \gamma \in \mathbb{R}$.*
4. *The set $G_\gamma^F = \{u \in U \mid F(u) > \gamma\}$ is open in $\sigma(U,U^*)$, $\forall \gamma \in \mathbb{R}$.*
5. $$\liminf_{v \rightharpoonup u} F(v) \geq F(u), \forall u \in U.$$

Proof. Assume that F is w-l.s.c..We will show that $Epi(F)^c$ is open in $\sigma(U,U^*)$ product with the usual topology in $\mathbb{R}$. Choose $(u,r) \in Epi(F)^c$. Then $(u,r) \notin Epi(F)$, so that $r < F(u)$. Select λ such that $r < \lambda < F(u)$. Since F is w-l.s.c., there exists a weak neighborhood $V_\lambda(u)$ such that

$$F(v) > \lambda, \forall v \in V_\lambda(u).$$

Thus

$$V_\lambda \times (-\infty, \lambda) \subset Epi(F)^c$$

so that (u,r) is an interior point of $Epi(F)^c$, and hence, since such a point in $Epi(F)^c$ is arbitrary, we may conclude that $Epi(F)^c$ is open so that $Epi(F)$ is closed in $\sigma(U,U^*)$ product with the usual topology in $\mathbb{R}$.

Now assume (2). Observe that

$$H_\gamma^F \times \{\gamma\} = Epi(F) \cap (U \times \{\gamma\}).$$

Since from the hypothesis $Epi(F)$ is closed, we have that $H_\gamma^F \times \{\gamma\}$ is closed and hence H_γ^F is closed.

Now assume (3). To obtain (4) just take the complement of H_γ^F. Suppose (4) is valid. Let $\gamma \in \mathbb{R}$ such that

$$\gamma < F(u).$$

Since G_γ^F is open in $\sigma(U,U^*)$ there exists a weak neighborhood $V(u)$ such that

$$V(u) \subset G_\gamma^F,$$

so that

$$F(v) > \gamma, \forall v \in V(u),$$

and hence

$$\inf_{v \in V(u)} F(v) \geq \gamma,$$

and hence in particular

$$\liminf_{v \rightharpoonup u} F(v) \geq \gamma.$$

Letting $\gamma \to F(u)$, we get

$$\liminf_{v \rightharpoonup u} F(v) \geq F(u).$$

Finally assume that

$$\liminf_{v \rightharpoonup u} F(v) \geq F(u).$$

Let $\lambda < F(u)$. Thus there exists a weak neighborhood $V(u)$ such that $F(v) \geq F(u) > \lambda, \forall v \in V(u)$. The proof is complete.

Similar result is valid for the strong topology of the Banach space U so that a functional $F : U \to \mathbb{R} \cup \{+\infty\}$ is strongly lower semicontinuous (l.s.c.) at $u \in U$, if

$$\liminf_{v \to u} F(v) \geq F(u). \tag{10.4}$$

Corollary 10.1.6. *Every convex l.s.c. function $F : U \to \overline{\mathbb{R}}$ is also w-l.s.c. (weakly lower semicontinuous).*

Proof. The result follows from the fact that the epigraph of F is convex and closed convex sets are weakly closed.

Definition 10.1.7 (Affine Continuous Function). Let U be a Banach space. A functional $F : U \to \mathbb{R}$ is said to be affine continuous if there exist $u^* \in U^*$ and $\alpha \in \mathbb{R}$ such that

$$F(u) = \langle u, u^* \rangle_U + \alpha, \forall u \in U. \tag{10.5}$$

Definition 10.1.8 ($\Gamma(U)$). Let U be a Banach space. We say that $F : U \to \bar{\mathbb{R}}$ belongs to $\Gamma(U)$ and write $F \in \Gamma(U)$ if F can be represented as the point-wise supremum of a family of affine continuous functions. If $F \in \Gamma(U)$ and $F(u) \in \mathbb{R}$ for some $u \in U$, then we write $F \in \Gamma_0(U)$.

The next result is proven in [25].

Proposition 10.1.9. *Let U be a Banach space, then $F \in \Gamma(U)$ if and only if F is convex and l.s.c., and if F takes the value $-\infty$, then $F \equiv -\infty$.*

Definition 10.1.10 (Convex Envelope). Let U be a Banach space. Given $F : U \to \bar{\mathbb{R}}$, we define its convex envelope, denoted by $CF : U \to \bar{\mathbb{R}}$ by

$$CF(u) = \sup_{(u^*, \alpha) \in A^*} \{\langle u, u^* \rangle + \alpha\}, \tag{10.6}$$

where

$$A^* = \{(u^*, \alpha) \in U^* \times \mathbb{R} \mid \langle v, u^* \rangle_U + \alpha \le F(v), \forall v \in U\} \tag{10.7}$$

Definition 10.1.11 (Polar Functionals). Given $F : U \to \bar{\mathbb{R}}$, we define the related polar functional, denoted by $F^* : U^* \to \bar{\mathbb{R}}$, by

$$F^*(u^*) = \sup_{u \in U}\{\langle u, u^* \rangle_U - F(u)\}, \forall u^* \in U^*. \tag{10.8}$$

Definition 10.1.12 (Bipolar Functional). Given $F : U \to \bar{\mathbb{R}}$, we define the related bipolar functional, denoted by $F^{**} : U \to \bar{\mathbb{R}}$, as

$$F^{**}(u) = \sup_{u^* \in U^*}\{\langle u, u^* \rangle_U - F^*(u^*)\}, \forall u \in U. \tag{10.9}$$

Proposition 10.1.13. *Given $F : U \to \bar{\mathbb{R}}$, then $F^{**}(u) = CF(u)$ and in particular if $F \in \Gamma(U)$, then $F^{**}(u) = F(u)$.*

Proof. By definition, the convex envelope of F is the supremum of all affine continuous minorants of F. We can consider only the maximal minorants, which functions of the form

$$u \mapsto \langle u, u^* \rangle_U - F^*(u^*). \tag{10.10}$$

Thus,

$$CF(u) = \sup_{u^* \in U^*}\{\langle u, u^* \rangle_U - F^*(u^*)\} = F^{**}(u). \tag{10.11}$$

Corollary 10.1.14. *Given $F : U \to \bar{\mathbb{R}}$, we have $F^* = F^{***}$.*

Proof. Since $F^{**} \le F$ we obtain

$$F^* \le F^{***}. \tag{10.12}$$

On the other hand, we have

$$F^{**}(u) \ge \langle u, u^* \rangle_U - F^*(u^*), \tag{10.13}$$

so that

$$F^{***}(u^*) = \sup_{u \in U}\{\langle u, u^* \rangle_U - F^{**}(u)\} \le F^*(u^*). \tag{10.14}$$

From (10.12) and (10.14) we obtain $F^*(u^*) = F^{***}(u^*)$.

Here we recall the definition of Gâteaux differentiability.

Definition 10.1.15 (Gâteaux Differentiability). A functional $F : U \to \bar{\mathbb{R}}$ is said to be Gâteaux differentiable at $u \in U$ if there exists $u^* \in U^*$ such that

$$\lim_{\lambda \to 0} \frac{F(u+\lambda h)-F(u)}{\lambda} = \langle h, u^* \rangle_U, \ \forall h \in U. \tag{10.15}$$

The vector u^* is said to be the Gâteaux derivative of $F : U \to \mathbb{R}$ at u and may be denoted as follows:

$$u^* = \frac{\partial F(u)}{\partial u} \ or\ u^* = \delta F(u) \tag{10.16}$$

Definition 10.1.16 (Sub-gradients). Given $F : U \to \bar{\mathbb{R}}$, we define the set of sub-gradients of F at u, denoted by $\partial F(u)$ as

$$\partial F(u) = \{u^* \in U^*, \ such \ \ that \ \langle v-u, u^* \rangle_U + F(u) \leq F(v), \forall v \in U\}. \tag{10.17}$$

Here we recall the definition of adjoint operator.

Definition 10.1.17 (Adjoint Operator). Let U and Y be Banach spaces and $\Lambda : U \to Y$ a continuous linear operator. The adjoint operator related to Λ, denoted by $\Lambda^* : Y^* \to U^*$, is defined through the equation

$$\langle u, \Lambda^* v^* \rangle_U = \langle \Lambda u, v^* \rangle_Y, \ \forall u \in U, \ v^* \in Y^*. \tag{10.18}$$

Lemma 10.1.18 (Continuity of Convex Functions). *If in a neighborhood of a point $u \in U$ a convex function F is bounded above by a finite constant, then F is continuous at u.*

Proof. By translation, we may reduce the problem to the case where $u = \theta$ and $F(u) = 0$. Let $\mathscr{V}$ be a neighborhood of origin such that $F(v) \leq a < +\infty, \forall v \in \mathscr{V}$. Define $\mathscr{W} = \mathscr{V} \cap (-\mathscr{V})$ (which is a symmetric neighborhood of origin). Pick $\varepsilon \in (0,1)$. If $v \in \varepsilon \mathscr{W}$, since F is convex and

$$\frac{v}{\varepsilon} \in \mathscr{V} \tag{10.19}$$

we may infer that

$$F(v) \leq (1-\varepsilon)F(\theta) + \varepsilon F(v/\varepsilon) \leq \varepsilon a. \tag{10.20}$$

Also

$$\frac{-v}{\varepsilon} \in \mathscr{V}. \tag{10.21}$$

Thus,

$$F(\theta) \leq \frac{F(v)}{1+\varepsilon} + \frac{\varepsilon}{1+\varepsilon} F(-v/\varepsilon),$$

so that

$$F(v) \geq (1+\varepsilon)F(\theta) - \varepsilon F(-v/\varepsilon) \geq -\varepsilon a. \tag{10.22}$$

Therefore

$$|F(v)| \leq \varepsilon a, \forall v \in \varepsilon \mathscr{W}, \tag{10.23}$$

that is, F is continuous at $u = \theta$.

Proposition 10.1.19. *Let $F : U \to \bar{\mathbb{R}}$ be a convex function finite and continuous at $u \in U$. Then $\partial F(u) \neq \emptyset$.*

Proof. Since F is convex, $Epi(F)$ is convex, as F is continuous at u, we have that $Epi(F)$ is nonempty. Observe that $(u, F(u))$ belongs to the boundary of $Epi(F)$, so that denoting $A = Epi(F)$, we may separate $(u, F(u))$ from $\mathring{A}$ by a closed hyperplane H, which may be written as

$$H = \{(v, a) \in U \times \mathbb{R} \mid \langle v, u^* \rangle_U + \alpha a = \beta\}, \tag{10.24}$$

for some fixed $\alpha, \beta \in \mathbb{R}$ and $u^* \in U^*$, so that

$$\langle v, u^* \rangle_U + \alpha a \geq \beta, \forall (v, a) \in Epi(F), \tag{10.25}$$

and

$$\langle u, u^* \rangle_U + \alpha F(u) = \beta, \tag{10.26}$$

where $(\alpha, \beta, u^*) \neq (0, 0, \theta)$. Suppose $\alpha = 0$. Thus we have

$$\langle v - u, u^* \rangle_U \geq 0, \forall v \in U, \tag{10.27}$$

and thus we obtain $u^* = \theta$ and $\beta = 0$. Therefore we may assume $\alpha > 0$ (considering (10.25)) so that $\forall v \in U$ we have

$$\frac{\beta}{\alpha} - \langle v, u^*/\alpha \rangle_U \leq F(v), \tag{10.28}$$

and

$$\frac{\beta}{\alpha} - \langle u, u^*/\alpha \rangle_U = F(u), \tag{10.29}$$

or

$$\langle v - u, -u^*/\alpha \rangle_U + F(u) \leq F(v), \forall v \in U, \tag{10.30}$$

so that

$$-u^*/\alpha \in \partial F(u). \tag{10.31}$$

Definition 10.1.20 (Carathéodory Mapping). Let $S \subset \mathbb{R}^n$ be an open set. We say that $g : S \times \mathbb{R}^l \to \mathbb{R}$ is a Carathéodory mapping if

$$\forall \xi \in \mathbb{R}^l,\ x \mapsto g(x,\xi) \text{ is a measurable function,}$$

and

$$\text{for almost all } x \in S,\ \xi \mapsto g(x,\xi) \text{ is a continuous function.}$$

The proof of next results may be found in Ekeland and Temam [25].

Proposition 10.1.21. *Let E and F be two Banach spaces, S a Borel subset of* $\mathbb{R}^n$ *and* $g : S \times E \to F$ *a Carathéodory mapping. For each measurable function* $u : S \to E$, *let* $G_1(u)$ *be the measurable function* $x \mapsto g(x,u(x)) \in F$.

If G_1 *maps* $L^p(S,E)$ *into* $L^r(S,F)$ *for* $1 \leq p,r < \infty$, *then* G_1 *is continuous in the norm topology.*

For the functional $G : U \to \mathbb{R}$, defined by $G(u) = \int_S g(x,u(x))dS$, where $U = U^* = [L^2(S)]^l$ (this is a special case of the more general hypothesis presented in [25]) we have the following result.

Proposition 10.1.22. *Considering the last proposition we can express* $G^* : U^* \to \bar{\mathbb{R}}$ *as*

$$G^*(u^*) = \int_S g^*(x,u^*(x))dS, \tag{10.32}$$

where $g^*(x,y) = \sup_{\eta \in \mathbb{R}^l} (y \cdot \eta - g(x,\eta))$, *almost everywhere in S.*

For non-convex functionals it may be sometimes difficult to express analytically conditions for a global extremum. This fact motivates the definition of Legendre transform, which is established through a local extremum.

Definition 10.1.23 (Legendre's Transform and Associated Functional). Consider a differentiable function $g : \mathbb{R}^n \to \mathbb{R}$. Its Legendre transform, denoted by $g_L^* : R_L^n \to \mathbb{R}$, is expressed as

$$g_L^*(y^*) = x_{0^i} \cdot y_i^* - g(x_0), \tag{10.33}$$

where x_0 is the solution of the system:

$$y_i^* = \frac{\partial g(x_0)}{\partial x_i}, \tag{10.34}$$

and $R_L^n = \{y^* \in \mathbb{R}^n \text{ such that (10.34) has a unique solution}\}$.

Furthermore, considering the functional $G : Y \to \mathbb{R}$ defined as $G(v) = \int_S g(v)dS$, we define the associated Legendre transform functional, denoted by $G_L^* : Y_L^* \to \mathbb{R}$ as

$$G_L^*(v^*) = \int_S g_L^*(v^*)dS, \tag{10.35}$$

where $Y_L^* = \{v^* \in Y^* \mid v^*(x) \in R_L^n, \text{ a.e. in } S\}$.

About the Legendre transform we still have the following results:

Proposition 10.1.24. *Considering the last definitions, suppose that for each $y^* \in R_L^n$ at least in a neighborhood (of y^*) it is possible to define a differentiable function by the expression*

$$x_0(y^*) = [\frac{\partial g}{\partial x}]^{-1}(y^*). \tag{10.36}$$

Then, $\forall\ i \in \{1,\ldots,n\}$ we may write

$$y_i^* = \frac{\partial g(x_0)}{\partial x_i} \Leftrightarrow x_{0^i} = \frac{\partial g_L^*(y^*)}{\partial y_i^*} \tag{10.37}$$

Proof. Suppose firstly that

$$y_i^* = \frac{\partial g(x_0)}{\partial x_i}, \forall i \in \{1,\ldots,n\}, \tag{10.38}$$

thus,

$$g_L^*(y^*) = y_i^* x_{0^i} - g(x_0) \tag{10.39}$$

and taking derivatives for this expression we have

$$\frac{\partial g_L^*(y^*)}{\partial y_i^*} = y_j^* \frac{\partial x_{0^j}}{\partial y_i^*} + x_{0^i} - \frac{\partial g(x_0)}{\partial x_j}\frac{\partial x_{0^j}}{\partial y_i^*}, \tag{10.40}$$

or

$$\frac{\partial g_L^*(y^*)}{\partial y_i^*} = (y_j^* - \frac{\partial g(x_0)}{\partial x_j})\frac{\partial x_{0^j}}{\partial y_i^*} + x_{0^i} \tag{10.41}$$

which from (10.38) implies that

$$\frac{\partial g_L^*(y^*)}{\partial y_i^*} = x_{0^i}, \forall i \in \{1,\ldots,n\}. \tag{10.42}$$

This completes the first half of the proof. Conversely, suppose now that

$$x_{0^i} = \frac{\partial g_L^*(y^*)}{\partial y_i^*}, \forall i \in \{1,\ldots,n\}. \tag{10.43}$$

As $y^* \in R_L^n$ there exists $\bar{x}_0 \in \mathbb{R}^n$ such that

$$y_i^* = \frac{\partial g(\bar{x}_0)}{\partial x_i} \forall i \in \{1,\ldots,n\}, \tag{10.44}$$

and

$$g_L^*(y^*) = y_i^* \bar{x}_{0^i} - g(\bar{x}_0) \tag{10.45}$$

and therefore taking derivatives for this expression we can obtain

$$\frac{\partial g_L^*(y^*)}{\partial y_i^*} = y_j^* \frac{\partial \bar{x}_{0j}}{\partial y_i^*} + \bar{x}_{0^i} - \frac{\partial g(\bar{x}_0)}{\partial x_j} \frac{\partial \bar{x}_{0j}}{\partial y_i^*}, \tag{10.46}$$

$\forall\, i \in \{1,\dots,n\}$, so that

$$\frac{\partial g_L^*(y^*)}{\partial y_i^*} = (y_j^* - \frac{\partial g(\bar{x}_0)}{\partial x_j}) \frac{\partial \bar{x}_{0j}}{\partial y_i^*} + \bar{x}_{0^i} \tag{10.47}$$

$\forall\, i \in \{1,\dots,n\}$, which from (10.43) and (10.44) implies that

$$\bar{x}_{0^i} = \frac{\partial g_L^*(y^*)}{\partial y_i^*} = x_{0^i}, \;\; \forall\; i \in \{1,\dots,n\}, \tag{10.48}$$

from this and (10.44) we have

$$y_i^* = \frac{\partial g(\bar{x}_0)}{\partial x_i} = \frac{\partial g(x_0)}{\partial x_i} \;\; \forall\; i \in \{1,\dots,n\}. \tag{10.49}$$

Theorem 10.1.25. *Consider the functional $J : U \to \bar{\mathbb{R}}$ defined as $J(u) = (G \circ \Lambda)(u) - \langle u, f\rangle_U$ where $\Lambda(= \{\Lambda_i\}) : U \to Y$ $(i \in \{1,\dots,n\})$ is a continuous linear operator and $G : Y \to \mathbb{R}$ is a functional that can be expressed as $G(v) = \int_S g(v)dS$, $\forall v \in Y$ (here $g : \mathbb{R}^n \to \mathbb{R}$ is a differentiable function that admits Legendre transform denoted by $g_L^* : R_L^n \to \mathbb{R}$. That is, the hypothesis mentioned at Proposition 10.1.24 is satisfied).*

Under these assumptions we have

$$\delta J(u_0) = \theta \Leftrightarrow \delta(-G_L^*(v_0^*) + \langle u_0, \Lambda^* v_0^* - f\rangle_U) = \theta, \tag{10.50}$$

where $v_0^* = \frac{\partial G(\Lambda(u_0))}{\partial v}$ is supposed to be such that $v_0^*(x) \in R_L^n$, a.e. in S and in this case

$$J(u_0) = -G_L^*(v_0^*). \tag{10.51}$$

Proof. Suppose first that $\delta J(u_0) = \theta$, that is,

$$\Lambda^* \frac{\partial G(\Lambda u_0)}{\partial v} - f = \theta \tag{10.52}$$

which, as $v_0^* = \frac{\partial G(\Lambda u_0)}{\partial v}$, implies that

$$\Lambda^* v_0^* - f = \theta, \tag{10.53}$$

and

$$v_{0^i}^* = \frac{\partial g(\Lambda u_0)}{\partial x_i}. \tag{10.54}$$

Thus from the last proposition we can write

$$\Lambda_i(u_0) = \frac{\partial g_L^*(v_0^*)}{\partial y_i^*}, \text{ for } i \in \{1,..,n\} \tag{10.55}$$

which means

$$\Lambda u_0 = \frac{\partial G_L^*(v_0^*)}{\partial v^*}. \tag{10.56}$$

Therefore from (10.53) and (10.56) we have

$$\delta(-G_L^*(v_0^*) + \langle u_0, \Lambda^* v_0^* - f\rangle_U) = \theta. \tag{10.57}$$

This completes the first part of the proof.

Conversely, suppose now that

$$\delta(-G_L^*(v_0^*) + \langle u_0, \Lambda^* v_0^* - f\rangle_U) = \theta, \tag{10.58}$$

that is,

$$\Lambda^* v_0^* - f = \theta \tag{10.59}$$

and

$$\Lambda u_0 = \frac{\partial G_L^*(v_0^*)}{\partial v^*}. \tag{10.60}$$

Clearly, from (10.60), the last proposition and (10.59), we can write

$$v_0^* = \frac{\partial G(\Lambda(u_0))}{\partial v} \tag{10.61}$$

and

$$\Lambda^* \frac{\partial G(\Lambda u_0)}{\partial v} - f = \theta, \tag{10.62}$$

which implies

$$\delta J(u_0) = \theta. \tag{10.63}$$

Finally, we have

$$J(u_0) = G(\Lambda u_0) - \langle u_0, f\rangle_U \tag{10.64}$$

From this, (10.59) and (10.61), we have

$$J(u_0) = G(\Lambda u_0) - \langle u_0, \Lambda^* v_0^*\rangle_U = G(\Lambda u_0) - \langle \Lambda u_0, v_0^*\rangle_Y \tag{10.65}$$

$$= -G_L^*(v_0^*). \tag{10.66}$$

10.2 Duality in Convex Optimization

Let U be a Banach space. Given $F : U \to \bar{\mathbb{R}}$ ($F \in \Gamma_0(U)$) we define the problem $\mathscr{P}$ as

$$\mathscr{P}: \text{ minimize } F(u) \text{ on } U. \tag{10.67}$$

We say that $u_0 \in U$ is a solution of problem $\mathscr{P}$ if $F(u_0) = \inf_{u \in U} F(u)$. Consider a function $\phi(u,p)$ ($\phi : U \times Y \to \bar{\mathbb{R}}$) such that

$$\phi(u,0) = F(u). \tag{10.68}$$

We define the problem $\mathscr{P}^*$ as

$$\mathscr{P}^*: \text{ maximize } -\phi^*(0,p^*) \text{ on } Y^*. \tag{10.69}$$

Observe that

$$\phi^*(0,p^*) = \sup_{(u,p) \in U \times Y} \{\langle 0,u\rangle_U + \langle p,p^*\rangle_Y - \phi(u,p)\} \geq -\phi(u,0), \tag{10.70}$$

or

$$\inf_{u \in U}\{\phi(u,0)\} \geq \sup_{p^* \in Y^*}\{-\phi^*(0,p^*)\}. \tag{10.71}$$

Proposition 10.2.1. *Consider $\phi \in \Gamma_0(U \times Y)$. If we define*

$$h(p) = \inf_{u \in U}\{\phi(u,p)\}, \tag{10.72}$$

then h is convex.

Proof. We have to show that given $p,q \in Y$ and $\lambda \in (0,1)$, we have

$$h(\lambda p + (1-\lambda)q) \leq \lambda h(p) + (1-\lambda)h(q). \tag{10.73}$$

If $h(p) = +\infty$ or $h(q) = +\infty$ we are done. Thus let us assume $h(p) < +\infty$ and $h(q) < +\infty$. For each $a > h(p)$ there exists $u \in U$ such that

$$h(p) \leq \phi(u,p) \leq a, \tag{10.74}$$

and if $b > h(q)$, there exists $v \in U$ such that

$$h(q) \leq \phi(v,q) \leq b. \tag{10.75}$$

Thus

$$\begin{aligned} h(\lambda p+(1-\lambda)q) &\leq \inf_{w\in U}\{\phi(w,\lambda p+(1-\lambda)q)\} \\ &\leq \phi(\lambda u+(1-\lambda)v,\lambda p+(1-\lambda)q) \leq \lambda\phi(u,p)+(1-\lambda)\phi(v,q) \\ &\leq \lambda a+(1-\lambda)b. \end{aligned} \tag{10.76}$$

Letting $a \to h(p)$ and $b \to h(q)$ we obtain

$$h(\lambda p+(1-\lambda)q) \leq \lambda h(p)+(1-\lambda)h(q). \tag{10.77}$$

Proposition 10.2.2. *For h as above, we have $h^*(p^*) = \phi^*(0,p^*)$, $\forall p^* \in Y^*$, so that*

$$h^{**}(0) = \sup_{p^*\in Y^*}\{-\phi^*(0,p^*)\}. \tag{10.78}$$

Proof. Observe that

$$h^*(p^*) = \sup_{p\in Y}\{\langle p,p^*\rangle_Y - h(p)\} = \sup_{p\in Y}\{\langle p,p^*\rangle_Y - \inf_{u\in U}\{\phi(u,p)\}\}, \tag{10.79}$$

so that

$$h^*(p^*) = \sup_{(u,p)\in U\times Y}\{\langle p,p^*\rangle_Y - \phi(u,p)\} = \phi^*(0,p^*). \tag{10.80}$$

Proposition 10.2.3. *The set of solutions of the problem $\mathscr{P}^*$ (the dual problem) is identical to $\partial h^{**}(0)$.*

Proof. Consider $p_0^* \in Y^*$ a solution of problem $\mathscr{P}^*$, that is,

$$-\phi^*(0,p_0^*) \geq -\phi^*(0,p^*), \forall p^* \in Y^*, \tag{10.81}$$

which is equivalent to

$$-h^*(p_0^*) \geq -h^*(p^*), \forall p^* \in Y^*, \tag{10.82}$$

which is equivalent to

$$\begin{aligned} -h(p_0^*) = \sup_{p^*\in Y^*}\{\langle 0,p^*\rangle_Y - h^*(p^*)\} &\Leftrightarrow -h^*(p_0^*) = h^{**}(0) \\ &\Leftrightarrow p_0^* \in \partial h^{**}(0). \end{aligned} \tag{10.83}$$

Theorem 10.2.4. *Consider $\phi : U\times Y \to \bar{\mathbb{R}}$ convex. Assume $\inf_{u\in U}\{\phi(u,0)\} \in \mathbb{R}$ and there exists $u_0 \in U$ such that $p \mapsto \phi(u_0,p)$ is finite and continuous at $0 \in Y$. Then*

$$\inf_{u\in U}\{\phi(u,0)\} = \sup_{p^*\in Y^*}\{-\phi^*(0,p^*)\}, \tag{10.84}$$

and the dual problem has at least one solution.

Proof. By hypothesis $h(0) \in \mathbb{R}$ and as was shown above, h is convex. As the function $p \mapsto \phi(u_0,p)$ is convex and continuous at $0 \in Y$, there exists a neighborhood $\mathscr{V}$ of zero in Y such that

$$\phi(u_0,p) \leq M < +\infty, \forall p \in \mathscr{V}, \tag{10.85}$$

for some $M \in \mathbb{R}$. Thus, we may write

$$h(p) = \inf_{u \in U}\{\phi(u,p)\} \leq \phi(u_0,p) \leq M, \forall p \in \mathscr{V}. \tag{10.86}$$

Hence, from Lemma 10.1.18, h is continuous at 0. Thus, by Proposition 10.1.19, h is sub-differentiable at 0, which means $h(0) = h^{**}(0)$. Therefore, by Proposition 10.2.3, the dual problem has solutions and

$$h(0) = \inf_{u \in U}\{\phi(u,0)\} = \sup_{p^* \in Y^*}\{-\phi^*(0,p^*)\} = h^{**}(0). \tag{10.87}$$

Now we apply the last results to $\phi(u,p) = G(\Lambda u + p) + F(u)$, where $\Lambda : U \to Y$ is a continuous linear operator whose adjoint operator is denoted by $\Lambda^* : Y^* \to U^*$. We may enunciate the following theorem.

Theorem 10.2.5. *Suppose U is a reflexive Banach space and define $J : U \to \mathbb{R}$ by*

$$J(u) = G(\Lambda u) + F(u) = \phi(u,0), \tag{10.88}$$

where $\lim J(u) = +\infty$ *as* $\|u\|_U \to \infty$ *and* $F \in \Gamma_0(U)$, $G \in \Gamma_0(Y)$. *Also suppose there exists $\hat{u} \in U$ such that $J(\hat{u}) < +\infty$ with the function $p \mapsto G(p)$ continuous at $\Lambda\hat{u}$. Under such hypothesis, there exist $u_0 \in U$ and $p_0^* \in Y^*$ such that*

$$\begin{aligned} J(u_0) &= \min_{u \in U}\{J(u)\} = \max_{p^* \in Y^*}\{-G^*(p^*) - F^*(-\Lambda^* p^*)\} \\ &= -G^*(p_0^*) - F^*(-\Lambda^* p_0^*). \end{aligned} \tag{10.89}$$

Proof. The existence of solutions for the primal problem follows from the direct method of calculus of variations. That is, considering a minimizing sequence, from above (coercivity hypothesis), such a sequence is bounded and has a weakly convergent subsequence to some $u_0 \in U$. Finally, from the lower semicontinuity of primal formulation, we may conclude that u_0 is a minimizer. The other conclusions follow from Theorem 10.2.4 just observing that

$$\begin{aligned} \phi^*(0,p^*) &= \sup_{u \in U, p \in Y}\{\langle p,p^*\rangle_Y - G(\Lambda u + p) - F(u)\} \\ &= \sup_{u \in U, q \in Y}\{\langle q,p^*\rangle - G(q) - \langle \Lambda u,p^*\rangle - F(u)\}, \end{aligned} \tag{10.90}$$

so that

$$\begin{aligned} \phi^*(0,p^*) &= G^*(p^*) + \sup_{u \in U}\{-\langle u,\Lambda^* p^*\rangle_U - F(u)\} \\ &= G^*(p^*) + F^*(-\Lambda^* p^*). \end{aligned} \tag{10.91}$$

Thus,

$$\inf_{u\in U}\{\phi(u,0)\} = \sup_{p^*\in Y^*}\{-\phi^*(0,p^*)\} \tag{10.92}$$

and solutions u_0 and p_0^* for the primal and dual problems, respectively, imply that

$$\begin{aligned} J(u_0) &= \min_{u\in U}\{J(u)\} = \max_{p^*\in Y^*}\{-G^*(p^*) - F^*(-\Lambda^* p^*)\} \\ &= -G^*(p_0^*) - F^*(-\Lambda^* p_0^*). \end{aligned} \tag{10.93}$$

10.3 The Min–Max Theorem

Our main objective in this section is to state and prove the min–max theorem.

Definition 10.1. Let U,Y be Banach spaces, $A \subset U$ and $B \subset Y$ and let $L : A\times B \to \mathbb{R}$ be a functional. We say that $(u_0,v_0)\in A\times B$ is a saddle point for L if

$$L(u_0,v) \leq L(u_0,v_0) \leq L(u,v_0),\ \forall u\in A,\ v\in B.$$

Proposition 10.1. *Let U,Y be Banach spaces, $A\subset U$ and $B\subset Y$. A functional $L : U\times Y\to\mathbb{R}$ has a saddle point if and only if*

$$\max_{v\in B}\inf_{u\in A} L(u,v) = \min_{u\in A}\sup_{v\in B} L(u,v).$$

Proof. Suppose $(u_0,v_0)\in A\times B$ is a saddle point of L.

Thus,

$$L(u_0,v)\leq L(u_0,v_0)\leq L(u,v_0), \forall u\in A,\ v\in B. \tag{10.94}$$

Define

$$F(u) = \sup_{v\in B} L(u,v).$$

Observe that

$$\inf_{u\in A} F(u) \leq F(u_0),$$

so that

$$\inf_{u\in A}\sup_{v\in B} L(u,v) \leq \sup_{v\in B} L(u_0,v). \tag{10.95}$$

Define

$$G(v) = \inf_{u\in A} L(u,v).$$

Thus

$$\sup_{v\in B} G(v) \geq G(v_0),$$

so that

$$\sup_{v\in B}\inf_{u\in A} L(u,v) \geq \inf_{u\in A} L(u,v_0). \tag{10.96}$$

From (10.94), (10.95), and (10.96) we obtain

$$\begin{aligned}\inf_{u\in A}\sup_{v\in B} L(u,v) &\leq \sup_{v\in B} L(u_0,v)\\ &\leq L(u_0,v_0)\\ &\leq \inf_{u\in A} L(u,v_0)\\ &\leq \sup_{v\in B}\inf_{u\in A} L(u,v).\end{aligned} \tag{10.97}$$

Hence

$$\begin{aligned}\inf_{u\in A}\sup_{v\in B} L(u,v) &\leq L(u_0,v_0)\\ &\leq \sup_{v\in B}\inf_{u\in A} L(u,v).\end{aligned} \tag{10.98}$$

On the other hand

$$\inf_{u\in A} L(u,v) \leq L(u,v), \forall u\in A,\ v\in B,$$

so that

$$\sup_{v\in B}\inf_{u\in A} L(u,v) \leq \sup_{vInB} L(u,v), \forall u\in A,$$

and hence

$$\sup_{v\in B}\inf_{u\in A} L(u,v) \leq \inf_{u\in A}\sup_{v\in B} L(u,v). \tag{10.99}$$

From (10.94), (10.98), and (10.99) we obtain

$$\begin{aligned}\inf_{u\in A}\sup_{v\in B} L(u,v) &= \sup_{v\in B} L(u_0,v)\\ &= L(u_0,v_0)\\ &= \inf_{u\in A} L(u,v_0)\\ &= \sup_{v\in B}\inf_{u\in A} L(u,v).\end{aligned} \tag{10.100}$$

Conversely suppose

$$\max_{v\in B}\inf_{u\in A} L(u,v) = \min_{u\in A}\sup_{v\in B} L(u,v)$$

As above defined,

$$F(u) = \sup_{v\in B} L(u,v),$$

and

$$G(v) = \inf_{u\in A} L(u,v).$$

From the hypotheses, there exists $(u_0, v_0) \in A \times B$ such that

$$\sup_{v \in B} G(v) = G(v_0) = F(u_0) = \inf_{u \in A} F(u).$$

so that

$$F(u_0) = \sup_{v \in B} L(u_0, v) = \inf_{u \in U} L(u, v_0) = G(v_0).$$

In particular

$$L(u_0, v_0) \leq \sup_{v \in B} L(u_0, v) = \inf_{u \in U} L(u, v_0) \leq L(u_0, v_0).$$

Therefore

$$\sup_{v \in B} L(u_0, v) = L(u_0, v_0) = \inf_{u \in U} L(u, v_0).$$

The proof is complete.

Proposition 10.2. *Let U, Y be Banach spaces, $A \subset U$, $B \subset Y$ and let $L : A \times B \to \mathbb{R}$ be a functional. Assume there exist $u_0 \in A$, $v_0 \in B$, and $\alpha \in \mathbb{R}$ such that*

$$L(u_0, v) \leq \alpha, \ \forall v \in B,$$

and

$$L(u, v_0) \geq \alpha, \ \forall u \in A.$$

Under such hypotheses (u_0, v_0) is a saddle point of L, that is,

$$L(u_0, v) \leq L(u_0, v_0) \leq L(u, v_0), \ \forall u \in A, \ v \in B.$$

Proof. Observe, from the hypotheses, that we have

$$L(u_0, v_0) \leq \alpha,$$

and

$$L(u_0, v_0) \geq \alpha,$$

so that

$$L(u_0, v) \leq \alpha = L(u_0, v_0) \leq L(u, v_0), \ \forall u \in A, \ v \in B.$$

This completes the proof.

In the next lines we state and prove the min-max theorem.

Theorem 10.1. *Let U, Y be reflexive Banach spaces, $A \subset U$, $B \subset Y$ and let $L : A \times B \to \mathbb{R}$ be a functional.*

Suppose that:

1. *$A \subset U$ is convex, closed, and nonempty.*
2. *$B \subset Y$ is convex, closed, and nonempty.*
3. *For each $u \in A$, $F_u(v) = L(u, v)$ is concave and upper semicontinuous.*

4. *For each $v \in B$, $G_v(u) = L(u,v)$ is convex and lower semicontinuous.*
5. *The sets A and B are bounded.*

Under such hypotheses L has at least one saddle point $(u_0, v_0) \in A \times B$ such that

$$\begin{aligned} L(u_0,v_0) &= \min_{u \in A} \max_{v \in B} L(u,v) \\ &= \max_{v \in B} \min_{u \in A} L(u,v). \end{aligned} \tag{10.101}$$

Proof. Fix $v \in B$. Observe that $G_v(u) = L(u,v)$ is convex and lower semicontinuous. Therefore it is weakly lower semicontinuous on the weak compact set A. At first we assume the additional hypothesis that $G_v(u)$ is strictly convex, $\forall v \in B$. Hence $G_v(u)$ attains a unique minimum on A. We denote the optimal $u \in A$ by $u(v)$.

Define

$$G(v) = \min_{u \in A} G_v(u) = \min_{u \in U} L(u,v).$$

Thus,

$$G(v) = L(u(v),v).$$

The function $G(v)$ is expressed as the minimum of a family of concave weakly upper semicontinuous functions, and hence it is also concave and upper semicontinuous.

Moreover, $G(v)$ is bounded above on the weakly compact set B, so that there exists $v_0 \in B$ such that

$$G(v_0) = \max_{v \in B} G(v) = \max_{v \in B} \min_{u \in A} L(u,v).$$

Observe that

$$G(v_0) = \min_{u \in A} L(u,v_0) \leq L(u,v_0), \; \forall u \in U.$$

Observe that from the concerned concavity, for $u \in A$, $v \in B$, and $\lambda \in (0,1)$, we have

$$L(u,(1-\lambda)v_0 + \lambda v) \geq (1-\lambda)L(u,v_0) + \lambda L(u,v).$$

In particular denote $u((1-\lambda)v_0 + \lambda v) = u_\lambda$, where u_λ is such that

$$\begin{aligned} G((1-\lambda)v_0 + \lambda v) &= \min_{u \in A} L(u,(1-\lambda)v_0 + \lambda v) \\ &= L(u_\lambda,(1-\lambda)v_0 + \lambda v). \end{aligned} \tag{10.102}$$

Therefore,

$$\begin{aligned} G(v_0) &= \max_{v \in B} G(v) \\ &\geq G((1-\lambda)v_0 + \lambda v) \\ &= L(u_\lambda,(1-\lambda)v_0 + \lambda v) \\ &\geq (1-\lambda)L(u_\lambda,v_0) + \lambda L(u_\lambda,v) \end{aligned}$$

$$\geq (1-\lambda)\min_{u\in A} L(u,v_0)+\lambda L(u_\lambda,v)$$
$$= (1-\lambda)G(v_0)+\lambda L(u_\lambda,v). \tag{10.103}$$

From this, we obtain

$$G(v_0) \geq L(u_\lambda,v). \tag{10.104}$$

Let $\{\lambda_n\} \subset (0,1)$ be such that $\lambda_n \to 0$.
Let $\{u_n\} \subset A$ be such that

$$\begin{aligned} G((1-\lambda_n)v_0+\lambda_n v) &= \min_{u\in A} L(u,(1-\lambda_n)v_0+\lambda_n v) \\ &= L(u_n,(1-\lambda_n)v_0+\lambda_n v). \end{aligned} \tag{10.105}$$

Since A is weakly compact, there exists a subsequence $\{u_{n_k}\} \subset \{u_n\} \subset A$ and $u_0 \in A$ such that

$$u_{n_k} \rightharpoonup u_0, \text{ weakly in } U, \text{ as } k \to \infty.$$

Observe that

$$\begin{aligned} (1-\lambda_{n_k})L(u_{n_k},v_0)+\lambda_{n_k}L(u_{n_k},v) &\leq L(u_{n_k},(1-\lambda_{n_k})v_0+\lambda_{n_k}v) \\ &= \min_{u\in A} L(u,(1-\lambda_{n_k})v_0+\lambda_{n_k}v) \\ &\leq L(u,(1-\lambda_{n_k})v_0+\lambda_{n_k}v), \end{aligned} \tag{10.106}$$

$\forall u \in A,\ k \in \mathbb{N}$.
Recalling that $\lambda_{n_k} \to 0$, from this and (10.106), we obtain

$$\begin{aligned} L(u_0,v_0) &\leq \liminf_{k\to\infty} L(u_{n_k},v_0) \\ &= \liminf_{k\to\infty}((1-\lambda_{n_k})L(u_{n_k},v_0)+\lambda_{n_k}L(u,v)) \\ &\leq \limsup_{k\to\infty} L(u,(1-\lambda_{n_k})v_0+\lambda_{n_k}v) \\ &\leq L(u,v_0),\ \forall u \in U. \end{aligned} \tag{10.107}$$

Hence, $L(u_0,v_0) = \min_{u\in A} L(u,v_0)$.
Observe that from (10.104) we have

$$G(v_0) \geq L(u_{n_k},v),$$

so that

$$G(v_0) \geq \liminf_{k\to\infty} L(u_{n_k},v) \geq L(u_0,v), \forall v \in B.$$

Denoting $\alpha = G(v_0)$ we have

$$\alpha = G(v_0) \geq L(u_0,v), \forall v \in B,$$

and

$$\alpha = G(v_0) = \min_{u\in U} L(u,v_0) \leq L(u,v_0),\ \forall u \in A.$$

From these last two results and Proposition 10.2 we have that (u_0, v_0) is a saddle point for L. Now assume that

$$G_v(u) = L(u,v)$$

is convex but not strictly convex $\forall v \in B$.

For each $n \in \mathbb{N}$ define L_n by

$$L_n(u,v) = L(u,v) + \|u\|_U/n.$$

In such a case

$$(G_v)_n(u) = L_n(u,v)$$

is strictly convex for all $n \in \mathbb{N}$.

From above we mainly obtain $(u_n, v_n) \in A \times B$ such that

$$\begin{aligned} L(u_n,v) + \|u_n\|_U/n &\leq L(u_n,v_n) + \|u_n\|_U/n \\ &\leq L(u,v_n) + \|u\|/n. \end{aligned} \tag{10.108}$$

Since $A \times B$ is weakly compact and $\{(u_n, v_n)\} \subset A \times B$, up to subsequence not relabeled, there exists $(u_0, v_0) \in A \times B$ such that

$$u_n \rightharpoonup u_0, \text{ weakly in } U,$$

$$v_n \rightharpoonup v_0, \text{ weakly in } Y,$$

so that

$$\begin{aligned} L(u_0,v) &\leq \liminf_{n\to\infty}(L(u_n,v) + \|u_n\|_U/n) \\ &\leq \limsup_{n\to\infty} L(u,v_n) + \|u\|_U/n \\ &\leq L(u,v_0). \end{aligned} \tag{10.109}$$

Hence,

$$L(u_0,v) \leq L(u,v_0), \; \forall u \in A, \; v \in B,$$

so that

$$L(u_0,v) \leq L(u_0,v_0) \leq L(u,v_0), \; \forall u \in A, \; v \in B.$$

This completes the proof.

In the next result we deal with more general situations.

Theorem 10.2. *Let U,Y be reflexive Banach spaces, $A \subset U$, $B \subset Y$ and let $L : A \times B \to \mathbb{R}$ be a functional.*

Suppose that

1. *$A \subset U$ is convex, closed, and nonempty.*
2. *$B \subset Y$ is convex, closed, and nonempty.*
3. *For each $u \in A$, $F_u(v) = L(u,v)$ is concave and upper semicontinuous.*

4. *For each $v \in B$, $G_v(u) = L(u,v)$ is convex and lower semicontinuous.*
5. *Either the set A is bounded or there exists $\tilde{v} \in B$ such that*

$$L(u,\tilde{v}) \to +\infty, \text{ as } \|u\| \to +\infty, \ u \in A.$$

6. *Either the set B is bounded or there exists $\tilde{u} \in A$ such that*

$$L(\tilde{u},v) \to -\infty, \text{ as } \|v\| \to +\infty, \ v \in B.$$

Under such hypotheses L has at least one saddle point $(u_0,v_0) \in A \times B$.

Proof. We prove the result just for the special case such that there exists $\tilde{v} \in B$ such that

$$L(u,\tilde{v}) \to +\infty, \text{ as } \|u\| \to +\infty, \ u \in A,$$

and B is bounded. The proofs of remaining cases are similar.

For each $n \in \mathbb{N}$ denote

$$A_n = \{u \in A \ : \ \|u\|_U \leq n\}.$$

Fix $n \in \mathbb{N}$. The sets A_n and B are closed, convex, and bounded, so that from the last Theorem 10.1 there exists a saddle point $(u_n,v_n) \in A_n \times B$ for

$$L : A_n \times B \to \mathbb{R}.$$

Hence,

$$L(u_n,v) \leq L(u_n,v_n) \leq L(u,v_n), \forall u \in A_n, \ v \in B.$$

For a fixed $\tilde{u} \in A_1$ we have

$$\begin{aligned} L(u_n,\tilde{v}) &\leq L(u_n,v_n) \\ &\leq L(\tilde{u},v_n) \\ &\leq \sup_{v \in B} L(\tilde{u},v) \equiv b \in \mathbb{R}. \end{aligned} \tag{10.110}$$

On the other hand, from the hypotheses,

$$G_{\tilde{v}}(u) = L(u,\tilde{v})$$

is convex, lower semicontinuous, and coercive, so that it is bounded below. Thus there exists $a \in \mathbb{R}$ such that

$$-\infty < a < G_{\tilde{v}}(u) = L(u,\tilde{v}), \ \forall u \in A.$$

Hence

$$a \leq L(u_n,\tilde{v}) \leq L(u_n,v_n) \leq b, \forall n \in \mathbb{N}.$$

Therefore $\{L(u_n,v_n)\}$ is bounded.

Moreover, from the coercivity hypotheses and

$$a \leq L(u_n,\tilde{v}) \leq b, \forall n \in \mathbb{N},$$

we may infer that $\{u_n\}$ is bounded.

Summarizing, $\{u_n\}, \{v_n\}$, and $\{L(u_n, v_n)\}$ are bounded sequences, and thus there exists a subsequence $\{n_k\}$, $u_0 \in A$, $v_0 \in B$, and $\alpha \in \mathbb{R}$ such that

$$u_{n_k} \rightharpoonup u_0, \text{ weakly in } U,$$

$$v_{n_k} \rightharpoonup v_0, \text{ weakly in } Y,$$

$$L(u_{n_k}, v_{n_k}) \to \alpha \in \mathbb{R},$$

as $k \to \infty$. Fix $(u,v) \in A \times B$. Observe that if $n_k > n_0 = \|u\|_U$, then

$$L(u_{n_k}, v) \leq L(u_{n_k}, v_{n_k}) \leq L(u, v_{n_k}),$$

so that letting $k \to \infty$, we obtain

$$\begin{aligned} L(u_0, v) &\leq \liminf_{k\to\infty} L(u_{n_k}, v) \\ &\leq \lim_{k\to\infty} L(u_{n_k}, v_{n_k}) = \alpha \\ &\leq \limsup_{k\to\infty} L(u, v_{n_k}) \\ &\leq L(u, v_0), \end{aligned} \tag{10.111}$$

that is,

$$L(u_0, v) \leq \alpha \leq L(u, v_0), \; \forall u \in A, \; v \in B.$$

From this and Proposition 10.2 we may conclude that (u_0, v_0) is a saddle point for $L : A \times B \to \mathbb{R}$.

The proof is complete.

10.4 Relaxation for the Scalar Case

In this section, $\Omega \subset \mathbb{R}^N$ denotes a bounded open set with a locally Lipschitz boundary. That is, for each point $x \in \partial\Omega$ there exists a neighborhood $\mathscr{U}_x$ whose intersection with $\partial\Omega$ is the graph of a Lipschitz continuous function.

We start with the following definition.

Definition 10.4.1. A function $u : \Omega \to \mathbb{R}$ is said to be affine if ∇u is constant on Ω. Furthermore, we say that $u : \Omega \to \mathbb{R}$ is piecewise affine if it is continuous and there exists a partition of Ω into a set of zero measure and finite number of open sets on which u is affine.

The proof of next result is found in [25].

Theorem 10.4.2. *Let $r \in \mathbb{N}$ and let u_k $1 < k < r$ be piecewise affine functions from Ω into $\mathbb{R}$ and $\{\alpha_k\}$ such that $\alpha_k > 0, \forall k \in \{1, \ldots, r\}$ and $\sum_{k=1}^r \alpha_k = 1$. Given $\varepsilon > 0$,* *there exists a locally Lipschitz function $u : \Omega \to \mathbb{R}$ and r disjoint open sets Ω_k, $1 \leq k \leq r$, such that*

$$|m(\Omega_k) - \alpha_k m(\Omega)| < \alpha_k \varepsilon, \ \forall k \in \{1, \ldots, r\}, \tag{10.112}$$

$$\nabla u(x) = \nabla u_k(x), \ a.e. \ on \ \Omega_k, \tag{10.113}$$

$$|\nabla u(x)| \leq \max_{1 \leq k \leq r} \{|\nabla u_k(x)|\}, \ a.e. \ on \ \Omega, \tag{10.114}$$

$$\left| u(x) - \sum_{k=1}^{r} \alpha_k u_k \right| < \varepsilon, \ \forall x \in \Omega, \tag{10.115}$$

$$u(x) = \sum_{k=1}^{r} \alpha_k u_k(x), \forall x \in \partial\Omega. \tag{10.116}$$

The next result is also found in [25].

Proposition 10.4.3. *Let $r \in \mathbb{N}$ and let u_k $1 \leq k \leq r$ be piecewise affine functions from Ω into $\mathbb{R}$. Consider a Carathéodory function $f : \Omega \times \mathbb{R}^N \to \mathbb{R}$ and a positive function $c \in L^1(\Omega)$ which satisfy*

$$c(x) \geq \sup\{|f(x,\xi)| \mid |\xi| \leq \max_{1 \leq k \leq r} \{\|\nabla u_k\|_\infty\}\}. \tag{10.117}$$

Given $\varepsilon > 0$, there exists a locally Lipschitz function $u : \Omega \to \mathbb{R}$ such that

$$\left| \int_\Omega f(x, \nabla u) dx - \sum_{k=1}^{r} \alpha_k \int_\Omega f(x, \nabla u_k) dx \right| < \varepsilon, \tag{10.118}$$

$$|\nabla u(x)| \leq \max_{1 \leq k \leq r} \{|\nabla u_k(x)|\}, \ a.e. \ in \ \Omega, \tag{10.119}$$

$$|u(x) - \sum_{k=1}^{r} \alpha_k u_k(x)| < \varepsilon, \forall x \in \Omega \tag{10.120}$$

$$u(x) = \sum_{k=1}^{r} \alpha_k u_k(x), \forall x \in \partial\Omega. \tag{10.121}$$

Proof. It is sufficient to establish the result for functions u_k affine over Ω, since Ω can be divided into pieces on which u_k are affine, and such pieces can be put together through (10.121). Let $\varepsilon > 0$ be given. We know that simple functions are dense in $L^1(\Omega)$, concerning the L^1 norm. Thus there exists a partition of Ω into a finite number of open sets $\mathscr{O}_i$, $1 \leq i \leq N_1$, and a negligible set, and there exists $\bar{f}_k$ constant functions over each $\mathscr{O}_i$ such that

$$\int_\Omega |f(x, \nabla u_k(x)) - \bar{f}_k(x)|dx < \varepsilon, \ 1 \leq k \leq r. \tag{10.122}$$

Now choose $\delta > 0$ such that

$$\delta \leq \frac{\varepsilon}{N_1(1 + \max_{1 \leq k \leq r}\{\|\bar{f}_k\|_\infty\})} \tag{10.123}$$

and if B is a measurable set

$$m(B) < \delta \Rightarrow \int_B c(x)dx \leq \varepsilon/N_1. \tag{10.124}$$

Now we apply Theorem 10.4.2, to each of the open sets $\mathscr{O}_i$; therefore there exists a locally Lipschitz function $u : \mathscr{O}_i \to \mathbb{R}$ and there exist r open disjoints spaces Ω_k^i, $1 \leq k \leq r$, such that

$$|m(\Omega_k^i) - \alpha_k m(\mathscr{O}_i)| \leq \alpha_k \delta, \text{ for } 1 \leq k \leq r, \tag{10.125}$$

$$\nabla u = \nabla u_k, \text{ a.e. in } \Omega_k^i, \tag{10.126}$$

$$|\nabla u(x)| \leq \max_{1 \leq k \leq r}\{|\nabla u_k(x)|\}, \text{ a.e. } \mathscr{O}_i, \tag{10.127}$$

$$\left| u(x) - \sum_{k=1}^r \alpha_k u_k(x) \right| \leq \delta, \forall x \in \mathscr{O}_i \tag{10.128}$$

$$u(x) = \sum_{k=1}^r \alpha_k u_k(x), \forall x \in \partial \mathscr{O}_i. \tag{10.129}$$

We can define $u = \sum_{k=1}^r \alpha_k u_k$ on $\Omega - \cup_{i=1}^{N_1} \mathscr{O}_i$. Therefore u is continuous and locally Lipschitz. Now observe that

$$\begin{aligned} &\int_{\mathscr{O}_i} f(x, \nabla u(x))dx - \sum_{k=1}^r \int_{\Omega_k^i} f(x, \nabla u_k(x))dx \\ &= \int_{\mathscr{O}_i - \cup_{k=1}^r \Omega_k^i} f(x, \nabla u(x))dx. \end{aligned} \tag{10.130}$$

From $|f(x, \nabla u(x))| \leq c(x)$, $m(\mathscr{O}_i - \cup_{k=1}^r \Omega_k^i) \leq \delta$ and (10.124) we obtain

$$\left|\int_{\mathscr{O}_i} f(x,\nabla u(x))dx - \sum_{k=1}^{r}\int_{\Omega_k^i} f(x,\nabla u_k(x)dx\right|$$
$$= \left|\int_{\mathscr{O}_i - \cup_{k=1}^{r}\Omega_k^i} f(x,\nabla u(x))dx\right| \leq \varepsilon/N_1. \tag{10.131}$$

Considering that $\bar{f}_k$ is constant in $\mathscr{O}_i$, from (10.123), (10.124), and (10.125), we obtain

$$\sum_{k=1}^{r}\left|\int_{\Omega_k^i}\bar{f}_k(x)dx - \alpha_k\int_{\mathscr{O}_i}\bar{f}_k(x)dx\right| < \varepsilon/N_1. \tag{10.132}$$

We recall that $\Omega_k = \cup_{i=1}^{N_1}\Omega_k^i$ so that

$$\left|\int_{\Omega} f(x,\nabla u(x))dx - \sum_{k=1}^{r}\alpha_k\int_{\Omega} f(x,\nabla u_k(x))dx\right|$$
$$\leq \left|\int_{\Omega} f(x,\nabla u(x))dx - \sum_{k=1}^{r}\int_{\Omega_k} f(x,\nabla u_k(x))dx\right|$$
$$+\sum_{k=1}^{r}\int_{\Omega_k}|f(x,\nabla u_k(x) - \bar{f}_k(x)|dx$$
$$+\sum_{k=1}^{r}\left|\int_{\Omega_k}\bar{f}_k(x)dx - \alpha_k\int_{\Omega}\bar{f}_k(x)dx\right|$$
$$+\sum_{k=1}^{r}\alpha_k\int_{\Omega}|\bar{f}_k(x) - f(x,\nabla u_k(x))|dx. \tag{10.133}$$

From (10.131), (10.122),(10.132), and (10.122) again, we obtain

$$\left|\int_{\Omega} f(x,\nabla u(x))dx - \sum_{k=1}^{r}\alpha_k\int_{\Omega} f(x,\nabla u_k)dx\right| < 4\varepsilon. \tag{10.134}$$

We do not prove the next result. It is a well-known result from the finite element theory.

Proposition 10.4.4. *If $u \in W_0^{1,p}(\Omega)$, there exists a sequence $\{u_n\}$ of piecewise affine functions over Ω, null on $\partial\Omega$, such that*

$$u_n \to u, \text{ in } L^p(\Omega) \tag{10.135}$$

and

$$\nabla u_n \to \nabla u, \text{ in } L^p(\Omega;\mathbb{R}^N). \tag{10.136}$$

Proposition 10.4.5. *For p such that $1 < p < \infty$, suppose that $f : \Omega \times \mathbb{R}^N \to \mathbb{R}$ is a Carathéodory function , for which there exist $a_1, a_2 \in L^1(\Omega)$ and constants $c_1 \geq c_2 > 0$ such that*

$$a_2(x) + c_2|\xi|^p \leq f(x,\xi) \leq a_1(x) + c_1|\xi|^p, \forall x \in \Omega,\ \xi \in \mathbb{R}^N. \tag{10.137}$$

Then, given $u \in W^{1,p}(\Omega)$ piecewise affine, $\varepsilon > 0$, and a neighborhood $\mathcal{V}$ of zero in the topology $\sigma(L^p(\Omega,\mathbb{R}^N), L^q(\Omega,\mathbb{R}^N))$, there exists a function $v \in W^{1,p}(\Omega)$ such that

$$\nabla v - \nabla u \in \mathcal{V}, \tag{10.138}$$

$$u = v \text{ on } \partial\Omega,$$

$$\|v - u\|_\infty < \varepsilon, \tag{10.139}$$

and

$$\left| \int_\Omega f(x, \nabla v(x))dx - \int_\Omega f^{**}(x, \nabla u(x))dx \right| < \varepsilon. \tag{10.140}$$

Proof. Suppose given $\varepsilon > 0$, $u \in W^{1,p}(\Omega)$ piecewise affine continuous, and a neighborhood $\mathcal{V}$ of zero, which may be expressed as

$$\mathcal{V} = \{w \in L^p(\Omega,\mathbb{R}^N) \mid \left| \int_\Omega h_m \cdot w dx \right| < \eta,$$
$$\forall m \in \{1,\ldots,M\}\}, \tag{10.141}$$

where $M \in \mathbb{N}$, $h_m \in L^q(\Omega,\mathbb{R}^N)$, $\eta \in \mathbb{R}^+$. By hypothesis, there exists a partition of Ω into a negligible set Ω_0 and open subspaces Δ_i, $1 \leq i \leq r$, over which $\nabla u(x)$ is constant. From standard results of convex analysis in $\mathbb{R}^N$, for each $i \in \{1,\ldots,r\}$, we can obtain $\{\alpha_k \geq 0\}_{1 \leq k \leq N+1}$ and ξ_k such that $\sum_{k=1}^{N+1} \alpha_k = 1$ and

$$\sum_{k=1}^{N+1} \alpha_k \xi_k = \nabla u, \forall x \in \Delta_i, \tag{10.142}$$

and

$$\sum_{k=1}^{N+1} \alpha_k f(x, \xi_k) = f^{**}(x, \nabla u(x)). \tag{10.143}$$

Define $\beta_i = \max_{k \in \{1,\ldots,N+1\}}\{|\xi_k| \quad on \quad \Delta_i\}$, and $\rho_1 = \max_{i \in \{1,\ldots,r\}}\{\beta_i\}$, and $\rho = \max\{\rho_1, \|\nabla u\|_\infty\}$. Now, observe that we can obtain functions $\hat{h}_m \in C_0^\infty(\Omega;\mathbb{R}^N)$ such that

$$\max_{m\in\{1,\ldots,M\}} \|\hat{h}_m - h_m\|_{L^q(\Omega,\mathbb{R}^N)} < \frac{\eta}{4\rho m(\Omega)}. \tag{10.144}$$

Define $C = \max_{m\in\{1,\ldots,M\}} \|div(\hat{h}_m)\|_{L^q(\Omega)}$ and we can also define

$$\varepsilon_1 = \min\{\varepsilon/4, 1/(m(\Omega))^{1/p}), \eta/(2Cm(\Omega))^{1/p}), 1/m(\Omega)\} \tag{10.145}$$

We recall that ρ does not depend on ε. Furthermore, for each $i \in \{1,\ldots,r\}$, there exists a compact subset $K_i \subset \Delta_i$ such that

$$\int_{\Delta_i - K_i} [a_1(x) + c_1(x) \max_{|\xi|\leq\rho}\{|\xi|^p\}]dx < \frac{\varepsilon_1}{r}. \tag{10.146}$$

Also, observe that the sets K_i may be obtained such that the restrictions of f and f^{**} to $K_i \times \rho B$ are continuous, so that from this and from the compactness of ρB, for all $x \in K_i$, we can find an open ball ω_x with center in x and contained in Ω, such that

$$|f^{**}(y, \nabla u(x)) - f^{**}(x, \nabla u(x))| < \frac{\varepsilon_1}{m(\Omega)}, \forall y \in \omega_x \cap K_i, \tag{10.147}$$

and

$$|f(y,\xi) - f(x,\xi)| < \frac{\varepsilon_1}{m(\Omega)}, \forall y \in \omega_x \cap K_i, \forall \xi \in \rho B. \tag{10.148}$$

Therefore, from this and (10.143), we may write

$$\left| f^{**}(y, \nabla u(x)) - \sum_{k=1}^{N+1} \alpha_k f(y, \xi_k) \right| < \frac{2\varepsilon_1}{m(\Omega)}, \forall y \in \omega_x \cap K_i. \tag{10.149}$$

We can cover the compact set K_i with a finite number of those open ball ω_x, denoted by ω_j, $1 \leq j \leq l$. Consider the open sets $\omega'_j = \omega_j - \cup_{i=1}^{j-1} \bar{\omega}_i$. We have that $\cup_{j=1}^{l} \bar{\omega}'_j = \cup_{j=1}^{l} \bar{\omega}_j$. Defining functions u_k, for $1 \leq k \leq N+1$ such that $\nabla u_k = \xi_k$ and $u = \sum_{k=1}^{N+1} \alpha_k u_k$, we may apply Proposition 10.4.3 to each of the open sets ω'_j, so that we obtain functions $v_i \in W^{1,p}(\Omega)$ such that

$$\left| \int_{\omega'_j} f(x, \nabla v_i(x)dx - \sum_{k=1}^{N+1} \alpha_k \int_{\omega'_j} f(x, \xi_k)dx \right| < \frac{\varepsilon_1}{rl}, \tag{10.150}$$

$$|\nabla v_i| < \rho, \forall x \in \omega'_j, \tag{10.151}$$

$$|v_i(x) - u(x)| < \varepsilon_1, \forall x \in \omega'_j, \tag{10.152}$$

and

$$v_i(x) = u(x), \forall x \in \partial \omega_j'. \tag{10.153}$$

Finally we set

$$v_i = u \text{ on } \Delta_i - \cup_{j=1}^{l} \omega_j. \tag{10.154}$$

We may define a continuous mapping $v : \Omega \to \mathbb{R}$ by

$$v(x) = v_i(x), \text{ if } x \in \Delta_i, \tag{10.155}$$

$$v(x) = u(x), \text{ if } x \in \Omega_0. \tag{10.156}$$

We have that $v(x) = u(x), \forall x \in \partial\Omega$, and $\|\nabla v\|_\infty < \rho$. Also, from (10.146)

$$\int_{\Delta_i - K_i} |f^{**}(x, \nabla u(x)| dx < \frac{\varepsilon_1}{r} \tag{10.157}$$

and

$$\int_{\Delta_i - K_i} |f(x, \nabla v(x)| dx < \frac{\varepsilon_1}{r}. \tag{10.158}$$

On the other hand, from (10.149) and (10.150)

$$\left| \int_{K_i \cap \omega_j'} f(x, \nabla v(x)) dx - \int_{K_i \cap \omega_j'} f^{**}(x, \nabla u(x)) dx \right| \\ \leq \frac{\varepsilon_1}{rl} + \frac{\varepsilon_1 m(\omega_j' \cap K_i)}{m(\Omega)} \tag{10.159}$$

so that

$$| \int_{K_i} f(x, \nabla v(x)) dx - \int_{K_i} f^{**}(x, \nabla u(x)) dx| \\ \leq \frac{\varepsilon_1}{r} + \frac{\varepsilon_1 m(K_i)}{m(\Omega)}. \tag{10.160}$$

Now summing up in i and considering (10.157) and (10.158) we obtain (10.140), that is,

$$| \int_\Omega f(x, \nabla v(x)) dx \quad - \quad \int_\Omega f^{**}(x, \nabla u(x)) dx| \quad < \quad 4\varepsilon_1 \quad \leq \quad \varepsilon. \tag{10.161}$$

Also, observe that from above, we have

$$\|v - u\|_\infty < \varepsilon_1, \tag{10.162}$$

and thus

$$\begin{aligned}\left|\int_\Omega \hat{h}_m\cdot(\nabla v(x)-\nabla u(x))dx\right| &= \left|-\int_\Omega div(\hat{h}_m)(v(x)-u(x))dx\right| \\ &\leq \|div(\hat{h}_m)\|_{L^q(\Omega)}\|v-u\|_{L^p(S)} \\ &\leq C\varepsilon_1 m(\Omega)^{1/p} \\ &< \frac{\eta}{2}.\end{aligned} \tag{10.163}$$

Also we have that

$$\begin{aligned}&\left|\int_\Omega (\hat{h}_m-h_m)\cdot(\nabla v-\nabla u)dx\right| \\ &\leq \|\hat{h}_m-h_m\|_{L^q(\Omega,\mathbb{R}^N)}\|\nabla v-\nabla u\|_{L^p(\Omega,\mathbb{R}^N)} \leq \frac{\eta}{2}.\end{aligned} \tag{10.164}$$

Thus

$$\left|\int_\Omega h_m\cdot(\nabla v-\nabla u)dx\right| < \eta, \forall m\in\{1,\ldots,M\}. \tag{10.165}$$

Theorem 10.4.6. *Assuming the hypothesis of last theorem, given a function $u\in W_0^{1,p}(\Omega)$, given $\varepsilon>0$, and a neighborhood of zero $\mathcal{V}$ in $\sigma(L^p(\Omega,\mathbb{R}^N),L^q(\Omega,\mathbb{R}^N))$, we have that there exists a function $v\in W_0^{1,p}(\Omega)$ such that*

$$\nabla v-\nabla u\in\mathcal{V}, \tag{10.166}$$

and

$$\left|\int_\Omega f(x,\nabla v(x))dx-\int_\Omega f^{**}(x,\nabla u(x))dx\right| < \varepsilon. \tag{10.167}$$

Proof. We can approximate u by a function w which is piecewise affine and null on the boundary. Thus, there exists $\delta>0$ such that we can obtain $w\in W_0^{1,p}(\Omega)$ piecewise affine such that

$$\|u-w\|_{1,p}<\delta \tag{10.168}$$

so that

$$\nabla w-\nabla u\in\frac{1}{2}\mathcal{V}, \tag{10.169}$$

and

$$\left|\int_\Omega f^{**}(x,\nabla w(x))dx-\int_\Omega f^{**}(x,\nabla u(x))dx\right| < \frac{\varepsilon}{2}. \tag{10.170}$$

From Proposition 10.4.5 we may obtain $v \in W_0^{1,p}(\Omega)$ such that

$$\nabla v - \nabla w \in \frac{1}{2}\mathscr{V}, \tag{10.171}$$

and

$$\left| \int_\Omega f^{**}(x, \nabla w(x))dx - \int_\Omega f(x, \nabla v(x))dx \right| < \frac{\varepsilon}{2}. \tag{10.172}$$

From (10.170) and (10.172)

$$\left| \int_\Omega f^{**}(x, \nabla u(x))dx - \int_\Omega f(x, \nabla v(x))dx \right| < \varepsilon. \tag{10.173}$$

Finally, from (10.169), (10.171), and from the fact that weak neighborhoods are convex, we have

$$\nabla v - \nabla u \in \mathscr{V}. \tag{10.174}$$

To finish this chapter, we present two theorems which summarize the last results.

Theorem 10.4.7. *Let f be a Carathéodory function from $\Omega \times \mathbb{R}^N$ into $\mathbb{R}$ which satisfies*

$$a_2(x) + c_2|\xi|^p \leq f(x, \xi) \leq a_1(x) + c_1|\xi|^p \tag{10.175}$$

where $a_1, a_2 \in L^1(\Omega)$, $1 < p < +\infty$, $b \geq 0$, and $c_1 \geq c_2 > 0$. Under such assumptions, defining $\hat{U} = W_0^{1,p}(\Omega)$, we have

$$\inf_{u \in \hat{U}} \left\{ \int_\Omega f(x, \nabla u)dx \right\} = \min_{u \in \hat{U}} \left\{ \int_\Omega f^{**}(x, \nabla u)dx \right\} \tag{10.176}$$

The solutions of relaxed problem are weak cluster points in $W_0^{1,p}(\Omega)$ of the minimizing sequences of primal problem.

Proof. The existence of solutions for the convex relaxed formulation is a consequence of the reflexivity of U and coercivity hypothesis, which allows an application of the direct method of calculus of variations. That is, considering a minimizing sequence, from above (coercivity hypothesis), such a sequence is bounded and has a weakly convergent subsequence to some $\hat{u} \subset W^{1,p}(\Omega)$. Finally, from the lower semicontinuity of relaxed formulation, we may conclude that $\hat{u}$ is a minimizer. The relation (10.176) follows from last theorem.

Theorem 10.4.8. *Let f be a Carathéodory function from $\Omega \times \mathbb{R}^N$ into $\mathbb{R}$ which satisfies*

$$a_2(x) + c_2|\xi|^p \leq f(x, \xi) \leq a_1(x) + c_1|\xi|^p \tag{10.177}$$

where $a_1,\ a_2 \in L^1(\Omega)$, $1 < p < +\infty$, $b \geq 0$ *and* $c_1 \geq c_2 > 0$. *Let* $u_0 \in W^{1,p}(\Omega)$. *Under such assumptions, defining* $\hat{U} = \{u \mid u - u_0 \in W_0^{1,p}(\Omega)\}$, *we have*

$$\inf_{u\in\hat{U}}\left\{\int_\Omega f(x,\nabla u)dx\right\} = \min_{u\in\hat{U}}\left\{\int_\Omega f^{**}(x,\nabla u)dx\right\} \qquad (10.178)$$

The solutions of relaxed problem are weak cluster points in $W^{1,p}(\Omega)$ *of the minimizing sequences of primal problem.*

Proof. Just apply the last theorem to the integrand $g(x,\xi) = f(x,\xi + \nabla u_0)$. For details see [25].

10.5 Duality Suitable for the Vectorial Case

10.5.1 The Ekeland Variational Principle

In this section we present and prove the Ekeland variational principle. This proof may be found in Giusti, [39], pp. 160–161.

Theorem 10.5.1 (Ekeland Variational Principle). *Let* (U,d) *be a complete metric space and let* $F : U \to \overline{\mathbb{R}}$ *be a lower semicontinuous bounded below functional taking a finite value at some point.*

Let $\varepsilon > 0$. *Assume for some* $u \in U$ *we have*

$$F(u) \leq \inf_{u\in U}\{F(u)\} + \varepsilon.$$

Under such hypotheses, there exists $v \in U$ *such that*

1. $d(u,v) \leq 1$,
2. $F(v) \leq F(u)$,
3. $F(v) \leq F(w) + \varepsilon d(v,w)$, $\forall w \in U$.

Proof. Define the sequence $\{u_n\} \subset U$ by

$$u_1 = u,$$

and having $u_1,\ldots,u_n$, select u_{n+1} as specified in the next lines. First, define

$$S_n = \{w \in U \mid F(w) \leq F(u_n) - \varepsilon d(u_n,w)\}.$$

Observe that $u_n \in S_n$ so that S_n in nonempty.

On the other hand, from the definition of infimum, we may select $u_{n+1} \in S_n$ such that

$$F(u_{n+1}) \leq \frac{1}{2}\left\{F(u_n) + \inf_{w\in S_n}\{F(w)\}\right\}. \qquad (10.179)$$

Since $u_{n+1} \in S_n$ we have

$$\varepsilon d(u_{n+1}, u_n) \leq F(u_n) - F(u_{n+1}). \tag{10.180}$$

and hence

$$\varepsilon d(u_{n+m}, u_n) \leq \sum_{i=1}^{m} d(u_{n+i}, u_{n+i-1}) \leq F(u_n) - F(u_{n+m}). \tag{10.181}$$

From (10.180) $\{F(u_n)\}$ is decreasing sequence bounded below by $\inf_{u \in U} F(u)$ so that there exists $\alpha \in \mathbb{R}$ such that

$$F(u_n) \to \alpha \text{ as } n \to \infty.$$

From this and (10.181), $\{u_n\}$ is a Cauchy sequence , converging to some $v \in U$. Since F is lower semicontinuous we get

$$\alpha = \liminf_{m \to \infty} F(u_{n+m}) \geq F(v),$$

so that letting $m \to \infty$ in (10.181) we obtain

$$\varepsilon d(u_n, v) \leq F(u_n) - F(v), \tag{10.182}$$

and, in particular, for $n = 1$ we get

$$0 \leq \varepsilon d(u, v) \leq F(u) - F(v) \leq F(u) - \inf_{u \in U} F(u) \leq \varepsilon.$$

Thus, we have proven 1 and 2.

Suppose, to obtain contradiction, that 3 does not hold.

Hence, there exists $w \in U$ such that

$$F(w) < F(v) - \varepsilon d(w, v).$$

In particular we have

$$w \neq v. \tag{10.183}$$

Thus, from this and (10.182), we have

$$F(w) < F(u_n) - \varepsilon d(u_n, v) - \varepsilon d(w, v) \leq F(u_n) - \varepsilon d(u_n, w), \forall n \in \mathbb{N}.$$

Now observe that $w \in S_n, \forall n \in \mathbb{N}$ so that

$$\inf_{w \in S_n} \{F(w)\} \leq F(w), \forall n \in \mathbb{N}.$$

From this and (10.179) we obtain

$$2F(u_{n+1}) - F(u_n) \leq F(w) < F(v) - \varepsilon d(v, w),$$

so that

$$2\liminf_{n\to\infty}\{F(u_{n+1})\} \leq F(v) - \varepsilon d(v,w) + \liminf_{n\to\infty}\{F(u_n)\}.$$

Hence,

$$F(v) \leq \liminf_{n\to\infty}\{F(u_{n+1})\} \leq F(v) - \varepsilon d(v,w),$$

so that

$$0 \leq -\varepsilon d(v,w),$$

which contradicts (10.183).

Thus 3 holds.

Remark 10.5.2. We may introduce in U a new metric given by $d_1 = \varepsilon^{1/2}d$. We highlight that the topology remains the same and also F remains lower semicontinuous. Under the hypotheses of the last theorem, if there exists $u \in U$ such that $F(u) < \inf_{u\in U} F(u) + \varepsilon$, then there exists $v \in U$ such that

1. $d(u,v) \leq \varepsilon^{1/2}$,
2. $F(v) \leq F(u)$,
3. $F(v) \leq F(w) + \varepsilon^{1/2}d(u,w), \forall w \in U$.

Remark 10.5.3. Observe that if U is a Banach space,

$$F(v) - F(v+tw) \leq \varepsilon^{1/2}t\|w\|_U, \forall t \in [0,1], \;\; w \in U, \tag{10.184}$$

so that if F is Gâteaux differentiable, we obtain

$$-\langle \delta F(v), w\rangle_U \leq \varepsilon^{1/2}\|w\|_U. \tag{10.185}$$

Similarly

$$F(v) - F(v+t(-w)) \leq \varepsilon^{1/2}t\|w\|_U \leq, \forall t \in [0,1], \;\; w \in U, \tag{10.186}$$

so that if F is Gâteaux differentiable, we obtain

$$\langle \delta F(v), w\rangle_U \leq \varepsilon^{1/2}\|w\|_U. \tag{10.187}$$

Thus

$$\|\delta F(v)\|_{U^*} \leq \varepsilon^{1/2}. \tag{10.188}$$

We have thus obtained, from the last theorem and remarks, the following result.

Theorem 10.5.4. *Let U be a Banach space. Let $F : U \to \mathbb{R}$ be a lower semicontinuous Gâteaux differentiable functional. Given $\varepsilon > 0$ suppose that $u \in U$ is such that*

$$F(u) \leq \inf_{u\in U}\{F(u)\} + \varepsilon. \tag{10.189}$$

Then there exists $v \in U$ such that

$$F(v) \leq F(u), \tag{10.190}$$

$$\|u - v\|_U \leq \sqrt{\varepsilon}, \tag{10.191}$$

and

$$\|\delta F(v)\|_{U^*} \leq \sqrt{\varepsilon}. \tag{10.192}$$

The next theorem easily follows from above results.

Theorem 10.5.5. *Let $J : U \to \mathbb{R}$ be defined by*

$$J(u) = G(\nabla u) - \langle f, u \rangle_{L^2(S;\mathbb{R}^N)}, \tag{10.193}$$

where

$$U = W_0^{1,2}(S;\mathbb{R}^N), \tag{10.194}$$

We suppose G is a l.s.c and Gâteaux differentiable so that J is bounded below. Then, given $\varepsilon > 0$, there exists $u_\varepsilon \subset U$ such that

$$J(u_\varepsilon) < \inf_{u \in U}\{J(u)\} + \varepsilon, \tag{10.195}$$

and

$$\|\delta J(u_\varepsilon)\|_{U^*} < \sqrt{\varepsilon}. \tag{10.196}$$

We finish this chapter with an important result for vectorial problems in the calculus of variations.

Theorem 10.5.6. *Let U be a reflexive Banach space. Consider $(G \circ \Lambda) : U \to \mathbb{R}$ and $(F \circ \Lambda_1) : U \to \mathbb{R}$ l.s.c. functionals such that $J : U \to \mathbb{R}$ defined as*

$$J(u) = (G \circ \Lambda)(u) - (F \circ \Lambda_1)(u) - \langle u, f \rangle_U$$

is below bounded. (Here $\Lambda : U \to Y$ and $\Lambda_1 : U \to Y_1$ are continuous linear operators whose adjoint operators are denoted by $\Lambda^* : Y^* \to U^*$ and $\Lambda_1^* : Y^* \to U^*$, respectively). Also we suppose the existence of $L : Y_1 \to Y$ continuous and linear operator such that L^* is onto and

$$\Lambda(u) - L(\Lambda_1(u)), \forall u \in U.$$

Under such assumptions, we have

$$\inf_{u \in U}\{J(u)\} \geq \sup_{v^* \in A^*}\{\inf_{z^* \in Y_1^*}\{F^*(L^*z^*) - G^*(v^* + z^*)\}\},$$

where

$$A^* = \{v^* \in Y^* \mid \Lambda^* v^* = f\}.$$

In addition we assume $(F \circ \Lambda_1) : U \to \mathbb{R}$ is convex and Gâteaux differentiable, and suppose there exists a solution (v_0^*, z_0^*) of the dual formulation, so that

$$L\left(\frac{\partial F^*(L^* z_0^*)}{\partial v^*}\right) \in \partial G^*(v_0^* + z_0^*),$$

$$\Lambda^* v_0^* - f = 0.$$

Suppose $u_0 \in U$ is such that

$$\frac{\partial F^*(L^* z_0^*)}{\partial v^*} = \Lambda_1 u_0,$$

so that

$$\Lambda u_0 \in \partial G^*(v_0^* + z_0^*).$$

Also we assume that there exists a sequence $\{u_n\} \subset U$ such that $u_n \rightharpoonup u_0$ weakly in U and

$$G(\Lambda u_n) \to G^{**}(\Lambda u_0) \text{ as } n \to \infty.$$

Under these additional assumptions we have

$$\begin{aligned}\inf_{u \in U}\{J(u)\} &= \max_{v^* \in A^*}\{\inf_{z^* \in Y_1^*}\{F^*(L^* z^*) - G^*(v^* + z^*)\}\} \\ &= F^*(L^* z_0^*) - G^*(v_0^* + z_0^*).\end{aligned}$$

Proof. Observe that

$$G^*(v^* + z^*) \geq \langle \Lambda u, v^* \rangle_Y + \langle \Lambda u, z^* \rangle_Y - G(\Lambda u), \forall u \in U,$$

that is,

$$\begin{aligned}&-F^*(L^* z^*) + G^*(v^* + z^*) \geq \langle u, f \rangle_U - F^*(L^* z^*) + \langle \Lambda_1 u, L^* z^* \rangle_{Y_1} \\ &- G(\Lambda u), \; \forall u \in U, v^* \in A^*\end{aligned}$$

so that

$$\begin{aligned}&\sup_{z^* \in Y_1^*}\{-F^*(L^* z^*) + G^*(v^* + z^*)\} \\ &\qquad \geq \sup_{z^* \in Y_1^*}\{\langle u, f \rangle_U - F^*(L^* z^*) + \langle \Lambda_1 u, L^* z^* \rangle_{Y_1} - G(\Lambda u)\},\end{aligned}$$

$\forall v^* \in A^*$, $u \in U$ and therefore

$$G(\Lambda u) - F(\Lambda_1 u) - \langle u, f \rangle_U \geq \inf_{z^* \in Y_1^*}\{F^*(L^* z^*) - G^*(v^* + z^*)\},$$

$$\forall v^* \in A^*, \; u \in U$$

which means

$$\inf_{u \in U}\{J(u)\} \geq \sup_{v^* \in A^*}\{\inf_{z^* \in Y_1^*}\{F^*(L^*z^*) - G^*(v^* + z^*)\}\},$$

where

$$A^* = \{v^* \in Y^* \mid \Lambda^* v^* = f\}.$$

Now suppose

$$L\left(\frac{\partial F^*(L^* z_0^*)}{\partial v^*}\right) \in \partial G^*(v_0^* + z_0^*),$$

and $u_0 \in U$ is such that

$$\frac{\partial F^*(L^* z_0^*)}{\partial v^*} = \Lambda_1 u_0.$$

Observe that

$$\Lambda u_0 = L(\Lambda_1 u_0) \in \partial G(v_0^* + z_0^*)$$

implies that

$$G^*(v_0^* + z_0^*) = \langle \Lambda u_0, v_0^* \rangle_Y + \langle \Lambda u_0, z_0^* \rangle_Y - G^{**}(\Lambda u_0).$$

From the hypothesis

$$u_n \rightharpoonup u_0 \text{ weakly in } U$$

and

$$G(\Lambda u_n) \to G^{**}(\Lambda u_0) \text{ as } n \to \infty.$$

Thus, given $\varepsilon > 0$, there exists $n_0 \in \mathbb{N}$ such that if $n \geq n_0$ then

$$G^*(v_0^* + z_0^*) - \langle \Lambda u_n, v_0^* \rangle_Y - \langle \Lambda u_n, z_0^* \rangle_Y + G(\Lambda u_n) < \varepsilon/2.$$

On the other hand, since $F(\Lambda_1 u)$ is convex and l.s.c., we have

$$\limsup_{n \to \infty}\{-F(\Lambda_1 u_n)\} \leq -F(\Lambda_1 u_0).$$

Hence, there exists $n_1 \in \mathbb{N}$ such that if $n \geq n_1$, then

$$\langle \Lambda u_n, z_0^* \rangle_Y - F(\Lambda_1 u_n) \leq \langle \Lambda u_0, z_0^* \rangle_Y - F(\Lambda_1 u_0) + \frac{\varepsilon}{2} = F^*(L^* z_0^*) + \frac{\varepsilon}{2},$$

so that for all $n \geq \max\{n_0, n_1\}$ we obtain

$$G^*(v_0^* + z_0^*) - F^*(L^* z_0^*) - \langle u_n, f \rangle_U - F(\Lambda_1 u_n) + G(\Lambda u_n) < \varepsilon.$$

Since ε is arbitrary, the proof is complete.

Chapter 11
Constrained Variational Optimization

11.1 Basic Concepts

For this chapter the most relevant reference is the excellent book of Luenberger [47], where more details may be found. Other relevant references are [15, 40–42]. We start with the definition of cone.

Definition 11.1.1 (Cone). Given a Banach space U, we say that $C \subset U$ is a cone with the vertex at the origin; if given $u \in C$, we have that $\lambda u \in C$, $\forall \lambda \geq 0$. By analogy we define a cone with the vertex at $p \in U$ as $P = p + C$, where C is any cone with the vertex at the origin. From now on we consider only cones with vertex at origin, unless otherwise indicated.

Definition 11.1.2. Let P be a convex cone in U. For $u, v \in U$ we write $u \geq v$ (with respect to P) if $u - v \in P$. In particular $u \geq \theta$ if and only if $u \in C$. Also

$$P^+ = \{u^* \in U^* \mid \langle u, u^* \rangle_U \geq 0, \forall u \in P\}. \tag{11.1}$$

If $u^* \in P^+$, we write $u^* \geq \theta^*$.

Proposition 11.1.3. *Let U be a Banach space and P be a closed cone in U. If $u \in U$ satisfies $\langle u, u^* \rangle_U \geq 0$, $\forall u^* \geq \theta^*$, then $u \geq \theta$.*

Proof. We prove the contrapositive. Assume $u \notin P$. Then by the separating hyperplane theorem there is an $u^* \in U^*$ such that $\langle u, u^* \rangle_U < \langle p, u^* \rangle_U, \forall p \in P$. Since P is a cone we must have $\langle p, u^* \rangle_U \geq 0$; otherwise we would have $\langle u, u^* \rangle > \langle \alpha p, u^* \rangle_U$ for some $\alpha > 0$. Thus $u^* \in P^+$. Finally, since $\inf_{p \in P}\{\langle p, u^* \rangle_U\} = 0$, we obtain $\langle u, u^* \rangle_U < 0$ which completes the proof.

Definition 11.1.4 (Convex Mapping). Let U, Z be vector spaces. Let $P \subset Z$ be a cone. A mapping $G : U \to Z$ is said to be convex if the domain of G is convex and

$$G(\alpha u_1 + (1-\alpha)u_2) \leq \alpha G(u_1) + (1-\alpha)G(u_2),$$
$$\forall u_1, u_2 \in U, \alpha \in [0,1]. \tag{11.2}$$

F. Botelho, *Functional Analysis and Applied Optimization in Banach Spaces: Applications to Non-Convex Variational Models*, DOI 10.1007/978-3-319-06074-3_11,

Consider the problem $\mathscr{P}$, defined as

$$\textit{Problem } \mathscr{P}: \text{ Minimize } F: U \to \mathbb{R} \text{ subject to } u \in \Omega, \text{ and } G(u) \leq \theta$$

Define

$$\omega(z) = \inf\{F(u) \mid u \in \Omega \text{ and } G(u) \leq z\}. \tag{11.3}$$

For such a functional we have the following result.

Proposition 11.1.5. *If F is a real convex functional and G is convex, then ω is convex.*

Proof. Observe that

$$\begin{aligned}\omega(\alpha z_1 + (1-\alpha)z_2) =& \inf\{F(u) \mid u \in \Omega \\ & \text{and } G(u) \leq \alpha z_1 + (1-\alpha)z_2\} \end{aligned} \tag{11.4}$$

$$\begin{aligned}\leq& \inf\{F(u) \mid u = \alpha u_1 + (1-\alpha)u_2 \; u_1, u_2 \in \Omega \\ & \text{and } G(u_1) \leq z_1, \; G(u_2) \leq z_2\}\end{aligned} \tag{11.5}$$

$$\begin{aligned}\leq& \alpha \inf\{F(u_1) \mid u_1 \in \Omega, \; G(u_1) \leq z_1\} \\ & + (1-\alpha)\inf\{F(u_2) \mid u_2 \in \Omega, \; G(u_2) \leq z_2\}\end{aligned} \tag{11.6}$$

$$\leq \alpha\omega(z_1) + (1-\alpha)\omega(z_2). \tag{11.7}$$

Now we establish the Lagrange multiplier theorem for convex global optimization.

Theorem 11.1.6. *Let U be a vector space, Z a Banach space, Ω a convex subset of U, and P a positive cone of Z. Assume that P contains an interior point. Let F be a real convex functional on Ω and G a convex mapping from Ω into Z. Assume the existence of $u_1 \in \Omega$ such that $G(u_1) < \theta$. Defining*

$$\mu_0 = \inf_{u \in \Omega}\{F(u) \mid G(u) \leq \theta\}, \tag{11.8}$$

then there exists $z_0^* \geq \theta$, $z_0^* \in Z^*$ such that

$$\mu_0 = \inf_{u \in \Omega}\{F(u) + \langle G(u), z_0^* \rangle_Z\}. \tag{11.9}$$

Furthermore, if the infimum in (11.8) is attained by $u_0 \in U$ such that $G(u_0) \leq \theta$, it is also attained in (11.9) by the same u_0 and also $\langle G(u_0), z_0^* \rangle_Z = 0$. We refer to z_0^* as the Lagrangian multiplier.

Proof. Consider the space $W = \mathbb{R} \times Z$ and the sets A, B where

$$A = \{(r,z) \in \mathbb{R} \times Z \mid r \geq F(u), \; z \geq G(u) \; \textit{for some} \; u \in \Omega\}, \tag{11.10}$$

and

$$B = \{(r,z) \in \mathbb{R} \times Z \mid r \leq \mu_0, \ z \leq \theta\}, \tag{11.11}$$

where $\mu_0 = \inf_{u \in \Omega}\{F(u) \mid G(u) \leq \theta\}$. Since F and G are convex, A and B are convex sets. It is clear that A contains no interior point of B, and since $N = -P$ contains an interior point, the set B contains an interior point. Thus, from the separating hyperplane theorem, there is a nonzero element $w_0^* = (r_0, z_0^*) \in W^*$ such that

$$r_0 r_1 + \langle z_1, z_0^* \rangle_Z \geq r_0 r_2 + \langle z_2, z_0^* \rangle_Z, \forall (r_1, z_1) \in A, \ (r_2, z_2) \in B. \tag{11.12}$$

From the nature of B it is clear that $w_0^* \geq \theta$. That is, $r_0 \geq 0$ and $z_0^* \geq \theta$. We will show that $r_0 > 0$. The point $(\mu_0, \theta) \in B$; hence

$$r_0 r + \langle z, z_0^* \rangle_Z \geq r_0 \mu_0, \forall (r,z) \in A. \tag{11.13}$$

If $r_0 = 0$, then $\langle G(u_1), z_0^* \rangle_Z \geq 0$ and $z_0^* \neq \theta$. Since $G(u_1) < \theta$ and $z_0^* \geq \theta$ we have a contradiction. Therefore $r_0 > 0$ and, without loss of generality, we may assume $r_0 = 1$. Since the point (μ_0, θ) is arbitrarily close to A and B, we have

$$\begin{aligned}\mu_0 = \inf_{(r,z) \in A} \{r + \langle z, z_0^* \rangle_Z\} &\leq \inf_{u \in \Omega} \{F(u) + \langle G(u), z_0^* \rangle_Z\} \\ &\leq \inf\{F(u) \mid u \in \Omega, \ G(u) \leq \theta\} = \mu_0. \end{aligned} \tag{11.14}$$

Also, if there exists u_0 such that $G(u_0) \leq \theta$, $\mu_0 = F(u_0)$, then

$$\mu_0 \leq F(u_0) + \langle G(u_0), z_0^* \rangle_Z \leq F(u_0) = \mu_0. \tag{11.15}$$

Hence

$$\langle G(u_0), z_0^* \rangle_Z = 0. \tag{11.16}$$

Corollary 11.1.7. *Let the hypothesis of the last theorem hold. Suppose*

$$F(u_0) = \inf_{u \in \Omega} \{F(u) \mid G(u) \leq \theta\}. \tag{11.17}$$

Then there exists $z_0^* \geq \theta$ such that the Lagrangian $L : U \times Z^* \to \mathbb{R}$ defined by

$$L(u, z^*) = F(u) + \langle G(u), z^* \rangle_Z \tag{11.18}$$

has a saddle point at (u_0, z_0^*). That is

$$L(u_0, z^*) \leq L(u_0, z_0^*) \leq L(u, z_0^*), \forall u \in \Omega, z^* \geq \theta. \tag{11.19}$$

Proof. For z_0^* obtained in the last theorem, we have

$$L(u_0, z_0^*) \leq L(u, z_0^*), \forall u \in \Omega. \tag{11.20}$$

As $\langle G(u_0), z_0^* \rangle_Z = 0$, we have

$$L(u_0, z^*) - L(u_0, z_0^*) = \langle G(u_0), z^* \rangle_Z - \langle G(u_0), z_0^* \rangle_Z$$
$$= \langle G(u_0), z^* \rangle_Z \leq 0. \quad (11.21)$$

We now prove two theorems relevant to develop the subsequent section.

Theorem 11.1.8. *Let $F : \Omega \subset U \to \mathbb{R}$ and $G : \Omega \to Z$. Let $P \subset Z$ be a cone. Suppose there exists $(u_0, z_0^*) \in U \times Z^*$ where $z_0^* \geq \theta$ and $u_0 \in \Omega$ are such that*

$$F(u_0) + \langle G(u_0), z_0^* \rangle_Z \leq F(u) + \langle G(u), z_0^* \rangle_Z, \forall u \in \Omega. \quad (11.22)$$

Then

$$F(u_0) + \langle G(u_0), z_0^* \rangle_Z$$
$$= \inf\{F(u) \mid u \in \Omega \text{ and } G(u) \leq G(u_0)\}. \quad (11.23)$$

Proof. Suppose there is a $u_1 \in \Omega$ such that $F(u_1) < F(u_0)$ and $G(u_1) \leq G(u_0)$. Thus

$$\langle G(u_1), z_0^* \rangle_Z \leq \langle G(u_0), z_0^* \rangle_Z \quad (11.24)$$

so that

$$F(u_1) + \langle G(u_1), z_0^* \rangle_Z < F(u_0) + \langle G(u_0), z_0^* \rangle_Z, \quad (11.25)$$

which contradicts the hypothesis of the theorem.

Theorem 11.1.9. *Let F be a convex real functional and $G : \Omega \to Z$ convex and let u_0 and u_1 be solutions to the problems $\mathscr{P}_0$ and $\mathscr{P}_1$ respectively, where*

$$\mathscr{P}_0 : \text{ minimize } F(u) \text{ subject to } u \in \Omega \text{ and } G(u) \leq z_0, \quad (11.26)$$

and

$$\mathscr{P}_1 : \text{ minimize } F(u) \text{ subject to } u \in \Omega \text{ and } G(u) \leq z_1. \quad (11.27)$$

Suppose z_0^* and z_1^* are the Lagrange multipliers related to these problems. Then

$$\langle z_1 - z_0, z_1^* \rangle_Z \leq F(u_0) - F(u_1) \leq \langle z_1 - z_0, z_0^* \rangle_Z. \quad (11.28)$$

Proof. For u_0, z_0^* we have

$$F(u_0) + \langle G(u_0) - z_0, z_0^* \rangle_Z \leq F(u) + \langle G(u) - z_0, z_0^* \rangle_Z, \forall u \in \Omega, \quad (11.29)$$

and, particularly for $u = u_1$ and considering that $\langle G(u_0) - z_0, z_0^* \rangle_Z = 0$, we obtain

$$F(u_0) - F(u_1) \leq \langle G(u_1) - z_0, z_0^* \rangle_Z \leq \langle z_1 - z_0, z_0^* \rangle_Z. \quad (11.30)$$

A similar argument applied to u_1, z_1^* provides us the other inequality.

11.2 Duality

Consider the basic convex programming problem:

$$\text{Minimize } F(u) \text{ subject to } G(u) \leq \theta,\ u \in \Omega, \tag{11.31}$$

where $F : U \to \mathbb{R}$ is a convex functional, $G : U \to Z$ is convex mapping, and Ω is a convex set. We define $\varphi : Z^* \to \mathbb{R}$ by

$$\varphi(z^*) = \inf_{u \in \Omega}\{F(u) + \langle G(u), z^* \rangle_Z\}. \tag{11.32}$$

Proposition 11.2.1. *φ is concave and*

$$\varphi(z^*) = \inf_{z \in \Gamma}\{\omega(z) + \langle z, z^* \rangle_Z\}, \tag{11.33}$$

where

$$\omega(z) = \inf_{u \in \Omega}\{F(u) \mid G(u) \leq z\}, \tag{11.34}$$

and

$$\Gamma = \{z \in Z \mid G(u) \leq z \text{ for some } u \in \Omega\}.$$

Proof. Observe that

$$\begin{aligned} \varphi(z^*) &= \inf_{u \in \Omega}\{F(u) + \langle G(u), z^* \rangle_Z\} \\ &\leq \inf_{u \in \Omega}\{F(u) + \langle z, z^* \rangle_Z \mid G(u) \leq z\} \\ &= \omega(z) + \langle z, z^* \rangle_Z, \forall z^* \geq \theta, z \in \Gamma. \end{aligned} \tag{11.35}$$

On the other hand, for any $u_1 \in \Omega$, defining $z_1 = G(u_1)$, we obtain

$$\begin{aligned} F(u_1) + \langle G(u_1), z^* \rangle_Z &\geq \inf_{u \in \Omega}\{F(u) + \langle z_1, z^* \rangle_Z \mid G(u) \leq z_1\} \\ &= \omega(z_1) + \langle z_1, z^* \rangle_Z, \end{aligned} \tag{11.36}$$

so that

$$\varphi(z^*) \geq \inf_{z \in \Gamma}\{\omega(z) + \langle z, z^* \rangle_Z\}. \tag{11.37}$$

Theorem 11.2.2 (Lagrange Duality). *Consider $F : \Omega \subset U \to \mathbb{R}$ is a convex functional, Ω a convex set, and $G : U \to Z$ a convex mapping. Suppose there exists a u_1 such that $G(u_1) < \theta$ and that $\inf_{u \in \Omega}\{F(u) \mid G(u) \leq \theta\} < \infty$. Under such assumptions, we have*

$$\inf_{u \in \Omega}\{F(u) \mid G(u) \leq \theta\} = \max_{z^* \geq \theta}\{\varphi(z^*)\}. \tag{11.38}$$

If the infimum on the left side in (11.38) is achieved at some $u_0 \in U$ and the max on the right side at $z_0^* \in Z^*$, then

$$\langle G(u_0), z_0^* \rangle_Z = 0 \tag{11.39}$$

and u_0 minimizes $F(u) + \langle G(u), z_0^* \rangle_Z$ on Ω.

Proof. For $z^* \geq \theta$ we have

$$\begin{aligned} \inf_{u \in \Omega} \{F(u) + \langle G(u), z^* \rangle_Z\} &\leq \inf_{u \in \Omega, G(u) \leq \theta} \{F(u) + \langle G(u), z^* \rangle_Z\} \\ &\leq \inf_{u \in \Omega, G(u) \leq \theta} F(u) \leq \mu_0. \end{aligned} \tag{11.40}$$

or

$$\varphi(z^*) \leq \mu_0. \tag{11.41}$$

The result follows from Theorem 11.1.6.

11.3 The Lagrange Multiplier Theorem

Remark 11.3.1. This section was published in similar form by the journal *Computational and Applied Mathematics*, SBMAC-Springer, reference [15].

In this section we develop a new and simpler proof of the Lagrange multiplier theorem in a Banach space context. In particular, we address the problem of minimizing a functional $F : U \to \mathbb{R}$ subject to $G(u) = \theta$, where θ denotes the zero vector and $G : U \to Z$ is a Fréchet differentiable transformation. Here U, Z are Banach spaces. General results on Banach spaces may be found in [1, 26], for example. For the theorem in question, among others, we would cite [13, 40, 47]. Specially the proof given in [47] is made through the generalized inverse function theorem. We emphasize such a proof is extensive and requires the continuous Fréchet differentiability of F and G. Our approach here is different and the results are obtained through other hypotheses.

The main result is summarized by the following theorem.

Theorem 11.3.2. *Let U and Z be Banach spaces. Assume u_0 is a local minimum of $F(u)$ subject to $G(u) = \theta$, where $F : U \to \mathbb{R}$ is a Gâteaux differentiable functional and $G : U \to Z$ is a Fréchet differentiable transformation such that $G'(u_0)$ maps U onto Z. Finally, assume there exist $\alpha > 0$ and $K > 0$ such that if $\|\varphi\|_U < \alpha$, then*

$$\|G'(u_0 + \varphi) - G'(u_0)\| \leq K \|\varphi\|_U.$$

Under such assumptions, there exists $z_0^* \in Z^*$ such that

$$F'(u_0) + [G'(u_0)]^*(z_0^*) = \theta,$$

that is,

$$\langle \varphi, F'(u_0)\rangle_U + \langle G'(u_0)\varphi, z_0^*\rangle_Z = 0, \forall \varphi \in U.$$

Proof. First observe that there is no loss of generality in assuming $0 < \alpha < 1$. Also from the generalized mean value inequality and our hypothesis, if $\|\varphi\|_U < \alpha$, then

$$\begin{aligned} &\|G(u_0+\varphi) - G(u_0) - G'(u_0)\cdot\varphi\| \\ &\quad \le \sup_{h\in[0,1]} \{\|G'(u_0+h\varphi) - G'(u_0)\|\}\, \|\varphi\|_U \\ &\quad \le K \sup_{h\in[0,1]} \{\|h\varphi\|_U\}\|\varphi\|_U \le K\|\varphi\|_U^2. \end{aligned} \tag{11.42}$$

For each $\varphi \in U$, define $H(\varphi)$ by

$$G(u_0+\varphi) = G(u_0) + G'(u_0)\cdot\varphi + H(\varphi),$$

that is,

$$H(\varphi) = G(u_0+\varphi) - G(u_0) - G'(u_0)\cdot\varphi.$$

Let $L_0 = N(G'(u_0))$ where $N(G'(u_0))$ denotes the null space of $G'(u_0)$. Observe that U/L_0 is a Banach space for which we define $A : U/L_0 \to Z$ by

$$A(u) = G'(u_0)\cdot u,$$

where $\bar{u} = \{u+v \mid v \in L_0\}$.

Since $G'(u_0)$ is onto, so is A, so that by the inverse mapping theorem A has a continuous inverse A^{-1}.

Let $\varphi \in U$ be such that $G'(u_0)\cdot\varphi = \theta$. For a given t such that $0 < |t| < \frac{\alpha}{1+\|\varphi\|_U}$, let $\psi_0 \in U$ be such that

$$G'(u_0)\cdot\psi_0 + \frac{H(t\varphi)}{t^2} = \theta,$$

Observe that from (11.42),

$$\|H(t\varphi)\| \le Kt^2\|\varphi\|_U^2,$$

and thus from the boundedness of A^{-1}, $\|\psi_0\|$ as a function of t may be chosen uniformly bounded relating t (i.e., despite the fact that ψ_0 may vary with t, there exists $K_1 > 0$ such that $\|\psi_0\|_U < K_1, \forall t$ such that $0 < |t| < \frac{\alpha}{1+\|\varphi\|_U}$).

Now choose $0 < r < 1/4$ and define $g_0 = \theta$.

Also define

$$\varepsilon = \frac{r}{4(\|A^{-1}\|+1)(K+1)(K_1+1)(\|\varphi\|_U+1)}.$$

Since from the hypotheses $G'(u)$ is continuous at u_0, we may choose $0 < \delta < \alpha$ such that if $\|v\|_U < \delta$ then

$$\|G'(u_0+v)-G'(u_0)\|<\varepsilon.$$

Fix $t\in\mathbb{R}$ such that

$$0<|t|<\frac{\delta}{2(1+\|\varphi\|_U+K_1)}.$$

Observe that $\psi\in U$ is such that $G(u_0+t\varphi+t^2\psi)=\theta$ if and only if

$$G'(u_0)\cdot\psi+\frac{H(t\varphi+t^2\psi)}{t^2}=\theta.$$

Define

$$L_1=A^{-1}\left[G'(u_0)\cdot(\psi_0-g_0)+\frac{H(t\varphi+t^2(\psi_0-g_0))}{t^2}\right],$$

so that

$$\begin{aligned}L_1&=A^{-1}[A(\overline{\psi_0-g_0})]+A^{-1}\left(\frac{H(t\varphi+t^2(\psi_0-g_0))}{t^2}\right)\\&=\overline{\psi_0-g_0}+\overline{w_1}\\&=\overline{\psi_0+w_1}\\&=\{\psi_0+w_1+v\mid v\in L_0\}.\end{aligned}$$

Here $w_1\in U$ is such that

$$\overline{w_1}=A^{-1}\left(\frac{H(t\varphi+t^2(\psi_0-g_0))}{t^2}\right),$$

that is,

$$A(\overline{w_1})=\frac{H(t\varphi+t^2(\psi_0-g_0))}{t^2},$$

so that

$$G'(u_0)\cdot w_1=\frac{H(t\varphi+t^2(\psi_0-g_0))}{t^2}.$$

Select $g_1\in L_1$ such that

$$\|g_1-g_0\|_U\le 2\|L_1-L_0\|.$$

This is possible since

$$\|L_1-L_0\|=\inf_{g\in L_1}\{\|g-g_0\|_U\}.$$

So we have that

$$L_1=A^{-1}\left[-\frac{H(t\varphi)}{t^2}+\frac{H(t\varphi+t^2(\psi_0-g_0))}{t^2}\right]. \tag{11.43}$$

However

$$\begin{aligned}
&H(t\varphi+t^2(\psi_0-g_0))-H(t\varphi)\\
&=G(u_0+t\varphi+t^2(\psi_0))-G(u_0)\\
&\quad-G'(u_0)\cdot(t\varphi+t^2(\psi_0))\\
&\quad-G(u_0+t\varphi)+G(u_0)\\
&\quad+G'(u_0)\cdot(t\varphi)\\
&=G(u_0+t\varphi+t^2(\psi_0))-G(u_0+t\varphi)\\
&\quad-G'(u_0)\cdot(t^2(\psi_0)),
\end{aligned} \tag{11.44}$$

so that by the generalized mean value inequality we may write

$$\begin{aligned}
&\|H(t\varphi+t^2(\psi_0-g_0))-H(t\varphi)\|\\
&\le \sup_{h\in[0,1]}\|G'(u_0+t\varphi+ht^2(\psi_0))-G'(u_0)\|\|t^2\psi_0\|_U\\
&<\varepsilon t^2\|\psi_0\|_U.
\end{aligned} \tag{11.45}$$

From this and (11.43) we get

$$\begin{aligned}
\|L_1\| &\le \|A^{-1}\|\|H(t\varphi+t^2(\psi_0-g_0))-H(t\varphi)\|/t^2\\
&<\|A^{-1}\|\varepsilon\|\psi_0\|_U\\
&<\|A^{-1}\|K_1\frac{r}{4(\|A^{-1}\|+1)(K+1)(K_1+1)(\|\varphi\|_U+1)}\\
&<\frac{r}{4}.
\end{aligned} \tag{11.46}$$

Hence

$$\|g_1\|_U<2\|L_1\|<r/2.$$

Now reasoning by induction, for $n\ge 2$, assume that $\|g_{n-1}\|_U<r$ and $\|g_{n-2}\|_U<r$ and define L_n by

$$L_n-L_{n-1}=A^{-1}\left[G'(u_0)\cdot(\psi_0-g_{n-1})+\frac{H(t\varphi+t^2(\psi_0-g_{n-1}))}{t^2}\right].$$

Observe that

$$\begin{aligned}
L_n&-A^{-1}\left[G'(u_0)\cdot(\psi_0-g_{n-1})+\frac{H(t\varphi+t^2(\psi_0-g_{n-1}))}{t^2}\right]+L_{n-1}\\
&=A^{-1}A(\overline{\psi_0-g_{n-1}})+A^{-1}\left[\frac{H(t\varphi+t^2(\psi_0-g_{n-1}))}{t^2}\right]+\overline{g}_{n-1}\\
&=\overline{\psi_0-g_{n-1}}+A^{-1}\left[\frac{H(t\varphi+t^2(\psi_0-g_{n-1}))}{t^2}\right]+\overline{g}_{n-1}
\end{aligned}$$

$$\begin{aligned}&= \overline{\psi_0} + A^{-1}\left[\frac{H(t\varphi + t^2(\psi_0 - g_{n-1}))}{t^2}\right]\\&= \{\psi_0 + w_n + v \mid v \in L_0\}.\end{aligned}$$

Here $w_n \in U$ is such that

$$\overline{w_n} = A^{-1}\left[\frac{H(t\varphi + t^2(\psi_0 - g_{n-1}))}{t^2}\right],$$

that is,

$$A(\overline{w_n}) = \left[\frac{H(t\varphi + t^2(\psi_0 - g_{n-1}))}{t^2}\right],$$

so that

$$G'(u_0) \cdot w_n = \left[\frac{H(t\varphi + t^2(\psi_0 - g_{n-1}))}{t^2}\right].$$

Choose $g_n \in L_n$ such that

$$\|g_n - g_{n-1}\|_U \leq 2\|L_n - L_{n-1}\|.$$

This is possible since

$$\|L_n - L_{n-1}\| = \inf_{g \in L_n}\{\|g - g_{n-1}\|_U\}.$$

Observe that we may write

$$L_{n-1} = A^{-1}[A(\bar{g}_{n-1})] = A^{-1}[G'(u_0) \cdot g_{n-1}].$$

Thus

$$L_n = A^{-1}\left[G'(u_0) \cdot (\psi_0 - g_{n-1}) + \frac{H(t\varphi + t^2(\psi_0 - g_{n-1}))}{t^2} + G'(u_0) \cdot g_{n-1}\right].$$

By analogy

$$L_{n-1} = A^{-1}\left[G'(u_0) \cdot (\psi_0 - g_{n-2}) + \frac{H(t\varphi + t^2(\psi_0 - g_{n-2}))}{t^2} + G'(u_0) \cdot g_{n-2}\right].$$

Observe that

$$\begin{aligned}&H(t\varphi + t^2(\psi_0 - g_{n-1})) - H(t\varphi + t^2(\psi_0 - g_{n-2}))\\&= G(u_0 + t\varphi + t^2(\psi_0 - g_{n-1})) - G(u_0)\\&\quad -G'(u_0) \cdot (t\varphi + t^2(\psi_0 - g_{n-1}))\\&\quad -G(u_0 + t\varphi + t^2(\psi_0 - g_{n-2})) + G(u_0)\\&\quad +G'(u_0) \cdot (t\varphi + t^2(\psi_0 - g_{n-2}))\end{aligned}$$

$$
\begin{aligned}
&= G(u_0+t\varphi+t^2(\psi_0-g_{n-1}))-G(u_0+t\varphi+t^2(\psi_0-g_{n-2})) \\
&\quad -G'(u_0)\cdot(t^2(-g_{n-1}+g_{n-2})), \qquad (11.47)
\end{aligned}
$$

so that by the generalized mean value inequality we may write

$$
\begin{aligned}
&\|H(t\varphi+t^2(\psi_0-g_{n-1}))-H(t\varphi+t^2(\psi_0-g_{n-2}))\| \\
&\quad \leq \sup_{h\in[0,1]} \|G'(u_0+t\varphi+t^2\psi_0-t^2(h(g_{n-1})+(1-h)g_{n-2}))-G'(u_0)\| \\
&\quad \times \|t^2(-g_{n-1}+g_{n-2})\|_U \\
&\quad < \varepsilon t^2\|g_{n-1}-g_{n-2}\|_U.
\end{aligned}
$$

Therefore, similarly as above,

$$
\begin{aligned}
\|L_n-L_{n-1}\| &\leq \frac{\|A^{-1}\|}{t^2}\|H(t\varphi+t^2(\psi_0-g_{n-1}))-H(t\varphi+t^2(\psi_0-g_{n-2}))\| \\
&< \varepsilon\|A^{-1}\|\,\|g_{n-1}-g_{n-2}\|_U \\
&< (r/4)\|g_{n-1}-g_{n-2}\|_U \\
&< \frac{1}{4}\|g_{n-1}-g_{n-2}\|_U. \qquad (11.48)
\end{aligned}
$$

Thus,

$$
\|g_n-g_{n-1}\|_U \leq 2\|L_n-L_{n-1}\| < \frac{1}{2}\|g_{n-1}-g_{n-2}\|_U.
$$

Finally

$$
\begin{aligned}
\|g_n\|_U &= \|g_n-g_{n-1}+g_{n-1}-g_{n-2}+g_{n-2}-\ldots+g_1-g_0\|_U \\
&\leq \|g_1\|_U\left(1+\frac{1}{2}+\ldots+\frac{1}{2^n}\right) < 2\|g_1\|_U < r. \qquad (11.49)
\end{aligned}
$$

Thus $\|g_n\|_U < r$ and

$$
\|g_n-g_{n-1}\|_U < \frac{1}{2}\|g_{n-1}-g_{n-2}\|_U, \forall n \in \mathbb{N},
$$

so that $\{g_n\}$ is a Cauchy sequence, and since U is a Banach space there exists $g \in U$ such that

$$
g_n \to g, \text{ in norm, as } n \to \infty.
$$

Hence

$$
L_n \to L = \bar{g}, \text{ in norm, as } n \to \infty,
$$

so that

$$
L_n - L_{n-1} \to L - L = \theta = A^{-1}\left[G'(u_0)\cdot(\psi_0-g)+\frac{H(t\varphi+t^2(\psi_0-g))}{t^2}\right].
$$

Since A^{-1} is a bijection, denoting $\tilde{\psi}_0 = (\psi_0 - g)$, we get

$$G'(u_0) \cdot \tilde{\psi}_0 + \frac{H(t\varphi + t^2(\tilde{\psi}_0))}{t^2} = \theta$$

Clarifying the dependence on t we denote $\tilde{\psi}_0 = \tilde{\psi}_0(t)$ where as above mentioned, $t \in \mathbb{R}$ is such that

$$0 < |t| < \frac{\delta}{2(1 + \|\varphi\|_U + K_1)}.$$

Therefore

$$G(u_0 + t\varphi + t^2\tilde{\psi}_0(t)) = \theta.$$

Observe also that $\|t^2\tilde{\psi}_0(t)\|_U = \|t^2(\psi_0(t) - g)\|_U \leq t^2(K_1 + r) \leq t^2(K_1 + 1)$ so that $t^2\tilde{\psi}_0(t) \to \theta$ as $t \to 0$. Thus, by defining $t^2\tilde{\psi}_0(t)|_{t=0} = \theta$ (observe that in principle such a function would not be defined at $t = 0$), we obtain

$$\frac{d(t^2\tilde{\psi}_0(t))}{dt}|_{t=0} = \lim_{t \to 0} \left(\frac{t^2\tilde{\psi}_0(t) - \theta}{t} \right) = \theta,$$

considering that

$$\|t\tilde{\psi}_0(t)\|_U \leq |t|(K_1 + 1) \to 0, \text{ as } t \to 0.$$

Finally, defining

$$\phi(t) = F(u_0 + t\varphi + t^2\tilde{\psi}_0(t)),$$

from the hypotheses, we have that there exists a suitable $\tilde{t}_2 > 0$ such that

$$\phi(0) = F(u_0) \leq F(u_0 + t\varphi + t^2\tilde{\psi}_0(t)) = \phi(t), \forall |t| < \tilde{t}_2,$$

also from the hypothesis we get

$$\phi'(0) = \delta F(u_0, \varphi) = 0,$$

that is,

$$\langle \varphi, F'(u_0) \rangle_U = 0, \forall \varphi \text{ such that } G'(u_0) \cdot \varphi = \theta.$$

In the next lines as usual $N[G'(u_0)]$ and $R[G'(u_0)]$ denote the null space and the range of $G'(u_0)$, respectively. Thus $F'(u_0)$ is orthogonal to the null space of $G'(u_0)$, which we denote by

$$F'(u_0) \perp N[G'(u_0)].$$

Since $R[G'(u_0)]$ is closed, we get $F'(u_0) \in R([G'(u_0)]^*)$, that is, there exists $z_0^* \in Z^*$ such that

$$F'(u_0) = [G'(u_0)]^*(-z_0^*).$$

The proof is complete.

11.4 Some Examples Concerning Inequality Constraints

In this section we assume the hypotheses of the last theorem for F and G specified below. As an application of this same result, consider the problem of locally minimizing $F(u)$ subject to $G_1(u) = \theta$ and $G_2(u) \leq \theta$, where $F: U \to \mathbb{R}$, U being a function Banach space, $G_1 : U \to [L^p(\Omega)]^{m_1}$, $G_2 : U \to [L^p(\Omega)]^{m_2}$ where $1 < p < \infty$, and Ω is an appropriate subset of $\mathbb{R}^N$. We refer to the simpler case in which the partial order in $[L^p(\Omega)]^{m_2}$ is defined by $u = \{u_i\} \geq \theta$ if and only if $u_i \in L^p(\Omega)$ and $u_i(x) \geq 0$ a.e. in $\Omega, \forall i \in \{1,\ldots,m_2\}$.

Observe that defining

$$\tilde{F}(u,v) = F(u),$$

$$G(u,v) = (\{(G_1)_i(u)\}_{m_1 \times 1}, \{(G_2)_i(u) + v_i^2\}_{m_2 \times 1})$$

it is clear that (locally) minimizing $\tilde{F}(u,v)$ subject to $G(u,v) = (\theta,\theta)$ is equivalent to the original problem. We clarify the domain of $\tilde{F}$ is denoted by $U \times Y$, where

$$Y = \{v \text{ measurable such that } v_i^2 \in L^p(\Omega), \ \forall i \in \{1,\ldots,m_2\}\}.$$

Therefore, if u_0 is a local minimum for the original constrained problem, then for an appropriate and easily defined v_0, we have that (u_0, v_0) is a point of local minimum for the extended constrained one, so that by the last theorem there exists a Lagrange multiplier $z_0^* = (z_1^*, z_2^*) \in [L^q(\Omega)]^{m_1} \times [L^q(\Omega)]^{m_2}$ where $1/p + 1/q = 1$ and

$$\tilde{F}'(u_0,v_0) + [G'(u_0,v_0)]^*(z_0^*) = (\theta,\theta),$$

that is,

$$F'(u_0) + [G_1'(u_0)]^*(z_1^*) + [G_2'(u_0)]^*(z_2^*) = \theta, \tag{11.50}$$

and

$$(z_2^*)_i v_{0^i} = \theta, \forall i \in \{1,\ldots,m_2\}.$$

In particular for almost all $x \in \Omega$, if x is such that $v_{0^i}(x)^2 > 0$, then $z_{2^i}^*(x) = 0$, and if $v_{0^i}(x) = 0$, then $(G_2)_i(u_0(x)) = 0$, so that $(z_2^*)_i(G_2)_i(u_0) = 0$, a.e. in $\Omega, \forall i \in \{1,\ldots,m_2\}$.

Furthermore, consider the problem of minimizing $F_1(v) = \tilde{F}(u_0,v) = F(u_0)$ subject $\{G_{2^i}(u_0) + v_i^2\} = \theta$. From the above such a local minimum is attained at v_0. Thus, from the stationarity of $F_1(v) + \langle z_2^*, \{(G_2)_i(u_0) + v_i^2\}\rangle_{[L^p(\Omega)]^{m_2}}$ at v_0 and the standard necessary conditions for the case of convex (in fact quadratic) constraints we get $(z_2^*)_i \geq 0$ a.e. in $\Omega, \forall i \in \{1,\ldots,m_2\}$, that is, $z_2^* \geq \theta$.

Summarizing, for the order in question, the first-order necessary optimality conditions are given by (12.37), $z_2^* \geq \theta$ and $(z_2^*)_i(G_2)_i(u_0) = \theta, \forall i \in \{1,\ldots,m_2\}$ (so that $\langle z_2^*, G_2(u_0)\rangle_{[L^p(\Omega)]^{m_2}} = 0$), $G_1(u_0) = \theta$, and $G_2(u_0) \leq \theta$.

Remark 11.4.1. For the case $U = \mathbb{R}^n$ and $\mathbb{R}^{m_k}$ replacing $[L^p(\Omega)]^{m_k}$, for $k \in \{1,2\}$, the conditions $(z_2^*)_i v_i = \theta$ mean that for the constraints not active (e.g., $v_i \neq 0$) the corresponding coordinate $(z_2^*)_i$ of the Lagrange multiplier is 0. If $v_i = 0$, then $(G_2)_i(u_0) = 0$, so that in any case $(z_2^*)_i(G_2)_i(u_0) = 0$.

Summarizing, for this last mentioned case, we have obtained the standard necessary optimality conditions: $(z_2^*)_i \geq 0$, and $(z_2^*)_i(G_2)_i(u_0) = 0, \forall i \in \{1,\ldots,m_2\}$.

11.5 The Lagrange Multiplier Theorem for Equality and Inequality Constraints

In this section we develop more rigorous results concerning the Lagrange multiplier theorem for the case involving equalities and inequalities.

Theorem 11.1. *Let U, Z_1, Z_2 be Banach spaces. Consider a cone C in Z_2 as specified above and such that if $z_1 \leq \theta$ and $z_2 < \theta$, then $z_1 + z_2 < \theta$, where $z \leq \theta$ means that $z \in -C$ and $z < \theta$ means that $z \in (-C)^\circ$. The concerned order is supposed to be also that if $z < \theta$, $z^* \geq \theta^*$ and $z^* \neq \theta$, then $\langle z, z^* \rangle_{Z_2} < 0$. Furthermore, assume $u_0 \in U$ is a point of local minimum for $F : U \to \mathbb{R}$ subject to $G_1(u) = \theta$ and $G_2(u_0) \leq \theta$, where $G_1 : U \to Z_1$, $G_2 : U \to Z_2$ and F are Fréchet differentiable at $u_0 \in U$. Suppose also $G_1'(u_0)$ is onto and that there exist $\alpha > 0, K > 0$ such that if $\|\varphi\|_U < \alpha$, then*

$$\|G_1'(u_0 + \varphi) - G_1'(u_0)\| \leq K \|\varphi\|_U.$$

Finally, suppose there exists $\varphi_0 \in U$ such that

$$G_1'(u_0) \cdot \varphi_0 = \theta$$

and

$$G_2'(u_0) \cdot \varphi_0 < \theta.$$

Under such hypotheses, there exists a Lagrange multiplier $z_0^* = (z_1^*, z_2^*) \in Z_1^* \times Z_2^*$ such that

$$F'(u_0) + [G_1'(u_0)]^*(z_1^*) + [G_2'(u_0)]^*(z_2^*) = \theta,$$

$$z_2^* \geq \theta^*,$$

and

$$\langle G_2(u_0), z_2^* \rangle_{Z_2} = 0.$$

Proof. Let $\varphi \in U$ be such that

$$G_1'(u_0) \cdot \varphi = \theta$$

and

$$G_2'(u_0) \cdot \varphi = v - \lambda G_2(u_0),$$

for some $v \leq \theta$ and $\lambda \geq 0$.

Select $\alpha \in (0,1)$ and define

$$\varphi_\alpha = \alpha \varphi_0 + (1 - \alpha) \varphi.$$

Observe that $G_1(u_0) = \theta$ and $G_1'(u_0) \cdot \varphi_\alpha = \theta$ so that as in the proof of the Lagrange multiplier Theorem 11.3.2 we may find $K_1 > 0$, $\varepsilon > 0$ and $\psi_0(t)$ such that

$$G_1(u_0 + t\varphi_\alpha + t^2\psi_0(t)) = \theta, \; \forall |t| < \varepsilon,$$

and

$$\|\psi_0(t)\|_U < K_1, \forall |t| < \varepsilon.$$

Observe that

$$\begin{aligned}
& G_2'(u_0) \cdot \varphi_\alpha \\
& \quad = \alpha G_2'(u_0) \cdot \varphi_0 + (1-\alpha) G_2'(u_0) \cdot \varphi \\
& \quad = \alpha G_2'(u_0) \cdot \varphi_0 + (1-\alpha)(v - \lambda G_2(u_0)) \\
& \quad = \alpha G_2'(u_0) \cdot \varphi_0 + (1-\alpha)v - (1-\alpha)\lambda G_2(u_0)) \\
& \quad = v_0 - \lambda_0 G_2(u_0),
\end{aligned} \tag{11.51}$$

where

$$\lambda_0 = (1-\alpha)\lambda,$$

and

$$v_0 = \alpha G_2'(u_0) \cdot \varphi_0 + (1-\alpha)v < \theta.$$

Hence, for $t > 0$,

$$G_2(u_0 + t\varphi_\alpha + t^2\psi_0(t)) = G_2(u_0) + G_2'(u_0) \cdot (t\varphi_\alpha + t^2\psi_0(t)) + r(t),$$

where

$$\lim_{t \to 0^+} \frac{\|r(t)\|}{t} = 0.$$

Therefore from (11.51) we obtain

$$G_2(u_0 + t\varphi_\alpha + t^2\psi_0(t)) = G_2(u_0) + tv_0 - t\lambda_0 G_2(u_0) + r_1(t),$$

where

$$\lim_{t \to 0^+} \frac{\|r_1(t)\|}{t} = 0.$$

Observe that there exists $\varepsilon_1 > 0$ such that if $0 < t < \varepsilon_1 < \varepsilon$, then

$$v_0 + \frac{r_1(t)}{t} < \theta,$$

and

$$G_2(u_0) - t\lambda_0 G_2(u_0) = (1 - t\lambda_0) G_2(u_0) \leq \theta.$$

Hence

$$G_2(u_0 + t\varphi_\alpha + t^2\psi_0(t)) < \theta, \text{ if } 0 < t < \varepsilon_1.$$

From this there exists $0 < \varepsilon_2 < \varepsilon_1$ such that

$$F(u_0 + t\varphi_\alpha + t^2\psi_0(t)) - F(u_0)$$
$$= \langle t\varphi_\alpha + t^2\psi_0(t), F'(u_0)\rangle_U + r_2(t) \geq 0, \tag{11.52}$$

where

$$\lim_{t\to 0^+} \frac{|r_2(t)|}{t} = 0.$$

Dividing the last inequality by $t > 0$ we get

$$\langle \varphi_\alpha + t\psi_0(t), F'(u_0)\rangle_U + r_2(t)/t \geq 0, \forall 0 < t < \varepsilon_2.$$

Letting $t \to 0^+$ we obtain

$$\langle \varphi_\alpha, F'(u_0)\rangle_U \geq 0.$$

Letting $\alpha \to 0^+$, we get

$$\langle \varphi, F'(u_0)\rangle_U \geq 0,$$

if

$$G_1'(u_0) \cdot \varphi = \theta,$$

and

$$G_2'(u_0) \cdot \varphi = v - \lambda G_2(u_0),$$

for some $v \leq \theta$ and $\lambda \geq 0$. Define

$$A = \{(\langle \varphi, F'(u_0)\rangle_U + r, G_1'(u_0)\cdot\varphi, G_2'(u_0)\varphi - v + \lambda G(u_0)),$$
$$\varphi \in U,\ r \geq 0, v \leq \theta, \lambda \geq 0\}. \tag{11.53}$$

Observe that A is a convex set with a nonempty interior.

If

$$G_1'(u_0) \cdot \varphi = \theta,$$

and

$$G_2'(u_0) \cdot \varphi - v + \lambda G_2(u_0) = \theta,$$

with $v \leq \theta$ and $\lambda \geq 0$ then

$$\langle \varphi, F'(u_0)\rangle_U \geq 0,$$

so that

$$\langle \varphi, F'(u_0)\rangle_U + r \geq 0.$$

Moreover, if

$$\langle \varphi, F'(u_0)\rangle + r = 0,$$

with $r \geq 0$,

$$G_1'(u_0) \cdot \varphi = \theta,$$

and

$$G_2'(u_0) \cdot \varphi - v + \lambda G_2(u_0) = \theta,$$

with $v \leq \theta$ and $\lambda \geq 0$, then we have

$$\langle \varphi, F'(u_0) \rangle_U \geq 0,$$

so that

$$\langle \varphi, F'(u_0) \rangle_U = 0,$$

and $r = 0$. Hence $(0, \theta, \theta)$ is on the boundary of A. Therefore, by the Hahn–Banach theorem, geometric form, there exists

$$(\beta, z_1^*, z_2^*) \in \mathbb{R} \times Z_1^* \times Z_2^*$$

such that

$$(\beta, z_1^*, z_2^*) \neq (0, \theta, \theta)$$

and

$$\beta(\langle \varphi, F'(u_0) \rangle_U + r) + \langle G_1'(u_0) \cdot \varphi, z_1^* \rangle_{Z_1} + \langle G_2'(u_0) \cdot \varphi - v + \lambda G_2(u_0), z_2^* \rangle_{Z_2} \geq 0, \quad (11.54)$$

$\forall \varphi \in U$, $r \geq 0$, $v \leq \theta$, $\lambda \geq 0$. Suppose $\beta = 0$. Fixing all variable except v we get $z_2^* \geq \theta$. Thus, for $\varphi = c\varphi_0$ with arbitrary $c \in \mathbb{R}$, $v = \theta, \lambda = 0$, if $z_2^* \neq \theta$, then $\langle G_2'(u_0) \cdot \varphi_0, z_2^* \rangle_{Z_2} < 0$, so that we get $z_2^* = \theta$. Since $G_1'(u_0)$ is onto, a similar reasoning lead us to $z_1^* = 0$, which contradicts $(\beta, z_1^*, z_2^*) \neq (0, 0, 0)$.

Hence, $\beta \neq 0$, and fixing all variables except r we obtain $\beta > 0$. There is no loss of generality in assuming $\beta = 1$.

Again fixing all variables except v, we obtain $z_2^* \geq \theta$. Fixing all variables except λ, since $G_2(u_0) \leq \theta$ we get

$$\langle G_2(u_0), z_2^* \rangle_{Z_2} = 0.$$

Finally, for $r = 0$, $v = \theta$, $\lambda = 0$, we get

$$\langle \varphi, F'(u_0) \rangle_U + \langle G_1'(u_0)\varphi, z_1^* \rangle_{Z_1} + \langle G_2'(u_0) \cdot \varphi, z_2^* \rangle_{Z_2} \geq 0, \ \forall \varphi \in U,$$

that is, since obviously such an inequality is valid also for $-\varphi$, $\forall \varphi \in U$, we obtain

$$\langle \varphi, F'(u_0) \rangle_U + \langle \varphi, [G_1'(u_0)]^*(z_1^*) \rangle_U + \langle \varphi, [G_2'(u_0)]^*(z_2^*) \rangle_U = 0, \ \forall \varphi \in U,$$

so that

$$F'(u_0) + [G_1'(u_0)]^*(z_1^*) + [G_2'(u_0)]^*(z_2^*) = \theta.$$

The proof is complete.

11.6 Second-Order Necessary Conditions

In this section we establish second-order necessary conditions for a class of constrained problems in Banach spaces. We highlight the next result is particularly applicable to optimization in $\mathbb{R}^n$.

Theorem 11.2. *Let U, Z_1, Z_2 be Banach spaces. Consider a cone C in Z_2 as above specified and such that if $z_1 \leq \theta$ and $z_2 < \theta$, then $z_1 + z_2 < \theta$, where $z \leq \theta$ means that $z \in -C$ and $z < \theta$ means that $z \in (-C)^\circ$. The concerned order is supposed to be also that if $z < \theta$, $z^* \geq \theta^*$ and $z^* \neq \theta$, then $\langle z, z^* \rangle_{Z_2} < 0$. Furthermore, assume $u_0 \in U$ is a point of local minimum for $F : U \to \mathbb{R}$ subject to $G_1(u) = \theta$ and $G_2(u_0) \leq \theta$, where $G_1 : U \to Z_1$, $G_2 : U \to (Z_2)^k$, and F are twice Fréchet differentiable at $u_0 \in U$. Assume $G_2(u) = \{(G_2)_i(u)\}$ where $(G_2)_i : U \to Z_2$, $\forall i \in \{1, \ldots, k\}$ and define*

$$A = \{i \in \{1, \ldots, k\} \; : \; (G_2)_i(u_0) = \theta\},$$

and also suppose that $(G_2)_i(u_0) < \theta$, if $i \notin A$. Moreover, suppose $\{G_1'(u_0), \{(G_2)_i'(u_0)\}_{i \in A}\}$ is onto and that there exist $\alpha > 0, K > 0$ such that if $\|\varphi\|_U < \alpha$, then

$$\|\tilde{G}'(u_0 + \varphi) - \tilde{G}'(u_0)\| \leq K \|\varphi\|_U,$$

where

$$\tilde{G}(u) = \{G_1(u), \{(G_2)_i(u)\}_{i \in A}\}.$$

Finally, suppose there exists $\varphi_0 \in U$ such that

$$G_1'(u_0) \cdot \varphi_0 = \theta$$

and

$$G_2'(u_0) \cdot \varphi_0 < \theta.$$

Under such hypotheses, there exists a Lagrange multiplier $z_0^* = (z_1^*, z_2^*) \in Z_1^* \times (Z_2^*)^k$ such that

$$F'(u_0) + [G_1'(u_0)]^*(z_1^*) + [G_2'(u_0)]^*(z_2^*) = \theta,$$

$$z_2^* \geq (\theta^*, \ldots, \theta^*) \equiv \theta_k^*,$$

and

$$\langle (G_2)_i(u_0), (z_2^*)_i \rangle_Z = 0, \forall i \in \{1, \ldots, k\},$$

$$(z_2^*)_i = \theta^*, \text{ if } i \notin A,$$

Moreover, defining

$$L(u, z_1^*, z_2^*) = F(u) + \langle G_1(u), z_1^* \rangle_{Z_1} + \langle G_2(u), z_2^* \rangle_{Z_2},$$

we have that

$$\delta_{uu}^2 L(u_0, z_1^*, z_2^*; \varphi) \geq 0, \forall \varphi \in \mathscr{V}_0,$$

where

$$\mathscr{V}_0 = \{\varphi \in U \,:\, G_1'(u_0)\cdot\varphi = \theta,\ (G_2)_i'(u_0)\cdot\varphi = \theta,\ \forall i \in A\}.$$

Proof. Observe that A is defined by

$$A = \{i \in \{1,\ldots,k\} \,:\, (G_2)_i(u_0) = \theta\}.$$

Observe also that $(G_2)_i(u_0) < \theta$, if $i \notin A$.

Hence the point $u_0 \in U$ is a local minimum for $F(u)$ under the constraints

$$G_1(u) = \theta, \text{ and } (G_2)_i(u) \leq \theta, \forall i \in A.$$

From the last Theorem 11.1 for such an optimization problem there exists a Lagrange multiplier $(z_1^*, \{(z_2^*)_{i\in A}\})$ such that $(z_2^*)_i \geq \theta^*$, $\forall i \in A$, and

$$F'(u_0) + [G_1'(u_0)]^*(z_1^*) + \sum_{i\in A}[(G_2)_i'(u_0)]^*((z_2^*)_i) = \theta. \tag{11.55}$$

The choice $(z_2^*)_i = \theta$, if $i \notin A$ leads to the existence of a Lagrange multiplier $(z_1^*, z_2^*) = (z_1^*, \{(z_2^*)_{i\in A}, (z_2^*)_{i\notin A}\})$ such that

$$z_2^* \geq 0_k^*$$

and

$$\langle (G_2)_i(u_0), (z_2^*)_i\rangle_Z - 0, \forall i \in \{1,\ldots,k\}.$$

Let $\varphi \in \mathscr{V}_0$, that is, $\varphi \in U$,

$$G_1'(u_0)\varphi = \theta$$

and

$$(G_2)_i'(u_0)\cdot\varphi = \theta,\ \forall i \in A.$$

Recall that $\tilde{G}(u) = \{G_1(u), (G_2)_{i\in A}(u)\}$ and therefore, similarly as in the proof of the Lagrange multiplier Theorem 11.3.2, we may obtain $\psi_0(t), K > 0$ and $\varepsilon > 0$ such that

$$\tilde{G}(u_0 + t\varphi + t^2\psi_0(t)) = \theta,\ \forall |t| < \varepsilon,$$

and

$$\|\psi_0(t)\| \leq K, \forall |t| < \varepsilon.$$

Also, if $i \notin A$, we have that $(G_2)_i(u_0) < \theta$, so that

$$(G_2)_i(u_0 + t\varphi + t^2\psi_0(t)) = (G_2)_i(u_0) + G_i'(u_0)\cdot(t\varphi + t^2\psi_0(t)) + r(t),$$

where

$$\lim_{t\to 0}\frac{\|r(t)\|}{t} = 0,$$

that is,

$$(G_2)_i(u_0 + t\varphi + t^2\psi_0(t)) = (G_2)_i(u_0) + t(G_2)_i'(u_0)\cdot\varphi + r_1(t),$$

where

$$\lim_{t\to 0}\frac{\|r_1(t)\|}{t}=0,$$

and hence there exists $0<\varepsilon_1<\varepsilon$ such that

$$(G_2)_i(u_0+t\varphi+t^2\psi_0(t))<\theta,\ \forall|t|<\varepsilon_1<\varepsilon.$$

Therefore, since u_0 is a point of local minimum under the constraint $G(u)\leq\theta$, there exists $0<\varepsilon_2<\varepsilon_1$ such that

$$F(u_0+t\varphi+t^2\psi_0(t))-F(u_0)\geq 0,\ \forall|t|<\varepsilon_2,$$

so that

$$\begin{aligned}
&F(u_0+t\varphi+t^2\psi_0(t))-F(u_0)\\
&=F(u_0+t\varphi+t^2\psi_0(t))-F(u_0)\\
&\quad+\langle G_1(u_0+t\varphi+t^2\psi_0(t)),z_1^*\rangle_{Z_1}+\sum_{i\in A}\{\langle (G_2)_i(u_0+t\varphi+t^2\psi_0(t)),(z_2^*)_i\rangle_{Z_2}\}\\
&\quad-\langle G_1(u_0),z_1^*\rangle_{Z_1}-\sum_{i\in A}\{\langle (G_2)_i(u_0),(z_2^*)_i\rangle_{Z_2}\}\\
&=F(u_0+t\varphi+t^2\psi_0(t))-F(u_0)\\
&\quad+\langle G_1(u_0+t\varphi+t^2\psi_0(t)),z_1^*\rangle_{Z_1}-\langle G_1(u_0),z_1^*\rangle_{Z_1}\\
&\quad+\langle G_2(u_0+t\varphi+t^2\psi_0(t)),z_2^*\rangle_{Z_2}-\langle G_2(u_0),z_2^*\rangle_{Z_2}\\
&=L(u_0+t\varphi+t^2\psi_0(t)),z_1^*,z_2^*)-L(u_0,z_1^*,z_2^*)\\
&=\delta_uL(u_0,z_1^*,z_2^*;t\varphi+t^2\psi_0(t))+\frac{1}{2}\delta^2_{uu}L(u_0,z_1^*,z_2^*;t\varphi+t^2\psi_0(t))+r_2(t)\\
&=\frac{t^2}{2}\delta^2_{uu}L(u_0,z_1^*,z_2^*;\varphi+t\psi_0(t))+r_2(t)\geq 0,\forall|t|<\varepsilon_2.
\end{aligned}$$

where

$$\lim_{t\to 0}|r_2(t)|/t^2=0.$$

To obtain the last inequality we have used

$$\delta_uL(u_0,z_1^*,z_2^*;t\varphi+t^2\psi_0(t))=0$$

Dividing the last inequality by $t^2>0$ we obtain

$$\frac{1}{2}\delta^2_{uu}L(u_0,z_1^*,z_2^*;\varphi+t\psi_0(t))+r_2(t)/t^2\geq 0,\forall 0<|t|<\varepsilon_2,$$

and finally, letting $t\to 0$, we get

$$\frac{1}{2}\delta^2_{uu}L(u_0,z_1^*,z_2^*;\varphi)\geq 0.$$

The proof is complete.

11.7 On the Banach Fixed Point Theorem

Now we recall a classical definition, namely, the Banach fixed theorem also known as the contraction mapping theorem.

Definition 11.7.1. Let C be a subset of a Banach space U and let $T : C \to C$ be an operator. Thus, T is said to be a contraction mapping if there exists $0 \leq \alpha < 1$ such that

$$\|T(u) - T(v)\|_U \leq \alpha\|u - v\|_U, \forall u, v \in C.$$

Remark 11.7.2. Observe that if $\|T'(u)\|_U \leq \alpha < 1$ on a convex set C, then T is a contraction mapping, since by the mean value inequality,

$$\|T(u) - T(v)\|_U \leq \sup_{u \in C}\{\|T'(u)\|\}\|u - v\|_U, \forall u, v \in C.$$

The next result is the base of our generalized method of lines.

Theorem 11.7.3 (Contraction Mapping Theorem). *Let C be a closed subset of a Banach space U. Assume T is contraction mapping on C, then there exists a unique $u_0 \in C$ such that $u_0 = T(u_0)$. Moreover, for an arbitrary $u_1 \in C$ defining the sequence*

$$u_2 = T(u_1) \text{ and } u_{n+1} = T(u_n), \forall n \in \mathbb{N}$$

we have

$$u_n \to u_0, \text{in norm, as } n \to +\infty.$$

Proof. Let $u_1 \in C$. Let $\{u_n\} \subset C$ be defined by

$$u_{n+1} = T(u_n), \forall n \in \mathbb{N}.$$

Hence, reasoning inductively

$$\begin{aligned}
\|u_{n+1} - u_n\|_U &= \|T(u_n) - T(u_{n-1})\|_U \\
&\leq \alpha\|u_n - u_{n-1}\|_U \\
&\leq \alpha^2\|u_{n-1} - u_{n-2}\|_U \\
&\leq \ldots\ldots \\
&\leq \alpha^{n-1}\|u_2 - u_1\|_U, \forall n \in \mathbb{N}.
\end{aligned} \tag{11.56}$$

Thus, for $p \in \mathbb{N}$, we have

$$\begin{aligned}
&\|u_{n+p} - u_n\|_U \\
&= \|u_{n+p} - u_{n+p-1} + u_{n+p-1} - u_{n+p-2} + \ldots - u_{n+1} + u_{n+1} - u_n\|_U \\
&\leq \|u_{n+p} - u_{n+p-1}\|_U + \|u_{n+p-1} - u_{n+p-2}\|_U + \ldots + \|u_{n+1} - u_n\|_U \\
&\leq (\alpha^{n+p-2} + \alpha^{n+p-3} + \ldots + \alpha^{n-1})\|u_2 - u_1\|_U \\
&\leq \alpha^{n-1}(\alpha^{p-1} + \alpha^{p-2} + \ldots + \alpha^0)\|u_2 - u_1\|_U
\end{aligned}$$

$$\leq \alpha^{n-1}\left(\sum_{k=0}^{\infty}\alpha^k\right)\|u_2-u_1\|_U$$
$$\leq \frac{\alpha^{n-1}}{1-\alpha}\|u_2-u_1\|_U \tag{11.57}$$

Denoting $n+p=m$, we obtain

$$\|u_m-u_n\|_U \leq \frac{\alpha^{n-1}}{1-\alpha}\|u_2-u_1\|_U, \forall m>n\in\mathbb{N}.$$

Let $\varepsilon>0$. Since $0\leq\alpha<1$, there exists $n_0\in\mathbb{N}$ such that if $n>n_0$ then

$$0\leq\frac{\alpha^{n-1}}{1-\alpha}\|u_2-u_1\|_U<\varepsilon,$$

so that

$$\|u_m-u_n\|_U<\varepsilon,\text{ if } m>n>n_0.$$

From this we may infer that $\{u_n\}$ is a Cauchy sequence, and since U is a Banach space, there exists $u_0\in U$ such that

$$u_n\to u_0,\text{ in norm, as } n\to\infty.$$

Observe that

$$\begin{aligned}\|u_0-T(u_0)\|_U &= \|u_0-u_n+u_n-T(u_0)\|_U\\ &\leq \|u_0-u_n\|_U+\|u_n-T(u_0)\|_U\\ &\leq \|u_0-u_n\|_U+\alpha\|u_{n-1}-u_0\|_U\\ &\to 0,\text{ as } n\to\infty.\end{aligned} \tag{11.58}$$

Thus $\|u_0-T(u_0)\|_U=0$.

Finally, we prove the uniqueness. Suppose $u_0,v_0\in C$ are such that

$$u_0=T(u_0)\text{ and } v_0=T(v_0).$$

Hence,

$$\begin{aligned}\|u_0-v_0\|_U &= \|T(u_0)-T(v_0)\|_U\\ &\leq \alpha\|u_0-v_0\|_U.\end{aligned} \tag{11.59}$$

From this we get

$$\|u_0-v_0\|_U\leq 0,$$

that is,

$$\|u_0-v_0\|_U=0.$$

The proof is complete.

11.8 Sensitivity Analysis

11.8.1 Introduction

In this section we state and prove the implicit function theorem for Banach spaces. A similar result may be found in Ito and Kunisch [40], page 31.

We emphasize the result found in [40] is more general; however, the proof present here is almost the same for a simpler situation. The general result found in [40] is originally from Robinson [53].

Theorem 11.3 (Implicit Function Theorem). *Let V,U,W be Banach spaces. Given a function $\hat{F}: V \times U \to W$, suppose $(x_0,u_0) \in V \times U$ is such that $\hat{F}(x_0,u_0) = \theta$. Assume $\hat{F}$ is Fréchet differentiable, $\hat{F}_x(x_0,u_0)$ is continuous, and $[\hat{F}_x(x_0,u_0)]^{-1}$ is a (single-valued) bounded linear operator so that we denote $\|[\hat{F}_x(x_0,u_0)]^{-1}\| = \rho > 0$. Under such hypotheses, for each $\varepsilon > 0$, there exist a neighborhood U_ε of u_0, a neighborhood V_ε of x_0, and a function $x: U_\varepsilon \to V_\varepsilon$ such that for each $u \in U_\varepsilon$, $x(u)$ is the unique solution of*

$$\hat{F}(x,u) = \theta,$$

that is,

$$\hat{F}(x(u),u) = \theta.$$

Moreover, for each $u,v \in U_\varepsilon$, we have

$$\|x(u) - x(v)\| \leq (\rho+\varepsilon)\|\hat{F}(x(v),u) - \hat{F}(x(v),v)\|.$$

Finally, if $\|\hat{F}(x,u) - \hat{F}(x,v)\| \leq K\|u-v\|, \forall (x,u) \in V_\varepsilon \times U_\varepsilon$, then

$$\|x(u) - x(v)\| \leq K_0\|u-v\|, \forall u,v \in U_\varepsilon,$$

where $K_0 = K(\rho+\varepsilon)$.

Proof. Let $\varepsilon > 0$. Choose $\delta > 0$ such that

$$\rho\delta < \frac{\varepsilon}{(\rho+\varepsilon)}.$$

Define

$$\begin{aligned} T(x) &= \hat{F}(x_0,u_0) + \hat{F}_x(x_0,u_0)(x-x_0) \\ &= \hat{F}_x(x_0,u_0)(x-x_0), \end{aligned} \tag{11.60}$$

and

$$\begin{aligned} h(x,u) &= \hat{F}(x_0,u_0) + \hat{F}_x(x_0,u_0)(x-x_0) - \hat{F}(x,u) \\ &= \hat{F}_x(x_0,u_0)(x-x_0) - \hat{F}(x,u). \end{aligned} \tag{11.61}$$

Select a ball U_ε about u_0 and a closed ball V_ε of radius $r > 0$ about x_0 such that for each $u \in U_\varepsilon$ and $x \in V_\varepsilon$ we have

$$\|\hat{F}_x(x,u) - \hat{F}_x(x_0,u_0)\| \leq \delta,$$

$$\rho\|\hat{F}(x_0,u) - \hat{F}(x_0,u_0)\| \leq (1-\rho\delta)r.$$

For each $u \in U_\varepsilon$ define

$$\phi_u(x) = T^{-1}(h(x,u)).$$

Fix $u \in U$. Observe that for $x_1, x_2 \in V_\varepsilon$ we have

$$\begin{aligned}
\|\phi_u(x_1) - \phi_u(x_2)\| &\leq \|T^{-1}\|\|h(x_1,u) - h(x_2,u)\| \\
&= \rho\|h(x_1,u) - h(x_2,u)\| \\
&= \rho\left\|\int_0^1 h_x(x_1 + t(x_2 - x_1),u)\cdot(x_1 - x_2)\,dt\right\| \\
&\leq \rho\delta\|x_1 - x_2\|, \qquad (11.62)
\end{aligned}$$

so that since $0 < \rho\delta < 1$ we may infer that $\phi_u(x)$ is a contractor. Observe also that $x_0 = T^{-1}(\theta)$, so that

$$\begin{aligned}
\|\phi_u(x_0) - x_0\| &\leq \rho\|h(x_0,u) - \theta\| \\
&= \rho\|\hat{F}(x_0,u) - \hat{F}(x_0,u_0)\| \\
&\leq (1-\rho\delta)r. \qquad (11.63)
\end{aligned}$$

Hence, for $x \in V_\varepsilon$, we obtain

$$\begin{aligned}
\|\phi_u(x) - x_0\| &\leq \|\phi_u(x) - \phi_u(x_0)\| \\
&\quad + \|\phi_u(x_0) - x_0\| \\
&\leq \rho\delta\|x - x_0\| + (1-\rho\delta)r \leq r. \qquad (11.64)
\end{aligned}$$

Therefore $\phi_u(x) \in V_\varepsilon, \forall x \in V_\varepsilon$ so that from this, (11.62) and the Banach fixed point theorem, ϕ_u has a unique fixed point in V_ε, which we denote by $x(u)$.

Thus,

$$\begin{aligned}
x(u) &= \phi_u(x(u)) \\
&= T^{-1}(h(x(u),u)) \\
&= T^{-1}(\hat{F}_x(x_0,u_0)(x(u) - x_0) - \hat{F}(x(u),u)) \\
&= [\hat{F}_x(x_0,u_0)]^{-1}(\hat{F}_x(x_0,u_0)(x(u) - x_0) - \hat{F}(x(u),u)) + x_0 \\
&= x(u) - x_0 + x_0 - [\hat{F}_x(x_0,u_0)]^{-1}(\hat{F}(x(u),u)) \\
&= x(u) - [\hat{F}_x(x_0,u_0)]^{-1}(\hat{F}(x(u),u)). \qquad (11.65)
\end{aligned}$$

From this,

$$[\hat{F}_x(x_0,u_0)]^{-1}(\hat{F}(x(u),u)) = \theta,$$

so that

$$\hat{F}(x(u),u) = \hat{F}_x(x_0,u_0)[\hat{F}_x(x_0,u_0)]^{-1}(\hat{F}(x(u),u)) = \hat{F}_x(x_0,u_0)\theta = \theta,$$

that is,

$$\hat{F}(x(u),u) = \theta.$$

Also as a consequence of the Banach fixed point theorem, we have that

$$\|x(u) - x\| \leq (1-\rho\delta)^{-1}\|\phi_u(x) - x\|.$$

Now observe that for $u, v \in U_\varepsilon$, with $x = x(v)$ in the last inequality, we get

$$\|x(u) - x(v)\| \leq (1-\rho\delta)^{-1}\|\phi_u(x(v)) - x(v)\|.$$

However, $x(v) = \phi_v(x(v))$, so that from this and the last inequality, we obtain

$$\begin{aligned}\|x(u) - x(v)\| &\leq (1-\rho\delta)^{-1}\|\phi_u(x(v)) - \phi_v(x(v))\| \\ &\leq (1-\rho\delta)^{-1}\rho\|h(x(v),u) - h(x(v),v)\| \\ &= \rho(1-\rho\delta)^{-1}\|\hat{F}(x(v),u) - \hat{F}(x(v),v)\|. \qquad (11.66)\end{aligned}$$

Since $\rho(1-\rho\delta)^{-1} \leq \rho + \varepsilon$, the proof is complete.

11.8.2 The Main Results About Gâteaux Differentiability

Again let V, U be Banach spaces and let $F : V \times U \to \mathbb{R}$ be a functional. Fix $u \in U$ and consider the problem of minimizing $F(x,u)$ subject to $G(x,u) \leq \theta$ and $H(x,u) = \theta$. Here the order and remaining details on the primal formulation are the same as those indicated in Section 11.4.

Hence, for the specific case in which

$$G : V \times U \to [L^p(\Omega)]^{m_1}$$

and

$$H : V \times U \to [L^p(\Omega)]^{m_2},$$

(the cases in which the co-domains of G and H are $\mathbb{R}^{m_1}$ and $\mathbb{R}^{m_2}$, respectively, are dealt similarly), we redefine the concerned optimization problem, again for a fixed $u \in U$, by minimizing $F(x,u)$ subject to

$$\{G_i(x,u) + v_i^2\} = \theta,$$

and

$$H(x,u) = \theta.$$

At this point we assume $F(x,u)$, $\tilde{G}(x,u,v) = \{G_i(x,u) + v_i^2\} \equiv G(u) + v^2$ (from now on we use this general notation) and $H(x,u)$ satisfy the hypotheses of the Lagrange multiplier Theorem 11.3.2.

Hence, for the fixed $u \in U$, we assume there exists an optimal $x \in V$ which locally minimizes $F(x,u)$ under the mentioned constraints.

From Theorem 11.3.2 there exist Lagrange multipliers λ_1, λ_2 such that denoting $[L^p(\Omega)]^{m_1}$ and $[L^p(\Omega)]^{m_2}$ simply by L^p and defining

$$\tilde{F}(x,u,\lambda_1,\lambda_2,v) = F(x,u) + \langle \lambda_1, G(u) + v^2 \rangle_{L^p} + \langle \lambda_2, H(x,u) \rangle_{L^p},$$

the following necessary conditions hold:

$$\tilde{F}_x(x,u) = F_x(x,u) + \lambda_1 \cdot G_x(x,u) + \lambda_2 \cdot H_x(x,u) = \theta, \tag{11.67}$$

$$G(x,u) + v^2 = \theta, \tag{11.68}$$

$$\lambda_1 \cdot v = \theta, \tag{11.69}$$

$$\lambda_1 \geq \theta, \tag{11.70}$$

$$H(x,u) = \theta. \tag{11.71}$$

Clarifying the dependence on u, we denote the solution $x, \lambda_1, \lambda_2, v$ by $x(u)$, $\lambda_1(u)$, $\lambda_2(u)$, $v(u)$, respectively. In particular, we assume that for a $u_0 \in U$, $x(u_0), \lambda_1(u_0), \lambda_2(u_0), v(u_0)$ satisfy the hypotheses of the implicit function theorem. Thus, for any u in an appropriate neighborhood of u_0, the corresponding $x(u), \lambda_1(u), \lambda_2(u), v(u)$ are uniquely defined.

We emphasize that from now on the main focus of our analysis is to evaluate variations of the optimal $x(u), \lambda_1(u), \lambda_2(u), v(u)$ with variations of u in a neighborhood of u_0.

For such an analysis, the main tool is the implicit function theorem and its main hypothesis is satisfied through the invertibility of the matrix of Fréchet second derivatives.

Hence, denoting $x_0 = x(u_0), (\lambda_1)_0 = \lambda_1(u_0), (\lambda_2)_0 = \lambda_2(u_0), v_0 = v(u_0)$, and

$$A_1 = F_x(x_0,u_0) + (\lambda_1)_0 \cdot G_x(x_0,u_0) + (\lambda_2)_0 \cdot H_x(x_0,u_0),$$

$$A_2 = G(x_0,u_0) + v_0^2$$

$$A_3 = H(x_0,u_0),$$

$$A_4 = (\lambda_1)_0 \cdot v_0,$$

we reiterate to assume that

$$A_1 = \theta,\ A_2 = \theta,\ A_3 = \theta,\ A_4 = \theta,$$

and M^{-1} to represent a bounded linear operator, where

$$M = \begin{bmatrix} (A_1)_x & (A_1)_{\lambda_1} & (A_1)_{\lambda_2} & (A_1)_v \\ (A_2)_x & (A_2)_{\lambda_1} & (A_2)_{\lambda_2} & (A_2)_v \\ (A_3)_x & (A_3)_{\lambda_1} & (A_3)_{\lambda_2} & (A_3)_v \\ (A_4)_x & (A_4)_{\lambda_1} & (A_4)_{\lambda_2} & (A_4)_v \end{bmatrix} \tag{11.72}$$

where the derivatives are evaluated at $(x_0, u_0, (\lambda_1)_0, (\lambda_2)_0, v_0)$ so that

$$M = \begin{bmatrix} A & G_x(x_0,u_0) & H_x(x_0,u_0) & \theta \\ G_x(x_0,u_0) & \theta & \theta & 2v_0 \\ H_x(x_0,u_0) & \theta & \theta & \theta \\ \theta & v_0 & \theta & (\lambda_1)_0 \end{bmatrix} \tag{11.73}$$

where

$$A = F_{xx}(x_0,u_0) + (\lambda_1)_0 \cdot G_{xx}(x_0,u_0) + (\lambda_2)_0 \cdot H_{xx}(x_0,u_0).$$

Moreover, also from the implicit function theorem,

$$\|(x(u),\lambda_1(u),\lambda_2(u),v(u)) - (x(u_0),\lambda_1(u_0),\lambda_2(u_0),v(u_0))\| \leq K\|u-u_0\|, \tag{11.74}$$

for some appropriate $K > 0$, $\forall u \in B_r(u_0)$, for some $r > 0$.

Beyond assuming $\tilde{F}$ to be twice Fréchet differentiable we suppose

$$\tilde{F}_{xx}$$

is continuous in a neighborhood of u_0 so that from (11.74) there exists $K_1 > 0$ such that

$$\|\tilde{F}_{xx}(x(u),u,\lambda(u),v(u))\| \leq K_1, \forall u \in B_{r_1}(u_0), \tag{11.75}$$

for some appropriate $K_1 > 0, r_1 > 0$. We highlight to have denoted $\lambda(u) = (\lambda_1(u),\lambda_2(u))$.

Let $\varphi \in [C^\infty(\Omega)]^k \cap U$, where k depends on the vectorial expression of U.

At this point we will be concerned with the following Gâteaux variation evaluation:

$$\delta_u \tilde{F}(x(u_0),u_0,\lambda(u_0),v(u_0);\varphi).$$

Observe that

$$\begin{aligned} &\delta_u \tilde{F}(x(u_0),u_0,\lambda(u_0),v(u_0);\varphi) \\ &= \lim_{\varepsilon \to 0} \left\{ \frac{\tilde{F}(x(u_0+\varepsilon\varphi),u_0+\varepsilon\varphi,\lambda(u_0+\varepsilon\varphi),v(u_0+\varepsilon\varphi))}{\varepsilon} \right. \\ &\left. - \frac{\tilde{F}(x(u_0),u_0,\lambda(u_0),v(u_0))}{\varepsilon} \right\}, \end{aligned}$$

so that

$$\delta_u \tilde{F}(x(u_0),u_0,\lambda(u_0),v(u_0);\varphi)$$

$$= \lim_{\varepsilon \to 0} \left\{ \frac{\tilde{F}(x(u_0+\varepsilon\varphi), u_0+\varepsilon\varphi, \lambda(u_0+\varepsilon\varphi), v(u_0+\varepsilon\varphi))}{\varepsilon} \right.$$
$$- \frac{\tilde{F}(x(u_0), u_0+\varepsilon\varphi, \lambda(u_0+\varepsilon\varphi), v(u_0+\varepsilon\varphi))}{\varepsilon}$$
$$+ \frac{\tilde{F}(x(u_0), u_0+\varepsilon\varphi, \lambda(u_0+\varepsilon\varphi), v(u_0+\varepsilon\varphi))}{\varepsilon}$$
$$\left. - \frac{\tilde{F}(x(u_0), u_0, \lambda(u_0), v(u_0))}{\varepsilon} \right\}.$$

However,

$$\left| \frac{\tilde{F}(x(u_0+\varepsilon\varphi), u_0+\varepsilon\varphi, \lambda(u_0+\varepsilon\varphi), v(u_0+\varepsilon\varphi))}{\varepsilon} \right.$$
$$\left. - \frac{\tilde{F}(x(u_0), u_0+\varepsilon\varphi, \lambda(u_0+\varepsilon\varphi), v(u_0+\varepsilon\varphi))}{\varepsilon} \right|$$
$$\leq \left\| \tilde{F}_x(x(u_0+\varepsilon\varphi), u_0+\varepsilon\varphi, \lambda(u_0+\varepsilon\varphi), v(u_0+\varepsilon\varphi)) \right\| K\|\varphi\|$$
$$+ \sup_{t\in[0,1]} \left\| \tilde{F}_{xx}(x(u_0+t\varepsilon\varphi), u_0+\varepsilon\varphi, \lambda(u_0+\varepsilon\varphi), v(u_0+\varepsilon\varphi)) \right\| K^2\|\varphi\|_4^2\varepsilon$$
$$\leq K_1 K^2 \|\varphi\|_4^2 \varepsilon$$
$$\to 0, \text{ as } \varepsilon \to 0.$$

In these last inequalities we have used

$$\limsup_{\varepsilon \to 0} \left\| \frac{x(u_0+\varepsilon\varphi) - x(u_0)}{\varepsilon} \right\| \leq K\|\varphi\|,$$

and

$$\tilde{F}_x(x(u_0+\varepsilon\varphi), u_0+\varepsilon\varphi, \lambda(u_0+\varepsilon\varphi), v(u_0+\varepsilon\varphi)) = \theta.$$

On the other hand,

$$\left\{ \frac{\tilde{F}(x(u_0), u_0+\varepsilon\varphi, \lambda(u_0+\varepsilon\varphi), v(u_0+\varepsilon\varphi))}{\varepsilon} \right.$$
$$\left. - \frac{\tilde{F}(x(u_0), u_0, \lambda(u_0), v(u_0))}{\varepsilon} \right\}$$
$$= \left\{ \frac{\tilde{F}(x(u_0), u_0+\varepsilon\varphi, \lambda(u_0+\varepsilon\varphi), v(u_0+\varepsilon\varphi))}{\varepsilon} \right.$$
$$- \frac{\tilde{F}(x(u_0), u_0+\varepsilon\varphi, \lambda(u_0), v(u_0))}{\varepsilon}$$
$$+ \frac{\tilde{F}(x(u_0), u_0+\varepsilon\varphi, \lambda(u_0), v(u_0))}{\varepsilon}$$
$$\left. - \frac{\tilde{F}(x(u_0), u_0, \lambda(u_0), v(u_0))}{\varepsilon} \right\}$$

Now observe that

$$
\begin{aligned}
&\frac{\tilde{F}(x(u_0),u_0+\varepsilon\varphi,\lambda(u_0+\varepsilon\varphi),v(u_0+\varepsilon\varphi))}{\varepsilon} \\
&\quad -\frac{\tilde{F}(x(u_0),u_0+\varepsilon\varphi,\lambda(u_0),v(u_0))}{\varepsilon} \\
&= \frac{\langle \lambda_1(u_0+\varepsilon\varphi),G(x(u_0),u_0+\varepsilon\varphi)+v(u_0+\varepsilon\varphi)^2\rangle_{L^p}}{\varepsilon} \\
&\quad -\frac{\langle \lambda_1(u_0),G(x(u_0),u_0+\varepsilon\varphi)+v(u_0)^2\rangle_{L^p}}{\varepsilon} \\
&\quad +\frac{\langle \lambda_2(u_0+\varepsilon\varphi)-\lambda_2(u_0),H(x(u_0),u_0+\varepsilon\varphi)\rangle_{L^p}}{\varepsilon}.
\end{aligned} \tag{11.76}
$$

Also,

$$
\begin{aligned}
&\left|\frac{\langle \lambda_1(u_0+\varepsilon\varphi),G(x(u_0),u_0+\varepsilon\varphi)+v(u_0+\varepsilon\varphi)^2\rangle_{L^p}}{\varepsilon}\right. \\
&\quad \left.-\frac{\langle \lambda_1(u_0),G(x(u_0),u_0+\varepsilon\varphi)+v(u_0)^2\rangle_{L^p}}{c}\right| \\
&\leq \left|\frac{\langle \lambda_1(u_0+\varepsilon\varphi),G(x(u_0),u_0+\varepsilon\varphi)+v(u_0+\varepsilon\varphi)^2\rangle_{L^p}}{\varepsilon}\right. \\
&\quad \left.-\frac{\langle \lambda_1(u_0),G(x(u_0),u_0+\varepsilon\varphi)+v(u_0+\varepsilon\varphi)^2\rangle_{L^p}}{\varepsilon}\right| \\
&\quad +\left|\frac{\langle \lambda_1(u_0),G(x(u_0),u_0+\varepsilon\varphi)+v(u_0+\varepsilon\varphi)^2\rangle_{L^p}}{\varepsilon}\right. \\
&\quad \left.-\frac{\langle \lambda_1(u_0),G(x(u_0),u_0+\varepsilon\varphi)+v(u_0)^2\rangle_{L^p}}{\varepsilon}\right| \\
&\leq \varepsilon\frac{K\|\varphi\|}{\varepsilon}\|G(x(u_0),u_0+\varepsilon\varphi)+v(u_0+\varepsilon\varphi)^2\| \\
&\quad +\|\lambda_1(u_0)(v(u_0+\varepsilon\varphi)+v(u_0))\|\,\frac{K\|\varphi\|\varepsilon}{\varepsilon} \\
&\to 0 \text{ as } \varepsilon\to 0.
\end{aligned}
$$

To obtain the last inequalities we have used

$$
\limsup_{\varepsilon\to 0}\left\|\frac{\lambda_1(u_0+\varepsilon\varphi)-\lambda_1(u_0)}{\varepsilon}\right\|\leq K\|\varphi\|,
$$

$$
\lambda_1(u_0)v(u_0)=\theta,
$$

$$
\lambda_1(u_0)v(u_0+\varepsilon\varphi)\to\theta,\ \text{as}\ \varepsilon\to 0,
$$

and

$$\begin{aligned}
&\left\| \frac{\lambda_1(u_0)(v(u_0+\varepsilon\varphi)^2 - v(u_0)^2)}{\varepsilon} \right\| \\
&= \left\| \frac{\lambda_1(u_0)(v(u_0+\varepsilon\varphi)+v(u_0))(v(u_0+\varepsilon\varphi)-v(u_0))}{\varepsilon} \right\| \\
&\leq \frac{\|\lambda_1(u_0)(v(u_0+\varepsilon\varphi)+v(u_0))\|K\|\varphi\|\varepsilon}{\varepsilon} \\
&\to 0, \text{ as } \varepsilon \to 0. \qquad (11.77)
\end{aligned}$$

Finally,

$$\begin{aligned}
&\left| \frac{\langle \lambda_2(u_0+\varepsilon\varphi) - \lambda_2(u_0), H(x(u_0), u_0+\varepsilon\varphi)\rangle_{L^p}}{\varepsilon} \right| \\
&\leq \frac{K\varepsilon\|\varphi\|}{\varepsilon}\|H(x(u_0), u_0+\varepsilon\varphi)\| \\
&\to 0, \text{ as } \varepsilon \to 0.
\end{aligned}$$

To obtain the last inequalities we have used

$$\limsup_{\varepsilon\to 0} \left\| \frac{\lambda_2(u_0+\varepsilon\varphi) - \lambda_2(u_0)}{\varepsilon} \right\| \leq K\|\varphi\|,$$

and

$$H(x(u_0), u_0+\varepsilon\varphi) \to \theta, \text{ as } \varepsilon \to 0.$$

From these last results, we get

$$\begin{aligned}
&\delta_u \tilde{F}(x(u_0), u_0, \lambda(u_0), v(u_0); \varphi) \\
&= \lim_{\varepsilon\to 0} \left\{ \frac{\tilde{F}(x(u_0), u_0+\varepsilon\varphi, \lambda(u_0), v(u_0))}{\varepsilon} \right. \\
&\quad \left. - \frac{\tilde{F}(x(u_0), u_0, \lambda(u_0), v(u_0))}{\varepsilon} \right\} \\
&= \langle F_u(x(u_0), u_0), \varphi\rangle_U + \langle \lambda_1(u_0)\cdot G_u(x(u_0), u_0), \varphi\rangle_{L^p} \\
&\quad + \langle \lambda_2(u_0)\cdot H_u(x(u_0), u_0), \varphi\rangle_{L^p}.
\end{aligned}$$

In the last lines we have proven the following corollary of the implicit function theorem.

Corollary 11.1. *Suppose $(x_0, u_0, (\lambda_1)_0, (\lambda_2)_0, v_0)$ is a solution of the system (11.67), (11.68),(11.69), (11.71), and assume the corresponding hypotheses of the implicit function theorem are satisfied. Also assume $\tilde{F}(x, u, \lambda_1, \lambda_2, v)$ is such that the Fréchet second derivative $\tilde{F}_{xx}(x, u, \lambda_1, \lambda_2)$ is continuous in a neighborhood of*

$$(x_0, u_0, (\lambda_1)_0, (\lambda_2)_0).$$

Under such hypotheses, for a given $\varphi \in [C^\infty(\Omega)]^k$, *denoting*

$$F_1(u) = \tilde{F}(x(u), u, \lambda_1(u), \lambda_2(u), v(u)),$$

we have

$$\begin{aligned}
&\delta(F_1(u); \varphi)|_{u=u_0} \\
&= \langle F_u(x(u_0), u_0), \varphi \rangle_U + \langle \lambda_1(u_0) \cdot G_u(x(u_0), u_0), \varphi \rangle_{L^p} \\
&\quad + \langle \lambda_2(u_0) \cdot H_u(x(u_0), u_0), \varphi \rangle_{L^p}.
\end{aligned}$$

Part III
Applications

Chapter 12
Duality Applied to Elasticity

12.1 Introduction

The first part of the present work develops a new duality principle applicable to nonlinear elasticity. The proof of existence of solutions for the model in question has been obtained in Ciarlet [21]. In earlier results (see [65] for details) the concept of complementary energy is equivalently developed under the hypothesis of positive definiteness of the stress tensor at a critical point. In more recent works, Gao [33, 34, 36] applied his triality theory to similar models obtaining duality principles for more general situations, including the case of negative definite optimal stress tensor.

We emphasize our main objective is to establish a new and different duality principle which allows the local optimal stress tensor to not be either positive or negative definite. Such a result is a kind of extension of a more basic one obtained in Toland [67]. Despite the fact we do not apply it directly, we follow a similar idea. The optimality conditions are also new. We highlight the basic tools on convex analysis here used may be found in [25, 54, 67] for example. For related results about the plate model presented in Ciarlet [22], see Botelho [11, 13].

In a second step, we present other two duality principles which qualitatively agree with the triality theory proposed by Gao (see again [33, 34], for details).

However, our proofs again are obtained through more traditional tools of convex analysis. Finally, in the last section, we provide a numerical example in which the optimal stress field is neither positive nor negative definite.

At this point we start to describe the primal formulation.

Consider $\Omega \subset \mathbb{R}^3$ an open, bounded, connected set, which represents the reference volume of an elastic solid under the loads $f \in L^2(\Omega;\mathbb{R}^3)$ and the boundary loads $\hat{f} \in L^2(\Gamma;\mathbb{R}^3)$, where Γ denotes the boundary of Ω. The field of displacements resulting from the actions of f and $\hat{f}$ is denoted by $u \equiv (u_1,u_2,u_3) \in U$, where u_1,u_2, and u_3 denote the displacements relating the directions x,y, and z, respectively, in the Cartesian system (x,y,z).

Here U is defined by

$$U = \{u = (u_1,u_2,u_3) \in W^{1,4}(\Omega;\mathbb{R}^3) \mid u = (0,0,0) \equiv \theta \text{ on } \Gamma_0\} \quad (12.1)$$

F. Botelho, *Functional Analysis and Applied Optimization in Banach Spaces: Applications to Non-Convex Variational Models*, DOI 10.1007/978-3-319-06074-3_12,

and $\Gamma = \Gamma_0 \cup \Gamma_1$, $\Gamma_0 \cap \Gamma_1 = \emptyset$ (for details about the Sobolev space U see [2]). We assume $|\Gamma_0| > 0$ where $|\Gamma_0|$ denotes the Lebesgue measure of Γ_0.

The stress tensor is denoted by $\{\sigma_{ij}\}$, where

$$\sigma_{ij} = H_{ijkl}\left(\frac{1}{2}(u_{k,l} + u_{l,k} + u_{m,k}u_{m,l})\right),$$
$$\{H_{ijkl}\} = \{\lambda\delta_{ij}\delta_{kl} + \mu(\delta_{ik}\delta_{jl} + \delta_{il}\delta_{jk})\}, \tag{12.2}$$

$\{\delta_{ij}\}$ is the Kronecker delta and $\lambda, \mu > 0$ are the Lamé constants (we assume they are such that $\{H_{ijkl}\}$ is a symmetric constant positive definite fourth-order tensor).

The boundary value form of the nonlinear elasticity model is given by

$$\begin{cases} \sigma_{ij,j} + (\sigma_{mj}u_{i,m})_{,j} + f_i = 0, & \text{in } \Omega, \\ u = \theta, & \text{on } \Gamma_0, \\ \sigma_{ij}n_j + \sigma_{mj}u_{i,m}\mathbf{n}_j = \hat{f}_i, & \text{on } \Gamma_1, \end{cases} \tag{12.3}$$

where $\mathbf{n}$ denotes the outward normal to the surface Γ.

The corresponding primal variational formulation is represented by $J : U \to \mathbb{R}$, where

$$J(u) = \frac{1}{2}\int_\Omega H_{ijkl}\left(\frac{1}{2}(u_{i,j} + u_{j,i} + u_{m,i}u_{m,j})\right)\left(\frac{1}{2}(u_{k,l} + u_{l,k} + u_{m,k}u_{m,l})\right) dx$$
$$-\langle u, f\rangle_{L^2(\Omega;\mathbb{R}^3)} - \int_{\Gamma_1} \hat{f}_i u_i \, d\Gamma \tag{12.4}$$

where

$$\langle u, f\rangle_{L^2(\Omega;\mathbb{R}^3)} = \int_\Omega f_i u_i \, dx.$$

Remark 12.1.1. Derivatives must be always understood in the distributional sense, whereas boundary conditions are in the sense of traces. Moreover, from now on by a regular boundary Γ of Ω, we mean regularity enough so that the standard Gauss–Green formulas of integrations by parts and the well-known Sobolev imbedding and trace theorems hold. Finally, we denote by θ the zero vector in appropriate function spaces, the standard norm for $L^2(\Omega)$ by $\|\cdot\|_2$, and $L^2(\Omega;\mathbb{R}^{3\times 3})$ simply by L^2.

12.2 The Main Duality Principle

Now we prove the main result.

Theorem 12.2.1. *Assume the statements of last section. In particular, let $\Omega \subset \mathbb{R}^3$ be an open, bounded, connected set with a regular boundary denoted by $\Gamma = \Gamma_0 \cup \Gamma_1$, where $\Gamma_0 \cap \Gamma_1 = \emptyset$ and $|\Gamma_0| > 0$. Consider the functional $(G \circ \Lambda) : U \to \mathbb{R}$ expressed by*

$$(G \circ \Lambda)(u)$$
$$= \frac{1}{2}\int_\Omega H_{ijkl}\left(\frac{u_{i,j} + u_{j,i}}{2} + \frac{u_{m,i}u_{m,j}}{2}\right)\left(\frac{u_{k,l} + u_{l,k}}{2} + \frac{u_{m,k}u_{m,l}}{2}\right) dx,$$

where $\Lambda : U \to Y \times Y$ *is given by*

$$\Lambda u = \{\Lambda_1 u, \Lambda_2 u\},$$

$$\Lambda_1 u = \left\{ \frac{u_{i,j} + u_{j,i}}{2} \right\}$$

and

$$\Lambda_2 u = \{u_{m,i}\}.$$

Here

$$U = \{u \in W^{1,4}(\Omega; \mathbb{R}^3) \mid u = (u_1, u_2, u_3) = \theta \text{ on } \Gamma_0\}.$$

Define $(F \circ \Lambda_2) : U \to \mathbb{R}$, $(G_K \circ \Lambda) : U \to \mathbb{R}$, *and* $(G_1 \circ \Lambda_2) : U \to \mathbb{R}$ *by*

$$(F \circ \Lambda_2)(u) = \frac{K}{2} \langle u_{m,i}, u_{m,i} \rangle_{L^2(\Omega)},$$

$$G_K(\Lambda u) = G_K(\Lambda_1 u, \Lambda_2 u) = G(\Lambda u) + \frac{K}{4} \langle u_{m,i}, u_{m,i} \rangle_{L^2(\Omega)},$$

and

$$(G_1 \circ \Lambda_2)(u) = \frac{K}{4} \langle u_{m,i}, u_{m,i} \rangle_{L^2(\Omega)},$$

respectively.

Also define

$$C = \{u \in U \mid G_K^{**}(\Lambda u) = G_K(\Lambda u)\},$$

where $K > 0$ *is an appropriate constant to be specified.*

For $f \in L^2(\Omega; \mathbb{R}^3)$, $\hat{f} \in L^2(\Gamma; \mathbb{R}^3)$, *let* $J : U \to \mathbb{R}$ *be expressed by*

$$J(u) = G(\Lambda u) - \langle u, f \rangle_{L^2(\Omega; \mathbb{R}^3)} - \int_{\Gamma_1} \hat{f}_i u_i \, d\Gamma. \tag{12.5}$$

Under such hypotheses, we have

$$\inf_{u \in C_1} \{J(u)\}$$
$$\geq \sup_{(\tilde{\sigma}, \sigma, v) \in \tilde{Y}} \left\{ \inf_{z^* \in Y^*} \{F^*(z^*) - \tilde{G}_K^*(\sigma, z^*, v) - \tilde{G}_1^*(\tilde{\sigma}, \sigma, z^*, v)\} \right\},$$

where $\tilde{Y} = A^* \times Y^* \times \hat{Y}^*$, $Y = Y^* = L^2(\Omega, \mathbb{R}^{3 \times 3}) = L^2$,

$$\hat{Y}^* = \{v \in Y^* \text{ such that } W^*(z^*) \text{ is positive definite in } \Omega\}, \tag{12.6}$$

and

$$W^*(z^*) = \frac{z^*_{mi} z^*_{mi}}{K} - \overline{H}_{ijkl} z^*_{ij} z^*_{kl} - \sum_{m,i=1}^{3} \frac{(z^*_{ij} v_{mj})^2}{K/2}. \tag{12.7}$$

Here $C_1 = C_2 \cap C$, *where*

$$C_2 = \{u \in U \mid \{u_{i,j}\} \in \hat{Y}^*\}.$$

Furthermore,

$$A^* = \{\tilde{\sigma} \in Y^* \mid \tilde{\sigma}_{ij,j} + f_i = 0 \text{ in } \Omega \text{ and } \tilde{\sigma}_{ij}\boldsymbol{n}_j = \hat{f}_i \text{ on } \Gamma_1\}.$$

Also

$$\begin{aligned} F^*(z^*) &= \sup_{v_2 \in Y}\{\langle v_2, z^*\rangle_Y - F(v_2)\} \\ &= \frac{1}{2K}\langle z^*_{mi}, z^*_{mi}\rangle_{L^2(\Omega)}, \end{aligned} \tag{12.8}$$

where we recall that $z^*_{ij} = z^*_{ji}$. *Through the relations*

$$Q_{mi} = (\sigma_{ij} + z^*_{ij})v_{mj} + (K/2)v_{mi},$$

we define

$$\begin{aligned} \tilde{G}^*_K(\sigma, z^*, v) &= G^*_K(\sigma + z^*, Q) \\ &= \sup_{(v_1, v_2) \in Y \times Y}\{\langle v_1, \sigma + z^*\rangle_Y + \langle v_2, Q\rangle_Y - G_K(v_1, v_2)\}, \end{aligned} \tag{12.9}$$

so that in particular,

$$\begin{aligned} \tilde{G}^*_K(\sigma, z^*, v) &= G^*_K(\sigma + z^*, Q) \\ &= \frac{1}{2}\int_\Omega \overline{H}_{ijkl}(\sigma_{ij} + z^*_{ij})(\sigma_{kl} + z^*_{kl})\, dx \\ &\quad + \frac{1}{2}\int_\Omega (\sigma_{ij} + z^*_{ij})v_{mi}v_{mj}\, dx + \frac{K}{4}\langle v_{mi}, v_{mi}\rangle_{L^2(\Omega)} \end{aligned}$$

if $(\tilde{\sigma}, \sigma, v, z^*) \in B^*$. *We emphasize to denote*

$$B^* = \{(\tilde{\sigma}, \sigma, v, z^*) \in [Y^*]^4 \mid \sigma_K(\sigma, z^*) \text{ is positive definite in } \Omega\},$$

$$\sigma_K(\sigma, z^*) = \left\{ \begin{matrix} \sigma_{11} + z^*_{11} + K/2 & \sigma_{12} + z^*_{12} & \sigma_{13} + z^*_{13} \\ \sigma_{21} + z^*_{21} & \sigma_{22} + z^*_{22} + K/2 & \sigma_{23} + z^*_{23} \\ \sigma_{31} + z^*_{31} & \sigma_{32} + z^*_{32} & \sigma_{33} + z^*_{33} + K/2 \end{matrix} \right\}, \tag{12.10}$$

and

$$\{\overline{H}_{ijkl}\} = \{H_{ijkl}\}^{-1}.$$

Moreover,

$$\begin{aligned}\tilde{G}_1^*(\tilde{\sigma},\sigma,z^*,v) &= G_1^*(\tilde{\sigma},-\sigma,-Q) \\ &= \sup_{v_2\in Y}\{\langle v_2,\tilde{\sigma}-\sigma-Q\rangle_Y - G_1(v_2)\} \\ &= \frac{1}{K}\sum_{m,i=1}^{3}\|\tilde{\sigma}_{mi}-\sigma_{mi}-Q_{mi}\|_2^2 \\ &= \frac{1}{K}\sum_{m,i=1}^{3}\|\tilde{\sigma}_{mi}-\sigma_{mi}-(\sigma_{ij}+z_{ij}^*)v_{mj}-(K/2)v_{mi}\|_2^2.\end{aligned}$$

Finally, if there exists a point $(u_0,\tilde{\sigma}_0,\sigma_0,v_0,z_0^*)\in C_1\times((\tilde{Y}\times Y^*)\cap B^*)$, *such that*

$$\delta\left\{\langle u_{0i},-\tilde{\sigma}_{0ij,j}-f_i\rangle_{L^2(\Omega)}-\int_{\Gamma_1}u_{0i}(\hat{f}_i-\tilde{\sigma}_{0ij}\boldsymbol{n}_j)\,d\Gamma \right.$$
$$\left. +F^*(z_0^*)-\tilde{G}_K^*(\sigma_0,z_0^*,v_0)-\tilde{G}_1^*(\tilde{\sigma}_0,\sigma_0,z_0^*,v_0)\right\}=\theta, \qquad (12.11)$$

we have

$$\begin{aligned}J(u_0) &= \min_{u\in C_1}\{J(u)\} \\ &= \sup_{(\tilde{\sigma},\sigma,v)\in\tilde{Y}}\left\{\inf_{z^*\in Y^*}\{F^*(z^*)-\tilde{G}_K^*(\sigma,z^*,v)-\tilde{G}_1^*(\tilde{\sigma},\sigma,z^*,v)\}\right\} \\ &= F^*(z_0^*)-\tilde{G}_K^*(\sigma_0,z_0^*,v_0)-\tilde{G}_1^*(\tilde{\sigma}_0,\sigma_0,z_0^*,v_0).\end{aligned} \qquad (12.12)$$

Proof. We start by proving that $G_K^*(\sigma+z^*,Q)=G_{K_L}^*(\sigma+z^*,Q)$ if $\sigma_K(\sigma,z^*)$ is positive definite in Ω, where

$$G_{K_L}^*(\sigma,Q)=\int_\Omega g_{K_L}^*(\sigma,Q)\,dx$$

is the Legendre transform of $G_K : Y\times Y\to\mathbb{R}$. To simplify the notation we denote $(\sigma,Q)=y^*=(y_1^*,y_2^*)$. We first formally calculate $g_{K_L}^*(y^*)$, the Legendre transform of $g_K(y)$, where

$$g_K(y)=H_{ijkl}\left(y_{1ij}+\frac{1}{2}y_{2mi}y_{2mj}\right)\left(y_{1kl}+\frac{1}{2}y_{2mk}y_{2ml}\right)$$
$$+\frac{K}{4}y_{2mi}y_{2mi}. \qquad (12.13)$$

We recall that

$$g_{K_L}^*(y^*)=\langle y,y^*\rangle_{\mathbb{R}^{18}}-g_K(y) \qquad (12.14)$$

where $y \in \mathbb{R}^{18}$ is the solution of equation

$$y^* = \frac{\partial g_K(y)}{\partial y}. \tag{12.15}$$

Thus

$$y^*_{1ij} = \sigma_{ij} = H_{ijkl}\left(y_{1kl} + \frac{1}{2}y_{2mk}y_{2ml}\right) \tag{12.16}$$

and

$$y^*_{2mi} = Q_{mi} = H_{ijkl}\left(y_{1kl} + \frac{1}{2}y_{2ok}y_{2ol}\right)y_{2mj} + (K/2)y_{2mi} \tag{12.17}$$

so that

$$Q_{mi} = \sigma_{ij}y_{2mj} + (K/2)y_{2mi}. \tag{12.18}$$

Inverting these last equations, we have

$$y_{2mi} = \overline{\sigma}^K_{ij}Q_{mj} \tag{12.19}$$

where $\{\overline{\sigma}^K_{ij}\} = \sigma_K^{-1}(\sigma)$,

$$\sigma_K(\sigma) = \left\{\begin{array}{lll} \sigma_{11} + K/2 & \sigma_{12} & \sigma_{13} \\ \sigma_{21} & \sigma_{22} + K/2 & \sigma_{23} \\ \sigma_{31} & \sigma_{32} & \sigma_{33} + K/2 \end{array}\right\} \tag{12.20}$$

and also

$$y_{1ij} = \overline{H}_{ijkl}\sigma_{kl} - \frac{1}{2}y_{2mi}y_{2mj}. \tag{12.21}$$

Finally

$$g^*_{K_L}(\sigma, Q) = \frac{1}{2}\overline{H}_{ijkl}\sigma_{ij}\sigma_{kl} + \frac{1}{2}\bar{\sigma}^K_{ij}Q_{mi}Q_{mj}. \tag{12.22}$$

Now we will prove that $g^*_{K_L}(y^*) = g^*_K(y^*)$ if $\sigma_K(y^*_1) = \sigma_K(\sigma)$ is positive definite. First observe that

$$\begin{aligned} g^*_K(y^*) &= \sup_{y \in \mathbb{R}^{18}} \{\langle y_1, \sigma\rangle_{\mathbb{R}^9} + \langle y_2, Q\rangle_{\mathbb{R}^9} - g_K(y)\} \\ &= \sup_{y \in \mathbb{R}^{18}} \Bigg\{ \langle y_1, \sigma\rangle_{\mathbb{R}^9} + \langle y_2, Q\rangle_{\mathbb{R}^9} \\ &\quad -\frac{1}{2}H_{ijkl}\left(y_{1ij} + \frac{1}{4}y_{2mi}y_{2mj}\right)\left(y_{1kl} + \frac{1}{2}y_{2mk}y_{2ml}\right) \\ &\quad -\frac{K}{4}y_{2mi}y_{2mi}\Bigg\} \end{aligned}$$

$$= \sup_{(\bar{y}_1,y_2)\in\mathbb{R}^9\times\mathbb{R}^9} \left\{ \langle \bar{y}_{1ij} - \frac{1}{2} y_{2mi} y_{2mj}, \sigma_{ij} \rangle_{\mathbb{R}} + \langle y_2, Q \rangle_{\mathbb{R}^9} \right.$$
$$\left. -\frac{1}{2} H_{ijkl}[\bar{y}_{1ij}][\bar{y}_{1kl}] - \frac{K}{4} y_{2mi} y_{2mi} \right\}.$$

The result follows just observing that

$$\sup_{\bar{y}_1\in\mathbb{R}^9} \left\{ \langle \bar{y}_{1ij}, \sigma_{ij} \rangle_{\mathbb{R}} - \frac{1}{2} H_{ijkl}[\bar{y}_{1ij}][\bar{y}_{1kl}] \right\} = \frac{1}{2} \overline{H}_{ijkl} \sigma_{ij} \sigma_{kl} \tag{12.23}$$

and

$$\sup_{y_2\in\mathbb{R}^9} \left\{ \langle -\frac{1}{2} y_{2mi} y_{2mj}, \sigma_{ij} \rangle_{\mathbb{R}} + \langle y_2, Q \rangle_{\mathbb{R}^9} - \frac{K}{4} y_{2mi} y_{2mi} \right\}$$
$$= \frac{1}{2} \overline{\sigma}^K_{ij} Q_{mi} Q_{mj} \tag{12.24}$$

if $\sigma_K(y_1^*) = \sigma_K(\sigma)$ is positive definite.

Now observe that using the relation

$$Q_{mi} = (\sigma_{ij} + z^*_{ij}) v_{mj} + (K/2) v_{mi},$$

we have

$$\tilde{G}^*_K(\sigma, z^*, v) = G^*_K(\sigma + z^*, Q)$$
$$= \int_\Omega g^*_{K_L}(\sigma + z^*, Q)\, dx, \tag{12.25}$$

if $\sigma_K(\sigma + z^*)$ is positive definite.

Also, considering the concerned symmetries, we may write

$$\tilde{G}^*_K(\sigma, z^*, v) + \tilde{G}^*_1(\tilde{\sigma}, \sigma, z^*, v) = G^*_K(\sigma + z^*, Q) + G^*_1(\tilde{\sigma}, -\sigma, -Q)$$
$$\geq \langle \Lambda_1 u, \sigma \rangle_{L^2} + \langle \Lambda_2 u, z^* + Q \rangle_{L^2}$$
$$+ \langle \Lambda_1 u, \tilde{\sigma} - \sigma \rangle_{L^2} - \langle \Lambda_2 u, Q \rangle_{L^2}$$
$$- G^{**}_K(\Lambda u) - G_1(\Lambda_2 u), \tag{12.26}$$

$\forall u \in U$, $z^* \in Y^*$, $(\tilde{\sigma}, \sigma, v) \in \tilde{Y}$, so that

$$\tilde{G}^*_K(\sigma, z^*, v) + \tilde{G}^*_1(\tilde{\sigma}, \sigma, z^*, v)$$
$$\geq \langle \Lambda_2 u, z^* \rangle_{L^2} + \langle \Lambda_1 u, \tilde{\sigma} \rangle_{L^2}$$
$$- G_K(\Lambda u) - G_1(\Lambda_2 u)$$
$$= \langle \Lambda_2 u, z^* \rangle_{L^2} + \langle u, f \rangle_{L^2(\Omega;\mathbb{R}^3)}$$
$$+ \int_{\Gamma_1} \hat{f}_i u_i\, d\Gamma - G_K(\Lambda u) - G_1(\Lambda_2 u), \tag{12.27}$$

$\forall u \in C_1$, $z^* \in Y^*$, $(\tilde{\sigma}, \sigma, v) \in \tilde{Y}$. Hence

$$\begin{aligned}
&-F^*(z^*) + \tilde{G}_K^*(\sigma, z^*, v) + \tilde{G}_1^*(\tilde{\sigma}, \sigma, z^*, v) \\
&\quad \geq -F^*(z^*) + \langle \Lambda_2 u, z^* \rangle_{L^2} + \langle u, f \rangle_{L^2(\Omega;\mathbb{R}^3)} \\
&\quad\quad + \int_{\Gamma_1} \hat{f}_i u_i \, d\Gamma - G_K(\Lambda u) - G_1(\Lambda_2 u),
\end{aligned} \tag{12.28}$$

$\forall u \in C_1$, $z^* \in Y^*$, $(\tilde{\sigma}, \sigma, v) \in \tilde{Y}$, and thus

$$\begin{aligned}
&\sup_{z^* \in Y^*} \{-F^*(z^*) + \tilde{G}_K^*(\sigma, z^*, v) + \tilde{G}_1^*(\tilde{\sigma}, \sigma, z^*, v)\} \\
&\quad \geq \sup_{z^* \in Y^*} \{-F^*(z^*) + \langle \Lambda_2 u, z^* \rangle_{L^2} + \langle u, f \rangle_{L^2(\Omega;\mathbb{R}^3)} \\
&\quad\quad + \int_{\Gamma_1} \hat{f}_i u_i \, d\Gamma - G_K(\Lambda u) - G_1(\Lambda_2 u)\},
\end{aligned} \tag{12.29}$$

$\forall u \in C_1, (\tilde{\sigma}, \sigma, v) \in \tilde{Y}$.

Therefore,

$$\begin{aligned}
&\sup_{z^* \in Y^*} \{-F^*(z^*) + \tilde{G}_K^*(\sigma, z^*, v) + \tilde{G}_1^*(\tilde{\sigma}, \sigma, z^*, v)\} \\
&\quad \geq F(\Lambda_2 u) + \langle u, f \rangle_{L^2(\Omega;\mathbb{R}^3)} + \int_{\Gamma_1} \hat{f}_i u_i \, d\Gamma \\
&\quad\quad - G_K(\Lambda u) - G_1(\Lambda_2 u),
\end{aligned} \tag{12.30}$$

$\forall u \in C_1, (\tilde{\sigma}, \sigma, v) \in \tilde{Y}$, that is,

$$\begin{aligned}
&\sup_{z^* \in Y^*} \{-F^*(z^*) + \tilde{G}_K^*(\sigma, z^*, v) + \tilde{G}_1^*(\tilde{\sigma}, \sigma, z^*, v)\} \\
&\quad \geq -J(u),
\end{aligned} \tag{12.31}$$

$\forall u \in C_1, (\tilde{\sigma}, \sigma, v) \in \tilde{Y}$. Finally,

$$\begin{aligned}
&\inf_{u \in C_1} \{J(u)\} \\
&\geq \sup_{(\tilde{\sigma}, \sigma, v) \in \tilde{Y}} \left\{ \inf_{z^* \in Y^*} \{F^*(z^*) - \tilde{G}_K^*(\sigma, z^*, v) - \tilde{G}_1^*(\tilde{\sigma}, \sigma, z^*, v)\} \right\}.
\end{aligned} \tag{12.32}$$

Now suppose there exists a point $(u_0, \tilde{\sigma}_0, \sigma_0, z_0^*, v_0) \in C_1 \times ((\tilde{Y} \times Y^*) \cap B^*)$, such that

$$\begin{aligned}
&\delta \left\{ \langle u_{0^i}, -\tilde{\sigma}_{0^{ij},j} - f_i \rangle_{L^2(\Omega)} - \int_{\Gamma_1} u_{0^i} (\hat{f}_i - \tilde{\sigma}_{0^{ij}} \mathbf{n}_j) \, d\Gamma \right. \\
&\left. + F^*(z_0^*) - \tilde{G}_K^*(\sigma_0, z_0^*, v_0) - \tilde{G}_1^*(\tilde{\sigma}_0, \sigma_0, z_0^*, v_0) \right\} = \theta,
\end{aligned} \tag{12.33}$$

that is,

$$\delta\left\{\langle u_{0i}, -\tilde{\sigma}_{0ij,j} - f_i\rangle_{L^2(\Omega)} - \int_{\Gamma_1} u_{0i}(\hat{f}_i - \tilde{\sigma}_{0ij}\mathbf{n}_j)\, d\Gamma \right.$$
$$+F^*(z_0^*) - \frac{1}{2}\int_\Omega \bar{H}_{ijkl}(\sigma_{0ij} + z_{0ij}^*)(\sigma_{0kl} + z_{0kl}^*)\, dx$$
$$-\frac{1}{2}\int_\Omega (\sigma_{0ij} + z_{0ij}^*) v_{0mi} v_{0mj}\, dx - \frac{K}{4}\langle v_{0mi}, v_{0mi}\rangle_{L^2(\Omega)}$$
$$\left. - \sum_{m,i=1}^{3} \frac{1}{K}\|\tilde{\sigma}_{0mi} - \sigma_{0mi} - (\sigma_{0ij} + z_{0ij}^*) v_{0mj} - K/2 v_{0mi}\|_2^2 \right\} = \theta.$$

Observe that the variation in $\tilde{\sigma}$ gives us

$$\tilde{\sigma}_{0mi} - \sigma_{0mi} - (\sigma_{0ij} + z_{0ij}^*) v_{0mj} - (K/2) v_{0mi} = (K/2) u_{0m,i} \text{ in } \Omega. \tag{12.34}$$

From this and recalling that $\tilde{\sigma}_{ij} = \tilde{\sigma}_{ji}$, so that we may use the replacement

$$\tilde{\sigma}_{ij} = \frac{\tilde{\sigma}_{ij} + \tilde{\sigma}_{ji}}{2} = \tilde{\sigma}_{ji}$$

(observe that a similar remark is valid for $\sigma_{0ij} + z_{0ij}^*$), the variation in σ gives us

$$-\bar{H}_{ijkl}(\sigma_{0kl} + z_{0kl}^*) - v_{0mi} v_{0mj}/2$$
$$+\frac{u_{0i,j} + u_{0j,i}}{2} + u_{0m,i} v_{0mj} = 0, \tag{12.35}$$

in Ω. From (12.34) and the variation in v we get

$$-(\sigma_{0ij} + z_{0ij}^*) v_{mj} - (K/2) v_{0mi}$$
$$+(\sigma_{0ij} + z_{0_{ij}}^*) u_{0m,j} + (K/2) u_{0m,i} = 0, \tag{12.36}$$

so that

$$\{v_{0ij}\} = \{u_{0i,j}\}, \text{ in } \Omega. \tag{12.37}$$

From this and (12.35) we get

$$\sigma_{0ij} + z_{0ij}^* = H_{ijkl}\left(\frac{u_{0k,l} + u_{0l,k}}{2} + \frac{u_{0m,k} u_{0m,l}}{2}\right). \tag{12.38}$$

Through such relations the variation in z^* gives us

$$z_{0ij}^* = \frac{K}{2}(u_{0i,j} + u_{0j,i}) \text{ in } \Omega. \tag{12.39}$$

Finally, from the variation in u, we get

$$\tilde{\sigma}_{0ij,j} + f_i = 0, \text{ in } \Omega, \tag{12.40}$$

$$u_0 = \theta \text{ on } \Gamma_0,$$

and

$$\tilde{\sigma}_{0ij}\mathbf{n}_j = \hat{f}_i \text{ on } \Gamma_1,$$

where from (12.34), (12.37), and (12.39), we have

$$\begin{aligned}\tilde{\sigma}_{0ij} &= H_{ijkl}\left(\frac{u_{0k,l}+u_{0l,k}}{2}+\frac{u_{0m,k}u_{0m,l}}{2}\right)\\ &+H_{mjkl}\left(\frac{u_{0k,l}+u_{0l,k}}{2}+\frac{u_{0p,k}u_{0p,l}}{2}\right)u_{0i,m}.\end{aligned} \tag{12.41}$$

Replacing such results in the dual formulation we obtain

$$J(u_0) = F^*(z_0^*) - \tilde{G}_K^*(\sigma_0, z_0^*, v_0) - \tilde{G}_1^*(\tilde{\sigma}_0, \sigma_0, z_0^*, v_0). \tag{12.42}$$

From the hypothesis indicated in (12.6), the extremal relation through which z_0^* is obtained is in fact a global one.

From this, (12.2) and (12.42), the proof is complete.

Remark 12.2.2. About the last theorem, there is no duality gap between the primal and dual problems, if K is big enough so that for the optimal dual point, $\sigma_K(\sigma_0, z_0^*)$ is positive definite in Ω, where

$$\sigma_K(\sigma, z^*) = \left\{\begin{array}{lll}\sigma_{11}+z_{11}^*+K/2 & \sigma_{12}+z_{12}^* & \sigma_{13}+z_{13}^*\\ \sigma_{21}+z_{21}^* & \sigma_{22}+z_{22}^*+K/2 & \sigma_{23}+z_{23}^*\\ \sigma_{31}+z_{31}^* & \sigma_{32}+z_{32}^* & \sigma_{33}+z_{33}^*+K/2\end{array}\right\}, \tag{12.43}$$

and

$$\sigma_{0ij} + z_{0ij}^* = H_{ijkl}\left(\frac{u_{0k,l}+u_{0l,k}}{2}+\frac{u_{0m,k}u_{0m,l}}{2}\right), \tag{12.44}$$

and, at the same time, K is small enough so that for the fixed point $\{v_{0mj}\} = \{u_{0m,j}\}$ the quadratic form (in z^*) $W^*(z^*)$ is also positive definite in Ω, where

$$W^*(z^*) = \frac{z_{mi}^* z_{mi}^*}{K} - \bar{H}_{ijkl} z_{ij}^* z_{kl}^* - \sum_{m,i=1}^{3} \frac{(z_{ij}^* v_{0mj})^2}{K/2}. \tag{12.45}$$

For $K \approx \mathscr{O}(\min\{H_{1111}/2, H_{2222}/2, H_{1212}/2\})$ there is a large class of external loads for which such a K satisfies the conditions above, including to some extent the large deformation context.

Finally, we have not formally proven, but one may obtain from the relation between the primal and dual variables that

$$\begin{aligned}C &= \{u \in U \mid G_K^{**}(\Lambda u) = G_K(\Lambda u)\}\\ &= \{u \in U \mid \sigma_K(\sigma(u), \theta) \text{ is positive definite in } \Omega\},\end{aligned} \tag{12.46}$$

where as above indicated

$$\sigma_{ij}(u) = H_{ijkl}\left(\frac{1}{2}(u_{k,l}+u_{l,k}+u_{m,k}u_{m,l})\right). \tag{12.47}$$

12.3 Other Duality Principles

At this point we present another main result, which is summarized by the following theorem.

Theorem 12.3.1. *Let $\Omega \subset \mathbb{R}^3$ be an open, bounded, connected set with a regular boundary denoted by $\Gamma = \Gamma_0 \cup \Gamma_1$, where $\Gamma_0 \cap \Gamma_1 = \emptyset$. Consider the functional $(G \circ \Lambda) : U \to \mathbb{R}$ expressed by*

$$\begin{aligned}&(G\circ\Lambda)(u)\\&= \frac{1}{2}\int_\Omega H_{ijkl}\left(\frac{u_{i,j}+u_{j,i}}{2}+\frac{u_{m,i}u_{m,j}}{2}\right)\left(\frac{u_{k,l}+u_{l,k}}{2}+\frac{u_{m,k}u_{m,l}}{2}\right)dx,\end{aligned}$$

where

$$U = \{u = (u_1,u_2,u_3) \in W^{1,4}(\Omega;\mathbb{R}^3) \mid u = (0,0,0) \equiv \theta \text{ on } \Gamma_0\}, \tag{12.48}$$

and $\Lambda : U \to Y = Y^* = L^2(\Omega;\mathbb{R}^{3\times 3}) \equiv L^2$ is given by

$$\Lambda u = \{\Lambda_{ij}(u)\} = \left\{\frac{1}{2}(u_{i,j}+u_{j,i}+u_{m,i}u_{m,j})\right\}.$$

Define $J : U \to \mathbb{R}$ by

$$J(u) = G(\Lambda u) - \langle u, f\rangle_{L^2(\Omega;\mathbb{R}^3)} - \int_{\Gamma_1} \hat{f}_i u_i\, d\Gamma. \tag{12.49}$$

Also define

$$J_K : U \times Y \to \mathbb{R}$$

by

$$J_K(u,p) = G(\Lambda u + p) + K\langle p,p\rangle_{L^2} - \langle u,f\rangle_{L^2(\Omega;\mathbb{R}^3)} - \int_{\Gamma_1} \hat{f}_i u_i\, d\Gamma - \frac{K}{2}\langle p,p\rangle_{L^2},$$

and assume that $K > 0$ is sufficiently big so that $J_K(u,p)$ is bounded below.

Also define

$$J_K^*(\sigma,u) = F_f(\sigma) - G^*(\sigma) + K\left\|\Lambda u - \frac{\partial G^*(\sigma)}{\partial \sigma}\right\|_{L^2}^2 + \frac{1}{2K}\langle \sigma,\sigma\rangle_{L^2}, \tag{12.50}$$

where

$$G^*(\sigma) = \sup_{v\in Y}\{\langle v,\sigma\rangle_{L^2} - G(v)\} = \frac{1}{2}\int_\Omega \overline{H}_{ijkl}\sigma_{ij}\sigma_{kl}\,dx, \tag{12.51}$$

$$\{\overline{H}_{ijkl}\} = \{H_{ijkl}\}^{-1}$$

and

$$F_f(\sigma) = \sup_{u\in U}\left\{\langle \Lambda u,\sigma\rangle_{L^2} - \langle u,f\rangle_{L^2(\Omega;\mathbb{R}^3)} - \int_{\Gamma_1}\hat{f}_i u_i\,d\Gamma\right\}.$$

Under such assumptions, we have

$$\inf_{(u,p)\in U}\{J_K(u,p)\} \leq \inf_{(\sigma,u)\in Y^*\times U}\{J_K^*(\sigma,u)\}. \tag{12.52}$$

Finally, assume that Γ_0, $f \in L^2(\Omega;\mathbb{R}^3)$ and $\hat{f} \in L^2(\Gamma;\mathbb{R}^3)$ are such that a local minimum of J_K over $V_0 = B_r(u_0)\times B_r(p_0)$ is attained at some $(u_0,p_0)\in U\times Y$ such that

$$\sigma_0 = \frac{\partial G(\Lambda u_0 + p_0)}{\partial v} \tag{12.53}$$

is negative definite.

Here

$$B_r(u_0) = \{u \in U \mid \|u-u_0\|_U < r\},$$

and

$$B_r(p_0) = \{p \in Y \mid \|p-p_0\|_Y < r\},$$

for some appropriate $r > 0$.

Under such hypotheses, there exists a set $\tilde{V}_0 \subset Y^*\times U$, such that

$$\begin{aligned} J_K(u_0,p_0) &= \inf_{(u,p)\in V_0}\{J_K(u,p)\} \\ &\leq \inf_{(\sigma,u)\in\tilde{V}_0}\{J_K^*(\sigma,u)\} \\ &\leq J_K^*(\sigma_0,u_0) \\ &= J_K(u_0,p_0) \\ &\approx J(u_0) + \mathscr{O}(1/K). \end{aligned} \tag{12.54}$$

Proof. Define

$$G_1(u,p) = G(\Lambda u + p) + K\langle p,p\rangle_{L^2},$$

and

$$G_2(u,p) = \langle u,f\rangle_{L^2(\Omega;\mathbb{R}^3)} + \int_{\Gamma_1}\hat{f}_i u_i\,d\Gamma + \frac{K}{2}\langle p,p\rangle_{L^2}.$$

Observe that $\alpha_K = \inf_{(u,p)\in U\times Y}\{J_K(u,p)\} \in \mathbb{R}$ is such that

$$J_K(u,p) = G_1(u,p) - G_2(u,p) \geq \alpha_K, \forall u \in U, p \in Y.$$

Thus,

$$-G_2(u,p) \geq -G_1(u,p) + \alpha_K, \forall u \in U, p \in Y,$$

so that

$$\langle \Lambda u + p, \sigma\rangle_{L^2} - G_2(u,p) \geq \langle \Lambda u + p, \sigma\rangle_{L^2} - G_1(u,p) + \alpha_K, \forall u \in U, p \in Y.$$

Hence,

$$\sup_{(u,p)\in U\times Y}\{\langle \Lambda u + p, \sigma\rangle_{L^2} - G_2(u,p)\} \geq \langle \Lambda u + p, \sigma\rangle_{L^2} - G_1(u,p) + \alpha_K, \forall u \in U, p \in Y. \tag{12.55}$$

In particular for u, p such that

$$\sigma = \frac{\partial G(\Lambda u + p)}{\partial v},$$

we get

$$p + \Lambda u = \frac{\partial G^*(\sigma)}{\partial \sigma},$$

that is,

$$p = \frac{\partial G^*(\sigma)}{\partial \sigma} - \Lambda u,$$

and

$$G^*(\sigma) = \langle \Lambda u + p, \sigma\rangle_{L^2} - G(\Lambda u + p).$$

Hence

$$\langle \Lambda u + p, \sigma\rangle_{L^2} - G_1(u,p) = G^*(\sigma) - K\left\|\frac{\partial G^*(\sigma)}{\partial \sigma} - \Lambda u\right\|_{L^2}^2.$$

On the other hand,

$$\sup_{(u,p)\in U\times Y}\{\langle \Lambda u + p, \sigma\rangle_{L^2} - G_2(u,p)\} = F_f(\sigma) + \frac{1}{2K}\langle \sigma, \sigma\rangle_{L^2}.$$

Replacing such results in (12.55), we get

$$F_f(\sigma) - G^*(\sigma) + K\left\|\frac{\partial G^*(\sigma)}{\partial \sigma} - \Lambda u\right\|_{L^2}^2 + \frac{1}{2K}\langle \sigma, \sigma\rangle_{L^2} > \alpha_K,$$

$\forall \sigma \in Y^*, u \in U$.

Thus,

$$\alpha_K = \inf_{(u,p)\in U\times Y}\{J_K(u,p)\} \leq \inf_{(\sigma,u)\in Y^*\times U}\{J_K^*(\sigma,u)\}. \tag{12.56}$$

Now, let $(u_0, p_0) \in U \times Y$ be such that

$$J(u_0, p_0) = \min_{(u,p) \in V_0} \{J_K(u,p)\}.$$

Defining

$$\sigma_0 = \frac{\partial G(\Lambda u_0 + p_0)}{\partial v}, \tag{12.57}$$

since for the extremal point, we have

$$\delta_u \left\{ G(\Lambda u + p_0) - \langle u, f \rangle_{L^2(\Omega;\mathbb{R}^3)} - \int_{\Gamma_1} \hat{f}_i u_i \, d\Gamma \right\} |_{u=u_0} = \theta,$$

from this and (12.57), we also have

$$\delta_u \left\{ \langle \Lambda u, \sigma_0 \rangle_{L^2} - \langle u, f \rangle_{L^2(\Omega;\mathbb{R}^3)} - \int_{\Gamma_1} \hat{f}_i u_i \, d\Gamma \right\} |_{u=u_0} = \theta,$$

and therefore, since σ_0 is negative definite, we obtain

$$F_f(\sigma_0) = \langle \Lambda u_0, \sigma_0 \rangle_{L^2} - \langle u_0, f \rangle_{L^2(\Omega;\mathbb{R}^3)} - \int_{\Gamma_1} \hat{f}_i u_{0_i} \, d\Gamma. \tag{12.58}$$

From (12.57), we get

$$G^*(\sigma_0) = \langle \Lambda u_0 + p_0, \sigma \rangle_{L^2} - G(\Lambda u_0 + p_0), \tag{12.59}$$

so that, from (12.58) and (12.59), we obtain

$$\begin{aligned} F_f(\sigma_0) - G^*(\sigma_0) + K \left\| \frac{\partial G^*(\sigma_0)}{\partial \sigma} - \Lambda u_0 \right\|_{L^2}^2 + \frac{1}{2K} \langle \sigma_0, \sigma_0 \rangle_{L^2} \\ = G(\Lambda u_0 + p_0) + \frac{K}{2} \langle p_0, p_0 \rangle_{L^2} - \langle u_0, f \rangle_{L^2(\Omega;\mathbb{R}^3)} \\ - \int_{\Gamma_1} \hat{f}_i u_{0_i} \, d\Gamma, \end{aligned} \tag{12.60}$$

that is,

$$J_K^*(\sigma_0, u_0) = J_K(u_0, p_0). \tag{12.61}$$

Observe that, from the hypotheses,

$$J_K(u,p) \geq J_K(u_0, p_0), \forall (u,p) \in V_0.$$

At this point we develop a reasoning similarly to the lines above but now for the specific case of a neighborhood around the local optimal point. We repeat some analogous details for the sake of clarity.

From above,

$$G_1(u,p) = G(\Lambda u + p) + K\langle p,p\rangle_{L^2},$$

and

$$G_2(u,p) = \langle u, f\rangle_{L^2(\Omega;\mathbb{R}^3)} + \int_{\Gamma_1} \hat{f}_i u_i \, d\Gamma + \frac{K}{2}\langle p,p\rangle_{L^2}.$$

Observe that $\alpha = \inf_{(u,p)\in V_0}\{J_K(u,p)\} \in \mathbb{R}$ is such that

$$J_K(u,p) = G_1(u,p) - G_2(u,p) \geq \alpha, \forall (u,p) \in V_0.$$

Thus,

$$-G_2(u,p) \geq -G_1(u,p) + \alpha, \forall (u,p) \in V_0,$$

so that

$$\langle \Lambda u + p, \sigma\rangle_{L^2} - G_2(u,p) \geq \langle \Lambda u + p, \sigma\rangle_{L^2} - G_1(u,p) + \alpha, \forall (u,p) \in V_0.$$

Hence,

$$\begin{aligned} &\sup_{(u,p)\in U} \{\langle \Lambda u + p, \sigma\rangle_{L^2} - G_2(u,p)\} \\ &\quad \geq \sup_{(u,p)\in V_0} \{\langle \Lambda u + p, \sigma\rangle_{L^2} - G_2(u,p)\} \\ &\quad \geq \langle \Lambda u + p, \sigma\rangle_{L^2} - G_1(u,p) + \alpha, \forall (u,p) \in V_0. \end{aligned} \tag{12.62}$$

In particular, if $(\sigma,u) \in \tilde{V}_0$, where such a set is defined by the points (σ,u) such that $u \in B_r(u_0)$ and for the σ in question there exists $p \in B_r(p_0)$ such that

$$\sigma = \frac{\partial G(\Lambda u + p)}{\partial v},$$

that is,

$$p + \Lambda u = \frac{\partial G^*(\sigma)}{\partial \sigma},$$

we get

$$p = \frac{\partial G^*(\sigma)}{\partial \sigma} - \Lambda u,$$

and

$$G^*(\sigma) = \langle \Lambda u + p, \sigma\rangle_{L^2} - G(\Lambda u + p).$$

Hence

$$\langle \Lambda u + p, \sigma\rangle_{L^2} - G_1(u,p) = G^*(\sigma) - K\left\| \frac{\partial G^*(\sigma)}{\partial \sigma} - \Lambda u \right\|_{L^2}^2. \tag{12.63}$$

On the other hand

$$
\begin{aligned}
&\sup_{(u,p)\in V_0}\{\langle \Lambda u+p,\sigma\rangle_{L^2}-G_2(u,p)\}\\
&\quad\leq \sup_{(u,p)\in U\times Y}\{\langle \Lambda u+p,\sigma\rangle_{L^2}-G_2(u,p)\}\\
&\quad= F_f(\sigma)+\frac{1}{2K}\langle \sigma,\sigma\rangle_{L^2}. \qquad (12.64)
\end{aligned}
$$

Observe that $\sigma_0\in\tilde{V}_0$. We do not provide details here, but from the generalized inverse function theorem, also an appropriate neighborhood of σ_0 belongs to $\tilde{V}_0$.

Replacing the last relations (12.63) and (12.64) into (12.62), we get

$$
\begin{aligned}
&F_f(\sigma)-G^*(\sigma)+K\left\|\frac{\partial G^*(\sigma)}{\partial\sigma}-\Lambda u\right\|_{L^2}^2\\
&\quad+\frac{1}{2K}\langle\sigma,\sigma\rangle_{L^2}\geq\alpha, \qquad (12.65)
\end{aligned}
$$

$\forall(\sigma,u)\in\tilde{V}_0$.

Thus,

$$
\alpha=\inf_{(u,p)\in V_0}\{J_K(u,p)\}\leq\inf_{(\sigma,u)\in\tilde{V}_0}\{J_K^*(\sigma,u)\}. \qquad (12.66)
$$

Finally, since

$$
p_0=-\frac{1}{K}\frac{\partial G(\Lambda u_0+p_0)}{\partial p}, \qquad (12.67)
$$

we get

$$
\|p_0\|_Y\approx\mathscr{O}\left(\frac{1}{K}\right),
$$

so that from this, (12.61), and (12.65), we may finally write

$$
\begin{aligned}
\alpha=J_K(u_0,p_0)&=\inf_{(u,p)\in V_0}\{J_K(u,p)\}\\
&\leq\inf_{(\sigma,u)\in\tilde{V}_0}\{J_K^*(\sigma,u)\}\\
&\leq J_K^*(\sigma_0,u_0)\\
&=J_K(u_0,p_0)\\
&\approx J(u_0)+\mathscr{O}(1/K). \qquad (12.68)
\end{aligned}
$$

The proof is complete.

Remark 12.3.2. Of particular interest is the model behavior as $K\to+\infty$. From (12.68) it seems to be clear that the duality gap between the original primal and dual formulations goes to zero as K goes to $+\infty$.

Our final result is summarized by the next theorem. It refers to a duality principle for the case of a local maximum for the primal formulation.

Theorem 12.3.3. *Let $\Omega \subset \mathbb{R}^3$ be an open, bounded, connected set with a regular boundary denoted by $\Gamma = \Gamma_0 \cup \Gamma_1$, where $\Gamma_0 \cap \Gamma_1 = \emptyset$. Consider the functional $(G \circ \Lambda) : U \to \mathbb{R}$ expressed by*

$$\begin{aligned}&(G \circ \Lambda)(u)\\&= \frac{1}{2}\int_\Omega H_{ijkl}\left(\frac{u_{i,j}+u_{j,i}}{2}+\frac{u_{m,i}u_{m,j}}{2}\right)\left(\frac{u_{k,l}+u_{l,k}}{2}+\frac{u_{m,k}u_{m,l}}{2}\right)\,dx,\end{aligned}$$

where

$$U = \{u = (u_1,u_2,u_3) \in W^{1,4}(\Omega;\mathbb{R}^3) \mid u = (0,0,0) \equiv \theta \text{ on } \Gamma_0\}, \tag{12.69}$$

and $\Lambda : U \to Y = Y^* = L^2(\Omega;\mathbb{R}^{3\times 3}) \equiv L^2$ is given by

$$\Lambda u = \{\Lambda_{ij}(u)\} = \left\{\frac{1}{2}(u_{i,j}+u_{j,i}+u_{m,i}u_{m,j})\right\}.$$

Define $J : U \to \mathbb{R}$ by

$$J(u) = G(\Lambda u) - \langle u, f\rangle_{L^2(\Omega;\mathbb{R}^3)} - \int_{\Gamma_1} \hat{f}_i u_i \, d\Gamma. \tag{12.70}$$

Assume that Γ_0, $f \in L^2(\Omega;\mathbb{R}^3)$, and $\hat{f} \in L^2(\Gamma;\mathbb{R}^3)$ are such that a local maximum of J over $V_0 = B_r(u_0)$ is attained at some $u_0 \in U$ such that

$$\sigma_0 = \frac{\partial G(\Lambda u_0)}{\partial v} \tag{12.71}$$

is negative definite.

Also define

$$J^*(\sigma) = F_f(\sigma) - G^*(\sigma), \tag{12.72}$$

where

$$\begin{aligned}G^*(\sigma) &= \sup_{v \in Y}\{\langle v,\sigma\rangle_{L^2} - G(v)\}\\&= \frac{1}{2}\int_\Omega \overline{H}_{ijkl}\sigma_{ij}\sigma_{kl}\,dx,\end{aligned} \tag{12.73}$$

$$\{\overline{H}_{ijkl}\} = \{H_{ijkl}\}^{-1}$$

and

$$F_f(\sigma) = \sup_{u\in U}\left\{\langle \Lambda u,\sigma\rangle_{L^2} - \langle u,f\rangle_{L^2(\Omega;\mathbb{R}^3)} - \int_{\Gamma_1}\hat{f}_i u_i\,d\Gamma\right\}.$$

Under such assumptions, there exists a set $\tilde{V}_0 \subset Y^*$ such that

$$-J^*(\sigma_0) = \max_{\sigma \in \tilde{V}_0}\{-J^*(\sigma)\} = \max_{u \in V_0}\{J(u)\} = J(u_0). \tag{12.74}$$

Proof. Define $\alpha = J(u_0)$.

Thus,

$$J(u) = G(\Lambda u) - \langle u, f\rangle_{L^2(\Omega;\mathbb{R}^3)} - \int_{\Gamma_1} \hat{f}_i u_i \, d\Gamma \leq J(u_0) = \alpha,$$

$\forall u \in V_0$.

Hence,

$$-\langle u, f\rangle_{L^2(\Omega;\mathbb{R}^3)} - \int_{\Gamma_1} \hat{f}_i u_i \, d\Gamma \leq -G(\Lambda u) + \alpha, \forall u \in V_0,$$

so that

$$\begin{aligned} &\langle \Lambda u, \sigma\rangle_{L^2} - \langle u, f\rangle_{L^2(\Omega;\mathbb{R}^3)} - \int_{\Gamma_1} \hat{f}_i u_i \, d\Gamma \\ &\quad \leq \langle \Lambda u, \sigma\rangle_{L^2} - G(\Lambda u) + \alpha, \forall u \in V_0, \sigma \in Y^*. \end{aligned} \tag{12.75}$$

Therefore,

$$\begin{aligned} &\sup_{u \in V_0}\left\{\langle \Lambda u, \sigma\rangle_{L^2} - \langle u, f\rangle_{L^2(\Omega;\mathbb{R}^3)} - \int_{\Gamma_1} \hat{f}_i u_i \, d\Gamma\right\} \\ &\quad \leq \sup_{v \in Y}\{\langle v, \sigma\rangle_{L^2} - G(v)\} + \alpha, \forall \sigma \in Y^*. \end{aligned} \tag{12.76}$$

We define $\tilde{V}_0$ by the points $\sigma \in Y^*$ such that

$$\begin{aligned} &\sup_{u \in V_0}\left\{\langle \Lambda u, \sigma\rangle_{L^2} - \langle u, f\rangle_{L^2(\Omega;\mathbb{R}^3)} - \int_{\Gamma_1} \hat{f}_i u_i \, d\Gamma\right\} \\ &\quad = \sup_{u \in U}\left\{\langle \Lambda u, \sigma\rangle_{L^2} - \langle u, f\rangle_{L^2(\Omega;\mathbb{R}^3)} - \int_{\Gamma_1} \hat{f}_i u_i \, d\Gamma\right\} \\ &\quad = F_f(\sigma). \end{aligned} \tag{12.77}$$

We highlight that $\sigma_0 \in \tilde{V}_0$, and from the generalized inverse function theorem, any σ in an appropriate neighborhood of σ_0 also belongs to $\tilde{V}_0$ (we do not provide the details here).

From this and (12.76), we get

$$F_f(\sigma) - G^*(\sigma) \leq \alpha = J(u_0), \forall \sigma \in \tilde{V}_0. \tag{12.78}$$

Finally, observe that

$$\begin{aligned} F_f(\sigma_0) - G^*(\sigma_0) &= G(\Lambda u_0) - \langle u_0, f\rangle_{L^2(\Omega;\mathbb{R}^3)} - \int_{\Gamma_1} \hat{f}_i u_{0i}\, d\Gamma \\ &= J(u_0). \end{aligned} \tag{12.79}$$

From this and (12.78), the proof is complete.

12.4 A Numerical Example

Consider the functional $J : U \to \mathbb{R}$ defined by

$$J(u) = \frac{H}{2}\int_0^1 \left(u_x + \frac{1}{2}u_x^2\right)^2 dx - \int_0^1 Pu\, dx,$$

where

$$U = \{u \in W^{1,4}([0,1]) \mid u(0) = u(1) = 0\} = W_0^{1,4}([0,1]),$$

$$H = 10^5$$

$$P = -1000$$

where the units refer to the international system. The condition indicated in (12.45) here stands for $W^*(z^*)$ to be positive definite in a critical point $u_0 \in U$, where

$$W^*(z^*) = \frac{(z^*)^2}{K} - \frac{(z^*)^2}{H} - \frac{(u_0'(x))^2(z^*)^2}{K/2},$$

which is equivalent to

$$\frac{\partial^2 W^*(z^*)}{\partial (z^*)^2} \geq 0,$$

so that, for $K = H/2$, we get

$$(u_0'(x))^2 \leq 0.25, \text{ a.e. in } [0,1],$$

that is,

$$|u_0'(x)| \leq 0.5, \text{ a.e. in } [0,1].$$

We have computed a critical point through the primal formulation, again denoted by $u_0 \in U$. Please see Fig. 12.1. For $u_0'(x)$, see Fig. 12.2.

We may observe that

$$|u_0'(x)| \leq 0.5,$$

in $[0,1]$, so that by the main duality, such a point is a local minimum on the set $C_1 = C \cap C_2$, where

$$\begin{aligned} C &= \{u \in U \mid G_K^{**}(u_x) = G_K(u_x)\} \\ &= \{u \in U \mid H(u_x + u_x^2/2) + K/2 > 0, \text{ in } [0,1]\}, \end{aligned} \tag{12.80}$$

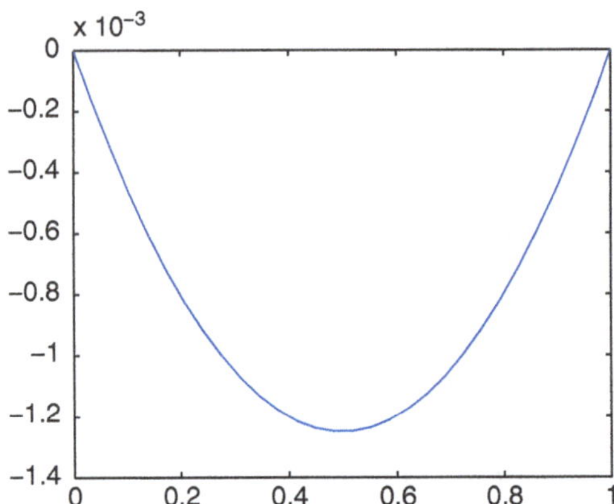

Fig. 12.1 The solution $u_0(x)$ through the primal formulation

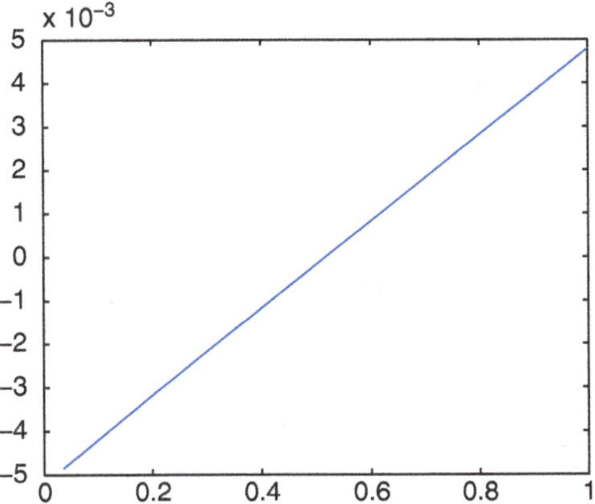

Fig. 12.2 The solution $u_0'(x)$ through the primal formulation

$C_2 = \{u \in U \mid u_x \in \hat{Y}^*\}$, where

$$G_K(u_x) = \frac{H}{2}\int_0^1 (u_x + u_x^2/2)^2\,dx + \frac{K}{4}\int_0^1 u_x^2\,dx,$$

and

$$\hat{Y}^* = \{v \in L^2([0,1]) \mid W^*(z^*) \text{ is positive definite in } [0,1]\}.$$

In fact, plotting the function $F(x) = H(x + x^2/2)^2/2$, we may observe that inside the set $[-0.5, 0.5]$ there is a local minimum, that is, in a close set, the Legendre necessary condition for a local minimum is satisfied. Please see Fig. 12.3.

We emphasize on the concerned sets there is no duality gap between the primal and dual formulations. Also, from the graphic of $u_0'(x)$, it is clear that the stress

$$H(u_0' + 1/2(u_0')^2)$$

is not exclusively positive or negative in $[0,1]$.

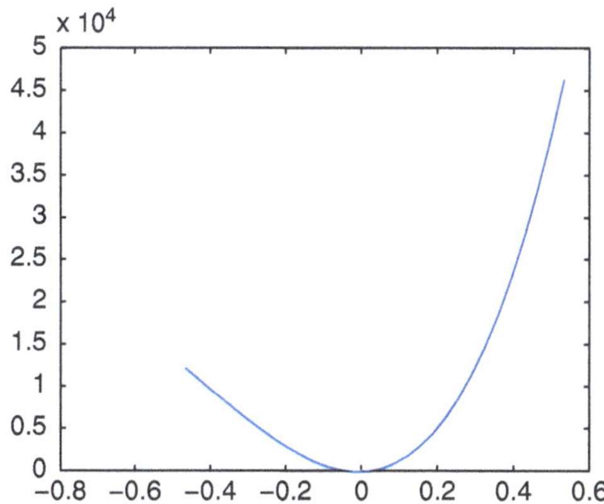

Fig. 12.3 The function $F(x) = H(x+x^2/2)^2/2$

12.5 Conclusion

In this chapter we develop new duality principles applicable to nonlinear finite elasticity. The results are obtained through the basic tools of convex analysis and include sufficient conditions of restricted optimality. It is worth mentioning that the methods developed here may be applied to many other situations, such as nonlinear models of plates and shells. Applications to related areas (specially to the shell model presented in [23]) are planned for future works.

Chapter 13
Duality Applied to a Plate Model

13.1 Introduction

In the present work we develop dual variational formulations for the Kirchhoff–Love thin plate model. Earlier results establish the complementary energy under the hypothesis of positive definiteness of the membrane force tensor at a critical point (please see [30–32, 36, 65] for details). In more recent works Gao has applied his triality theory to models in elasticity (see [33–35] for details) obtaining duality principles which allow the optimal stress tensor to be negative definite. Here for the present case we have obtained a dual variational formulation which allows the global optimal point in question to be not only positive definite (for related results see Botelho [11, 13]) but also not necessarily negative definite. The approach developed also includes sufficient conditions of optimality for the primal problem. Moreover, a numerical example concerning the main duality principle application is presented in the last section.

It is worth mentioning that the standard tools of convex analysis used in this text may be found in [13, 25, 54], for example. Another relating result may be found in [14].

At this point we start to describe the primal formulation.

Let $\Omega \subset \mathbb{R}^2$ be an open, bounded, connected set which represents the middle surface of a plate of thickness h. The boundary of Ω, which is assumed to be regular, is denoted by $\partial\Omega$. The vectorial basis related to the Cartesian system $\{x_1, x_2, x_3\}$ is denoted by $(\mathbf{a}_\alpha, \mathbf{a}_3)$, where $\alpha = 1, 2$ (in general Greek indices stand for 1 or 2) and where $\mathbf{a}_3$ is the vector normal to Ω, whereas $\mathbf{a}_1$ and $\mathbf{a}_2$ are orthogonal vectors parallel to Ω. Also, $\mathbf{n}$ is the outward normal to the plate surface.

The displacements will be denoted by

$$\hat{\mathbf{u}} = \{\hat{u}_\alpha, \hat{u}_3\} = \hat{u}_\alpha \mathbf{a}_\alpha + \hat{u}_3 \mathbf{a}_3.$$

The Kirchhoff–Love relations are

$$\hat{u}_\alpha(x_1, x_2, x_3) = u_\alpha(x_1, x_2) - x_3 w(x_1, x_2)_{,\alpha} \text{ and } \hat{u}_3(x_1, x_2, x_3) = w(x_1, x_2).$$

F. Botelho, *Functional Analysis and Applied Optimization in Banach Spaces: Applications to Non-Convex Variational Models*, DOI 10.1007/978-3-319-06074-3_13,

Here $-h/2 \leq x_3 \leq h/2$ so that we have $u = (u_\alpha, w) \in U$ where

$$U = \left\{ (u_\alpha, w) \in W^{1,2}(\Omega;\mathbb{R}^2) \times W^{2,2}(\Omega),\ u_\alpha = w = \frac{\partial w}{\partial \mathbf{n}} = 0 \text{ on } \partial\Omega \right\}$$
$$= W_0^{1,2}(\Omega;\mathbb{R}^2) \times W_0^{2,2}(\Omega).$$

It is worth emphasizing that the boundary conditions here specified refer to a clamped plate.

We define the operator $\Lambda : U \to Y \times Y$, where $Y = Y^* = L^2(\Omega;\mathbb{R}^{2\times 2})$, by

$$\Lambda(u) = \{\gamma(u), \kappa(u)\},$$

$$\gamma_{\alpha\beta}(u) = \frac{u_{\alpha,\beta} + u_{\beta,\alpha}}{2} + \frac{w_{,\alpha} w_{,\beta}}{2},$$

$$\kappa_{\alpha\beta}(u) = -w_{,\alpha\beta}.$$

The constitutive relations are given by

$$N_{\alpha\beta}(u) = H_{\alpha\beta\lambda\mu}\gamma_{\lambda\mu}(u), \tag{13.1}$$

$$M_{\alpha\beta}(u) = h_{\alpha\beta\lambda\mu}\kappa_{\lambda\mu}(u), \tag{13.2}$$

where $\{H_{\alpha\beta\lambda\mu}\}$ and $\{h_{\alpha\beta\lambda\mu} = \frac{h^2}{12} H_{\alpha\beta\lambda\mu}\}$ are symmetric positive definite fourth-order tensors. From now on, we denote $\{\overline{H}_{\alpha\beta\lambda\mu}\} = \{H_{\alpha\beta\lambda\mu}\}^{-1}$ and $\{\overline{h}_{\alpha\beta\lambda\mu}\} = \{h_{\alpha\beta\lambda\mu}\}^{-1}$.

Furthermore $\{N_{\alpha\beta}\}$ denote the membrane force tensor and $\{M_{\alpha\beta}\}$ the moment one. The plate stored energy, represented by $(G \circ \Lambda) : U \to \mathbb{R}$, is expressed by

$$(G \circ \Lambda)(u) = \frac{1}{2}\int_\Omega N_{\alpha\beta}(u)\gamma_{\alpha\beta}(u)\,dx + \frac{1}{2}\int_\Omega M_{\alpha\beta}(u)\kappa_{\alpha\beta}(u)\,dx, \tag{13.3}$$

and the external work, represented by $F : U \to \mathbb{R}$, is given by

$$F(u) = \langle w, P\rangle_{L^2(\Omega)} + \langle u_\alpha, P_\alpha\rangle_{L^2(\Omega)}, \tag{13.4}$$

where $P, P_1, P_2 \in L^2(\Omega)$ are external loads in the directions $\mathbf{a}_3$, $\mathbf{a}_1$, and $\mathbf{a}_2$, respectively. The potential energy, denoted by $J : U \to \mathbb{R}$, is expressed by

$$J(u) = (G \circ \Lambda)(u) - F(u).$$

It is important to emphasize that the existence of a minimizer (here denoted by u_0) related to $J(u)$ has been proven in Ciarlet [22]. Some inequalities of Sobolev type are necessary to prove the above result. In particular, we assume the boundary $\partial\Omega$ of Ω is regular enough so that the standard Gauss–Green formulas of integration by parts and the well-known Sobolev imbedding and trace theorems hold. Details about such results may be found in [1].

Finally, we also emphasize from now on, as their meaning is clear, we may denote $L^2(\Omega)$ and $L^2(\Omega;\mathbb{R}^{2\times 2})$ simply by L^2 and the respective norms by $\|\cdot\|_2$. Moreover derivatives are always understood in the distributional sense, θ denotes the zero vector in appropriate Banach spaces, and the following and relating notations are used:

$$w_{,\alpha\beta} = \frac{\partial^2 w}{\partial x_\alpha \partial x_\beta},$$

$$u_{\alpha,\beta} = \frac{\partial u_\alpha}{\partial x_\beta},$$

$$N_{\alpha\beta,1} = \frac{\partial N_{\alpha\beta}}{\partial x_1},$$

and

$$N_{\alpha\beta,2} = \frac{\partial N_{\alpha\beta}}{\partial x_2}.$$

13.2 The Main Duality Principle

In this section, we develop a duality principle presented in similar form in [13, 14]. The novelty here is its suitability for the Kirchhoff–Love plate model.

Theorem 13.2.1. *Let $\Omega \subset \mathbb{R}^2$ be an open, bounded, connected set with a regular boundary denoted by Γ. Suppose $(G \circ \Lambda) : U \to \mathbb{R}$ is defined by*

$$\begin{aligned} G(\Lambda u) &= \frac{1}{2}\int_\Omega H_{\alpha\beta\lambda\mu}\gamma_{\alpha\beta}(u)\gamma_{\lambda\mu}(u)\,dx \\ &+\frac{1}{2}\int_\Omega h_{\alpha\beta\lambda\mu}\kappa_{\alpha\beta}(u)\kappa_{\lambda\mu}(u)\,dx \\ &+\frac{K}{2}\int_\Omega (w_x)^2\,dx + \frac{K}{2}\int_\Omega (w_y)^2\,dx, \end{aligned} \tag{13.5}$$

and let $(F \circ \Lambda_2) : U \to \mathbb{R}$ be expressed by

$$F(\Lambda_2 u) = \frac{K}{2}\int_\Omega (w_x)^2\,dx + \frac{K}{2}\int_\Omega (w_y)^2\,dx,$$

where

$$\gamma_{\alpha\beta}(u) = \Lambda_{1\alpha\beta}(u) + \frac{1}{2}\Lambda_{2\alpha}(u)\Lambda_{2\beta}(u),$$

$$\{\Lambda_{1\alpha\beta}(u)\} = \left\{\frac{u_{\alpha,\beta}+u_{\beta,\alpha}}{2}\right\},$$

$$\{\Lambda_{2\alpha}(u)\} = \{w_{,\alpha}\},$$

$$\{\kappa_{\alpha\beta}(u)\} = \{-\Lambda_{3\alpha\beta}(u)\} = \{-w_{,\alpha\beta}\},$$

where $u = (u_\alpha, w) \in U = W_0^{1,2}(\Omega;\mathbb{R}^2) \times W_0^{2,2}(\Omega)$.

Also, define $F_1 : U \to \mathbb{R}$ *by*

$$F_1(u) = \langle w, P\rangle_{L^2} + \langle u_\alpha, P_\alpha\rangle_{L^2} \equiv \langle u, \tilde{P}\rangle_{L^2},$$

where $\tilde{P} = (P, P_\alpha)$, *and let* $J : U \to \mathbb{R}$ *be expressed by*

$$J(u) = (G \circ \Lambda)(u) - F(\Lambda_2 u) - F_1(u).$$

Under such hypotheses, we have

$$\inf_{u\in U}\{J(u)\} \geq \sup_{v^*\in A^*}\left\{\inf_{z^*\in Y_0^*}\{\tilde{F}^*(z^*) - G^*(v^*, z^*)\}\right\}, \tag{13.6}$$

where, when the meaning is clear, denoting $Y = Y^* = L^2(\Omega, \mathbb{R}^{2\times 2}) \equiv L^2$, $Y_1 = Y_1^* = L^2(\Omega, \mathbb{R}^2) \equiv L^2$, $v^* = (v_1^*, v_2^*, v_3^*)$, *and* $v_1^* = \{N_{\alpha\beta}\}$, $v_2^* = \{Q_\alpha\}$, *and* $v_3^* = \{M_{\alpha\beta}\}$, *we have*

$$A^* = \{v^* \in Y^* \mid \Lambda^* v^* = \tilde{P}\},$$

that is, recalling that $\tilde{P} = (P, P_\alpha)$, *we may write*

$$A^* = A_1 \cap A_2.$$

Here

$$A_1 = \{v^* \in Y^* \mid M_{\alpha\beta,\alpha\beta} + Q_{\alpha,\alpha} + P = 0, \text{ in } \Omega\},$$
$$A_2 = \{v^* \in Y^* \mid N_{\alpha\beta,\beta} + P_\alpha = 0, \text{ in } \Omega\},$$

and

$$Y_0^* = \{z^* \in Y_1^* \mid z_{11}^* = z_{22}^* = 0 \text{ and } (z_{11}^*)_x \boldsymbol{n}_1 + (z_{22}^*)_y \boldsymbol{n}_2 = 0 \text{ on } \Gamma\}.$$

Also,

$$\begin{aligned}\tilde{F}^*(z^*) &= \sup_{v\in Y_1}\{-\langle v_1, (z_{11}^*)_x\rangle_{L^2(\Omega)} - \langle v_2, (z_{22}^*)_y\rangle_{L^2(\Omega)} \\ &\quad -\frac{K}{2}\int_\Omega (v_1)^2\,dx - \frac{K}{2}\int_\Omega (v_2)^2\,dx\} \\ &= \frac{1}{2K}\int_\Omega ((z_{11}^*)_x)^2\,dx + \frac{1}{2K}\int_\Omega ((z_{22}^*)_y)^2\,dx, \end{aligned} \tag{13.7}$$

and

$$\begin{aligned}G^*(v^*, z^*) &= \sup_{v\in Y}\{\langle v_1, v_1^*\rangle_{L^2} + \langle v_2, v_2^*\rangle_{L^2} \\ &\quad + \langle v_3, v_3^* + z^*\rangle_{L^2} - G(v)\} \\ &= \frac{1}{2}\int_\Omega \overline{h}_{\alpha\beta\lambda\mu}(M_{\alpha\beta} + z_{\alpha\beta}^*)(M_{\lambda\mu} + z_{\lambda\mu}^*)\,dx \\ &\quad + \frac{1}{2}\int_\Omega \overline{H}_{\alpha\beta\lambda\mu} N_{\alpha\beta} N_{\lambda\mu}\,dx \\ &\quad + \frac{1}{2}\int_\Omega \overline{N}_{\alpha\beta}^K Q_\alpha Q_\beta\,dx, \end{aligned} \tag{13.8}$$

if N^K is positive definite, where

$$N^K = \left\{ \begin{matrix} N_{11} + K & N_{12} \\ N_{21} & N_{22} + K \end{matrix} \right\}, \tag{13.9}$$

$$\{\overline{N}^K_{\alpha\beta}\} = \{N^K_{\alpha\beta}\}^{-1}.$$

Moreover,

$$\{\overline{H}_{\alpha\beta\lambda\mu}\} = \{H_{\alpha\beta\lambda\mu}\}^{-1},$$

and

$$\{\overline{h}_{\alpha\beta\lambda\mu}\} = \{h_{\alpha\beta\lambda\mu}\}^{-1}.$$

Here we recall that $z^*_{12} = z^*_{21} = 0$ in Ω.

Finally, if there is a point $(u_0, v^*_0, z^*_0) \in U \times A^* \times Y^*_0$ such that

$$\delta\{\tilde{F}^*(z^*_0) - G^*(v^*_0, z^*_0) + \langle u_0, \Lambda^* v^*_0 - \tilde{P}\rangle_{L^2}\} = \theta,$$

where $K > 0$ is such that

$$\tilde{F}^*(z^*) - G^*(z^*) > 0, \forall z^* \in Y^*_0 \text{ such that } z^* \neq \theta, \tag{13.10}$$

we have that

$$\begin{aligned} J(u_0) &= \min_{u \in U}\{J(u)\} \\ &= \sup_{v^* \in A^*}\left\{ \inf_{z^* \in Y^*_0}\{\tilde{F}^*(z^*) - G^*(v^*, z^*)\}\right\} \\ &= \tilde{F}^*(z^*_0) - G^*(v^*_0, z^*_0). \end{aligned} \tag{13.11}$$

Proof. Observe that

$$\begin{aligned} G^*(v^*, z^*) \geq{}& \langle \Lambda_3 u, v^*_3 + z^*\rangle_{L^2} + \langle \Lambda_1 u, v^*_1\rangle_{L^2} \\ &+ \langle \Lambda_1 u, v^*_2\rangle_{L^2} - G(\Lambda u), \end{aligned} \tag{13.12}$$

$\forall u \in U, v^* \in A^*, z^* \in Y^*_0$.

Thus,

$$\begin{aligned} -\tilde{F}^*(z^*) + G^*(v^*, z^*) \geq{}& -\tilde{F}^*(z^*) + \langle \Lambda_3 u, z^*\rangle_{L^2} \\ &- G(\Lambda u) + \langle u, \tilde{P}\rangle_{L^2}, \end{aligned} \tag{13.13}$$

$\forall u \in U, v^* \in A^*, z^* \in Y^*_0$, so that, taking the supremum in z^* at both sides of last inequality, we obtain

$$\begin{aligned} &\sup_{z^* \in Y^*_0}\{-\tilde{F}^*(z^*) + G^*(v^*, z^*)\} \\ &\quad \geq F(\Lambda_2 u) - G(\Lambda u) + \langle u, \tilde{P}\rangle_{L^2} \end{aligned}$$

$$
\begin{aligned}
&= F(\Lambda_2 u) - G(\Lambda u) + F_1(u) \\
&= J(u),
\end{aligned} \tag{13.14}
$$

$\forall u \in U, v^* \in A^*$. Hence,

$$
\inf_{u \in U} \{J(u)\} \geq \sup_{v^* \in A^*} \left\{ \inf_{z^* \in Y_0^*} \{\tilde{F}^*(z^*) - G^*(v^*, z^*)\} \right\}. \tag{13.15}
$$

Finally, suppose that $(u_0, v_0^*, z_0^*) \in U \times A^* \times Y_0^*$ is such that

$$
\delta\{\tilde{F}^*(z_0^*) - G^*(v_0^*, z_0^*) + \langle u_0, \Lambda^* v_0^* - \tilde{P} \rangle_{L^2}\} = \theta.
$$

From the variation in v^* we obtain

$$
\frac{\partial G^*(v_0^*, z_0^*)}{\partial v^*} - \Lambda(u_0) = \theta, \tag{13.16}
$$

that is,

$$
v_{03}^* + z_0^* = \frac{\partial G(\Lambda u_0)}{\partial v_3},
$$

$$
v_{01}^* = \frac{\partial G(\Lambda u_0)}{\partial v_1},
$$

$$
v_{02}^* = \frac{\partial G(\Lambda u_0)}{\partial v_2},
$$

and

$$
\begin{aligned}
G^*(v_0^*, z_0^*) = {} & \langle \Lambda_3 u_0, v_{03}^* + z_0^* \rangle_{L^2} + \langle \Lambda_1 u_0, v_{01}^* \rangle_{L^2} \\
& \langle \Lambda_2 u_0, v_{02}^* \rangle_{L^2} - G(\Lambda u_0).
\end{aligned} \tag{13.17}
$$

From the variation in z^* we get

$$
-((z_0)_{11}^*)_{xx}/K - \frac{\partial G^*(v_0^*, z_0^*)}{\partial z_{11}^*} = \theta, \tag{13.18}
$$

$$
-((z_0)_{22}^*)_{yy}/K - \frac{\partial G^*(v_0^*, z_0^*)}{\partial z_{22}^*} = \theta, \tag{13.19}
$$

that is, from this and (13.16) we get

$$
-((z_0)_{11}^*)_x = K(w_0)_x,
$$

$$
-((z_0)_{22}^*)_y = K(w_0)_y,
$$

so that

$$
\tilde{F}^*(z_0^*) = \langle (z_0)_{11}^*, (w_0)_{xx} \rangle_{L^2(\Omega)} + \langle (z_0)_{22}^*, (w_0)_{yy} \rangle_{L^2(\Omega)} - F(\Lambda_2 u_0). \tag{13.20}
$$

From the variation in u, we get

$$\Lambda^* v_0^* - \tilde{P} = 0, \tag{13.21}$$

so that $v_0^* \in A^*$. Therefore, from (13.17), (13.20), and (13.21) we have

$$\begin{aligned}
&\tilde{F}^*(z_0^*) - G^*(v_0^*, z_0^*) \\
&= G(\Lambda u_0) - F_2(\Lambda u_0) - \langle u_0, \tilde{P} \rangle_{L^2} \\
&= J(u_0).
\end{aligned} \tag{13.22}$$

To complete the proof just observe that from the condition indicated in (13.10), the extremal relations (13.18) and (13.19) refer to a global optimization (in z^*, for a fixed v_0^*), so that the infimum indicated in the dual formulation is attained for the z_0^* in question.

From this and (13.22), the proof is complete.

13.3 Another Duality Principle

In this section we present another result, which is summarized by the following theorem.

Theorem 13.3.1. *Considering the introduction statements, let $(G \circ \Lambda) : U \to \mathbb{R}$ be expressed by*

$$\begin{aligned}
G(\Lambda u) &= \frac{1}{2} \int_\Omega H_{\alpha\beta\lambda\mu} \gamma_{\alpha\beta}(u) \gamma_{\lambda\mu}(u)\, dx \\
&+ \frac{1}{2} \int_\Omega h_{\alpha\beta\lambda\mu} \kappa_{\alpha\beta}(u) \kappa_{\lambda\mu}(u)\, dx,
\end{aligned} \tag{13.23}$$

where

$$\Lambda(u) = \{\gamma(u), \kappa(u)\},$$

$$\gamma_{\alpha\beta}(u) = \frac{u_{\alpha,\beta} + u_{\beta,\alpha}}{2} + \frac{w_{,\alpha} w_{,\beta}}{2},$$

$$\kappa_{\alpha\beta}(u) = -w_{,\alpha\beta},$$

where $u = (u_\alpha, w) \in U = W_0^{1,2}(\Omega; \mathbb{R}^2) \times W_0^{2,2}(\Omega)$.

As above, define $F : U \to \mathbb{R}$ by

$$F(u) = \langle w, P \rangle_{L^2} + \langle u_\alpha, P_\alpha \rangle_{L^2},$$

and $J : U \to \mathbb{R}$ by

$$J(u) = (G \circ \Lambda)(u) - F(u).$$

Under such assumptions, we have

$$\inf_{u \in U}\{J(u)\} \geq \sup_{N \in A^*}\{-G_1^*(N) - \tilde{G}_2^*(-N)\}, \tag{13.24}$$

where

$$G_1(\gamma(u)) = \frac{1}{2}\int_\Omega H_{\alpha\beta\lambda\mu}\gamma_{\alpha\beta}(u)\gamma_{\lambda\mu}(u)\,dx,$$

$$G_2(u) = \frac{1}{2}\int_\Omega h_{\alpha\beta\lambda\mu}\kappa_{\alpha\beta}(u)\kappa_{\lambda\mu}(u)\,dx - F(u),$$

so that

$$J(u) = G_1(\gamma(u)) + G_2(u).$$

Moreover,

$$\begin{aligned} G_1^*(N) &= \sup_{v \in Y}\{\langle v, N\rangle_{L^2} - G_1(v)\} \\ &= \frac{1}{2}\int_\Omega \overline{H}_{\alpha\beta\lambda\mu}N_{\alpha\beta}N_{\lambda\mu}\,dx, \end{aligned} \tag{13.25}$$

and

$$\begin{aligned} \tilde{G}_2^*(-N) &= \sup_{u \in U}\{\langle \gamma(u), -N\rangle_{L^2} - G_2(u)\}, \\ &= \frac{1}{2}\int_\Omega h_{\alpha\beta\lambda\mu}\hat{w}_{,\alpha\beta}\hat{w}_{,\lambda\mu}\,dx + \frac{1}{2}\int_\Omega N_{\alpha\beta}\hat{w}_{,\alpha}\hat{w}_{,\beta}\,dx, \end{aligned}$$

if

$$N = \{N_{\alpha\beta}\} \in A^* = A_1 \cap A_2,$$

where $\hat{w} \in W_0^{2,2}(\Omega)$ is the solution of equation

$$(h_{\alpha\beta\lambda\mu}\hat{w}_{,\lambda\mu})_{,\alpha\beta} - (N_{\alpha\beta}\hat{w}_{,\alpha})_{,\beta} - P = 0, \text{ in } \Omega.$$

Also,

$$A_1 = \{N \in Y^* \mid \tilde{J}(w) > 0, \forall w \in W_0^{2,2}(\Omega) \text{ such that } w \neq \theta\},$$

$$\tilde{J}(w) = \frac{1}{2}\int_\Omega h_{\alpha\beta\lambda\mu}w_{,\alpha\beta}w_{,\lambda\mu}\,dx + \frac{1}{2}\int_\Omega N_{\alpha\beta}w_{,\alpha}w_{,\beta}\,dx,$$

and

$$A_2 = \{N \in Y^* \mid N_{\alpha\beta,\beta} + P_\alpha = 0 \text{ in } \Omega\}.$$

Finally, if there exists $u_0 \in U$ such that $\delta J(u_0) = \theta$ and $N_0 = \{N_{\alpha\beta}(u_0)\} \in A_1$, where $N_{\alpha\beta}(u_0) = H_{\alpha\beta\lambda\mu}\gamma_{\lambda\mu}(u_0)$, then

$$\begin{aligned} J(u_0) &= \min_{u \in U}\{J(u)\} \\ &= \max_{N \in A^*}\{-G_1^*(N) - \tilde{G}_2^*(-N)\} \\ &= -G_1^*(N_0) - \tilde{G}_2^*(-N_0)\}. \end{aligned} \tag{13.26}$$

Proof. Clearly

$$
\begin{aligned}
J(u) &= G_1(\gamma(u)) + G_2(u) \\
&= -\langle \gamma(u), N\rangle_{L^2} + G_1(\gamma(u)) + \langle \gamma(u), N\rangle_{L^2} + G_2(u) \\
&\geq \inf_{v\in Y}\{-\langle v, N\rangle_{L^2} + G_1(v)\} + \inf_{u\in U}\{-\langle \gamma(u), -N\rangle_{L^2} + G_2(u)\} \\
&= -G_1^*(N) - \tilde{G}_2^*(-N), \forall u \in U,\ N \in Y^*.
\end{aligned} \tag{13.27}
$$

Hence,

$$
\inf_{u\in U}\{J(u)\} \geq \sup_{N\in A^*}\{-G_1^*(N) - \tilde{G}_2^*(-N)\}. \tag{13.28}
$$

Now suppose there exists $u_0 \in U$ such that $\delta J(u_0) = \theta$ and $N_0 = \{N_{\alpha\beta}(u_0)\} \in A_1$.

First, note that from $\delta J(u_0) = \theta$, the following extremal equation is satisfied:

$$
(N_0)_{\alpha\beta,\beta} + P_\alpha = 0 \text{ in } \Omega,
$$

that is, $N_0 \in A_2$, so that $N_0 \in A_1 \cap A_2 = A^*$.

Thus, from $N_0 \in A_1$, we obtain

$$
\begin{aligned}
\tilde{G}_2^*(-N_0) &= \sup_{u\in U}\{\langle \gamma(u), -N_0\rangle_{L^2} - G_2(u)\}, \\
&= \langle \gamma(\hat{u}), -N_0\rangle_{L^2} - G_2(\hat{u}) \\
&= \left\langle \frac{\hat{w}_{,\alpha}\hat{w}_{,\beta}}{2}, -(N_0)_{\alpha\beta} \right\rangle_{L^2} - \frac{1}{2}\int_\Omega h_{\alpha\beta\lambda\mu}\hat{w}_{,\alpha\beta}\hat{w}_{,\lambda\mu}\,dx \\
&\quad + \langle \hat{w}, P\rangle_{L^2},
\end{aligned} \tag{13.29}
$$

where $\hat{w} \in W_0^{2,2}(\Omega)$ is the solution of equation

$$
\left((N_0)_{\alpha\beta}\hat{w}_{,\alpha}\right)_{,\beta} - (h_{\alpha\beta\lambda\mu}\hat{w}_{,\lambda\mu})_{,\alpha\beta} + P = 0, \text{ in } \Omega. \tag{13.30}
$$

Replacing such a relation in (13.29), we obtain

$$
\tilde{G}_2^*(-N_0) = \frac{1}{2}\int_\Omega h_{\alpha\beta\lambda\mu}\hat{w}_{,\alpha\beta}\hat{w}_{,\lambda\mu}\,dx + \frac{1}{2}\int_\Omega (N_0)_{\alpha\beta}\hat{w}_{,\alpha}\hat{w}_{,\beta}\,dx.
$$

Hence, also from the equation $\delta J(u_0) = \theta$ and (13.30), we may get $\hat{w} = w_0$, so that from this and (13.29), we obtain

$$
\tilde{G}_2^*(-N_0) = \langle \gamma(u_0), -N_0\rangle_{L^2} - G_2(u_0). \tag{13.31}
$$

Finally, considering that

$$
(N_0)_{\alpha\beta} = H_{\alpha\beta\lambda\mu}\gamma_{\lambda\mu}(u_0),
$$

we get

$$
G_1^*(N_0) = \langle \gamma(u_0), N_0\rangle_{L^2} - G_1(\gamma(u_0)),
$$

so that

$$\begin{aligned} -G_1^*(N_0) - \tilde{G}_2^*(-N_0) &= -\langle \gamma(u_0), N_0 \rangle_{L^2} + G_1(\gamma(u_0)) \\ &\quad -\langle \gamma(u_0), -N_0 \rangle_{L^2} + G_2(u_0) \\ &= G_1(\gamma(u_0)) + G_2(u_0) \\ &= J(u_0). \end{aligned} \tag{13.32}$$

From this and (13.28) and also from the fact that $N_0 \in A^*$, the proof is complete.

Remark 13.3.2. From the last duality principle, we may write

$$\inf_{u \in U}\{J(u)\} \geq \sup_{(M,N,u)\in \hat{A}} \{-\tilde{J}^*(M,N,u)\},$$

where

$$\begin{aligned} \tilde{J}^*(M,N,u) &= G_1^*(N) + \tilde{G}_2^*(-N) \\ &= \frac{1}{2}\int_\Omega \overline{H}_{\alpha\beta\lambda\mu} N_{\alpha\beta} N_{\lambda\mu}\,dx + \frac{1}{2}\int_\Omega \overline{h}_{\alpha\beta\lambda\mu} M_{\alpha\beta} M_{\lambda\mu}\,dx \\ &\quad + \frac{1}{2}\int_\Omega N_{\alpha\beta} w_{,\alpha} w_{,\beta}\,dx, \end{aligned} \tag{13.33}$$

$\hat{A} = A_1 \cap A_2 \cap A_3 \cap A_4$,

$$A_3 = \{(M,N,u) \in Y^* \times Y^* \times U \mid M_{\alpha\beta,\alpha\beta} + (N_{\alpha\beta} w_{,\alpha})_{,\beta} + P = 0, \text{ in } \Omega\},$$

$$A_4 = \{(M,N,u) \in Y^* \times Y^* \times U \mid \{M_{\alpha\beta}\} = \{h_{\alpha\beta\lambda\mu}(-w_{,\lambda\mu})\}, \text{ in } \Omega\},$$

and A_1 and A_2 as above specified, that is,

$$A_1 = \{N \in Y^* \mid \tilde{J}(w) > 0, \forall w \in W_0^{2,2}(\Omega) \text{ such that } w \neq \theta\},$$

where

$$\tilde{J}(w) = \frac{1}{2}\int_\Omega h_{\alpha\beta\lambda\mu} w_{,\alpha\beta} w_{,\lambda\mu}\,dx + \frac{1}{2}\int_\Omega N_{\alpha\beta} w_{,\alpha} w_{,\beta}\,dx,$$

and

$$A_2 = \{(M,N,u) \in Y^* \times Y^* \times U \mid N_{\alpha\beta,\beta} + P_\alpha = 0 \text{ in } \Omega\}.$$

Finally, we could suggest as a possible approximate dual formulation the problem of maximizing $-J_K^*(M,N,u)$ on $A_1 \cap A_2 \cap A_3$, where $K > 0$ and

$$\begin{aligned} J_K^*(M,N,u) &= \frac{1}{2}\int_\Omega \overline{H}_{\alpha\beta\lambda\mu} N_{\alpha\beta} N_{\lambda\mu}\,dx + \frac{1}{2}\int_\Omega \overline{h}_{\alpha\beta\lambda\mu} M_{\alpha\beta} M_{\lambda\mu}\,dx \\ &\quad + \frac{1}{2}\int_\Omega N_{\alpha\beta} w_{,\alpha} w_{,\beta}\,dx \\ &\quad + \frac{K}{2}\sum_{\alpha,\beta=1}^{2} \|M_{\alpha\beta} - h_{\alpha\beta\lambda\mu}(-w_{,\lambda\mu})\|_2^2. \end{aligned} \tag{13.34}$$

A study about the system behavior as $K \to +\infty$ is planned for a future work. Anyway, big values for $K > 0$ allow the gap function $\frac{1}{2}\int_\Omega N_{\alpha\beta} w_{,\alpha} w_{,\beta}\, dx$ to be nonpositive at a possible optimal point inside the region of convexity of J_K^*.

13.4 An Algorithm for Obtaining Numerical Results

In this section we develop an algorithm which we prove, under certain mild hypotheses; it is convergent up to a subsequence (the result stated in the next lines must be seen as an existence one and, of course, it is not the full proof of convergence from a numerical analysis point of view). Such a result is summarized by the following theorem.

Theorem 13.4.1. *Consider the system of equations relating the boundary value form of the Kirchhoff–Love plate model, namely*

$$\begin{cases} M_{\alpha\beta,\alpha\beta} + (N_{\alpha\beta} w_{,\alpha})_{,\beta} + P = 0, & \text{in } \Omega \\ N_{\alpha\beta,\beta} + P_\alpha = 0 & \text{in } \Omega \\ u_\alpha = w = \frac{\partial w}{\partial n} = 0 & \text{on } \partial\Omega \end{cases} \tag{13.35}$$

where

$$N_{\alpha\beta}(u) = H_{\alpha\beta\lambda\mu}\gamma_{\lambda\mu}(u), \tag{13.36}$$

$$M_{\alpha\beta}(u) = h_{\alpha\beta\lambda\mu}\kappa_{\lambda\mu}(u). \tag{13.37}$$

Define, as above,

$$(G \circ \Lambda)(u) = \frac{1}{2}\int_\Omega N_{\alpha\beta}(u)\gamma_{\alpha\beta}(u)\, dx + \frac{1}{2}\int_\Omega M_{\alpha\beta}(u)\kappa_{\alpha\beta}(u)\, dx, \tag{13.38}$$

$$F(u) = \langle w, P\rangle_{L^2} + \langle u_\alpha, P_\alpha\rangle_{L^2} \tag{13.39}$$

and $J : U \to \mathbb{R}$ by

$$J(u) = G(\Lambda u) - F(u), \forall u \in U. \tag{13.40}$$

Assume $\{\|P_\alpha\|_2\}$ are small enough so that (from [22] pages 285–287) if either

$$\|u_\alpha\|_{W^{1,2}(\Omega)} \to \infty$$

or

$$\|w\|_{W^{2,2}(\Omega)} \to \infty,$$

then

$$J(u) \to +\infty.$$

Let $\{u_n = ((u_n)_\alpha, w_n)\} \subset U$ *be the sequence obtained through the following algorithm:*

1. *Set* $n = 1$.
2. *Choose* $(z_1^*)_1, (z_2^*)_1 \in L^2(\Omega)$.
3. *Compute* u_n *by*

$$\begin{aligned} u_n = argmin_{u \in U} \Bigg\{ & G(\Lambda u) + \frac{K}{2}\int_\Omega (w_x)^2\,dx + \frac{K}{2}\int_\Omega (w_y)^2\,dx \\ & - \langle w_x, (z_1^*)_n \rangle_{L^2} - \langle w_y, (z_2^*)_n \rangle_{L^2} \\ & + \frac{1}{2K}\int_\Omega (z_1^*)_n^2\,dx + \frac{1}{2K}\int_\Omega (z_2^*)_n^2\,dx - F(u) \Bigg\}, \end{aligned}$$

which means to solve the equation

$$\begin{cases} M_{\alpha\beta,\alpha\beta} + (N_{\alpha\beta} w_{,\alpha})_{,\beta} + P + K w_{,\alpha\alpha} - (z_n^*)_{\alpha,\alpha} = 0, & in\ \Omega \\ N_{\alpha\beta,\beta} + P_\alpha = 0 & in\ \Omega \\ u_\alpha = w = \frac{\partial w}{\partial n} = 0 & on\ \partial\Omega \end{cases} \tag{13.41}$$

4. *Compute* $z_{n+1}^* = ((z_1^*)_{n+1}, (z_2^*)_{n+1})$ *by*

$$\begin{aligned} z_{n+1}^* = argmin_{z^* \in L^2 \times L^2} \Bigg\{ & G(\Lambda u_n) + \frac{K}{2}\int_\Omega (w_n)_x^2\,dx + \frac{K}{2}\int_\Omega (w_n)_y^2\,dx \\ & - \langle (w_n)_x, z_1^* \rangle_{L^2} - \langle (w_n)_y, z_2^* \rangle_{L^2} \\ & + \frac{1}{2K}\int_\Omega (z_1^*)^2\,dx + \frac{1}{2K}\int_\Omega (z_2^*)^2\,dx - F(u_n) \Bigg\}, \end{aligned}$$

that is,

$$(z_1^*)_{n+1} = K(w_n)_x,$$

and

$$(z_2^*)_{n+1} = K(w_n)_y.$$

5. *Set* $n \to n+1$ *and go to step 3 till the satisfaction of a suitable approximate convergence criterion.*

Assume $\{u_n = ((u_n)_\alpha, w_n)\} \subset U$ *is such that for a sufficiently big* $K > 0$ *we have*

$$\begin{aligned} & N_{11}(u_n) + K > 0,\ N_{22}(u_n) + K > 0, \\ & and\ (N_{11}(u_n) + K)(N_{22}(u_n) + K) - N_{12}^2(u_n) > 0, \\ & in\ \Omega,\ \forall n \in \mathbb{N}. \end{aligned} \tag{13.42}$$

Under such assumptions, the sequence $\{u_n\}$ *is uniquely defined (depending only on* $(z^*)_1$*), and such that, up to a subsequence not relabeled, for some* $u_0 = ((u_0)_\alpha, w_0) \in U$*, we have*

$$(u_n)_\alpha \rightharpoonup (u_0)_\alpha, \text{ weakly in } W_0^{1,2}(\Omega),$$

$$(u_n)_\alpha \to (u_0)_\alpha, \text{ strongly in } L^2(\Omega),$$

$$w_n \rightharpoonup w_0, \text{ weakly in } W_0^{2,2}(\Omega),$$

and

$$w_n \to w_0, \text{ strongly in } W_0^{1,2}(\Omega),$$

where

$$u_0 \in U$$

is a solution for the system of equations indicated in (13.35).

Proof. Since $J : U \to \mathbb{R}$ is defined by

$$J(u) = G(\Lambda u) - F(u), \tag{13.43}$$

we have

$$\begin{aligned}
J(u) &= G(\Lambda u) + \frac{K}{2}\int_\Omega (w_x)^2\,dx + \frac{K}{2}\int_\Omega (w_y)^2\,dx \\
&\quad -\langle w_x, z_1^*\rangle_{L^2} - \langle w_y, z_2^*\rangle_{L^2} \\
&\quad -\frac{K}{2}\int_\Omega (w_x)^2\,dx - \frac{K}{2}\int_\Omega (w_y)^2\,dx \\
&\quad +\langle w_x, z_1^*\rangle_{L^2} + \langle w_y, z_2^*\rangle_{L^2} - F(u) \\
&\leq G(\Lambda u) + \frac{K}{2}\int_\Omega (w_x)^2\,dx \frac{K}{2}\int_\Omega (w_y)^2\,dx \\
&\quad -\langle w_x, z_1^*\rangle_{L^2} - \langle w_y, z_2^*\rangle_{L^2} \\
&\quad + \sup_{v \in L^2 \times L^2}\left\{ -\frac{K}{2}\int_\Omega v_1^2\,dx - \frac{K}{2}\int_\Omega v_2^2\,dx \right. \\
&\quad \left. +\langle v_1, z_1^*\rangle_{L^2} + \langle v_2, z_2^*\rangle_{L^2} \right\} - F(u) \\
&= G(\Lambda u) + \frac{K}{2}\int_\Omega (w_x)^2\,dx + \frac{K}{2}\int_\Omega (w_y)^2\,dx + \\
&\quad -\langle w_x, z_1^*\rangle_{L^2} - \langle w_y, z_2^*\rangle_{L^2} \\
&\quad + \frac{1}{2K}\int_\Omega (z_1^*)^2\,dx + \frac{1}{2K}\int_\Omega (z_2^*)^2\,dx - F(u),
\end{aligned} \tag{13.44}$$

$\forall u \in U,\ z^* \in L^2(\Omega) \times L^2(\Omega)$.

From the hypotheses, $\{u_n\}$ is inside the region of strict convexity of the functional in U (for z^* fixed) in question, so that it is uniquely defined for each z_n^*

(through the general results in [11] we may infer the region of convexity of the functional

$$\begin{aligned}\bar{J}(u) &= G(\Lambda u) + \frac{K}{2}\int_\Omega (w_x)^2\,dx + \frac{K}{2}\int_\Omega (w_y)^2\,dx + \\ &-\langle w_x, (z_1^*)_n\rangle_{L^2} - \langle w_y, (z_2^*)_n\rangle_{L^2} \\ &+\frac{1}{2K}\int_\Omega (z_1^*)_n^2\,dx + \frac{1}{2K}\int_\Omega (z_2^*)_n^2\,dx - F(u), \end{aligned} \tag{13.45}$$

corresponds to the satisfaction of constraints

$$\begin{aligned}&N_{11}(u)+K>0,\ N_{22}(u)+K>0, \\ &\text{and } (N_{11}(u)+K)(N_{22}(u)+K) - N_{12}^2(u) > 0, \text{ in } \Omega). \end{aligned} \tag{13.46}$$

Denoting

$$\begin{aligned}\alpha_n = &\ G(\Lambda u_n) + \frac{K}{2}\int_\Omega (w_n)_x^2\,dx + \frac{K}{2}\int_\Omega (w_n)_y^2\,dx + \\ &-\langle (w_n)_x, (z_1^*)_n\rangle_{L^2} - \langle (w_n)_y, (z_2^*)_n\rangle_{L^2} \\ &+\frac{1}{2K}\int_\Omega (z_1^*)_n^2\,dx + \frac{1}{2K}\int_\Omega (z_2^*)_n^2\,dx - F(u_n), \end{aligned} \tag{13.47}$$

we may easily verify that $\{\alpha_n\}$ is a real nonincreasing sequence bounded below by $\inf_{u\in U}\{J(u)\}$; therefore, there exists $\alpha\in\mathbb{R}$ such that

$$\lim_{n\to\infty} \alpha_n = \alpha. \tag{13.48}$$

From the hypotheses

$$J(u) \to +\infty,$$

if either $\|u_\alpha\|_{W_0^{1,2}(\Omega)} \to \infty$ or $\|w\|_{W_0^{2,2}(\Omega)} \to \infty$.

From this, (13.44), (13.47), and (13.48) we may infer there exists $C>0$ such that

$$\|w_n\|_{W_0^{2,2}(\Omega)} < C, \forall n \in \mathbb{N},$$

and

$$\|(u_\alpha)_n\|_{W_0^{1,2}(\Omega)} < C, \forall n \in \mathbb{N}.$$

Thus, from the Rellich–Kondrachov theorem, up to a subsequence not relabeled, there exists $u_0 = ((u_0)_\alpha, w_0) \in U$ such that

$$(u_n)_\alpha \rightharpoonup (u_0)_\alpha, \text{ weakly in } W_0^{1,2}(\Omega),$$

$$(u_n)_\alpha \to (u_0)_\alpha, \text{ strongly in } L^2(\Omega),$$

$$w_n \rightharpoonup w_0, \text{ weakly in } W_0^{2,2}(\Omega),$$

and

$$w_n \to w_0, \text{ strongly in } W_0^{1,2}(\Omega),$$

so that, considering the algorithm in question,

$$z_n \to z_0^* \text{ strongly in } L^2(\Omega;\mathbb{R}^2),$$

where

$$(z_0^*)_\alpha = K(w_0)_{,\alpha}.$$

From these last results, the Sobolev imbedding theorem and relating results (more specifically, Korn's inequality and its consequences; details may be found in [21]), we have that there exist $K_1, K_2 > 0$ such that

$$\|(u_n)_{\alpha,\beta} + (u_n)_{\beta,\alpha}\|_2 < K_1, \forall n \in \mathbb{N},\ \alpha,\beta \in \{1,2\},$$

and

$$\|(w_n)_{,\alpha}\|_4 < K_2, \forall n \in \mathbb{N},\ \alpha \in \{1,2\}.$$

On the other hand, $u_n \in U$ such that

$$\begin{aligned} u_n = \operatorname{argmin}_{u\in U} \Big\{ & G(\Lambda u) + \frac{K}{2}\int_\Omega (w_x)^2\,dx + \frac{K}{2}\int_\Omega (w_y)^2\,dx \\ & - \langle w_x, (z_1^*)_n\rangle_{L^2} - \langle w_y, (z_2^*)_n\rangle_{L^2} \\ & + \frac{1}{2K}\int_\Omega (z_1^*)_n^2\,dx + \frac{1}{2K}\int_\Omega (z_2^*)_n^2\,dx - F(u)\Big\} \end{aligned} \tag{13.49}$$

is also such that

$$\begin{aligned} & \left(h_{\alpha\beta\lambda\mu}(w_n)_{,\lambda\mu}\right)_{,\alpha\beta} \\ & - \left(H_{\alpha\beta\lambda\mu}\left(\frac{(u_n)_{\lambda,\mu} + (u_n)_{\lambda,\mu}}{2} + \frac{(w_n)_{,\lambda}(w_n)_{,\mu}}{2}\right)(w_n)_{,\beta}\right)_{,\alpha} \\ & - K(w_n)_{,\alpha\alpha} + (z_n^*)_{\alpha,\alpha} - P = 0 \text{ in } \Omega, \end{aligned} \tag{13.50}$$

and

$$\begin{aligned} & \left(H_{\alpha\beta\lambda\mu}\left(\frac{(u_n)_{\lambda,\mu} + (u_n)_{\lambda,\mu}}{2} + \frac{(w_n)_{,\lambda}(w_n)_{,\mu}}{2}\right)\right)_{,\beta} \\ & + P_\alpha = 0 \text{ in } \Omega, \end{aligned} \tag{13.51}$$

in the sense of distributions (theoretical details about similar results may be found in [25]).

Fix $\phi \in C_c^\infty(\Omega)$. In the next lines, we will prove that

$$\left\langle \left(\frac{(u_n)_{\alpha,\beta} + (u_n)_{\beta,\alpha}}{2} + (w_n)_{,\alpha}(w_n)_{,\beta} \right) (w_n)_{,\alpha}, \phi_{,\beta} \right\rangle_{L^2}$$
$$\to \left\langle \left(\frac{(u_0)_{\alpha,\beta} + (u_0)_{\beta,\alpha}}{2} + (w_0)_{,\alpha}(w_0)_{,\beta} \right) (w_0)_{,\alpha}, \phi_{,\beta} \right\rangle_{L^2}, \tag{13.52}$$

as $n \to \infty$, $\forall \alpha, \beta \in \{1,2\}$ (here the repeated indices do not sum).

Observe that, since

$$(u_n)_\alpha \rightharpoonup (u_0)_\alpha, \text{ weakly in } W_0^{1,2}(\Omega),$$

from the Hölder inequality, we obtain

$$\begin{aligned}
&\left| \left\langle \left(\frac{(u_n)_{\alpha,\beta} + (u_n)_{\beta,\alpha}}{2} \right) (w_n)_{,\alpha}, \phi_{,\beta} \right\rangle_{L^2} \right. \\
&\quad \left. - \left\langle \left(\frac{(u_0)_{\alpha,\beta} + (u_0)_{\beta,\alpha}}{2} \right) (w_0)_{,\alpha}, \phi_{,\beta} \right\rangle_{L^2} \right| \\
&= \left| \left\langle \left(\frac{(u_n)_{\alpha,\beta} + (u_n)_{\beta,\alpha}}{2} \right) (w_n)_{,\alpha} - \left(\frac{(u_n)_{\alpha,\beta} + (u_n)_{\beta,\alpha}}{2} \right) (w_0)_{,\alpha} \right. \right. \\
&\quad \left. \left. + \left(\frac{(u_n)_{\alpha,\beta} + (u_n)_{\beta,\alpha}}{2} \right) (w_0)_{,\alpha} - \left(\frac{(u_0)_{\alpha,\beta} + (u_0)_{\beta,\alpha}}{2} \right) (w_0)_{,\alpha}, \phi_{,\beta} \right\rangle_{L^2} \right| \\
&\leq \left\| \frac{(u_n)_{\alpha,\beta} + (u_n)_{\beta,\alpha}}{2} \right\|_2 \|(w_n)_{,\alpha} - (w_0)_{,\alpha}\|_2 \|\phi_{,\beta}\|_\infty \\
&\quad + \left| \left\langle \left(\frac{(u_n)_{\alpha,\beta} + (u_n)_{\beta,\alpha}}{2} \right) (w_0)_{,\alpha} - \left(\frac{(u_0)_{\alpha,\beta} + (u_0)_{\beta,\alpha}}{2} \right) (w_0)_{,\alpha}, \phi_{,\beta} \right\rangle_{L^2} \right| \\
&\leq K_1 \|(w_n)_{,\alpha} - (w_0)_{,\alpha}\|_2 \|\phi_{,\beta}\|_\infty \\
&\quad + \left| \left\langle \left(\frac{(u_n)_{\alpha,\beta} + (u_n)_{\beta,\alpha}}{2} \right) - \left(\frac{(u_0)_{\alpha,\beta} + (u_0)_{\beta,\alpha}}{2} \right), (w_0)_{,\alpha} \phi_{,\beta} \right\rangle_{L^2} \right| \\
&\to 0, \text{ as } n \to \infty.
\end{aligned}$$

Moreover, from the generalized Hölder inequality, we get

$$\begin{aligned}
&|\langle (w_n)_{,\alpha}(w_n)_{,\beta}(w_n)_{,\alpha}, \phi_{,\beta} \rangle_{L^2} - \langle (w_0)_{,\alpha}(w_0)_{,\beta}(w_0)_{,\alpha}, \phi_{,\beta} \rangle_{L^2}| \\
&= \left| \langle (w_n)_{,\alpha}^2 (w_n)_{,\beta} - (w_n)_{,\alpha}^2 (w_0)_{,\beta} + (w_n)_{,\alpha}^2 (w_0)_{,\beta} - (w_0)_{,\alpha}^2 (w_0)_{,\beta}, \phi_{,\beta} \rangle_{L^2} \right| \\
&\leq \left| \langle (w_n)_{,\alpha}^2 ((w_n)_{,\beta} - (w_0)_{,\beta}), \phi_{,\beta} \rangle_{L^2} \right| + \left| \langle ((w_n)_{,\alpha}^2 - (w_0)_{,\alpha}^2)(w_0)_{,\beta}, \phi_{,\beta} \rangle_{L^2} \right| \\
&\leq \left| \langle (w_n)_{,\alpha}^2 ((w_n)_{,\beta} - (w_0)_{,\beta}), \phi_{,\beta} \rangle_{L^2} \right| \\
&\quad + |\langle ((w_n)_{,\alpha} + (w_0)_{,\alpha})((w_n)_{,\alpha} - (w_0)_{,\alpha})(w_0)_{,\beta}, \phi_{,\beta} \rangle_{L^2}|
\end{aligned}$$

$$\begin{aligned}
&\leq \|(w_n)_{,\alpha}\|_4^2 \|(w_n)_{,\alpha} - (w_0)_{,\alpha}\|_2 \|\phi_{,\beta}\|_\infty \\
&\quad + \|(w_n)_{,\alpha} + (w_0)_{,\alpha}\|_4 \|(w_n)_{,\alpha} - (w_0)_{,\alpha}\|_2 \|(w_0)_{,\beta}\|_4 \|\phi_{,\beta}\|_\infty \\
&\leq (K_2^2 + 2K_2K_2) \|(w_n)_{,\alpha} - (w_0)_{,\alpha}\|_2 \|\phi_{,\beta}\|_\infty \\
&\to 0, \text{ as } n \to \infty,
\end{aligned}$$

where also up to here the repeated indices do not sum.

Thus (13.52) has been proven, so that we may infer that

$$\begin{aligned}
&\Bigg\langle \left(h_{\alpha\beta\lambda\mu}(w_0)_{,\lambda\mu}\right)_{,\alpha\beta} \\
&\quad - \left(H_{\alpha\beta\lambda\mu}\left(\frac{(u_0)_{\lambda,\mu} + (u_0)_{\lambda,\mu}}{2} + \frac{(w_0)_{,\lambda}(w_0)_{,\mu}}{2}\right)(w_0)_{,\beta}\right)_{,\alpha} - P, \phi \Bigg\rangle_{L^2} \\
&= \lim_{n\to\infty} \Bigg\{ \Bigg\langle \left(h_{\alpha\beta\lambda\mu}(w_n)_{,\lambda\mu}\right)_{,\alpha\beta} \\
&\quad - \left(H_{\alpha\beta\lambda\mu}\left(\frac{(u_n)_{\lambda,\mu} + (u_n)_{\lambda,\mu}}{2} + \frac{(w_n)_{,\lambda}(w_n)_{,\mu}}{2}\right)(w_n)_{,\beta}\right)_{,\alpha} \\
&\quad - K(w_n)_{,\alpha\alpha} + (z_n^*)_{\alpha,\alpha} - P, \phi \Bigg\rangle_{L^2} \Bigg\} \\
&= \lim_{n\to\infty} 0 = 0.
\end{aligned}$$

Since $\phi \in C_c^\infty(\Omega)$ is arbitrary, we obtain

$$\begin{aligned}
&\left(h_{\alpha\beta\lambda\mu}(w_0)_{,\lambda\mu}\right)_{,\alpha\beta} \\
&\quad - \left(H_{\alpha\beta\lambda\mu}\left(\frac{(u_0)_{\lambda,\mu} + (u_0)_{\lambda,\mu}}{2} + \frac{(w_0)_{,\lambda}(w_0)_{,\mu}}{2}\right)(w_0)_{,\beta}\right)_{,\alpha} - P = 0, \text{ in } \Omega
\end{aligned}$$

in the distributional sense.

Similarly,

$$\begin{aligned}
&\left(H_{\alpha\beta\lambda\mu}\left(\frac{(u_0)_{\lambda,\mu} + (u_0)_{\lambda,\mu}}{2} + \frac{(w_0)_{,\lambda}(w_0)_{,\mu}}{2}\right)\right)_{,\beta} \\
&\quad + P_\alpha = 0, \text{ in } \Omega,
\end{aligned} \tag{13.53}$$

for $\alpha \in \{1, 2\}$, also in the distributional sense.

From the convergence in question, we also get in a weak sense

$$(u_0)_\alpha = w_0 = \frac{\partial w_0}{\partial \mathbf{n}} = 0, \text{ on } \partial\Omega.$$

The proof is complete.

Remark 13.4.2. We emphasize that for each $n \in \mathbb{N}$, from the condition indicated in (13.42), $\{u_n\}$ is obtained through the minimization of a convex functional. Therefore the numerical procedure translates into the solution of a sequence of convex optimization problems.

13.5 Numerical Results

In this section we present some numerical results. Let $\Omega = [0,1] \times [0,1]$ and consider the problem of minimizing $J : U \to \mathbb{R}$ where

$$J(u) = G(\Lambda u) - F(u),$$

$$(G \circ \Lambda)(u) = \frac{1}{2}\int_\Omega N_{\alpha\beta}(u)\gamma_{\alpha\beta}(u)\,dx + \frac{1}{2}\int_\Omega M_{\alpha\beta}(u)\kappa_{\alpha\beta}(u)\,dx, \tag{13.54}$$

$\{N_{\alpha\beta}\}$ denote the membrane force tensor and $\{M_{\alpha\beta}\}$ the moment one, so that from the constitutive relations,

$$N_{\alpha\beta}(u) = H_{\alpha\beta\lambda\mu}\gamma_{\lambda\mu}(u), \tag{13.55}$$

$$M_{\alpha\beta}(u) = h_{\alpha\beta\lambda\mu}\kappa_{\lambda\mu}(u). \tag{13.56}$$

Also, $F : U \to \mathbb{R}$ is given by

$$F(u) = \langle w, P\rangle_{L^2} + \langle u_\alpha, P_\alpha\rangle_{L^2}. \tag{13.57}$$

Here

$$U = \{(u_\alpha, w) \in W^{1,2}(\Omega;\mathbb{R}^2) \times W^{2,2}(\Omega) \mid u_\alpha = 0 \text{ on } \Gamma_0,\ w = 0 \text{ on } \partial\Omega\},$$

$\Gamma_0 = \{[0,y] \cup [x,0],\ 0 \le x,y \le 1\}$, and $P, P_1, P_2 \in L^2$ denote the external loads in the directions $\mathbf{a}_3$, $\mathbf{a}_1$, and $\mathbf{a}_2$, respectively.

We consider the particular case where all entries of $\{H_{\alpha\beta\lambda\mu}\}$ and $\{h_{\alpha\beta\lambda\mu}\}$ are zero, except for $H_{1111} = H_{2222} = H_{1212} = 10^5$ and $h_{1111} = h_{2222} = h_{1212} = 10^4$. Moreover $P = 1000$, $P_1 = -100$, and $P_2 = -100$ (units refer to the international system). In a first moment, define the trial functions $w : \Omega \to \mathbb{R}$, $u_1 : \Omega \to \mathbb{R}$, and $u_2 : \Omega \to \mathbb{R}$ by

$$w(x,y) = a_1 \sin(\pi x)\sin(\pi y) + a_2 \sin(2\pi x)\sin(2\pi y),$$

$$u_1(x,y) = a_3 \sin(\pi x/2)\sin(\pi y/2),$$

$$u_2(x,y) = a_4 \sin(\pi x/2)\sin(\pi y/2),$$

respectively.

The coefficients $\{a_1, a_2, a_3, a_4\}$ will be found through the extremal points of J.

We have obtained only one real critical point, namely,

$$(a_0)_1 = 0.000832$$

$$(a_0)_2 = -1.038531 * 10^{-8}$$

$$(a_0)_3 = -0.000486$$

$$(a_0)_4 = -0.000486$$

so that the candidate to optimal point is $((u_0)_1, (u_0)_2, w_0)$ where

$$w_0(x,y) = (a_0)_1 \sin(\pi x)\sin(\pi y) + (a_0)_2 \sin(2\pi x)\sin(2\pi y),$$

$$(u_0)_1(x,y) = (a_0)_3 \sin(\pi x/2)\sin(\pi y/2),$$

$$(u_0)_2 = (a_0)_4 \sin(\pi x/2)\sin(\pi y/2).$$

With such values for the coefficients, it is clear that $N_{11}(u_0)$ and $N_{22}(u_0)$ are negative in Ω so that $\{N_{\alpha\beta}(u_0)\}$ is not positive definite. Even so, as we shall see in the next lines, the optimality criterion of the second duality principle developed may be applied. Let

$$w(x,y) = a_1 \sin(\pi x)\sin(\pi y) + a_2 \sin(2\pi x)\sin(2\pi y).$$

We have that

$$\begin{aligned} W(a_1,a_2) &= \frac{1}{2}\int_\Omega h_{\alpha\beta\lambda\mu} w_{,\alpha\beta} w_{,\lambda\mu}\, dx + \frac{1}{2}\int_\Omega (N(u_0))_{\alpha\beta} w_{,\alpha} w_{,\beta}\, dx \\ &= 360319.a_1^2 + 191.511 a_1 a_2 + 5.7668 * 10^6 a_2^2. \end{aligned} \tag{13.58}$$

Now observe that

$$s_{11} = \frac{\partial^2 W(a_1,a_2)}{\partial a_1^2} = 720638.0,$$

$$s_{22} = \frac{\partial^2 W(a_1,a_2)}{\partial a_2^2} = 1.15336 * 10^7,$$

$$s_{12} = \frac{\partial^2 W(a_1,a_2)}{\partial a_1 \partial a_2} = 191.511.$$

Therefore $s_{11} > 0$, $s_{22} > 0$, and $s_{11}s_{22} - s_{12}^2 = 8.31155 * 10^{12} > 0$ so that $W(a_1,a_2)$ is a positive definite quadratic form.

Hence, from the second duality principle, we may conclude that $((u_0)_1, (u_0)_2, w_0)$ is indeed the optimal solution (approximate global minimizer for J).

Refining the results through finite differences using the algorithm of last section, we obtain again the field of displacements $w_0(x,y)$ (please see Fig. 13.1).

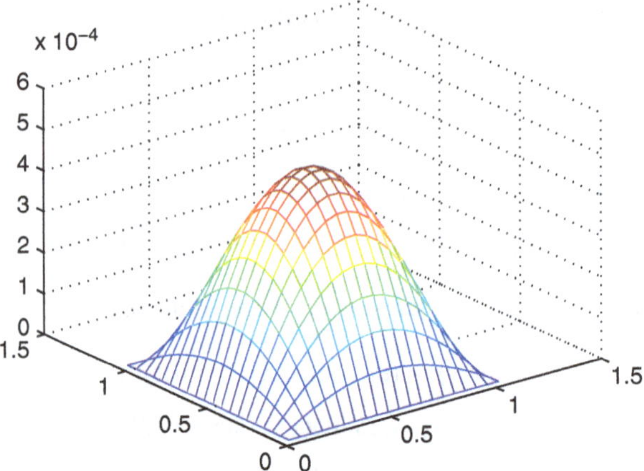

Fig. 13.1 Vertical axis: $w_0(x,y)$-field of displacements

13.6 Conclusion

In this chapter, we develop duality principles for the Kirchhoff–Love plate model. The results are obtained through the basic tools of convex analysis and include sufficient conditions of optimality. It is worth mentioning that earlier results require the membrane force tensor to be positive definite in a critical point in order to guarantee global optimality, whereas from the new results here presented, we are able to guarantee global optimality for a critical point such that $N_{11}(u_0) < 0$ and $N_{22}(u_0) < 0$, in Ω. Finally, the methods developed here may be applied to many other nonlinear models of plates and shells. Applications to related areas (specially to the shell models found in [23]) are planned for future works.

Chapter 14
About Ginzburg–Landau-Type Equations: The Simpler Real Case

14.1 Introduction

In this chapter, our first objectives are to show existence and develop dual formulations concerning the real semi-linear Ginzburg–Landau equation. We start by describing the primal formulation.

By $S \subset \mathbb{R}^3$ we denote an open connected bounded set with a sufficiently regular boundary $\Gamma = \partial S$ (regular enough so that the Sobolev imbedding theorem holds). The Ginzburg–Landau equation is given by

$$\begin{cases} -\nabla^2 u + \alpha(\frac{u^2}{2} - \beta)u - f = 0 \text{ in } S, \\ u = 0 \text{ on } \Gamma, \end{cases} \tag{14.1}$$

where $u : S \to \mathbb{R}$ denotes the primal field and $f \in L^2(S)$. Moreover, α, β are real positive constants.

Remark 14.1.1. The complex Ginzburg–Landau equation plays a fundamental role in the theory of superconductivity (see [4], for details). In the present work we deal with the simpler real form; however, the results obtained may be easily extended to the complex case.

The corresponding variational formulation is given by the functional $J : U \to \mathbb{R}$, where

$$J(u) = \frac{1}{2}\int_S |\nabla u|^2\, dx + \frac{\alpha}{2}\int_S (\frac{u^2}{2} - \beta)^2\, dx - \int_S fu\, dx \tag{14.2}$$

where $U = \{u \in W^{1,2}(S) \mid u = 0 \text{ on } \Gamma\} = W_0^{1,2}(S)$.

We are particularly concerned with the fact that equations indicated in (14.1) are necessary conditions for the solution of problem $\mathscr{P}$, where

$$\text{Problem } \mathscr{P}: \text{ to find } u_0 \in U \text{ such that } J(u_0) = \min_{u \in U}\{J(u)\}.$$

F. Botelho, *Functional Analysis and Applied Optimization in Banach Spaces: Applications to Non-Convex Variational Models*, DOI 10.1007/978-3-319-06074-3_14,

14.1.1 Existence of Solution for the Ginzburg–Landau Equation

We start with a remark.

Remark 14.1.2. From the Sobolev imbedding theorem for

$$mp < n, \ \ n - mp < n, \ \ p \leq q \leq p^* = np/(n-mp),$$

we have

$$W^{j+m,p}(\Omega) \hookrightarrow W^{j,q}(\Omega).$$

Therefore, considering $n = 3$, $m = 1$, $j = 0$, $p = 2$, and $q = 4$, we obtain

$$W^{1,2}(\Omega) \subset L^4(\Omega) \subset L^2(\Omega)$$

and thus

$$\|u\|_{L^4(\Omega)} \to +\infty \Rightarrow \|u\|_{W^{1,2}(\Omega)} \to +\infty.$$

Furthermore, from the above and the Poincaré inequality, it is clear that for J given by (14.2), we have

$$J(u) \to +\infty \text{ as } \|u\|_{W^{1,2}(S)} \to +\infty,$$

that is, J is coercive.

Now we establish the existence of a minimizer for $J : U \to \mathbb{R}$. It is a well-known procedure (the direct method of calculus of variations). We present it here for the sake of completeness.

Theorem 14.1.3. *For $\alpha, \beta \in \mathbb{R}^+$, $f \in L^2(S)$ there exists at least one $u_0 \in U$ such that*

$$J(u_0) = \min_{u \in U}\{J(u)\}$$

where

$$J(u) = \frac{1}{2}\int_S |\nabla u|^2\,dx + \frac{\alpha}{2}\int_S (\frac{u^2}{2} - \beta)^2\,dx - \int_S fu\,dx$$

and $U = \{u \in W^{1,2}(S) \mid u = 0 \text{ on } \Gamma\} = W_0^{1,2}(S)$.

Proof. From Remark 14.1.2 we have

$$J(u) \to +\infty \text{ as } \|u\|_U \to +\infty.$$

Also from the Poincaré inequality, there exists $\alpha_1 \in \mathbb{R}$ such that $\alpha_1 = \inf_{u \in U}\{J(u)\}$ so that for $\{u_n\}$ minimizing sequence, in the sense that

$$J(u_n) \to \alpha_1 \text{ as } n \to +\infty \tag{14.3}$$

we have that $\|u_n\|_U$ is bounded, and thus, as $W_0^{1,2}(S)$ is reflexive, there exists $u_0 \in W_0^{1,2}(S)$ and a subsequence $\{u_{nj}\} \subset \{u_n\}$ such that

$$u_{nj} \rightharpoonup u_0, \text{ weakly in } W_0^{1,2}(S). \tag{14.4}$$

From (14.4), by the Rellich–Kondrachov theorem, up to a subsequence, which is also denoted by $\{u_{nj}\}$, we have

$$u_{nj} \to u_0, \text{ strongly in } L^2(S). \tag{14.5}$$

Furthermore, defining $J_1 : U \to \mathbb{R}$ as

$$J_1(u) = \frac{1}{2}\int_S |\nabla u|^2\, dx + \frac{\alpha}{8}\int_S u^4\, dx - \int_S fu\, dx$$

we have that $J_1 : U \to \mathbb{R}$ is convex and strongly continuous, therefore weakly lower semicontinuous, so that

$$\liminf_{j\to+\infty}\{J_1(u_{nj})\} \geq J_1(u_0). \tag{14.6}$$

On the other hand, from (14.5),

$$\int_S (u_{nj})^2 dx \to \int_S u_0^2 dx, \text{ as } j \to +\infty \tag{14.7}$$

and thus, from (14.6) and (14.7), we may write

$$\alpha_1 = \inf_{u\in U}\{J(u)\} = \liminf_{j\to+\infty}\{J(u_{nj})\} \geq J(u_0).$$

14.2 A Concave Dual Variational Formulation

We start this section by enunciating the following theorem which has been proven in [11].

Theorem 14.2.1. *Let U be a reflexive Banach space, $(G\circ\Lambda) : U \to \bar{\mathbb{R}}$ a convex Gâteaux differentiable functional, and $(F\circ\Lambda_1) : U \to \bar{\mathbb{R}}$ convex, coercive, and lower semicontinuous (l.s.c.) such that the functional*

$$J(u) = (G\circ\Lambda)(u) - F(\Lambda_1 u) - \langle u, u_0^*\rangle_U$$

is bounded from below, where $\Lambda : U \to Y$ and $\Lambda_1 : U \to Y_1$ are continuous linear operators.

Then we may write

$$\inf_{z^*\in Y_1^*}\ \sup_{v^*\in B^*(z^*)}\{F^*(z^*) - G^*(v^*)\} \geq \inf_{u\in U}\{J(u)\}$$

where $B^*(z^*) = \{v^* \in Y^*$ such that $\Lambda^* v^* - \Lambda_1^* z^* - u_0^* = 0\}$.

Our next result refers to a convex dual variational formulation, through which we obtain sufficient conditions for optimality.

Theorem 14.2.2. *Consider $J : U \to \mathbb{R}$, where*

$$J(u) = \int_S \frac{1}{2}|\nabla u|^2\,dx + \int_S \frac{\alpha}{2}(\frac{u^2}{2} - \beta)^2\,dx - \int_S fu\,dx,$$

and $U = W_0^{1,2}(S)$. For $K = 1/K_0$, where K_0 stands for the constant related to the Poincaré inequality, we have the following duality principle:

$$\inf_{u \in U}\{J(u)\} \geq \sup_{(z^*, v_1^*, v_0^*) \in B^*}\{-G_L^*(v^*, z^*)\}$$

where

$$G_L^*(v^*, z^*) = \frac{1}{2K^2}\int_S |\nabla z^*|^2\,dx - \frac{1}{2K}\int_S (z^*)^2\,dx + \frac{1}{2}\int_S \frac{(v_1^*)^2}{v_0^* + K}dx$$
$$+\frac{1}{2\alpha}\int_S (v_0^*)^2\,dx + \beta\int_S v_0^*\,dx, \tag{14.8}$$

and

$$B^* = \{(z^*, v_1^*, v_0^*) \in L^2(S;\mathbb{R}^3) \mid$$
$$-\frac{1}{K}\nabla^2 z^* + v_1^* - z^* = f,\ v_0^* + K > 0,\ \text{a.e. in } S,\ z^* = 0 \text{ on } \Gamma\}.$$

If in addition there exists $u_0 \in U$ such that $\delta J(u_0) = \theta$ and $\bar{v}_0^* + K = (\alpha/2)u_0^2 - \beta + K > 0$, a.e. in S, then

$$J(u_0) = \min_{u \in U}\{J(u)\} = \max_{(z^*, v_1^*, v_0^*) \in B^*}\{-G_L^*(v^*, z^*)\} = -G_L^*(\bar{v}^*, \bar{z}^*),$$

where

$$\bar{v}_0^* = \frac{\alpha}{2}u_0^2 - \beta,$$
$$\bar{v}_1^* = (\bar{v}_0^* + K)u_0$$

and

$$\bar{z}^* = Ku_0.$$

Proof. Observe that we may write

$$J(u) = G(\Lambda u) - F(\Lambda_1 u) - \int_S fu\,dx,$$

where

$$G(\Lambda u) = \int_S \frac{1}{2}|\nabla u|^2 dx + \int_S \frac{\alpha}{2}(\frac{u^2}{2} - \beta + 0)^2 dx + \frac{K}{2}\int_S u^2 dx,$$
$$F(\Lambda_1 u) = \frac{K}{2}\int_S u^2\,dx,$$

where

$$\Lambda u = \{\Lambda_0 u, \Lambda_1 u, \Lambda_2 u\},$$

and

$$\Lambda_0 u = 0,\ \Lambda_1 u = u,\ \Lambda_2 u = \nabla u.$$

From Theorem 14.2.1 (here this is an auxiliary theorem through which we obtain A^*, below indicated), we have

$$\inf_{u \in U}\{J(u)\} = \inf_{z^* \in Y_1^*} \sup_{v^* \in A^*}\{F^*(z^*) - G^*(v^*)\},$$

where

$$F^*(z^*) = \frac{1}{2K}\int_S (z^*)^2\, dx,$$

and

$$G^*(v^*) = \frac{1}{2}\int_S |v_2^*|^2\, dx + \frac{1}{2}\int_S \frac{(v_1^*)^2}{v_0^* + K}\, dx + \frac{1}{2\alpha}\int_S (v_0^*)^2\, dx + \beta\int_S v_0^*\, dx,$$

if $v_0^* + K > 0$, a.e. in S, and

$$A^* = \{v^* \in Y^* \mid \Lambda^* v^* - \Lambda_1^* z^* - f = 0\},$$

or

$$A^* = \{(z^*, v^*) \in L^2(S) \times L^2(S;\mathbb{R}^5) \mid \\ -div(v_2^*) + v_1^* - z^* - f = 0, \text{ a.e. in } S\}.$$

Observe that

$$G^*(v^*) \geq \langle \Lambda u, v^* \rangle_Y - G(\Lambda u),\ \forall u \in U,\ v^* \in A^*,$$

and thus

$$-F^*(z^*) + G^*(v^*) \geq -F^*(z^*) + \langle \Lambda_1 u, z^* \rangle_{L^2(S)} + \langle u, f \rangle_U - G(\Lambda u), \tag{14.9}$$

and hence, making z^* an independent variable through A^*, from (14.9), we may write

$$\sup_{z^* \in L^2(S)}\{-F^*(z^*) + G^*(v_2^*(v_1^*, z^*), v_1^*, v_0^*)\} > \sup_{z^* \in L^2(S)}\left\{-F^*(z^*) \right. \\ \left. + \langle \Lambda_1 u, z^* \rangle_{L^2(S)} + \int_S fu\, dx - G(\Lambda u)\right\}, \tag{14.10}$$

so that

$$\sup_{z^* \in L^2(S)} \left\{ -\frac{1}{2K} \int_S (z^*)^2 \, dx + \frac{1}{2} \int_S (v_2^*(z^*, v_1^*))^2 \, dx + \frac{1}{2} \int_S \frac{(v_1^*)^2}{v_0^* + K} \, dx \right.$$
$$\left. + \frac{1}{2\alpha} \int_S (v_0^*)^2 \, dx + \beta \int_S v_0^* \, dx \right\}$$
$$\geq F(\Lambda_1 u) + \int_S f u dx - G(\Lambda u). \tag{14.11}$$

Therefore, if $K \leq 1/K_0$ (here K_0 denotes the constant concerning the Poincaré inequality), the supremum in the left side of (14.11) is attained through the relations

$$v_2^* = \frac{\nabla z^*}{K} \text{ and } z^* = 0 \text{ on } \Gamma,$$

so that the final format of our duality principle is given by

$$\inf_{u \in U} \{J(u)\} \geq \sup_{(z^*, v_1^*, v_0^*) \in B^*} \left\{ -\frac{1}{2K^2} \int_S |\nabla z^*|^2 \, dx + \frac{1}{2K} \int_S (z^*)^2 \, dx \right.$$
$$\left. - \frac{1}{2} \int_S \frac{(v_1^*)^2}{v_0^* + K} \, dx - \frac{1}{2\alpha} \int_S (v_0^*)^2 \, dx - \beta \int_S v_0^* \, dx \right\}, \tag{14.12}$$

where

$$B^* = \{(z^*, v_1^*, v_0^*) \in L^2(S; \mathbb{R}^3) \mid$$
$$-\frac{1}{K} \nabla^2 z^* + v_1^* - z^* = f, \; v_0^* + K > 0, \text{ a.e. in } S, \; z^* = 0 \; on \; \Gamma\}.$$

The remaining conclusions follow from an application (with little changes) of Theorem 10.1.25.

Remark 14.2.3. The relations

$$v_2^* = \frac{\nabla z^*}{K} \text{ and } z^* = 0 \text{ on } \Gamma,$$

are sufficient for the attainability of the supremum indicated in (14.11) but just partially necessary; however, we assume them because the expression of dual problem is simplified without violating inequality (14.12) (in fact the difference between the primal and dual functionals even increases under such relations).

14.3 A Numerical Example

In this section we present numerical results for a one-dimensional example originally due to Bolza (see [50] for details about the primal formulation).

Consider $J : U \to \mathbb{R}$ expressed as

$$J(u) = \frac{1}{2}\int_0^1 ((u_{,x})^2 - 1)^2 dx + \frac{1}{2}\int_0^1 (u - f)^2 dx$$

or, defining $S = [0,1]$,

$$G(\Lambda u) = \frac{1}{2}\int_0^1 ((u_{,x})^2 - 1)^2 dx$$

and

$$F(u) = \frac{1}{2}\int_0^1 (u - f)^2 dx$$

we may write

$$J(u) = G(\Lambda u) + F(u)$$

where, for convenience, we define $\Lambda : U \to Y \equiv L^4(S) \times L^2(S)$ as

$$\Lambda u = \{u_{,x}, 0\}.$$

Furthermore, we have

$$U = \{u \in W^{1,4}(S) \mid u(0) = 0 \text{ and } u(1) = 0.5\}$$

For $Y = Y^* = L^4(S) \times L^2(S)$, defining

$$G(\Lambda u + p) = \frac{1}{2}\int_S ((u_{,x} + p_1)^2 - 1.0 + p_0)^2 dx$$

for $v_0^* > 0$, we obtain

$$G(\Lambda u) + F(u) \geq \inf_{p \in Y}\{-\langle p_0, v_0^*\rangle_{L^2(S)} - \langle p_1, v_1^*\rangle_{L^2(S)} + G(\Lambda u + p) + F(u)\}$$

or

$$G(\Lambda u) + F(u) \geq \inf_{p \in Y}\{-\langle q_0, v_0^*\rangle_{L^2(S)} - \langle q_1, v_1^*\rangle_{L^2(S)} + G(q) + \langle 0, v_0^*\rangle_{L^2(S)} + \langle u', v_1^*\rangle_{L^2(S)} + F(u)\}.$$

Here $q = \Lambda u + p$ so that

$$G(\Lambda u) + F(u) \geq -G^*(v^*) + \langle 0, v_0^*\rangle_{L^2(S)} + \langle u_{,x}, v_1^*\rangle_{L^2(S)} + F(u).$$

That is,

$$G(\Lambda u)+F(u)\geq -G^*(v^*)+\inf_{u\in U}\{\langle 0,v_0^*\rangle_{L^2(S)}+\langle u_{,x},v_1^*\rangle_{L^2(S)}+F(u)\},$$

or

$$\inf_{u\in U}\{G(\Lambda u)+F(u)\}\geq \sup_{v^*\in A^*}\{-G^*(v^*)-F^*(-\Lambda^*v^*)\}$$

where

$$G^*(v^*)=\frac{1}{2}\int_S\frac{(v_1^*)^2}{v_0^*}dx+\frac{1}{2}\int_S(v_0^*)^2dx,$$

if $v_0^*>0,\ \ a.e.\ \ in\ \ S$. Also

$$F^*(-\Lambda^*v^*)=\frac{1}{2}\int_S[(v_1^*)_{,x}]^2dx+\langle f,(v_1^*)_{,x}\rangle_{L^2(S)}-v_1^*(1)u(1)$$

and

$$A^*=\{v^*\in Y^*\ \mid\ v_0^*>0,\ \text{a.e. } in\ S\}.$$

Remark 14.3.1. Through the extremal condition $v_0^*=((u_{,x})^2-1)$ and Weierstrass condition $(u_{,x})^2-1.0\geq 0$ we can see that the dual formulation is convex for $v_0^*>0$; however, it is possible that the primal formulation has no minimizers, and we could expect a microstructure formation through $v_0^*=0$ (i.e., $u_{,x}=\pm 1$, depending on f(x)). To allow $v_0^*=0$ we will redefine the primal functional as indicated below.

Define $G_1:U\to\mathbb{R}$ and $F_1:U\to\mathbb{R}$ by

$$G_1(u)=G(\Lambda u)+F(u)+\frac{K}{2}\int_S(u_{,x})^2dx$$

and

$$F_1(u)=\frac{K}{2}\int_S(u_{,x})^2dx.$$

Also defining $\hat{G}(\Lambda u)=G(\Lambda u)+\frac{K}{2}\int_S(u_{,x})^2dx$, from Theorem 14.2.1, we can write

$$\inf_{u\in U}\{J(u)\}\leq \inf_{z^*\in Y^*}\sup_{v^*\in B^*(z^*)}\{F_1^*(z^*)-\hat{G}^*(v_0^*,v_2^*)-F^*(v_1^*)\} \tag{14.13}$$

where

$$F_1^*(z^*)=\frac{1}{2K}\int_S(z^*)^2dx,$$

$$\hat{G}^*(v_0^*,v_2^*)=\frac{1}{2}\int_S\frac{(v_2^*)^2}{v_0^*+K}dx+\frac{1}{2}\int_S(v_0^*)^2dx,$$

$$F^*(v_1^*)=\frac{1}{2}\int_S(v_1^*)^2dx+\langle f,v_1^*\rangle_{L^2(S)}-v_2^*(1)u(1)$$

and

$$B^*(z^*) = \{v^* \in Y^* \mid -(v_2^*)_{,x} + v_1^* - z^* = 0 \ and \ v_0^* \geq 0 \ a.e. \ in \ S\}.$$

We developed an algorithm based on the dual formulation indicated in (14.13). It is relevant to emphasize that such a dual formulation is convex if the supremum indicated is evaluated under the constraint $v_0^* \geq 0$ *a.e. in S* (this result follows from the traditional Weierstrass condition, so that there is no duality gap between the primal and dual formulations and the inequality indicated in (14.13) is in fact an equality).

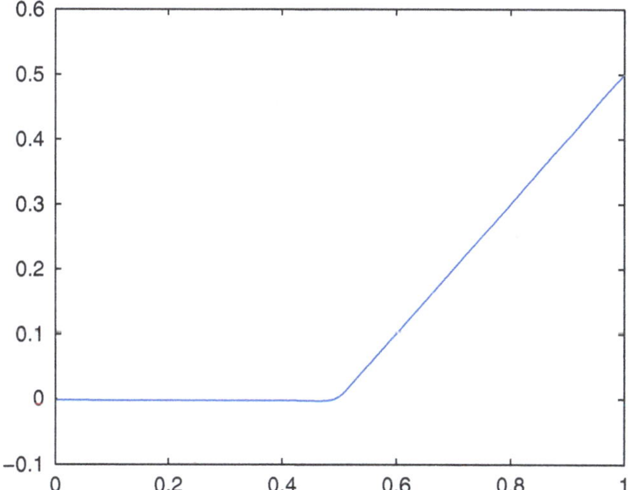

Fig. 14.1 Vertical axis: $u_0(x)$-weak limit of minimizing sequences for f(x)=0

We present numerical results for $f(x) = 0$ (see Fig. 14.1), $f(x) = 0.3 * Sin(\pi * x)$ (Fig. 14.2), and $f(x) = 0.3 * Cos(\pi * x)$ (Fig. 14.3). The solutions indicated as optimal through the dual formulations (denoted by u_0) are in fact weak cluster points of minimizing sequences for the primal formulations.

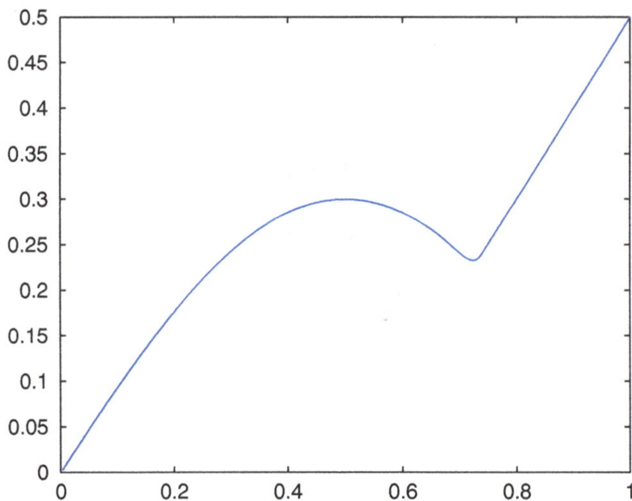

Fig. 14.2 Vertical axis: $u_0(x)$-weak limit of minimizing sequences for $f(x) = 0.3 * Sin(\pi * x)$

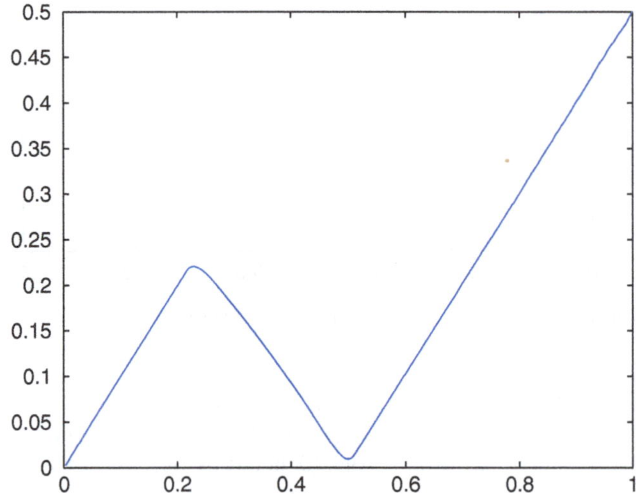

Fig. 14.3 Vertical axis: $u_0(x)$-weak limit of minimizing sequences for $f(x) = 0.3 * Cos(\pi * x)$

Chapter 15
The Full Complex Ginzburg–Landau System

15.1 Introduction

Remark 15.1.1. This chapter was published in an article form by *Applied Mathematics and Computation-Elsevier, reference [12].*

We recall that about the year 1950 Ginzburg and Landau introduced a theory to model the superconducting behavior of some types of materials below a critical temperature T_c, which depends on the material in question. They postulated that the free-energy density may be written close to T_c as

$$F_s(T) = F_n(T) + \frac{\hbar}{4m}\int_\Omega |\nabla\psi|_2^2\,dx + \frac{\alpha(T)}{4}\int_\Omega |\psi|^4\,dx - \frac{\beta(T)}{2}\int_\Omega |\psi|^2\,dx,$$

where ψ is a complex parameter and $F_n(T)$ and $F_s(T)$ are the normal and superconducting free-energy densities, respectively. (see [4, 9, 45, 46] for details). Here $\Omega \subset \mathbb{R}^3$ denotes the superconducting sample with a boundary denoted by $\partial\Omega = \Gamma$. The complex function $\psi \in W^{1,2}(\Omega;\mathbb{C})$ is intended to minimize $F_s(T)$ for a fixed temperature T.

Denoting $\alpha(T)$ and $\beta(T)$ simply by α and β, the corresponding Euler–Lagrange equations are given by

$$\begin{cases} -\frac{\hbar}{2m}\nabla^2\psi + \alpha|\psi|^2\psi - \beta\psi = 0, & \text{in } \Omega \\ \frac{\partial\psi}{\partial \mathbf{n}} = 0, & \text{on } \partial\Omega. \end{cases} \tag{15.1}$$

This last system of equations is well known as the Ginzburg–Landau (G-L) one. In the physics literature, it is also well known the G-L energy in which a magnetic potential here denoted by **A** is included. The functional in question is given by

F. Botelho, *Functional Analysis and Applied Optimization in Banach Spaces: Applications to Non-Convex Variational Models*, DOI 10.1007/978-3-319-06074-3_15,

$$J(\psi,\mathbf{A}) = \frac{1}{8\pi}\int_{\mathbb{R}^3} |\operatorname{curl}\mathbf{A}-\mathbf{B}_0|_2^2\,dx + \frac{\hbar^2}{4m}\int_\Omega \left|\nabla\psi - \frac{2ie}{\hbar c}\mathbf{A}\psi\right|_2^2 dx$$
$$+\frac{\alpha}{4}\int_\Omega |\psi|^4\,dx - \frac{\beta}{2}\int_\Omega |\psi|^2\,dx \qquad (15.2)$$

Considering its minimization on the space U, where

$$U = W^{1,2}(\Omega;\mathbb{C}) \times W^{1,2}(\mathbb{R}^3;\mathbb{R}^3),$$

through the physics notation, the corresponding Euler–Lagrange equations are

$$\begin{cases} \frac{1}{2m}\left(-i\hbar\nabla - \frac{2e}{c}\mathbf{A}\right)^2\psi + \alpha|\psi|^2\psi - \beta\psi = 0, & \text{in } \Omega \\ \left(i\hbar\nabla\psi + \frac{2e}{c}\mathbf{A}\psi\right)\cdot\mathbf{n} = 0, & \text{on } \partial\Omega, \end{cases} \qquad (15.3)$$

and

$$\begin{cases} \operatorname{curl}(\operatorname{curl}\mathbf{A}) = \operatorname{curl}\mathbf{B}_0 + \frac{4\pi}{c}\tilde{J}, & \text{in } \Omega \\ \operatorname{curl}(\operatorname{curl}\mathbf{A}) = \operatorname{curl}\mathbf{B}_0, & \text{in } \mathbb{R}^3 - \overline{\Omega}, \end{cases} \qquad (15.4)$$

where

$$\tilde{J} = -\frac{ie\hbar}{2m}(\psi^*\nabla\psi - \psi\nabla\psi^*) - \frac{2e^2}{mc}|\psi|^2\mathbf{A}.$$

and

$$\mathbf{B}_0 \in L^2(\mathbb{R}^3;\mathbb{R}^3)$$

is a known applied magnetic field.

15.2 Global Existence for the Ginzburg–Landau System

The existence of a global minimizer for the Ginzburg–Landau energy for a system in superconductivity in the presence of a magnetic field is proven in the next lines. The key hypothesis is the boundedness of infinity norm of the magnetic potential. It is worth emphasizing that such a hypothesis is physically observed. We start with the following remark:

Remark 15.2.1. For an open bounded subset $\Omega \subset \mathbb{R}^3$, we denote the $L^2(\Omega)$ norm by $\|\cdot\|_{L^2(\Omega)}$ or simply by $\|\cdot\|_2$. A similar remark is valid for the $L^2(\Omega;\mathbb{R}^3)$ norm, which is denoted by $\|\cdot\|_{L^2(\Omega;\mathbb{R}^3)}$ or simply by $\|\cdot\|_2$, when its meaning is clear, and for the $L^4(\Omega)$ one, which is denoted by $\|\cdot\|_{L^4(\Omega)}$ or simply by $\|\cdot\|_4$. On the other hand, by $|\cdot|_2$, we denote the standard Euclidean norm in $\mathbb{R}^3$ or $\mathbb{C}^3$, $|\Omega|$ denotes the Lebesgue measure of Ω, and $\mathbf{n}$ is the outward normal to its boundary.

Moreover derivatives are always understood in the distributional sense. Finally, by a regular boundary $\partial\Omega = \Gamma$ of Ω, we mean regularity enough so that the standard

Sobolev imbedding theorems, the trace theorem, and the Gauss–Green formulas of integration by parts hold. Details about such results may be found in [1, 26].

Theorem 15.2.2. *Let $\Omega \subset \mathbb{R}^3$ be an open, bounded, and connected set with a regular boundary denoted by $\partial\Omega = \Gamma$. Consider the functional $J : U \to \mathbb{R}$ given by*

$$J(\psi,\mathbf{A}) = \frac{1}{8\pi}\int_{\mathbb{R}^3} |\,curl\,\mathbf{A} - \mathbf{B}_0|_2^2\,dx + \frac{\hbar^2}{4m}\int_\Omega \left|\nabla\psi - \frac{2ie}{\hbar c}\mathbf{A}\psi\right|_2^2 dx$$
$$+\frac{\alpha}{4}\int_\Omega |\psi|^4\,dx - \frac{\beta}{2}\int_\Omega |\psi|^2\,dx \tag{15.5}$$

where $\hbar, m, c, e, \alpha, \beta$ are positive constants, i is the imaginary unit, and

$$U = W^{1,2}(\Omega;\mathbb{C}) \times W^{1,2}(\mathbb{R}^3;\mathbb{R}^3).$$

Assume there exists a minimizing sequence $\{(\psi_n,\mathbf{A}_n)\} \subset U$ such that

$$\|\mathbf{A}_n\|_\infty < K, \forall n \in \mathbb{N}$$

for some finite $K > 0$. Under such a hypothesis, there exists $(\psi_0,\mathbf{A}_0) \in U$ such that

$$J(\psi_0,\mathbf{A}_0) = \min_{(\psi,\mathbf{A})\in U}\{J(\psi,\mathbf{A})\}.$$

Proof. Suppose $\{(\psi_n,\mathbf{A}_n)\} \subset U$ is a minimizing sequence for J, that is,

$$\lim_{n\to\infty} J(\psi_n,\mathbf{A}_n) = \inf_{(u,\mathbf{A})\in U}\{J(\psi,\mathbf{A})\}, \tag{15.6}$$

such that

$$\|\mathbf{A}_n\|_\infty < K, \forall n \in \mathbb{N}$$

for some finite $K > 0$.

Observe that

$$J(\psi_n,\mathbf{A}_n) \geq \frac{1}{8\pi}\int_{\mathbb{R}^3} |\,\text{curl}\,\mathbf{A}_n - \mathbf{B}_0|_2^2\,dx$$
$$+\frac{\hbar^2}{4m}\int_\Omega |\nabla\psi_n|_2^2\,dx - K\frac{\hbar^2}{2m}\left|\frac{2ie}{\hbar c}\right| \|\nabla\psi_n\|_2\|\psi_n\|_2$$
$$+\frac{\hbar^2}{4m}\int_\Omega \left|\frac{2ie}{\hbar c}\mathbf{A}_n\psi_n\right|_2^2 dx + \frac{\alpha}{4}\int_\Omega |\psi_n|^4\,dx$$
$$-\frac{\beta}{2}\int_\Omega |\psi_n|^2\,dx,\ \forall n \in \mathbb{N}. \tag{15.7}$$

Suppose, to obtain contradiction, that there exists a subsequence $\{\psi_{n_k}\}$ such that either $\|\psi_{n_k}\|_4 \to \infty$ or $\|\nabla\psi_{n_k}\|_2 \to \infty$, as $k \to \infty$. In such a case from (15.7) we would obtain $J(\psi_{n_k},\mathbf{A}_{n_k}) \to +\infty$, as $k \to \infty$, which contradicts (15.6).

Therefore, there exists $K_1 > 0$ such that

$$\|\psi_n\|_4 < K_1 \text{ and } \|\nabla\psi_n\|_2 < K_1,$$

$\forall n \in \mathbb{N}$. From this we may conclude that there exists $K_2 > 0$ such that $\|\psi_n\|_2 < K_2, \forall n \in \mathbb{N}$. Hence by the Rellich–Kondrachov theorem, there exists $\psi_0 \in W^{1,2}(\Omega;\mathbb{C})$ and a subsequence not relabeled such that

$$\nabla\psi_n \rightharpoonup \nabla\psi_0, \text{ weakly in } L^2(\Omega;\mathbb{C}^3)$$

and

$$\psi_n \to \psi_0 \text{ strongly in } L^2(\Omega;\mathbb{C}). \tag{15.8}$$

On the other hand, since $\{\|\mathbf{A}_n\|_\infty\}$ is uniformly bounded, there exists $\mathbf{A}_0 \in L^\infty(\Omega;\mathbb{R}^3)$ such that up to a subsequence not relabeled we have

$$\mathbf{A}_n \rightharpoonup \mathbf{A}_0, \text{ weakly star in } L^\infty(\Omega;\mathbb{R}^3).$$

Fix $v \in L^2(\Omega,\mathbb{C}^3)$, since

$$\int_\Omega |v\psi_0|\, dx \leq \|v\|_2\|\psi_0\|_2$$

we have that $v\psi_0 \in L^1(\Omega;\mathbb{C}^3)$, so that

$$\begin{aligned}
&\left|\int_\Omega (\mathbf{A}_n\psi_n - \mathbf{A}_0\psi_0)\cdot v\, dx\right| \\
&\quad = \left|\int_\Omega (\mathbf{A}_n\psi_n - \mathbf{A}_n\psi_0 + \mathbf{A}_n\psi_0 - \mathbf{A}_0\psi_0)\cdot v\, dx\right| \\
&\quad \leq \int_\Omega |(\mathbf{A}_n\psi_n - \mathbf{A}_n\psi_0)\cdot v|\, dx \\
&\qquad + \left|\int_\Omega (\mathbf{A}_n\psi_0 - \mathbf{A}_0\psi_0)\cdot v\, dx\right| \\
&\quad \leq \|\mathbf{A}_n\cdot v\|_2\|\psi_n - \psi_0\|_2 + \left|\int_\Omega (\mathbf{A}_n - \mathbf{A}_0)\cdot v\psi_0\, dx\right| \\
&\quad \leq K\|v\|_2\|\psi_n - \psi_0\|_2 + \left|\int_\Omega (\mathbf{A}_n - \mathbf{A}_0)\cdot v\psi_0\, dx\right| \\
&\to 0 \text{ as } n \to \infty.
\end{aligned} \tag{15.9}$$

Thus, since $v \in L^2(\Omega,\mathbb{C}^3)$ is arbitrary, we obtain

$$\nabla\psi_n - \frac{2ie}{\hbar c}\mathbf{A}_n\psi_n \rightharpoonup \nabla\psi_0 - \frac{2ie}{\hbar c}\mathbf{A}_0\psi_0, \text{ weakly in } L^2(\Omega;\mathbb{C}^3)$$

so that

$$\liminf_{n\to\infty} \int_\Omega \left|\nabla\psi_n - \frac{2ie}{\hbar c}\mathbf{A}_n\psi_n\right|_2^2 dx \geq \int_\Omega \left|\nabla\psi_0 - \frac{2ie}{\hbar c}\mathbf{A}_0\psi_0\right|_2^2 dx. \tag{15.10}$$

Also it is clear that

$$\int_{\mathbb{R}^3} |\operatorname{curl} \mathbf{A}_n|_2^2\, dx < K_3, \ \forall n \in \mathbb{N},$$

for some $K_3 > 0$, so that there exists $\mathbf{v}_0 \in L^2(\mathbb{R}^3;\mathbb{R}^3)$ such that up to a subsequence not relabeled, we have

$$\text{curl } \mathbf{A}_n \rightharpoonup \mathbf{v}_0, \text{ weakly in } L^2(\mathbb{R}^3;\mathbb{R}^3).$$

Fix $\phi \in C_c^\infty(\mathbb{R}^3;\mathbb{R}^3)$. Hence, we get

$$\begin{aligned} \langle \mathbf{A}_0, \text{curl}^* \phi \rangle_{L^2(\mathbb{R}^3;\mathbb{R}^3)} &= \lim_{n\to\infty} \langle \mathbf{A}_n, \text{curl}^* \phi \rangle_{L^2(\mathbb{R}^3;\mathbb{R}^3)} \\ &= \lim_{n\to\infty} \langle \text{curl } \mathbf{A}_n, \phi \rangle_{L^2(\mathbb{R}^3;\mathbb{R}^3)} \\ &= \langle \mathbf{v}_0, \phi \rangle_{L^2(\mathbb{R}^3;\mathbb{R}^3)}. \end{aligned} \tag{15.11}$$

Since $\phi \in C_c^\infty(\mathbb{R}^3;\mathbb{R}^3)$ is arbitrary we have that

$$\mathbf{v}_0 = \text{curl } \mathbf{A}_0,$$

in the distributional sense, so that

$$\text{curl } \mathbf{A}_n \rightharpoonup \text{curl } \mathbf{A}_0, \text{ weakly in } L^2(\mathbb{R}^3;\mathbb{R}^3).$$

Therefore, considering the convexity of the functional in question, we obtain

$$\begin{aligned} &\liminf_{n\to\infty} \left\{ \frac{1}{8\pi} \int_{\mathbb{R}^3} |\text{ curl } \mathbf{A}_n - \mathbf{B}_0|_2^2 \, dx + \frac{\alpha}{4} \int_\Omega |\psi_n|^4 \, dx \right\} \\ &\geq \frac{1}{8\pi} \int_{\mathbb{R}^3} |\text{ curl } \mathbf{A}_0 - \mathbf{B}_0|_2^2 \, dx + \frac{\alpha}{4} \int_\Omega |\psi_0|^4 \, dx. \end{aligned} \tag{15.12}$$

From this, (15.10) and (15.8) we get

$$\inf_{(\psi,\mathbf{A})\in U} \{J(\psi,\mathbf{A})\} = \liminf_{n\to\infty} \{J(\psi_n,\mathbf{A}_n)\} \geq J(\psi_0,\mathbf{A}_0).$$

The proof is complete.

15.3 A Related Optimal Control Problem

In this section we study the existence of solutions for a closely related optimal control problem. In particular the state equation is of Ginzburg–Landau type. It is worth mentioning that the present case refers to the simpler real one. In the next lines we describe such a problem.

Let $\Omega \subset \mathbb{R}^3$ be an open, bounded, and connected set with a regular boundary denoted by $\partial\Omega = \Gamma$. Let $\psi_d : \Omega \to \mathbb{R}$ be a function such that $\psi_d \in L^2(\Omega)$. Consider the problem $\mathscr{P}$, that is, the problem of minimizing $J : U \to \mathbb{R}$ given by

$$J(\psi, u) = \frac{1}{2} \int_\Omega |\psi - \psi_d|^2 \, dx + \frac{1}{2} \int_{\partial\Omega} |u|^2 \, d\Gamma$$

subject to

$$\begin{cases} -\nabla^2\psi + \alpha\psi^3 - \beta\psi = f, & \text{in } \Omega \\ \frac{\partial\psi}{\partial\mathbf{n}} = u, & \text{on } \partial\Omega. \end{cases} \tag{15.13}$$

Here $U = W^{1,2}(\Omega) \times L^2(\partial\Omega)$ and $f \in L^2(\Omega)$.

We say that the set of admissible fields for problem $\mathscr{P}$ is nonempty if there exists $(\psi, u) \in U$ satisfying (15.13).

A similar problem is studied in [40] through a different approach. We will prove that such a problem has a solution. We start with the following proposition:

Proposition 15.3.1. *The set of admissible fields for problem $\mathscr{P}$ is nonempty.*

Proof. From reference [13], Chap. 13, there exists $\tilde{\psi} \in W_0^{1,2}(\Omega)$ which minimizes $\tilde{J}$ on $W_0^{1,2}(\Omega)$, where

$$\tilde{J}(\psi) = \frac{1}{2}\int_\Omega |\nabla\psi|_2^2\,dx + \frac{\alpha}{4}\int_\Omega |\psi|^4\,dx - \frac{\beta}{2}\int_\Omega |\psi|^2\,dx - \langle\psi, f\rangle_{L^2(\Omega)},$$

so that

$$\begin{cases} -\nabla^2\tilde{\psi} + \alpha\tilde{\psi}^3 - \beta\tilde{\psi} = f, & \text{in } \Omega \\ \tilde{\psi} = 0, & \text{on } \partial\Omega. \end{cases} \tag{15.14}$$

Therefore $(\psi, u) = (\tilde{\psi}, \frac{\partial\tilde{\psi}}{\partial\mathbf{n}})$ is an admissible field for problem $\mathscr{P}$.

Theorem 15.3.2. *Problem $\mathscr{P}$ has at least one solution.*

Proof. Let $\{(\psi_n, u_n)\} \subset U$ be a minimizing sequence for problem $\mathscr{P}$. Clearly there exists $K > 0$ such that

$$\|\psi_n\|_2 < K \text{ and } \|u_n\|_{L^2(\partial\Omega)} < K,$$

$\forall n \in \mathbb{N}$. Therefore, there exist $\psi_0 \in L^2(\Omega)$ and $u_0 \in L^2(\partial\Omega)$ such that up to a subsequence not relabeled we have

$$\psi_n \rightharpoonup \psi_0, \text{ weakly in } L^2(\Omega),$$

and

$$u_n \rightharpoonup u_0, \text{ weakly in } L^2(\partial\Omega),$$

We claim that there exists $K_1 > 0$ such that

$$\|\psi_n\|_4 < K_1, \forall n \in \mathbb{N}.$$

Suppose, to obtain contradiction, that there exists a subsequence $\{\psi_{n_k}\}$ such that

$$\|\psi_{n_k}\|_4 \to \infty, \text{ as } k \to \infty. \tag{15.15}$$

Observe that for each $k \in \mathbb{N}$ we have

$$\begin{cases} -\nabla^2\psi_{n_k} + \alpha\psi_{n_k}^3 - \beta\psi_{n_k} = f, & \text{in } \Omega \\ \frac{\partial\psi_{n_k}}{\partial\mathbf{n}} = u_{n_k}, & \text{on } \partial\Omega, \end{cases} \tag{15.16}$$

so that

$$\int_\Omega |\nabla \psi_{n_k}|_2^2\,dx + \alpha \int_\Omega |\psi_{n_k}|^4\,dx - \beta \int_\Omega |\psi_{n_k}|^2\,dx$$
$$-\langle \psi_{n_k}, f\rangle_{L^2(\Omega)} - \left\langle \frac{\partial \psi_{n_k}}{\partial \mathbf{n}}, \psi_{n_k}\right\rangle_{L^2(\partial\Omega)} = 0, \tag{15.17}$$

Hence

$$\begin{aligned}\beta \int_\Omega |\psi_{n_k}|^2\,dx &\geq \int_\Omega |\nabla \psi_{n_k}|_2^2\,dx + \alpha \int_\Omega |\psi_{n_k}|^4\,dx \\ &\quad - \|\psi_{n_k}\|_2 \|f\|_2 - \|u_{n_k}\|_{L^2(\partial\Omega)} \|\psi_{n_k}\|_{L^2(\partial\Omega)} \\ &\geq \int_\Omega |\nabla \psi_{n_k}|_2^2\,dx + \alpha \int_\Omega |\psi_{n_k}|^4\,dx \\ &\quad - K\|f\|_{L^2(\Omega)} - K\|\psi_{n_k}\|_{L^2(\partial\Omega)} \end{aligned} \tag{15.18}$$

and thus, from the trace theorem, there exists $C_1 > 0$ such that

$$\begin{aligned}\beta \int_\Omega |\psi_{n_k}|^2\,dx + \|\psi_{n_k}\|_2^2 &\geq \int_\Omega |\nabla \psi_{n_k}|_2^2\,dx + \alpha \int_\Omega |\psi_{n_k}|^4\,dx \\ &\quad + \|\psi_{n_k}\|_2^2 - K\|f\|_2 \\ &\quad - KC_1 \|\psi_{n_k}\|_{W^{1,2}(\Omega)}. \end{aligned} \tag{15.19}$$

From this and (15.15) we obtain

$$\int_\Omega |\psi_{n_k}|^2\,dx \to \infty \text{ as } k \to \infty,$$

which is a contradiction. Hence, there exists $K_1 > 0$ such that

$$\|\psi_n\|_4 < K_1, \forall n \in \mathbb{N}.$$

Thus, up to a subsequence not relabeled, there exists $\tilde{\psi}$ such that

$$\psi_n \rightharpoonup \tilde{\psi}, \text{ weakly in } L^4(\Omega),$$

so that from

$$\psi_n \rightharpoonup \psi_0, \text{ weakly in } L^2(\Omega),$$

we get

$$\tilde{\psi} = \psi_0.$$

From the last results and (15.17) we may obtain

$$\begin{aligned}\int_\Omega |\nabla \psi_n|_2^2\,dx &\leq \alpha K_1^4 + \beta K^2 + K\|f\|_2 + K\,C_1 \|\psi_n\|_{W^{1,2}(\Omega)} \\ &< \alpha K_1^4 + \beta K^2 + K\|f\|_2 + K\,C_1 \sqrt{\|\psi_n\|_2^2 + \|\nabla \psi_n\|_2^2} \\ &\leq \alpha K_1^4 + \beta K^2 + K\|f\|_2 + K\,C_1 \sqrt{K^2 + \|\nabla \psi_n\|_2^2}, \end{aligned}$$

so that there exists $K_2 > 0$ such that

$$\|\nabla \psi_n\|_2 < K_2, \forall n \in \mathbb{N}.$$

So, we may conclude that there exists $K_3 > 0$ such that

$$\|\psi_n\|_{W^{1,2}(\Omega)} < K_3, \forall n \in \mathbb{N}.$$

Therefore, from the Rellich–Kondrachov theorem, up to a subsequence not relabeled, we may infer that there exists $\hat{\psi} \in W^{1,2}(\Omega)$ such that

$$\nabla \psi_n \rightharpoonup \nabla \hat{\psi}, \text{ weakly in } L^2(\Omega)$$

and

$$\psi_n \to \hat{\psi}, \text{ strongly in } L^2(\Omega)$$

so that as

$$\psi_n \rightharpoonup \psi_0, \text{ weakly in } L^2(\Omega)$$

we can get

$$\hat{\psi} = \psi_0,$$

that is,

$$\nabla \psi_n \rightharpoonup \nabla \psi_0, \text{ weakly in } L^2(\Omega)$$

and

$$\psi_n \to \psi_0, \text{ strongly in } L^2(\Omega).$$

Choose $\phi \in C_c^\infty(\Omega)$. Clearly we have

$$\langle \psi_n, -\nabla^2 \phi \rangle_{L^2(\Omega)} \to \langle \psi_0, -\nabla^2 \phi \rangle_{L^2(\Omega)}, \tag{15.20}$$

and

$$\langle \psi_n, \phi \rangle_{L^2(\Omega)} \to \langle \psi_0, \phi \rangle_{L^2(\Omega)}, \tag{15.21}$$

and in the next lines, we will prove that

$$\langle \psi_n^3, \phi \rangle_{L^2(\Omega)} \to \langle \psi_0^3, \phi \rangle_{L^2(\Omega)}, \tag{15.22}$$

as $n \to \infty$. Observe that

$$\begin{aligned} \left| \int_\Omega (\psi_n^3 - \psi_0^3)\phi \, dx \right| &\le \left| \int_\Omega (\psi_n^3 - \psi_n^2 \psi_0 + \psi_n^2 \psi_0 - \psi_0^3)\phi \, dx \right| \\ &\le \int_\Omega \left| \psi_n^2 (\psi_n - \psi_0)\phi \right| \, dx \\ &\quad + \int_\Omega \left| \psi_0 (\psi_n^2 - \psi_0^2)\phi \right| \, dx. \end{aligned} \tag{15.23}$$

Also observe that

$$\begin{aligned}\int_\Omega \left|\psi_n^2(\psi_n-\psi_0)\phi\right|\,dx &\leq \|\psi_n\|_{L^4(\Omega)}^2\|(\psi_n-\psi_0)\phi\|_{L^2(\Omega)}\\ &\leq K_1^2\|\psi_n-\psi_0\|_{L^2(\Omega)}\|\phi\|_\infty\\ &\to 0,\text{ as } n\to\infty. \end{aligned}\tag{15.24}$$

On the other hand, from the generalized Hölder inequality, we get

$$\begin{aligned}\int_\Omega \left|\psi_0(\psi_n^2-\psi_0^2)\phi\right|\,dx &= \int_\Omega \left|\psi_0(\psi_n+\psi_0)(\psi-\psi_0)\phi\right|\,dx\\ &\leq \|\psi_0\|_4\|\psi_n+\psi_0\|_4\|(\psi_n-\psi_0)\phi\|_2\\ &\leq \|\psi_0\|_4(\|\psi_n\|_4+\|\psi_0\|_4)\|(\psi_n-\psi_0)\|_2\|\phi\|_\infty\\ &\leq K_1(K_1+K_1)\|\phi\|_\infty\|\psi_n-\psi_0\|_2\\ &\to 0,\text{ as } n\to\infty. \end{aligned}\tag{15.25}$$

Summarizing the last results we get

$$\langle\psi_n^3,\phi\rangle_{L^2(\Omega)}\to\langle\psi_0^3,\phi\rangle_{L^2(\Omega)},\tag{15.26}$$

as $n\to\infty$. Therefore

$$\begin{aligned}&\langle\psi_0,-\nabla^2\phi\rangle_{L^2(\Omega)}+\langle\psi_0^3-\psi_0-f,\phi\rangle_{L^2(\Omega)}\\ &=\lim_{n\to\infty}\langle\psi_n,-\nabla^2\phi\rangle_{L^2(\Omega)}+\langle\psi_n^3-\psi_n-f,\phi\rangle_{L^2(\Omega)}\\ &=\lim_{n\to\infty}0=0.\end{aligned}\tag{15.27}$$

Since $\phi\in C_c^\infty(\Omega)$ is arbitrary we get

$$-\nabla^2\psi_0+\alpha\psi_0^3-\beta\psi_0-f=0,\text{in }\Omega,$$

in the distributional sense. From the weak convergence and

$$\frac{\partial\psi_n}{\partial\mathbf{n}}=u_n,\text{ on }\partial\Omega,\ \forall n\in\mathbb{N},$$

we may also obtain

$$\frac{\partial\psi_0}{\partial\mathbf{n}}=u_0,\text{ on }\partial\Omega.$$

Finally, from the convexity of the functional in question,

$$\liminf_{n\to\infty}J(\psi_n,u_n)\geq J(\psi_0,u_0).$$

Therefore (ψ_0,u_0) is a solution of Problem $\mathscr{P}$.

15.4 The Generalized Method of Lines

In this section we prepare a route to obtain numerical results. We reintroduce the generalized method of lines, originally presented in Botelho [13]. In the present context we add new theoretical and applied results to the original presentation. Specially the computations are almost all completely new. Consider first the equation

$$\nabla^2 u = 0, \text{ in } \Omega \subset \mathbb{R}^2, \tag{15.28}$$

with the boundary conditions

$$u = 0 \text{ on } \Gamma_0 \text{ and } u = u_f, \text{ on } \Gamma_1.$$

From now on we assume that u_f is a smooth function, unless otherwise specified. Here Γ_0 denotes the internal boundary of Ω and Γ_1 the external one. Consider the simpler case where

$$\Gamma_1 = 2\Gamma_0,$$

and suppose there exists $r(\theta)$, a smooth function such that

$$\Gamma_0 = \{(\theta, r(\theta)) \mid 0 \leq \theta \leq 2\pi\},$$

being $r(0) = r(2\pi)$.

Also assume $(0,0) \notin \Omega$ and

$$\min_{\theta \in [0,2\pi]} \{r(\theta)\} \gtrapprox \mathcal{O}(1).$$

We emphasize this is a crucial assumption for the application of the contraction mapping theorem, which is the base of this method.

In polar coordinates the above equation may be written as

$$\frac{\partial^2 u}{\partial r^2} + \frac{1}{r}\frac{\partial u}{\partial r} + \frac{1}{r^2}\frac{\partial^2 u}{\partial \theta^2} = 0, \text{ in } \Omega, \tag{15.29}$$

and

$$u = 0 \text{ on } \Gamma_0 \text{ and } u = u_f, \text{ on } \Gamma_1.$$

Define the variable t by

$$t = \frac{r}{r(\theta)}.$$

Also defining $\bar{u}$ by

$$u(r,\theta) = \bar{u}(t,\theta),$$

dropping the bar in $\bar{u}$, (15.28) is equivalent to

$$\begin{aligned} &\frac{\partial^2 u}{\partial t^2} + \frac{1}{t} f_2(\theta)\frac{\partial u}{\partial t} \\ &\quad + \frac{1}{t} f_3(\theta)\frac{\partial^2 u}{\partial \theta \partial t} + \frac{f_4(\theta)}{t^2}\frac{\partial^2 u}{\partial \theta^2} = 0, \end{aligned} \tag{15.30}$$

in Ω. Here $f_2(\theta)$, $f_3(\theta)$, and $f_4(\theta)$ are known functions.

More specifically, denoting

$$f_1(\theta)=\frac{-r'(\theta)}{r(\theta)},$$

we have

$$f_2(\theta)=1+\frac{f_1'(\theta)}{1+f_1(\theta)^2},$$

$$f_3(\theta)=\frac{2f_1(\theta)}{1+f_1(\theta)^2},$$

and

$$f_4(\theta)=\frac{1}{1+f_1(\theta)^2}.$$

Observe that $t \in [1,2]$ in Ω. Discretizing in t (N equal pieces which will generate N lines) we obtain the equation

$$\begin{aligned}&\frac{u_{n+1}-2u_n+u_{n-1}}{d^2}+\frac{(u_n-u_{n-1})}{d}\frac{1}{t_n}f_2(\theta)\\&+\frac{\partial(u_n-u_{n-1})}{\partial\theta}\frac{1}{t_nd}f_3(\theta)+\frac{\partial^2 u_n}{\partial\theta^2}\frac{f_4(\theta)}{t_n^2}=0,\end{aligned} \tag{15.31}$$

$\forall n \in \{1,\ldots,N-1\}$. Here, $u_n(\theta)$ corresponds to the solution on the line n. Thus we may write

$$u_n=T(u_{n-1},u_n,u_{n+1}),$$

where

$$\begin{aligned}T(u_{n-1},u_n,u_{n+1})=&\frac{u_{n+1}+u_{n-1}}{2}+\frac{d^2}{2}\left(\frac{(u_n-u_{n-1})}{d}\frac{1}{t_n}f_2(\theta)\right.\\&\left.+\frac{\partial(u_n-u_{n-1})}{\partial\theta}\frac{1}{t_nd}f_3(\theta)+\frac{\partial^2 u_n}{\partial\theta^2}\frac{f_4(\theta)}{t_n^2}\right).\end{aligned} \tag{15.32}$$

Now we recall a classical definition.

Definition 15.4.1. Let C be a subset of a Banach space U and let $T : C \to C$ be an operator. Thus T is said to be a contraction mapping if there exists $0 \leq \alpha < 1$ such that

$$\|T(x_1)-T(x_2)\|_U \leq \alpha\|x_1-x_2\|_U, \forall x_1,x_2 \in C.$$

Remark 15.4.2. Observe that if $\|T'(x)\|_U \leq \alpha < 1$ on a convex set C, then T is a contraction mapping, since by the mean value inequality,

$$\|T(x_1)-T(x_2)\|_U \leq \sup_{x\in C}\{\|T'(x)\|\}\|x_1-x_2\|_U, \forall x_1,x_2 \in C.$$

The next result is the base of our generalized method of lines. For a proof see [47].

Theorem 15.4.3 (Contraction Mapping Theorem). *Let C be a closed subset of a Banach space U. Assume T is contraction mapping on C, then there exists a unique $\tilde{x} \in C$ such that $\tilde{x} = T(\tilde{x})$. Moreover, for an arbitrary $x_0 \in C$ defining the sequence*

$$x_1 = T(x_0) \text{ and } x_{k+1} = T(x_k), \forall k \in \mathbb{N}$$

we have

$$x_k \to \tilde{x}, \text{in norm, as } k \to +\infty.$$

From (15.32), if $d = 1/N$ is small enough and if $u_{n-1} \approx u_n$, it is clear that for a fixed u_{n+1}, $G(u_n) = T(u_{n-1}, u_n, u_{n+1})$ is a contraction mapping, considering that d may be chosen so that $\|G'(u_n)\| \leq \alpha < 1$, for some $0 < \alpha < 1$ in a set that contains the solution of the equation in question.

In particular for $n = 1$ we have

$$u_1 = T(0, u_1, u_2).$$

We may use the contraction mapping theorem to calculate u_1 as a function of u_2. The procedure would be

1. set $x_0 = u_2$,
2. obtain $x_1 = T(0, x_0, u_2)$,
3. obtain recursively
$$x_{k+1} = T(0, x_k, u_2), \text{and}$$
4. finally get
$$u_1 = \lim_{k \to \infty} x_k = g_1(u_2).$$

We have obtained thus

$$u_1 = g_1(u_2).$$

We can repeat the process for $n = 2$, that is, we can solve the equation

$$u_2 = T(u_1, u_2, u_3),$$

which from the above stands for

$$u_2 = T(g_1(u_2), u_2, u_3).$$

The procedure would be :

1. set $x_0 = u_3$,
2. calculate
$$x_{k+1} = T(g_1(x_k), x_k, u_3),$$
3. obtain
$$u_2 = \lim_{k \to \infty} x_k = g_2(u_3).$$

We proceed in this fashion until obtaining

$$u_{N-1} = g_{N-1}(u_N) = g_{N-1}(u_f).$$

u_f being known, we have obtained u_{N-1}. We may then calculate

$$u_{N-2} = g_{N-2}(u_{N-1}),$$

$$u_{N-3} = g_{N-3}(u_{N-2}),$$

and so on, up to finding

$$u_1 = g_1(u_2).$$

Thus the problem is solved.

Remark 15.4.4. Here we consider some points concerning the convergence of the method.

In the next lines the norm indicated as in $\|x_k\|$ refers to $W^{2,2}([0,2\pi])$. In particular for $n=1$ from the above we have

$$u_1 = T(0,u_1,u_2).$$

We will construct the sequence x_k (in a little different way as above) by defining

$$x_1 = u_2/2,$$

and

$$x_{k+1} = T(0,x_k,u_2) = u_2/2 + d\tilde{T}(x_k),$$

where the operator $\tilde{T}$ is properly defined from the expression of T. Observe that

$$\|x_{k+2} - x_{k+1}\| \leq d\|\tilde{T}\|\,\|x_{k+1} - x_k\|,$$

and if

$$0 \leq \alpha = d\|\tilde{T}\| < 1,$$

we have that $\{x_k\}$ is (Cauchy) convergent. Through a standard procedure for this kind of sequence, we may obtain

$$\|x_{k+1} - x_1\| \leq \frac{1}{1-\alpha}\|x_2 - x_1\|,$$

so that denoting $u_1 = \lim_{k\to\infty} x_k$, we get

$$\|u_1 - u_2/2\| \leq \frac{1}{1-\alpha} d\|\tilde{T}\|\|u_2/2\|,$$

Having such an estimate, we may similarly obtain

$$u_2 \approx u_3 + \mathcal{O}(d),$$

and generically

$$u_n \approx u_{n+1} + \mathcal{O}(d), \forall n \in \{1,\ldots,N-1\}.$$

This last calculation is just to clarify that the procedure of obtaining the relation between consecutive lines through the contraction mapping theorem is well defined.

15.4.1 About the Approximation Error

Consider again the equation in finite differences for the example in question:

$$\frac{u_{n+1} - 2u_n + u_{n-1}}{d^2} + \frac{(u_n - u_{n-1})}{d}\frac{1}{t}f_2(\theta) + \frac{\partial(u_n - u_{n-1})}{\partial\theta}\frac{1}{td}f_3(\theta) + \frac{\partial^2 u_n}{\partial\theta^2}\frac{f_4(\theta)}{t^2} = 0, \tag{15.33}$$

$\forall n \in \{1,\ldots,N-1\}$. Here, $u_n(\theta)$ corresponds to the solution on the line n. Thus, as above, we may write

$$u_n = T(u_{n-1}, u_n, u_{n+1}),$$

where

$$T(u_{n-1}, u_n, u_{n+1}) = \frac{u_{n+1} + u_{n-1}}{2} + \frac{d^2}{2}\left(\frac{(u_n - u_{n-1})}{d}\frac{1}{t}f_2(\theta) + \frac{\partial(u_n - u_{n-1})}{\partial\theta}\frac{1}{td}f_3(\theta) + \frac{\partial^2 u_n}{\partial\theta^2}\frac{f_4(\theta)}{t^2}\right). \tag{15.34}$$

For $n = 1$, we evaluate $u_1 = g_1(u_2)$ through the contraction mapping theorem obtaining

$$u_1(x) \approx 0.5u_2(x) + 0.25du_2(x)f_2(x) + 0.25df_3(x)u_2'(x) + 0.25d^2 f_4(x)u_2''(x). \tag{15.35}$$

We can also obtain $u_n(x) = \tilde{g}_n(u_{n+1}, u_{n-1})$, that is,

$$\begin{aligned} u_n(x) \approx\ & 0.5u_{n-1}(x) + 0.5u_{n+1}(x) - 0.25d\, u_{n-1}(x)f_2(x)/t_n \\ & + 0.25d\, u_{n+1}(x)f_2(x)/t_n - 0.25d\, f_3(x)u_{n-1}'(x)/t_n \\ & + 0.25d\, f_3(x)u_{n+1}'(x)/t_n + 0.25d^2 f_4(x)u_{n-1}''(x)/t_n^2 \\ & + 0.25d^2 f_4(x)u_{n+1}''(x)/t_n^2. \end{aligned} \tag{15.36}$$

The approximation error in (15.35) is of order $\mathscr{O}(d^3)$ plus the error concerning the application of the contraction mapping theorem, which is well known and, if d is small enough, may be made arbitrarily small in a reasonable number of iterations. Also we may infer that the approximation error in (15.36) is also of order $\mathscr{O}(d^3)$. The discretization error in this case is known to be of order $\mathscr{O}(d)$ (see [63] for details).

15.4.2 The Solution of Laplace Equation for a Special Class of Domains

As an example, we compute by the generalized method of lines the solution of the equation

$$\nabla^2 u = 0, \text{ in } \Omega \subset \mathbb{R}^2, \tag{15.37}$$

with the boundary conditions

$$u = u_0 \text{ on } \Gamma_0 \text{ and } u = u_f, \text{ on } \Gamma_1.$$

We assume u_0 and u_f are smooth functions. As above Γ_0 denotes the internal boundary of Ω and Γ_1 the external one. We consider the simpler case where

$$\Gamma_1 = 2\Gamma_0.$$

Suppose there exists $r(\theta)$, a smooth function such that

$$\Gamma_0 = \{(\theta, r(\theta)) \mid 0 \leq \theta \leq 2\pi\},$$

being $r(0) = r(2\pi)$.

Also assume $(0,0) \notin \Omega$ and

$$\min_{\theta \in [0,2\pi]} \{r(\theta)\} \gtrapprox \mathcal{O}(1).$$

Denoting $x = \theta$, particularly for $N = 10$, truncating the series up the terms in d^2, we obtain the following expression for the lines:

Line 1

$$\begin{aligned} u_1(x) = {} & 0.1u_f(x) + 0.9u_0(x) - 0.034u_0(x)f_2(x) + 0.034f_2(x)u_f(x) \\ & -0.034f_3(x)u_0'(x) + 0.034f_3(x)u_f'(x) \\ & +0.018f_4(x)u_0''(x) + 0.008f_4(x)u_f''(x) \end{aligned}$$

Line 2

$$\begin{aligned} u_2(x) = {} & 0.2u_f(x) + 0.8u_0(x) - 0.058u_0(x)f_2(x) + 0.058f_2(x)u_f(x) \\ & -0.058f_3(x)u_0'(x) + 0.058f_3(x)u_f'(x) \\ & +0.029f_4(x)u_0''(x) + 0.015f_4(x)u_f''(x) \end{aligned}$$

Line 3

$$\begin{aligned} u_3(x) = {} & 0.3u_f(x) + 0.7u_0(x) - 0.075u_0(x)f_2(x) + 0.075f_2(x)u_f(x) \\ & -0.075f_3(x)u_0'(x) + 0.075f_3(x)u_f'(x) \\ & +0.034f_4(x)u_0''(x) + 0.020f_4(x)u_f''(x) \end{aligned}$$

Line 4

$$\begin{aligned} u_4(x) &= 0.4u_f(x) + 0.6u_0(x) - 0.083u_0(x)f_2(x) + 0.083f_2(x)u_f(x) \\ &\quad -0.083f_3(x)u_0'(x) + 0.083f_3(x)u_f'(x) \\ &\quad +0.035f_4(x)u_0''(x) + 0.024f_4(x)u_f''(x) \end{aligned}$$

Line 5

$$\begin{aligned} u_5(x) &= 0.5u_f(x) + 0.5u_0(x) - 0.085u_0(x)f_2(x) + 0.085f_2(x)u_f(x) \\ &\quad -0.085f_3(x)u_0'(x) + 0.085f_3(x)u_f'(x) \\ &\quad +0.033f_4(x)u_0''(x) + 0.026f_4(x)u_f''(x) \end{aligned}$$

Line 6

$$\begin{aligned} u_6(x) &= 0.6u_f(x) + 0.4u_0(x) - 0.080u_0(x)f_2(x) + 0.080f_2(x)u_f(x) \\ &\quad -0.080f_3(x)u_0'(x) + 0.080f_3(x)u_f'(x) \\ &\quad +0.028f_4(x)u_0''(x) + 0.026f_4(x)u_f''(x) \end{aligned}$$

Line 7

$$\begin{aligned} u_7(x) &= 0.7u_f(x) + 0.3u_0(x) - 0.068u_0(x)f_2(x) + 0.068f_2(x)u_f(x) \\ &\quad -0.068f_3(x)u_0'(x) + 0.068f_3(x)u_f''(x) \\ &\quad +0.023f_4(x)u_0''(x) + 0.023f_4(x)u_f''(x) \end{aligned}$$

Line 8

$$\begin{aligned} u_8(x) &= 0.8u_f(x) + 0.2u_0(x) - 0.051u_0(x)f_2(x) + 0.051f_2(x)u_f(x) \\ &\quad -0.051f_3(x)u_0'(x) + 0.051f_3(x)u_f'(x) \\ &\quad +0.015f_4(x)u_0''(x) + 0.018f_4(x)u_f''(x) \end{aligned}$$

Line 9

$$\begin{aligned} u_9(x) &= 0.9u_f(x) + 0.1u_0(x) - 0.028u_0(x)f_2(x) + 0.028f_2(x)u_f(x) \\ &\quad -0.028f_3(x)u_0'(x) + 0.028f_3(x)u_f'(x) \\ &\quad +0.008f_4(x)u_0''(x) + 0.010f_4(x)u_f''(x) \end{aligned}$$

Remark 15.4.5. Here a word of caution is necessary. Consider for example the equation

$$\varepsilon \nabla^2 u + G(u) = 0, \text{ in } \Omega \subset \mathbb{R}^2, \tag{15.38}$$

with the boundary conditions

$$u = u_0 \text{ on } \Gamma_0 \text{ and } u = u_f, \text{ on } \Gamma_1.$$

We assume G, u_0, and u_f are smooth functions.

If ε is too small, for example, about 0.001 or 0.0001, the error just truncating the series up the order d^2 is big. It seems that higher-order approximations or even discretizing more does not solve the problem. However, for example, for $G(u) = u$, by solving the equation with $\varepsilon = 1$, we can infer that the solution at each line has the general format

$$\begin{aligned} u_n(x) \approx\ & a_1[n]u_f(x) + a_2[n]u_0(x) + a_3[n]u_0(x)f_2(x) + a_4[n]f_2(x)u_f(x) \\ & a_5[n]f_3(x)u_0'(x) + a_6[n]f_3(x)u_f'(x) \\ & + a_7[n]f_4(x)u_0''(x) + a_8[n]f_4(x)u_f''(x) \\ & + a_9[n]f_5(x)u_0(x) + a_{10}[n]f_5(x)u_f(x), \end{aligned}$$

where $f_5(x) = r^2(x)f_4(x)$.

This expression we get from the series that would represent the exact solution obtained through an application of contraction mapping theorem for the concerned inversions (which the first terms are qualitatively known up to the exact coefficient values).

Thus, we just have to calculate the optimal real coefficients $\{a_k[n]\}$ which minimize the error concerning the original differential equation. Here derivatives must be understood as matrices acting on vectors. A similar remark is valid as

$$\max_{\theta \in [0,2\pi]} \{r(\theta)\} \gg \mathcal{O}(2).$$

So, to summarize, we emphasize that through the problem solution with $\varepsilon = 1$ we may discover its general format for smaller values of ε, up to constants which may be easily evaluated (e.g., by the error minimization). Such a procedure has worked very well in all examples we have so far developed. Of course, for this specific example, other procedures are possible.

15.5 A First Numerical Example

Just to illustrate the possibilities of the generalized method of lines, we apply it to the equation

$$\nabla^2 u = \nabla^2 \bar{u}, \text{ in } \Omega,$$

where

$$\Omega = \{(r,\theta) \mid 1 \le r \le 2,\ 0 \le \theta \le 2\pi\},$$

$$u = \bar{u} \text{ on } \Gamma_0 \text{ and } \Gamma_1,$$

where Γ_0 and Γ_1 are boundaries of the circles with centers at the origin and radius 1 and 2, respectively. Finally, in polar coordinates (here x stands for θ),

$$\bar{u} = r^2 \cos(x).$$

See below the approximate values for the 9 lines ($N = 10$) obtained by the generalized method of lines ($u_n(x)$) and the exact values ($\bar{u}_n(x)$ for the same lines):

Line 1

$$u_1(x) = 1.21683\cos(x), \quad \bar{u}_1(x) = 1.21\cos(x)$$

Line 2

$$u_2(x) = 1.44713\cos(x), \quad \bar{u}_2(x) = 1.44\cos(x)$$

Line 3

$$u_3(x) = 1.69354\cos(x), \quad \bar{u}_3(x) = 1.69\cos(x)$$

Line 4

$$u_4(x) = 1.95811\cos(x) \quad \bar{u}_4(x) = 1.96\cos(x)$$

Line 5

$$u_5(x) = 2.24248\cos(x), \quad \bar{u}_5(x) = 2.25\cos(x)$$

Line 6

$$u_6(x) = 2.54796\cos(x), \quad \bar{u}_6(x) = 2.56\cos(x)$$

Line 7

$$u_7(x) = 2.87563\cos(x), \quad \bar{u}_7(x) = 2.89\cos(x)$$

Line 8

$$u_8(x) = 3.22638\cos(x), \quad \bar{u}_8(x) = 3.24\cos(x)$$

Line 9

$$u_9(x) = 3.60096\cos(x), \quad \bar{u}_9(x) = 3.61\cos(x)$$

15.6 A Numerical Example Concerning the Optimal Control Problem

We compute the solution of problem $\mathscr{P}$, that is, the problem of minimizing $J : U \to \mathbb{R}$, which similarly as above stated, is given by

$$J(\psi, u) = \frac{1}{2}\int_{\Omega} |\psi - \psi_d|^2 \, dx + \frac{1}{2}\int_{\Gamma_1} |u|^2 \, d\Gamma$$

subject to

$$\begin{cases} -\nabla^2 \psi + \alpha \psi^3 - \beta \psi = 0, & \text{in } \Omega \\ \psi = 0, & \text{on } \Gamma_0, \\ \frac{\partial \psi}{\partial \mathbf{n}} = u, & \text{on } \Gamma_1. \end{cases} \tag{15.39}$$

Also $U = W^{1,2}(\Omega) \times L^2(\Gamma_1)$. In this example we consider in polar coordinates

$$\Omega = \{(r, \theta) \mid 1 \leq r \leq 2,\ 0 \leq \theta \leq 2\pi\},$$

$$\Gamma_0 = \{(1, \theta) \mid 0 \leq \theta \leq 2\pi\},$$
$$\Gamma_1 = \{(2, \theta) \mid 0 \leq \theta \leq 2\pi\},$$

and

$$\psi_d(r, \theta) = (r-1)^2 \sin\theta.$$

We discretize the domain in lines (in fact curves). We divide the interval $[1,2]$ into 10 pieces (corresponding to the discretization in r) obtaining the following system of equations:

$$\frac{\psi_{n+1} - 2\psi_n + \psi_{n-1}}{d^2} + \frac{1}{r_n} \frac{\psi_n - \psi_{n-1}}{d} + \frac{1}{r_n^2} \frac{\partial^2 \psi_n}{\partial \theta^2}$$
$$- \alpha \psi_n^3 + \beta \psi_n = 0, \forall n \in \{1, \ldots, N-1\}, \tag{15.40}$$

where $N = 10$, $d = 1/N$, and $r_n = 1 + nd$.

Thus ψ_n corresponds to the solution on the line n.

For $\alpha = \beta = 1$ our procedure was first to compute ψ through the generalized method of lines as a function of its value on the boundary, which we have denoted by $u_f(x)$ (where as above x stands for θ), obtaining the following approximate expressions for the lines (we have truncated the series up the terms in d^2):

Line 1

$\psi_1(x) = 0.150254u_f(x) - 0.004917u_f(x)^3 + 0.0078404u_f''(x)$

Line 2

$\psi_2(x) = 0.290418u_f(x) - 0.009824u_f(x)^3 + 0.0148543u_f''(x)$

Line 3

$\psi_3(x) = 0.420248u_f(x) - 0.014651u_f(x)^3 + 0.0204794u_f''(x)$

Line 4

$\psi_4(x) = 0.539385u_f(x) - 0.019208u_f(x)^3 + 0.0243293u_f''(x)$

Line 5

$\psi_5(x) = 0.64738u_f(x) - 0.023125u_f(x)^3 + 0.0261384u_f''(x)$

Line 6

$$\psi_6(x) = 0.743709u_f(x) - 0.025792u_f(x)^3 + 0.0257253u_f''(x)$$

Line 7

$$\psi_7(x) = 0.827787u_f(x) - 0.026299u_f(x)^3 + 0.0229684u_f''(x)$$

Line 8

$$\psi_8(x) = 0.898983u_f(x) - 0.023376u_f(x)^3 + 0.0177894u_f''(x)$$

Line 9

$$\psi_9(x) = 0.956623u_f(x) - 0.015333u_f(x)^3 + 0.0101412u_f''(x)$$

The second step is to replace the field ψ obtained in J and then to compute through a numerical minimization of J the optimal u_f. For the candidate to optimal $u_f(x)$ see Fig. 15.1. Finally, we have computed a critical point, but we cannot guarantee it is the global optimal solution.

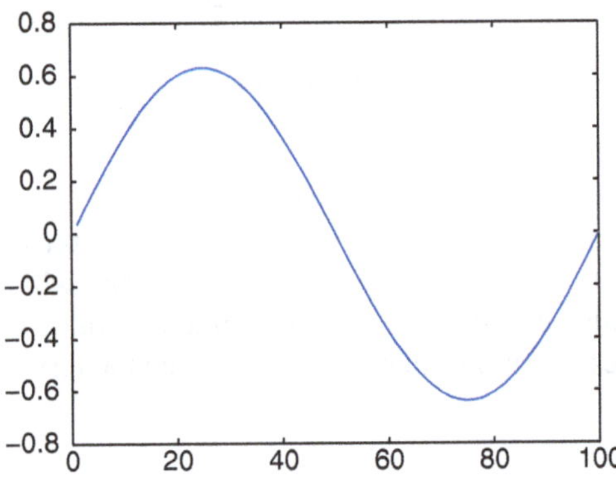

Fig. 15.1 The optimal (candidate) $u_f(x)$-units in x: $2\pi/100$

15.7 Conclusion

In this chapter, we have presented global existence results concerning the Ginzburg–Landau system in the presence of a magnetic field. In the second step we prove the existence of solution for a closely related optimal control problem, introducing the generalized method of lines (or briefly the GMOL) as an efficient tool for computing its solution. It seems to be clear that the generalized method of lines may used for solving a large class of nonlinear problems, specially when we apply its matrix version. It is our objective in the future to develop applications of GMOL to three-dimensional and time-dependent problems.

Chapter 16
More on Duality and Computation for the Ginzburg–Landau System

16.1 Introduction

We recall again here (more details may be found in the introduction at Chap. 15) that close to a critical temperature T_c, the Ginzburg–Landau energy would be expressed by

$$F_s(T) = F_n(T) + \frac{\hbar}{4m}\int_\Omega |\nabla\psi|_2^2\,dx + \frac{\alpha(T)}{4}\int_\Omega |\psi|^4\,dx - \frac{\beta(T)}{2}\int_\Omega |\psi|^2\,dx,$$

where ψ is a complex parameter and $F_n(T)$ and $F_s(T)$ are the normal and superconducting free-energy densities, respectively (see [4, 9, 45, 46] for details). Here $\Omega \subset \mathbb{R}^3$ denotes the superconducting sample with a boundary denoted by $\partial\Omega = \Gamma$. The complex function $\psi \in W^{1,2}(\Omega;\mathbb{C})$ is intended to minimize $F_s(T)$ for a fixed temperature T.

Denoting $\alpha(T)$ and $\beta(T)$ simply by α and β, the corresponding Euler–Lagrange equations are given by

$$\begin{cases} -\frac{\hbar}{2m}\nabla^2\psi + \alpha|\psi|^2\psi - \beta\psi = 0, & \text{in } \Omega \\ \frac{\partial\psi}{\partial\mathbf{n}} = 0, & \text{on } \partial\Omega. \end{cases} \tag{16.1}$$

This last system of equations is well known as the Ginzburg–Landau (G-L) one. In the physics literature, it is also well known the G-L energy in which a magnetic potential here denoted by $\mathbf{A}$ is included. The functional in question is given by

$$\begin{aligned} J(\psi,\mathbf{A}) = {} & \frac{1}{8\pi}\int_{\mathbb{R}^3} |\operatorname{curl}\mathbf{A} - \mathbf{B}_0|_2^2\,dx + \frac{\hbar^2}{4m}\int_\Omega \left|\nabla\psi - \frac{2ie}{\hbar c}\mathbf{A}\psi\right|_2^2 dx \\ & + \frac{\alpha}{4}\int_\Omega |\psi|^4\,dx - \frac{\beta}{2}\int_\Omega |\psi|^2\,dx \end{aligned} \tag{16.2}$$

F. Botelho, *Functional Analysis and Applied Optimization in Banach Spaces: Applications to Non-Convex Variational Models*, DOI 10.1007/978-3-319-06074-3_16,

Considering its minimization on the space U, where

$$U = W^{1,2}(\Omega;\mathbb{C}) \times W^{1,2}(\mathbb{R}^3;\mathbb{R}^3),$$

through the physics notation, the corresponding Euler–Lagrange equations are

$$\begin{cases} \frac{1}{2m}\left(-i\hbar\nabla - \frac{2e}{c}\mathbf{A}\right)^2 \psi + \alpha|\psi|^2\psi - \beta\psi = 0, & \text{in } \Omega \\ \left(i\hbar\nabla\psi + \frac{2e}{c}\mathbf{A}\psi\right)\cdot \mathbf{n} = 0, & \text{on } \partial\Omega, \end{cases} \tag{16.3}$$

and

$$\begin{cases} \text{curl (curl } \mathbf{A}) = \text{ curl } \mathbf{B}_0 + \frac{4\pi}{c}\tilde{J}, & \text{in } \Omega \\ \text{curl (curl } \mathbf{A}) = \text{ curl } \mathbf{B}_0, & \text{in } \mathbb{R}^3 \setminus \overline{\Omega}, \end{cases} \tag{16.4}$$

where

$$\tilde{J} = -\frac{ie\hbar}{2m}(\psi^*\nabla\psi - \psi\nabla\psi^*) - \frac{2e^2}{mc}|\psi|^2\mathbf{A}.$$

and

$$\mathbf{B}_0 \in L^2(\mathbb{R}^3;\mathbb{R}^3)$$

is a known applied magnetic field.

16.2 The Duality Principle

In this section we develop a duality principle for the Ginzburg–Landau system in the presence of a magnetic field. Such a result includes sufficient conditions of global optimality and is summarized by the next theorem.

Theorem 16.2.1. *Let $\Omega \subset \mathbb{R}^3$ be an open, bounded, connected set with a regular boundary denoted by $\partial\Omega = \Gamma$. Consider the functional $J : U \to \mathbb{R}$ given by*

$$\begin{aligned} J(\psi,\boldsymbol{A}) = & \frac{1}{8\pi}\int_{\mathbb{R}^3} |\, curl\, \boldsymbol{A} - \boldsymbol{B}_0|_2^2\, dx + \frac{\hbar^2}{4m}\int_\Omega \left|\nabla\psi - \frac{2ie}{\hbar c}\boldsymbol{A}\psi\right|_2^2 dx \\ & + \frac{\alpha}{4}\int_\Omega |\psi|^4\, dx - \frac{\beta}{2}\int_\Omega |\psi|^2\, dx \end{aligned} \tag{16.5}$$

where $\hbar, m, c, e, \alpha, \beta$ are positive constants, i is the imaginary unit, and

$$U = W^{1,2}(\Omega;\mathbb{C}) \times W_0^{1,2}(\mathbb{R}^3;\mathbb{R}^3).$$

Under such a hypothesis, we have

$$\inf_{(\psi,\boldsymbol{A})\in U}\{J(\psi,\boldsymbol{A})\} \geq \sup_{v^*\in A^*}\{-J^*(v^*)\},$$

where $v^* = (v_1^*, v_2^*, \sigma)$ *and*

$$J^*(v^*) = G_0^*(v_2^*) + G_1^*(v_1^*) + G_2^*(\sigma) + G_3^*(v_1^*, v_2^*, \sigma). \tag{16.6}$$

Furthermore,

$$G_1(v_1) = \frac{\hbar^2}{4m} \int_\Omega |v_1|_2^2 \, dx,$$

$$G_2(v_3) = \frac{\alpha}{4} \int_\Omega (v_3)^2 \, dx - \frac{\beta}{2} \int_\Omega v_3 \, dx$$

and

$$G_3(\psi, \boldsymbol{A}) = -\langle \psi, div(v_1^*) \rangle_{L^2(\Omega;\mathbb{C})} - \left\langle \frac{2ie}{\hbar c} \boldsymbol{A} \psi, v_1^* \right\rangle_{L^2(\Omega;\mathbb{C}^3)}. \tag{16.7}$$

Also,

$$G_0^*(v_2^*) = \sup_{(\psi,\boldsymbol{A}) \in U} \left\{ \langle \boldsymbol{A}_k^2, v_{2k}^* \rangle_{L^2(\Omega)} - \frac{1}{8\pi} \int_{\mathbb{R}^3} |curl\, \boldsymbol{A} - \boldsymbol{B}_0|_2^2 \, dx \right\},$$

$$\begin{aligned} G_1^*(v_1^*) &= \sup_{v_1 \in L^2(\Omega;\mathbb{C}^3)} \{ \langle v_1, v_1^* \rangle_{L^2(\Omega;\mathbb{C}^3)} - G_1(v_1) \} \\ &= \frac{m}{\hbar^2} \langle v_1^*, v_1^* \rangle_{L^2(\Omega;\mathbb{C}^3)}. \end{aligned} \tag{16.8}$$

Despite that we are dealing with complex and real variables, we highlight the functionals in question are real, so that we denote, for $u, v \in L^2(\Omega; \mathbb{C})$,

$$\langle u, v \rangle_{L^2(\Omega;\mathbb{C})} = \int_\Omega u_1 v_1 \, dx + \int_\Omega u_2 v_2 \, dx,$$

where u_1, v_1 *are the real parts and* u_2, v_2 *are the imaginary ones of* u, v, *respectively. A similar remark is valid for* $L^2(\Omega; \mathbb{C}^3)$.

Moreover,

$$\begin{aligned} G_2^*(\sigma) &= \sup_{v_3 \in L^2(\Omega)} \{ \langle v_3, \sigma \rangle_{L^2} - G_2(v_3) \} \\ &= \frac{1}{\alpha} \int_\Omega (\sigma + \beta/2)^2 \, dx \end{aligned} \tag{16.9}$$

and

$$\begin{aligned} G_3^*(v_1^*, v_2^*, \sigma) &= \sup_{(\psi,\boldsymbol{A}) \in U} \{ -\langle |\psi|^2, \sigma \rangle_{L^2(\Omega)} - \langle \boldsymbol{A}_k^2, v_{2k}^* \rangle_{L^2(\Omega)} - G_3(\psi, \boldsymbol{A}) \} \\ &= \sup_{(\psi,\boldsymbol{A}) \in U} \Big\{ -\langle |\psi|^2, \sigma \rangle_{L^2(\Omega)} - \langle \boldsymbol{A}_k^2, v_{2k}^* \rangle_{L^2(\Omega)} \\ &\quad + \langle \psi, div(v_1^*) \rangle_{L^2(\Omega,\mathbb{C})} + \left\langle \frac{2ie}{\hbar c} \boldsymbol{A} \psi, v_1^* \right\rangle_{L^2(\Omega;\mathbb{C}^3)} \Big\} \end{aligned}$$

so that if $v^* \in A^*$, *we have*

$$\begin{aligned} G_3^*(v_1^*, v_2^*, \sigma) &= -\langle |\tilde{\psi}|^2, \sigma \rangle_{L^2(\Omega)} \\ &\quad -\langle \tilde{A}_k^2, v_{2k}^* \rangle_{L^2(\Omega)} - G_3(\tilde{\psi}, \tilde{A}), \end{aligned} \tag{16.10}$$

where $(\tilde{\psi}, \tilde{A}) \in U$ *is the only critical point of the quadratic functional indicated in (16.10).*

Here,

$$Y^* = W^{1,2}(\Omega; \mathbb{C}^3) \times L^2(\Omega; \mathbb{R}^3) \times L^2(\Omega),$$

and defining

$$\tilde{G}(\psi, A) = \langle |\psi|^2, \sigma \rangle_{L^2(\Omega)} + \langle A_k^2, v_{2k}^* \rangle_{L^2(\Omega)} - \left\langle \frac{2ie}{\hbar c} A\psi, v_1^* \right\rangle_{L^2(\Omega;\mathbb{C}^3)},$$

we also define

$$A^* = A_1 \cap A_2,$$

$$\begin{aligned} A_1 = \{ v^* \in Y^* \mid \tilde{G}(\psi, A) > 0 \, \forall (\psi, A) \in U \text{ such that} \\ (\psi, A) \neq (\theta, \theta), \text{ and } v_1^* \cdot \boldsymbol{n} = 0 \text{ on } \partial\Omega \}. \end{aligned}$$

Furthermore,

$$A_2 = \{ v^* \in Y^* \mid \tilde{J}(A) > 0, \forall A \in W_0^{1,2}(\mathbb{R}^3; \mathbb{R}^3) \text{ such that } A \neq \theta \},$$

and

$$\tilde{J}(A) = \frac{1}{8\pi} \int_{\mathbb{R}^3} |curl\, A|_2^2 \, dx - \langle A_k^2, v_{2k}^* \rangle_{L^2(\Omega)}.$$

Finally, define $Ind_0 : U \to \mathbb{R} \cup \{+\infty\}$ *by*

$$Ind_0(\psi, A) = \begin{cases} 0, & \text{if } \left(i\hbar\nabla\psi + \frac{2e}{c} A\psi\right) \cdot \boldsymbol{n} = 0, \text{ on } \partial\Omega, \\ +\infty, & \text{otherwise.} \end{cases} \tag{16.11}$$

Assume $(\psi_0, A_0) \in U$ *is such that* $\delta J(\psi_0, A_0) = \theta$ *and* $Ind_0(\psi_0, A_0) = 0$ *and also such that*

$$v_0^* = (v_{01}^*, v_{02}^*, \sigma_0) \in A^*,$$

where such a point is the solution of the following relations:

$$\begin{aligned} v_{01}^* &= \frac{\partial G_1(\nabla\psi_0 - \frac{2ie}{\hbar c} A_0 \psi_0)}{\partial v_1} \\ &= \frac{\hbar^2}{2m} \left(\nabla\psi_0 - \frac{2ie}{\hbar c} A_0 \psi_0 \right), \text{ in } \Omega, \end{aligned} \tag{16.12}$$

$$-\sum_{k=1}^{3} 2(v_{02}^*)_k A_{0k} \boldsymbol{e}_k + \frac{curl\, curl\, A_0 - curl\, \boldsymbol{B}_0}{4\pi} = \theta \text{ in } \Omega, \tag{16.13}$$

where $\{\mathbf{e}_1, \mathbf{e}_2, \mathbf{e}_3\}$ *denotes the canonical basis of* $\mathbb{R}^3$ *and*

$$\sigma_0 = \frac{\partial G_2(|\psi_0|^2)}{\partial v_3} = \frac{\alpha}{2}|\psi_0|^2 - \frac{\beta}{2}, \text{ in } \Omega. \tag{16.14}$$

Under such hypotheses, we have

$$J(\psi_0, \mathbf{A}_0) = \min_{(\psi,\mathbf{A})\in U} J(\psi,\mathbf{A}) = \max_{v^*\in A^*}\{-J^*(v^*)\} = -J^*(v_0^*).$$

Proof. Observe that

$$\begin{aligned} J(\psi,\mathbf{A}) &= \frac{1}{8\pi}\int_{\mathbb{R}^3} |\text{curl}\,\mathbf{A} - \mathbf{B}_0|_2^2\,dx \\ &+ \frac{\hbar^2}{4m}\int_\Omega \left|\nabla\psi - \frac{2ie}{\hbar c}\mathbf{A}\psi\right|_2^2 dx \\ &+ \frac{\alpha}{4}\int_\Omega |\psi|^4\,dx - \frac{\beta}{2}\int_\Omega |\psi|^2\,dx \\ &+ \left\langle \nabla\psi - \frac{2ie}{\hbar c}\mathbf{A}\psi, v_1^*\right\rangle_{L^2(\Omega;\mathbb{C}^3)} - \left\langle \nabla\psi - \frac{2ie}{\hbar c}\mathbf{A}\psi, v_1^*\right\rangle_{L^2(\Omega;\mathbb{C}^3)} \\ &+ \langle \mathbf{A}_k^2, v_{2k}^*\rangle_{L^2(\Omega)} - \langle \mathbf{A}_k^2, v_{2k}^*\rangle_{L^2(\Omega)} \\ &+ \langle |\psi|^2, \sigma\rangle_{L^2(\Omega)} - \langle |\psi|^2, \sigma\rangle_{L^2(\Omega)}, \end{aligned} \tag{16.15}$$

$\forall(\psi,\mathbf{A}) \in U, v^* \in A^*$.

Hence,

$$\begin{aligned} J(\psi,\mathbf{A}) &= -\langle \mathbf{A}_k^2, v_{2k}^*\rangle_{L^2(\Omega)} + \frac{1}{8\pi}\int_{\mathbb{R}^3} |\text{ curl}\,\mathbf{A} - \mathbf{B}_0|_2^2\,dx \\ &- \left\langle \nabla\psi - \frac{2ie}{\hbar c}\mathbf{A}\psi, v_1^*\right\rangle_{L^2(\Omega;\mathbb{C}^3)} + \frac{\hbar^2}{4m}\int_\Omega \left|\nabla\psi - \frac{2ie}{\hbar c}\mathbf{A}\psi\right|_2^2 dx \\ &- \langle |\psi|^2, \sigma\rangle_{L^2(\Omega)} + \frac{\alpha}{4}\int_\Omega |\psi|^4\,dx - \frac{\beta}{2}\int_\Omega |\psi|^2\,dx \\ &+ \langle \mathbf{A}_i^2, v_{2i}^*\rangle_{L^2(\Omega)} + \langle |\psi|^2, \sigma\rangle_{L^2(\Omega)} + \left\langle \nabla\psi - \frac{2ie}{\hbar c}\mathbf{A}\psi, v_1^*\right\rangle_{L^2(\Omega;\mathbb{C}^3)} \end{aligned} \tag{16.16}$$

$\forall(\psi,\mathbf{A}) \in U, v^* \in A^*$, so that

$$\begin{aligned} J(\psi,\mathbf{A}) \geq &\inf_{\mathbf{A}\in U}\left\{-\langle \mathbf{A}_k^2, v_{2k}^*\rangle_{L^2(\Omega)} + \frac{1}{8\pi}\int_{\mathbb{R}^3} |\text{ curl}\,\mathbf{A} - \mathbf{B}_0|_2^2\,dx\right\} \\ &+ \inf_{v_1\in L^2(\Omega;\mathbb{C}^3)}\left\{-\langle v_1, v_1^*\rangle_{L^2(\Omega;\mathbb{C}^3)} + \frac{\hbar^2}{4m}\int_\Omega |v_1|_2^2\,dx\right\} \end{aligned}$$

$$+\inf_{v_3\in L^2(\Omega)}\left\{-\langle v_3,\sigma\rangle_{L^2(\Omega)}+\frac{\alpha}{4}\int_\Omega (v_3)^2\,dx-\frac{\beta}{2}\int_\Omega v_3\,dx\right\}$$
$$+\inf_{(\psi,\mathbf{A})\in U}\left\{-\langle \psi, div(v_1^*)\rangle_{L^2(\Omega;\mathbb{C})}-\left\langle \frac{2ie}{\hbar c}\mathbf{A}\psi, v_1^*\right\rangle_{L^2(\Omega;\mathbb{C}^3)}\right.$$
$$\left.+\langle|\psi|^2,\sigma\rangle_{L^2(\Omega)}+\langle \mathbf{A}_k^2, v_{2k}^*\rangle_{L^2(\Omega)}\right\},$$

$\forall(\psi,\mathbf{A})\in U,\ v^*\in A^*$.

Therefore,

$$\begin{aligned}J(\psi,\mathbf{A}) &\geq -G_0^*(v_2^*)-G_1^*(v_1^*)-G_2^*(\sigma)-G_3^*(v_1^*,v_2^*,\sigma)\\ &= -J^*(v^*),\end{aligned} \tag{16.17}$$

$\forall(\psi,\mathbf{A})\in U,\ v^*\in A^*$.

Now observe that since

$$\delta J(\psi_0,\mathbf{A}_0)=\theta,$$

from the variation in ψ, we get

$$-div\left(\frac{\hbar^2}{2m}\left(\nabla\psi_0-\frac{2ie}{\hbar c}\mathbf{A}_0\psi_0\right)\right)+\frac{\hbar^2}{2m}\left(\nabla\psi_0-\frac{2ie}{\hbar c}\mathbf{A}_0\psi_0\right)\cdot\left(\frac{-2ie\mathbf{A}_0}{\hbar c}\right)$$
$$+\alpha|\psi_0|^2\psi_0-\beta\psi_0=\theta,\ \text{in } \Omega.$$

Hence, from this, (16.12) and (16.14), we obtain

$$-div(v_{01}^*)-v_{01}^*\cdot\left(\frac{2ie\mathbf{A}_0}{\hbar c}\right)+2\sigma_0\psi_0=\theta,\ \text{in } \Omega. \tag{16.18}$$

On the other hand, the variation in $\mathbf{A}$ gives us

$$\begin{aligned}&\frac{\text{curl curl }\mathbf{A}_0}{4\pi}-\frac{\text{curl }\mathbf{B}_0}{4\pi}\\ &+Re\left[\frac{\hbar^2}{2m}\left(\nabla\psi_0-\frac{2ie}{\hbar c}\mathbf{A}_0\psi_0\right)\left(\frac{-2ie\psi_0}{\hbar c}\right)\right]=\theta,\ \text{in } \Omega,\end{aligned} \tag{16.19}$$

where $Re[v]$ denotes the real part of v and

$$\frac{\text{curl curl }\mathbf{A}_0}{4\pi}-\frac{\text{curl }\mathbf{B}_0}{4\pi}=\theta,\ \text{in } \mathbb{R}^3\setminus\Omega. \tag{16.20}$$

From (16.12), (16.13), and (16.19), we obtain

$$\sum_{k=1}^3 2(v_{02}^*)_k(\mathbf{A}_0)_k\mathbf{e}_k+Re\left[v_{01}^*\left(\frac{-2ie\psi_0}{\hbar c}\right)\right]=\theta,\ \text{in } \Omega. \tag{16.21}$$

From (16.18) and (16.21) we get

$$G_3^*((v_0^*)_1,(v_0^*)_2,\sigma_0) = -\langle|\psi_0|^2,\sigma\rangle_{L^2(\Omega)} \\ -\langle(\mathbf{A}_0)_k^2,(v_0^*)_{2k}\rangle_{L^2(\Omega)} - G_3(\psi_0,\mathbf{A}_0), \qquad (16.22)$$

From (16.12) we obtain

$$\begin{aligned} G_1^*((v_0^*)_1) &= \left\langle \nabla\psi_0 - \frac{2ie}{\hbar c}\mathbf{A}_0\psi_0,(v_0^*)_1 \right\rangle_{L^2(\Omega;\mathbb{C}^3)} \\ &\quad -G_1(\nabla\psi_0 - \frac{2ie}{\hbar c}\mathbf{A}_0\psi_0) \\ &= \left\langle \nabla\psi_0 - \frac{2ie}{\hbar c}\mathbf{A}_0\psi_0,(v_0^*)_1 \right\rangle_{L^2(\Omega;\mathbb{C}^3)} \\ &\quad -\frac{\hbar^2}{4m}\int_\Omega \left|\nabla\psi_0 - \frac{2ie}{\hbar c}\mathbf{A}_0\psi_0\right|_2^2 dx. \end{aligned} \qquad (16.23)$$

By (16.14) we may infer that

$$G_2^*(\sigma_0) = \langle|\psi_0|^2,\sigma_0\rangle_{L^2} - G_2(|\psi_0|^2). \qquad (16.24)$$

From (16.13) and (16.20) we get

$$G_0^*((v_0^*)_2) = \langle(\mathbf{A}_0)_k^2,(v_0^*)_{2k}\rangle_{L^2(\Omega)} \\ -\frac{1}{8\pi}\int_{\mathbb{R}^3}|\text{curl}\,\mathbf{A}_0 - \mathbf{B}_0|_2^2\,dx. \qquad (16.25)$$

Finally, by (16.22), (16.23), (16.24), (16.25), and from the fact that $v_0^* \in A^*$, we obtain

$$\begin{aligned} &G_0^*((v_0^*)_2) + G_1^*((v_0^*)_1) + G_2^*(\sigma_0) + G_3^*((v_0^*)_1,(v_0^*)_2,\sigma_0) \\ &= \langle(\mathbf{A}_0)_k^2,(v_0^*)_{2k}\rangle_{L^2(\Omega)} \\ &\quad -\frac{1}{8\pi}\int_{\mathbb{R}^3}|\text{curl}\,\mathbf{A}_0 - \mathbf{B}_0|_2^2\,dx \\ &\quad +\left\langle \nabla\psi_0 - \frac{2ie}{\hbar c}\mathbf{A}_0\psi_0,(v_0^*)_1 \right\rangle_{L^2(\Omega;\mathbb{C}^3)} - G_1\left(\nabla\psi_0 - \frac{2ie}{\hbar c}\mathbf{A}_0\psi_0\right) \\ &\quad +\langle|\psi_0|^2,\sigma_0\rangle_{L^2} - G_2(|\psi_0|^2) \\ &\quad -\left\langle \nabla\psi_0 - \frac{2ie}{\hbar c}\mathbf{A}_0\psi_0,(v_0^*)_1 \right\rangle_{L^2(\Omega;\mathbb{C}^3)} \\ &\quad -\langle|\psi_0|^2,\sigma_0\rangle_{L^2(\Omega)} - \langle(\mathbf{A}_0)_k^2,(v_0^*)_{2k}\rangle_{L^2(\Omega)} \\ &= -\frac{1}{8\pi}\int_{\mathbb{R}^3}|\text{curl}\,\mathbf{A}_0 - \mathbf{B}_0|_2^2\,dx - G_1\left(\nabla\psi_0 - \frac{2ie}{\hbar c}\mathbf{A}_0\psi_0\right) - G_2(|\psi_0|^2) \\ &= -J(\psi_0,\mathbf{A}_0), \end{aligned}$$

that is,

$$J(\psi_0, \mathbf{A}_0) = -J^*(v_0^*).$$

From this and (16.17), the proof is complete.

16.3 On the Numerical Procedures for Ginzburg–Landau-Type Equations

We first apply Newton's method. The solution here is obtained similarly as for the generalized method of lines procedure. See the next sections for details on such a method for PDEs.

Consider again the equation.

$$\begin{cases} u'' + f(u) + g = 0, \text{ in } [0,1] \\ u(0) = u_0, \ \ u(1) = u_f, \end{cases} \tag{16.26}$$

As above, in finite differences, we have

$$u_{n+1} - 2u_n + u_{n-1} + f(u_n)d^2 + g_n d^2 = 0.$$

Assume such an equation is nonlinear. Linearizing it about a first solution $\{\tilde{u}\}$, we have (in fact this is an approximation)

$$u_{n+1} - 2u_n + u_{n-1} + f(\tilde{u}_n)d^2 + f'(\tilde{u}_n)(u_n - \tilde{u}_n)d^2 + g_n d^2 = 0.$$

Thus we may write

$$u_{n+1} - 2u_n + u_{n-1} + A_n u_n d^2 + B_n d^2 = 0,$$

where

$$A_n = f'(\tilde{u}_n),$$

and

$$B_n = f(\tilde{u}_n) - f'(\tilde{u}_n)\tilde{u}_n + g_n.$$

In particular for $n = 1$ we get

$$u_2 - 2u_1 + u_0 + A_1 u_1 d^2 + B_1 d^2 = 0.$$

Solving such an equation for u_1, we get

$$u_1 = a_1 u_2 + b_1 u_0 + c_1,$$

where

$$a_1 = (2 - A_1 d^2)^{-1}, \ \ b_1 = a_1, \ \ c_1 = a_1 B_1.$$

Reasoning inductively, having

$$u_{n-1} = a_{n-1}u_n + b_{n-1}u_0 + c_{n-1},$$

and

$$u_{n+1} - 2u_n + u_{n-1} + A_n u_n d^2 + B_n d^2 = 0,$$

we get

$$u_{n+1} - 2u_n + a_{n-1}u_n + b_{n-1}u_0 + c_{n-1} + A_n u_n d^2 + B_n d^2 = 0,$$

so that

$$u_n = a_n u_{n+1} + b_n u_0 + c_n,$$

where

$$a_n = (2 - a_{n-1} - A_n d^2)^{-1},$$

$$b_n = a_n b_{n-1},$$

and

$$c_n = a_n(c_{n-1} + B_n d^2),$$

$\forall n \in 1,\ldots,N-1$.

We have thus obtained

$$u_n = a_n u_{n+1} + b_n u_0 + c_n \equiv H_n(u_{n+1}), \forall n \in \{1,\ldots,N-1\},$$

and in particular

$$u_{N-1} = H_{N-1}(u_f),$$

so that we may calculate

$$u_{N-2} = H_{N-2}(u_{N-1}),$$

$$u_{N-3} = H_{N-3}(u_{N-2}),$$

and so on, up to finding

$$u_1 = H_1(u_2).$$

The next step is to replace $\{\tilde{u}_n\}$ by the $\{u_n\}$ calculated and repeat the process up to the satisfaction of an appropriate convergence criterion. We present numerical results for the equation

$$\begin{cases} u'' - \frac{u^3}{\varepsilon} + \frac{u}{\varepsilon} + g = 0, \text{ in } [0,1] \\ u(0) = 0, \;\; u(1) = 0, \end{cases} \tag{16.27}$$

where

$$g(x) = \frac{1}{\varepsilon},$$

The results are obtained for $\varepsilon = 1.0$, $\varepsilon = 0.1$, $\varepsilon = 0.01$, and $\varepsilon = 0.001$. Please see Figs. 16.1, 16.2, 16.3, and 16.4, respectively. For the other two solutions for $\varepsilon = 0.01$ see Figs. 16.5 and 16.6.

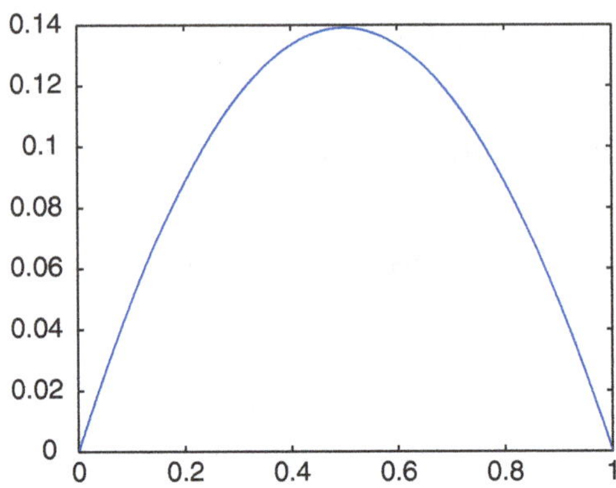

Fig. 16.1 The solution $u(x)$ by Newton's method for $\varepsilon = 1$

Other solutions through Newton's method are also shown.

Remark 16.3.1. We highlight that the results obtained through Newton's method are consistent with problem physics.

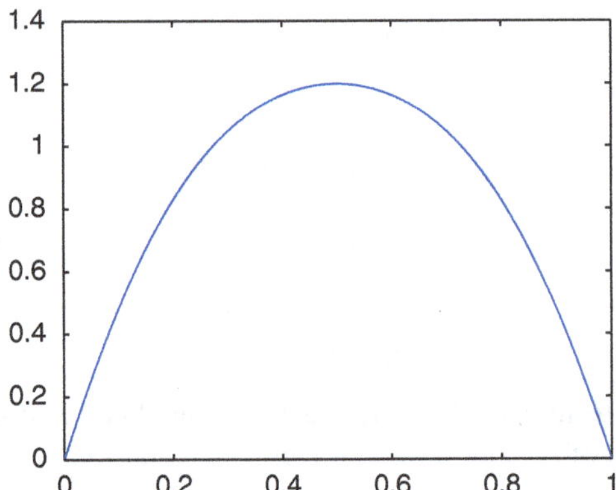

Fig. 16.2 The solution $u(x)$ by Newton's method for $\varepsilon = 0.1$

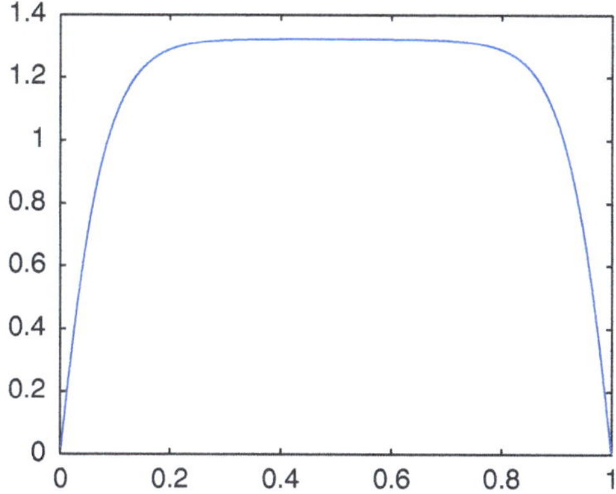

Fig. 16.3 The solution $u(x)$ by Newton's method for $\varepsilon = 0.01$

16.4 Numerical Results for Related PDEs

16.4.1 A Related PDE on a Special Class of Domains

We start by describing a similar equation, but now in a two-dimensional context. Let $\Omega \subset \mathbb{R}^2$ be an open, bounded, connected set with a regular boundary denoted by $\partial\Omega$. Consider a real Ginzburg–Landau-type equation (see [4, 9, 45, 46] for details about such an equation), given by

$$\begin{cases} \varepsilon \nabla^2 u - \alpha u^3 + \beta u = f, & \text{in } \Omega \\ u = 0, & \text{on } \partial\Omega, \end{cases} \tag{16.28}$$

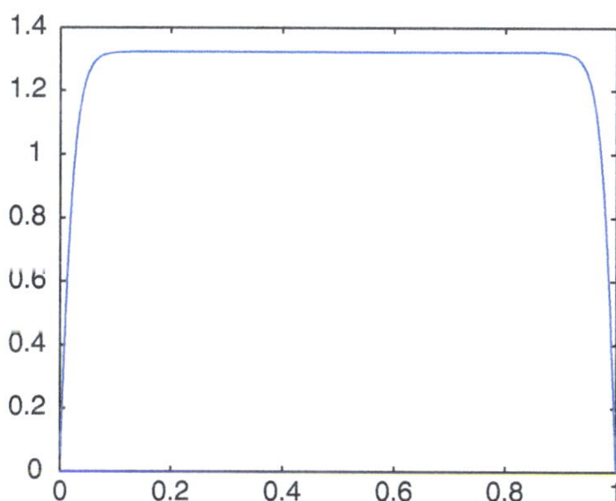

Fig. 16.4 The solution $u(x)$ by Newton's method for $\varepsilon = 0.001$

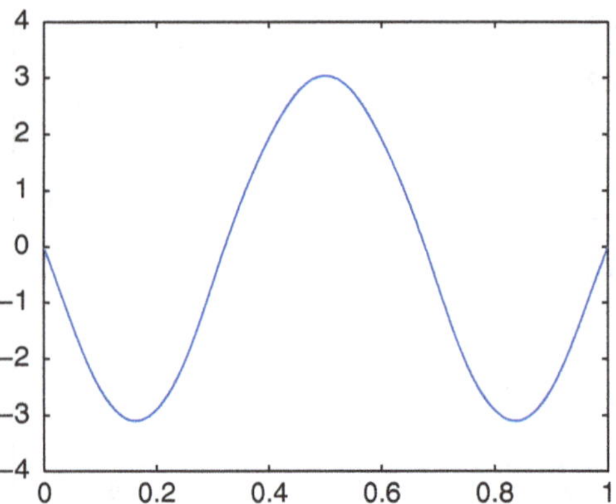

Fig. 16.5 Other solution $u(x)$ by Newton's method for $\varepsilon = 0.01$

where $\alpha,\ \beta,\ \varepsilon > 0$, $u \in U = W_0^{1,2}(\Omega)$, and $f \in L^2(\Omega)$. The corresponding primal variational formulation is represented by $J : U \to \mathbb{R}$, where

$$J(u) = \frac{\varepsilon}{2}\int_\Omega \nabla u \cdot \nabla u\, dx + \frac{\alpha}{4}\int_\Omega u^4\, dx - \frac{\beta}{2}\int_\Omega u^2\, dx + \int_\Omega f u\, dx.$$

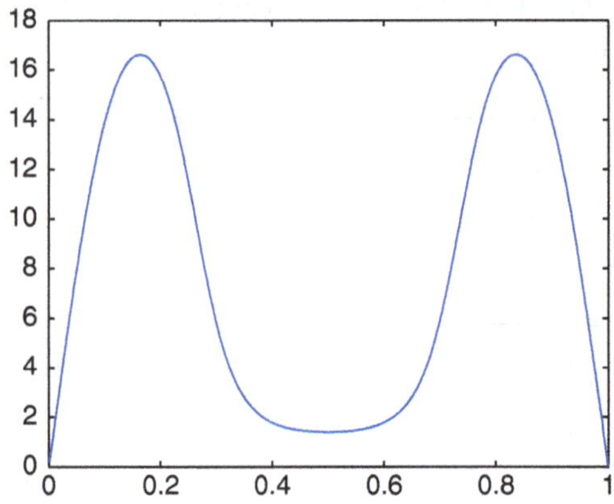

Fig. 16.6 Other solution $u(x)$ by Newton's method for $\varepsilon = 0.01$

16.4.2 About the Matrix Version of GMOL

The generalized method of lines was originally developed in [13]. In this work we address its matrix version. Consider the simpler case where $\Omega = [0,1] \times [0,1]$. We discretize the domain in x, that is, in $N+1$ vertical lines obtaining the following equation in finite differences (see [63] for details about finite differences schemes).

$$\frac{\varepsilon(u_{n+1}-2u_n+u_{n-1})}{d^2}+\varepsilon M_2 u_n/d_1^2-\alpha u_n^3+\beta u_n=f_n, \tag{16.29}$$

$\forall n \in \{1,\ldots,N-1\}$, where $d = 1/N$ and u_n corresponds to the solution on the line n. The idea is to apply Newton's method. Thus choosing an initial solution $\{(u_0)_n\}$ we linearize (16.29) about it, obtaining the linear equation

$$\begin{aligned} u_{n+1}-2u_n+u_{n-1} &+ \tilde{M}_2 u_n - \frac{3\alpha d^2}{\varepsilon}(u_0)_n^2 u_n \\ &+ \frac{2\alpha}{\varepsilon}(u_0)_n^3 d^2 + \frac{\beta d^2}{\varepsilon}u_n - f_n\frac{d^2}{\varepsilon} = 0, \end{aligned} \tag{16.30}$$

where $\tilde{M}_2 = M_2\frac{d^2}{d_1^2}$ and

$$M_2 = \begin{bmatrix} -2 & 1 & 0 & 0 & \ldots & 0 \\ 1 & -2 & 1 & 0 & \ldots & 0 \\ 0 & 1 & -2 & 1 & 0.. & 0 \\ \ldots & \ldots & \ldots & \ldots & \ldots & \ldots \\ 0 & 0 & \ldots & 1 & -2 & 1 \\ 0 & 0 & \ldots & \ldots & 1 & -2 \end{bmatrix}, \tag{16.31}$$

with N_1 lines corresponding to the discretization in the y axis. Furthermore $d_1 = 1/N_1$.

In particular for $n = 1$ we get

$$\begin{aligned} u_2-2u_1 &+ \tilde{M}_2 u_1 - \frac{3\alpha d^2}{\varepsilon}(u_0)_1^2 u_1 \\ &+ \frac{2\alpha}{\varepsilon}(u_0)_1^3 d^2 + \frac{\beta d^2}{\varepsilon}u_1 - f_1\frac{d^2}{\varepsilon} = 0. \end{aligned} \tag{16.32}$$

Denoting

$$M_{12}[1] = 2I_d - \tilde{M}_2 + 3\frac{\alpha d^2}{\varepsilon}(u_0)_1^2 I_d - \frac{\beta d^2}{\varepsilon}I_d,$$

where I_d denotes the $(N_1-1)\times(N_1-1)$ identity matrix,

$$Y_0[1] = \frac{2\alpha d^2}{\varepsilon}(u_0)_1^3 - f_1\frac{d^2}{\varepsilon},$$

and $M_{50}[1] = M_{12}[1]^{-1}$, we obtain

$$u_1 = M_{50}[1]u_2 + z[1].$$

where

$$z[1] = M_{50}[1]\cdot Y_0[1]$$

Now for $n = 2$ we get

$$u_3 - 2u_2 + u_1 + \tilde{M}_2 u_2 - \frac{3\alpha d^2}{\varepsilon}(u_0)_2^2 u_2 + \frac{2\alpha}{\varepsilon}(u_0)_2^3 d^2 + \frac{\beta d^2}{\varepsilon} u_2 - f_2 \frac{d^2}{\varepsilon} = 0, \quad (16.33)$$

that is,

$$u_3 - 2u_2 + M_{50}[1]u_2 + z[1] + \tilde{M}_2 u_2 - \frac{3\alpha d^2}{\varepsilon}(u_0)_2^2 u_2 + \frac{2\alpha}{\varepsilon}(u_0)_2^3 d^2 + \frac{\beta d^2}{\varepsilon} u_2 - f_2 \frac{d^2}{\varepsilon} = 0, \quad (16.34)$$

so that denoting

$$M_{12}[2] = 2I_d - \tilde{M}_2 - M_{50}[1] + 3\frac{\alpha d^2}{\varepsilon}(u_0)_2^2 I_d - \frac{\beta d^2}{\varepsilon} I_d,$$

$$Y_0[2] = \frac{2\alpha d^2}{\varepsilon}(u_0)_2^3 - f_2 \frac{d^2}{\varepsilon},$$

and $M_{50}[2] = M_{12}[2]^{-1}$, we obtain

$$u_2 = M_{50}[2]u_3 + z[2],$$

where

$$z[2] = M_{50}[2] \cdot (Y_0[2] + z[1]).$$

Proceeding in this fashion, for the line n, we obtain

$$u_{n+1} - 2u_n + M_{50}[n-1]u_n + z[n-1] + \tilde{M}_2 u_n - \frac{3\alpha d^2}{\varepsilon}(u_0)_n^2 u_n + \frac{2\alpha}{\varepsilon}(u_0)_n^3 d^2 + \frac{\beta d^2}{\varepsilon} u_n - f_n \frac{d^2}{\varepsilon} = 0, \quad (16.35)$$

so that denoting

$$M_{12}[n] = 2I_d - \tilde{M}_2 - M_{50}[n-1] + 3\frac{\alpha d^2}{\varepsilon}(u_0)_n^2 I_d - \frac{\beta d^2}{\varepsilon} I_d,$$

and also denoting

$$Y_0[n] = \frac{2\alpha d^2}{\varepsilon}(u_0)_n^3 - f_n \frac{d^2}{\varepsilon},$$

and $M_{50}[n] = M_{12}[n]^{-1}$, we obtain

$$u_n = M_{50}[n]u_{n+1} + z[n],$$

where

$$z[n] = M_{50}[n] \cdot (Y_0[n] + z[n-1]).$$

Observe that we have

$$u_N = \theta,$$

where θ denotes the zero matrix $(N_1 - 1) \times 1$, so that we may calculate

$$u_{N-1} = M_{50}[N-1] \cdot u_N + z[N-1],$$

and

$$u_{N-2} = M_{50}[N-2] \cdot u_{N-1} + z[N-2],$$

and so on, up to obtaining

$$u_1 = M_{50}[1] \cdot u_2 + z[1].$$

The next step is to replace $\{(u_0)_n\}$ by $\{u_n\}$ and thus to repeat the process until convergence is achieved.

This is Newton's method; what seems to be relevant is the way we inverted the big matrix $((N_1-1)\cdot(N-1)) \times ((N_1-1)\cdot(N-1))$, and in fact instead of inverting it directly we have inverted $N-1$ matrices $(N_1-1) \times (N_1-1)$ through an application of the generalized method of lines.

So far we cannot guarantee convergence; however, through the next theorem, we describe a procedure that always leads to a solution. Anyway, we highlight the next result is not a formal proof of convergence in a numerical analysis context. In fact, such a result must be seen as an existence of one of the critical points for the equation in question.

Theorem 16.4.1. *Let $\Omega \subset \mathbb{R}^2$ be an open, bounded, connected set with a regular boundary denoted by $\partial\Omega$. Consider the real Ginzburg–Landau-type equation, given by*

$$\begin{cases} \varepsilon\nabla^2 u - \alpha u^3 + \beta u = f, & \text{in } \Omega \\ u = 0, & \text{on } \partial\Omega, \end{cases} \tag{16.36}$$

where $\alpha,\ \beta,\ \varepsilon > 0$, $u \in U = W_0^{1,2}(\Omega)$, and $f \in L^2(\Omega)$.

Consider the sequence obtained through the algorithm:

1. *Set $n = 1$.*
2. *Choose $z_1^* \in L^2(\Omega)$*
3. *Compute u_n by*

$$u_n = argmin_{u\in U}\left\{\frac{\varepsilon}{2}\int_\Omega \nabla u\cdot\nabla u\,dx + \frac{\alpha}{4}\int_\Omega u^4\,dx \right.$$
$$\left. -\langle u, z_n^*\rangle_{L^2} + \frac{1}{2\beta}\int_\Omega (z_n^*)^2\,dx + \int_\Omega fu\,dx\right\}, \tag{16.37}$$

which means to solve the equation

$$\begin{cases} \varepsilon\nabla^2 u - \alpha u^3 + z_n^* = f, & in\ \Omega \\ u = 0, & on\ \partial\Omega. \end{cases} \tag{16.38}$$

4. *Compute z_{n+1}^* by*

$$z_{n+1}^* = argmin_{z^*\in L^2(\Omega)} \left\{ \frac{\varepsilon}{2}\int_\Omega \nabla u_n\cdot\nabla u_n\,dx + \frac{\alpha}{4}\int_\Omega u_n^4\,dx \right.$$
$$\left. -\langle u_n, z^*\rangle_{L^2} + \frac{1}{2\beta}\int_\Omega (z^*)^2\,dx + \int_\Omega f u_n\,dx \right\}, \tag{16.39}$$

that is,

$$z_{n+1}^* = \beta u_n.$$

5. *Set $n \to n+1$ and go to step 3 (up to the satisfaction of an appropriate convergence criterion).*

The sequence $\{u_n\}$ is such that up to a subsequence not relabeled

$$u_n \to u_0,\ strongly\ in\ L^2(\Omega),$$

where

$$u_0 \in W_0^{1,2}(\Omega)$$

is a solution of equation (16.36).

Proof. Observe that defining $J : U \to \mathbb{R}$, by

$$J(u) = \left\{ \frac{\varepsilon}{2}\int_\Omega \nabla u\cdot\nabla u\,dx + \frac{\alpha}{4}\int_\Omega u^4\,dx \right.$$
$$\left. -\frac{\beta}{2}\int_\Omega u^2\,dx + \int_\Omega f u\,dx \right\}, \tag{16.40}$$

we have

$$\begin{aligned} J(u) &= \frac{\varepsilon}{2}\int_\Omega \nabla u\cdot\nabla u\,dx + \frac{\alpha}{4}\int_\Omega u^4\,dx - \frac{\beta}{2}\int_\Omega u^2\,dx \\ &\quad + \langle u, z^*\rangle_{L^2} - \langle u, z^*\rangle_{L^2} + \int_\Omega f u\,dx \\ &\leq \frac{\varepsilon}{2}\int_\Omega \nabla u\cdot\nabla u\,dx + \frac{\alpha}{4}\int_\Omega u^4\,dx \\ &\quad -\langle u, z^*\rangle_{L^2} + \sup_{u\in U}\left\{ \langle u, z^*\rangle_{L^2} - \frac{\beta}{2}\int_\Omega u^2\,dx \right\} + \int_\Omega f u\,dx \\ &\leq \frac{\varepsilon}{2}\int_\Omega \nabla u\cdot\nabla u\,dx + \frac{\alpha}{4}\int_\Omega u^4\,dx \\ &\quad -\langle u, z^*\rangle_{L^2} + \frac{1}{2\beta}\int_\Omega (z^*)^2\,dx + \int_\Omega f u\,dx. \end{aligned} \tag{16.41}$$

Denoting

$$\alpha_n = \frac{\varepsilon}{2}\int_\Omega \nabla u_n \cdot \nabla u_n \, dx + \frac{\alpha}{4}\int_\Omega u_n^4 \, dx$$
$$-\langle u_n, z_n^* \rangle_{L^2} + \frac{1}{2\beta}\int_\Omega (z_n^*)^2 \, dx + \int_\Omega f u_n \, dx, \qquad (16.42)$$

we may easily verify that $\{\alpha_n\}$ is a real nonincreasing sequence bounded below by $\inf_{u\in U}\{J(u)\}$; therefore there exists $\alpha \in \mathbb{R}$ such that

$$\lim_{n\to\infty} \alpha_n = \alpha. \qquad (16.43)$$

From the Poincaré inequality (see [1] for details) we have that $J(u) \to +\infty$ if $\|u\|_{W^{1,2}(\Omega)} \to \infty$. From this, (16.41), (16.42), and (16.43), we may infer that

$$\|u_n\|_{W^{1,2}(\Omega)} < C, \forall n \in \mathbb{N}$$

for some $C > 0$.

Thus, from the Rellich–Kondrachov theorem, up to a not relabeled subsequence, there exists $u_0 \subset W^{1,2}(\Omega)$ such that

$$\nabla u_n \rightharpoonup \nabla u_0 \text{ weakly in } L^2(\Omega),$$

$$u_n \to u_0 \text{ strongly in } L^2(\Omega),$$

so that considering the algorithm in question

$$z_n \to z_0^* \text{ strongly in } L^2(\Omega),$$

where

$$z_0^* = \beta u_0.$$

Observe that the unique $u_n \in U$ such that

$$u_n = \text{argmin}_{u\in U}\left\{\frac{\varepsilon}{2}\int_\Omega \nabla u \cdot \nabla u \, dx + \frac{\alpha}{4}\int_\Omega u^4 \, dx\right.$$
$$\left. -\langle u, z_n^* \rangle_{L^2} + \frac{1}{2\beta}\int_\Omega (z_n^*)^2 \, dx + \int_\Omega f u \, dx\right\}, \qquad (16.44)$$

is also such that

$$\varepsilon\nabla^2 u_n - \alpha u_n^3 + z_n^* + f - 0 \text{ in } \Omega,$$

in the sense of distributions (details about this result may be found in [25]).

Fix $\phi \in C_c^\infty(\Omega)$. In the next lines, we will prove that

$$\langle u_n^3, \phi \rangle_{L^2(\Omega)} \to \langle u_0^3, \phi \rangle_{L^2(\Omega)}, \qquad (16.45)$$

as $n \to \infty$. First observe that from (16.41), (16.42), and (16.43), it is clear that there exists $K_1 > 0$ such that

$$\|u_n\|_4 < K_1, \forall n \in \mathbb{N}.$$

Observe also that

$$\begin{aligned} \left| \int_\Omega (u_n^3 - u_0^3)\phi \, dx \right| &\leq \left| \int_\Omega (u_n^3 - u_n^2 u_0 + u_n^2 u_0 - u_0^3)\phi \, dx \right| \\ &\leq \int_\Omega \left| u_n^2 (u_n - u_0)\phi \right| \, dx \\ &\quad + \int_\Omega \left| u_0 (u_n^2 - u_0^2)\phi \right| \, dx. \end{aligned} \tag{16.46}$$

Furthermore,

$$\begin{aligned} \int_\Omega \left| u_n^2 (u_n - u_0)\phi \right| \, dx &\leq \|u_n\|_4^2 \|(u_n - u_0)\phi\|_2 \\ &\leq K_1^2 \|u_n - u_0\|_2 \|\phi\|_\infty \\ &\to 0, \text{ as } n \to \infty. \end{aligned} \tag{16.47}$$

On the other hand, from the generalized Hölder inequality, we get

$$\begin{aligned} \int_\Omega \left| u_0 (u_n^2 - u_0^2)\phi \right| \, dx &= \int_\Omega \left| u_0 (u_n + u_0)(u_n - u_0)\phi \right| \, dx \\ &\leq \|u_0\|_4 \|u_n + u_0\|_4 \|(u_n - u_0)\phi\|_2 \\ &\leq \|u_0\|_4 (\|u_n\|_4 + \|u_0\|_4) \|(u_n - u_0)\|_2 \|\phi\|_\infty \\ &\leq K_1 (K_1 + K_1) \|\phi\|_\infty \|u_n - u_0\|_2 \\ &\to 0, \text{ as } n \to \infty. \end{aligned} \tag{16.48}$$

Summarizing the last results we get

$$\langle u_n^3, \phi \rangle_{L^2(\Omega)} \to \langle u_0^3, \phi \rangle_{L^2(\Omega)}, \tag{16.49}$$

as $n \to \infty$. So, we may write

$$\begin{aligned} 0 &= \lim_{n \to \infty} \{ \langle u_n, \varepsilon \nabla^2 \phi \rangle_{L^2} + \langle -\alpha u_n^3 + z_n^* - f, \phi \rangle_{L^2} \} \\ &= \langle u_0, \varepsilon \nabla^2 \phi \rangle_{L^2} + \langle -\alpha u_0^3 + z_0^* - f, \phi \rangle_{L^2} \\ &= \langle u_0, \varepsilon \nabla^2 \phi \rangle_{L^2} + \langle -\alpha u_0^3 + \beta u_0 - f, \phi \rangle_{L^2} \end{aligned} \tag{16.50}$$

that is,

$$\varepsilon \nabla^2 u_0 - \alpha u_0^3 + \beta u_0 = f \text{ in } \Omega,$$

in the sense of distributions. From $u_n = 0$ on $\partial\Omega, \forall n \in \mathbb{N}$ we also obtain in a weak sense

$$u_0 = 0, \text{ on } \partial\Omega.$$

The proof is complete.

Remark 16.4.2. Observe that for each n, the procedure of evaluating u_n stands for the solution of a convex optimization problem with unique solution, given by the one of equation

$$\varepsilon \nabla^2 u_n - \alpha u_n^3 + z_n^* + f = 0 \text{ in } \Omega,$$

which may be easily obtained, due to convexity, through the generalized method of lines (matrix version) associated with Newton's method as above described.

16.5 Numerical Results

We solve the equation

$$\begin{cases} \varepsilon \nabla^2 u - \alpha u^3 + \beta u + 1 = 0, & \text{in } \Omega = [0,1] \times [0,1] \\ u = 0, & \text{on } \partial \Omega, \end{cases} \tag{16.51}$$

through the algorithm specified in the last theorem. We consider $\alpha = \beta = 1$. For $\varepsilon = 1.0$ see Fig. 16.7, and for $\varepsilon = 0.0001$ see Fig. 16.8.

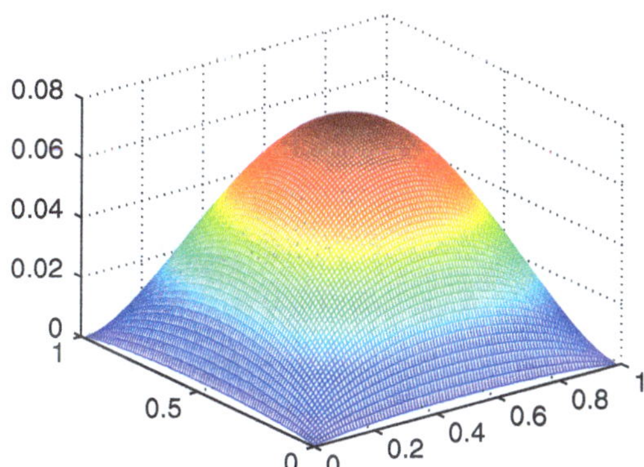

Fig. 16.7 The solution $u(x,y)$ for $\varepsilon = 1.0$

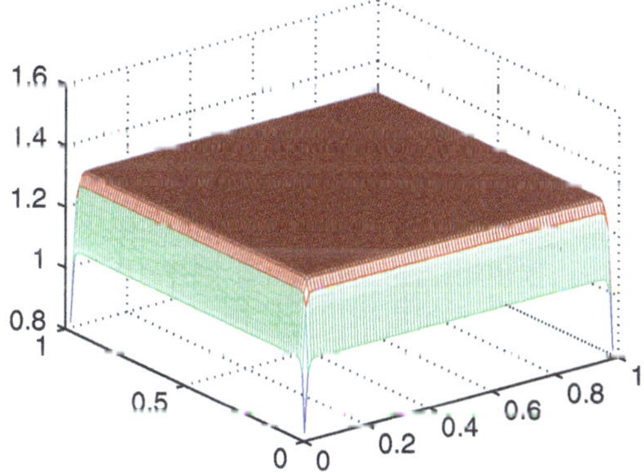

Fig. 16.8 The solution $u(x,y)$ for $\varepsilon = 0.0001$

16.6 A New Procedure to Obtain Approximate PDE Solutions

In this section we describe a procedure to obtain an approximate solution of a class of PDE. We start with the following theorem.

Theorem 16.6.1. *Consider the equation given by*

$$\begin{cases} \nabla^2 u + G(u) = f, & \text{in } \Omega \\ u = 0, & \text{on } \partial\Omega, \end{cases} \tag{16.52}$$

where $G : \mathbb{R} \to \mathbb{R}$ is a smooth function with bounded first derivatives in bounded sets, $u \in U = W_0^{1,2}(\Omega)$, and $f \in L^2(\Omega)$. Consider the simpler case where $\Omega = [0,1] \times [0,1]$. We discretize the domain in x, that is, in $N+1$ vertical lines obtaining the following equation in finite differences:

$$\frac{(u_{n+1} - 2u_n + u_{n-1})}{d^2} + M_2 u_n / d_1^2 + G(u_n) = f_n, \tag{16.53}$$

$\forall n \in \{1, \ldots, N-1\}$, where $d = 1/N$ and u_n corresponds to the solution on the line n. We rewrite equation (16.53), obtaining

$$u_{n+1} - 2u_n + u_{n-1} + \tilde{M}_2 u_n + G(u_n) d^2 - f_n d^2 = 0, \tag{16.54}$$

where $\tilde{M}_2 = M_2 \frac{d^2}{d_1^2}$ and

$$M_2 = \begin{bmatrix} -2 & 1 & 0 & 0 & \ldots & 0 \\ 1 & -2 & 1 & 0 & \ldots & 0 \\ 0 & 1 & -2 & 1 & 0.. & 0 \\ \ldots & \ldots & \ldots & \ldots & \ldots & \ldots \\ 0 & 0 & \ldots & 1 & -2 & 1 \\ 0 & 0 & \ldots & \ldots & 1 & -2 \end{bmatrix}, \tag{16.55}$$

with N_1 lines corresponding to the discretization in the y axis. Furthermore, $d_1 = 1/N_1$. Then, for such a system, we have the following relations:

$$u_n = M_{50}[n] u_{n+1} + M_{60}[n] G(u_{n+1}) d^2 + z[n] + Er[n],$$

where

$$M_{12}[1] = 2I_d - \tilde{M}_2,$$

I_d denotes the $(N_1 - 1) \times (N_1 - 1)$ identity matrix, $M_{50}[1] = M_{12}[1]^{-1}$,

$$M_{60}[1] = M_{50}[1],$$

$$z[1] = M_{50}[1] \cdot (-f_1 d^2),$$

$$Er[1] = M_{50}[1](G(u_1) - G(u_2))d^2,$$

$$M_{12}[n] = 2I_d - \tilde{M}_2 - M_{50}[n-1],$$

$M_{50}[n] = M_{12}[n]^{-1}$,

$$M_{60}[n] = M_{50}[n] \cdot (M_{60}[n-1] + I_d),$$

$$z[n] = M_{50}[n] \cdot (z[n-1] - f_n d^2),$$

and

$$Er[n] = M_{50}[n](Er[n-1]) + M_{60}[n](G(u_n) - G(u_{n+1}))d^2,$$

$\forall n \in \{1, \ldots, N-1\}$.

Proof. In particular for $n = 1$ we get

$$u_2 - 2u_1 + \tilde{M}_2 u_1 + G(u_1)d^2 - f_1 d^2 = 0. \tag{16.56}$$

Denoting

$$M_{12}[1] = 2I_d - \tilde{M}_2,$$

where I_d denotes the $(N_1 - 1) \times (N_1 - 1)$ identity matrix and $M_{50}[1] = M_{12}[1]^{-1}$, we obtain

$$u_1 = M_{50}[1](u_2 + G(u_1)d^2 - f_1 d^2),$$

so that

$$u_1 = M_{50}[1](u_2) + M_{60}[1]G(u_2)d^2 + z[1] + Er[1],$$

where

$$M_{60}[1] = M_{50}[1],$$

$$z[1] = M_{50}[1] \cdot (-f_1 d^2),$$

and

$$Er[1] = M_{50}[1](G(u_1) - G(u_2))d^2.$$

Now for $n = 2$ we get

$$u_3 - 2u_2 + u_1 + \tilde{M}_2 u_2 + G(u_2)d^2 - f_2 d^2 = 0, \tag{16.57}$$

that is,

$$\begin{aligned} u_3 - 2u_2 + M_{50}[1]u_2 + M_{60}[1]G(u_2)d^2 + z[1] \\ + Er[1] + \tilde{M}_2 u_2 + G(u_2)d^2 - f_2 d^2 = 0, \end{aligned} \tag{16.58}$$

so that denoting

$$M_{12}[2] = 2I_d - \tilde{M}_2 - M_{50}[1],$$

and also denoting $M_{50}[2] = M_{12}[2]^{-1}$, we obtain

$$u_2 = M_{50}[2]u_3 + M_{60}[2]G(u_3)d^2 + z[2] + Er[2],$$

where

$$M_{60}[2] = M_{50}[2] \cdot (M_{60}[1] + I_d),$$
$$z[2] = M_{50}[2] \cdot (z[1] - f_2 d^2).$$

and

$$Er[2] = M_{50}[2](Er[1]) + M_{60}[2](G(u_2) - G(u_3))d^2.$$

Proceeding in this fashion, for the line n, we obtain

$$\begin{aligned} u_{n+1} &- 2u_n + M_{50}[n-1]u_n + M_{60}[n-1]G(u_n)d^2 \\ &+ z[n-1] + Er[n-1] + \tilde{M}_2 u_n + G(u_n)d^2 - f_n d^2 = 0, \end{aligned} \tag{16.59}$$

so that denoting

$$M_{12}[n] = 2I_d - \tilde{M}_2 - M_{50}[n-1],$$

and $M_{50}[n] = M_{12}[n]^{-1}$, we obtain

$$u_n = M_{50}[n]u_{n+1} + M_{60}[n]G(u_{n+1})d^2 + z[n] + Er[n],$$

where

$$M_{60}[n] = M_{50}[n] \cdot (M_{60}[n-1] + I_d),$$
$$z[n] = M_{50}[n] \cdot (z[n-1] - f_n d^2),$$

and

$$Er[n] = M_{50}[n](Er[n-1]) + M_{60}[n](G(u_n) - G(u_{n+1}))d^2.$$

Remark 16.6.2. We may use, as a first approximation for the solution, the relations

$$u_n \approx M_{50}[n]u_{n+1} + M_{60}[n]G(u_{n+1})d^2 + z[n].$$

Observe that we have

$$u_N = \theta,$$

where θ denotes the zero matrix $(N_1 - 1) \times 1$, so that we may calculate

$$u_{N-1} \approx M_{50}[N-1] \cdot u_N + M_{60}[N-1] \cdot G(u_N)d^2 + z[N-1],$$

and

$$u_{N-2} \approx M_{50}[N-2] \cdot u_{N-1} + M_{60}[N-2] \cdot G(u_{N-1})d^2 + z[N-2],$$

and so on, up to obtaining

$$u_1 \approx M_{50}[1] \cdot u_2 + M_{60}[1] \cdot G(u_2)d^2 + z[1].$$

The next step is to use the $\{u_n\}$ obtained as the initial solution for Newton's method.

What is relevant is that in general, the first approximation is a good one for the exact solution.

We have computed the first approximation using such a method, for

$$G(u) = \frac{-u^3}{\varepsilon} + \frac{u}{\varepsilon},$$

$$f(x,y) = -\frac{1}{\varepsilon}, \forall (x,y) \in \Omega$$

and $\varepsilon = 0.01$. Please see Fig. 16.9.

This first approximation is close to the solution obtained through Newton's method. For the solution through the earlier approach, see Fig. 16.10.

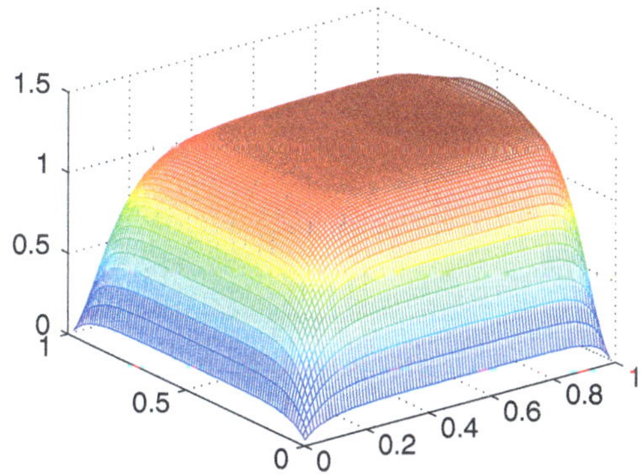

Fig. 16.9 The first approximation for $u(x,y)$ for $\varepsilon = 0.01$

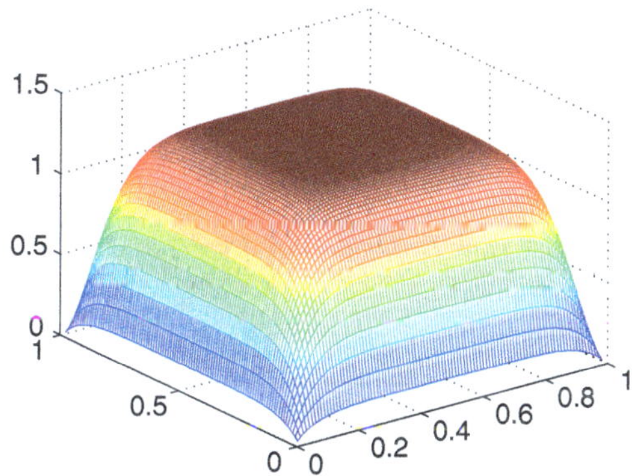

Fig. 16.10 The solution $u(x,y)$ by Newton's method for $\varepsilon = 0.01$

16.7 Final Results, Newton's Method for a First-Order System

Consider the first-order system and respective boundary conditions

$$\begin{cases} u' + f_1(u,v) + g_1 = 0, \text{ in } [0,1] \\ v' + f_2(u,v) + g_2 = 0, \text{ in } [0,1] \\ u(0) = u_0, \ \ v(1) = v_f, \end{cases} \tag{16.60}$$

Linearizing the equations about the first solutions $\tilde{u}$, and $\tilde{v}$, we obtain

$$\begin{aligned} &u' + f_1(\tilde{u},\tilde{v}) + \frac{\partial f_1(\tilde{u},\tilde{v})}{\partial u}(u-\tilde{u}) \\ &+\frac{\partial f_1(\tilde{u},\tilde{v})}{\partial v}(v-\tilde{v}) + g_1 = 0, \end{aligned} \tag{16.61}$$

$$\begin{aligned} &v' + f_2(\tilde{u},\tilde{v}) + \frac{\partial f_2(\tilde{u},\tilde{v})}{\partial u}(u-\tilde{u}) \\ &+\frac{\partial f_2(\tilde{u},\tilde{v})}{\partial v}(v-\tilde{v}) + g_2 = 0. \end{aligned} \tag{16.62}$$

In finite differences, we could write

$$\begin{aligned} &u_n - u_{n-1} + f_1(\tilde{u}_{n-1},\tilde{v}_{n-1})d + \frac{\partial f_1(\tilde{u}_{n-1},\tilde{v}_{n-1})}{\partial u}(u_{n-1}-\tilde{u}_{n-1})d \\ &+\frac{\partial f_1(\tilde{u}_{n-1},\tilde{v}_{n-1})}{\partial v}(v_{n-1}-\tilde{v}_{n-1})d + (g_1)_{n-1}d = 0, \end{aligned} \tag{16.63}$$

$$\begin{aligned} &v_n - v_{n-1} + f_2(\tilde{u}_{n-1},\tilde{v}_{n-1})d + \frac{\partial f_2(\tilde{u}_{n-1},\tilde{v}_{n-1})}{\partial u}(u_{n-1}-\tilde{u}_{n-1})d \\ &+\frac{\partial f_2(\tilde{u}_{n-1},\tilde{v}_{n-1})}{\partial v}(v_{n-1}-\tilde{v}_{n-1})d + (g_2)_{n-1}d = 0. \end{aligned} \tag{16.64}$$

Hence, we may write

$$\begin{aligned} u_n &= a_n u_{n-1} + b_n v_{n-1} + c_n, \\ v_n &= d_n u_{n-1} + e_n v_{n-1} + f_n, \end{aligned}$$

where

$$\begin{aligned} a_n &= -\frac{\partial f_1(\tilde{u}_{n-1},\tilde{v}_{n-1})}{\partial u}d + 1, \\ b_n &= -\frac{\partial f_1(\tilde{u}_{n-1},\tilde{v}_{n-1})}{\partial v}d, \end{aligned}$$

$$\begin{aligned} c_n = & -f_1(\tilde{u}_{n-1},\tilde{v}_{n-1})d + \frac{\partial f_1(\tilde{u}_{n-1},\tilde{v}_{n-1})}{\partial u}\tilde{u}_{n-1}d \\ & + \frac{\partial f_1(\tilde{u}_{n-1},\tilde{v}_{n-1})}{\partial v}\tilde{v}_{n-1}d - (g_1)_{n-1}d, \end{aligned} \tag{16.65}$$

and

$$d_n = -\frac{\partial f_2(\tilde{u}_{n-1},\tilde{v}_{n-1})}{\partial u}d,$$

$$e_n = -\frac{\partial f_2(\tilde{u}_{n-1},\tilde{v}_{n-1})}{\partial v}d + 1,$$

$$\begin{aligned} f_n = & -f_2(\tilde{u}_{n-1},\tilde{v}_{n-1})d + \frac{\partial f_2(\tilde{u}_{n-1},\tilde{v}_{n-1})}{\partial u}\tilde{u}_{n-1}d \\ & + \frac{\partial f_2(\tilde{u}_{n-1},\tilde{v}_{n-1})}{\partial v}\tilde{v}_{n-1}d - (g_2)_{n-1}d. \end{aligned} \tag{16.66}$$

In particular, for $n = 1$, we get

$$u_1 = a_1 u_0 + b_1 v_0 + c_1, \tag{16.67}$$

and

$$v_1 = d_1 u_0 + e_1 v_0 + f_1. \tag{16.68}$$

From this last equation,

$$v_0 = (v_1 - d_1 u_0 - f_1)/e_1,$$

so that from this and Eq. (16.67), we get

$$u_1 = a_1 u_0 + b_1 (v_1 - d_1 u_0 - f_1)/e_1 + c_1 = F_1 v_1 + G_1,$$

where

$$F_1 = b_1/e_1, \;\; G_1 = a_1 u_0 - b_1(d_1 u_0 + f_1)/e_1 + c_1.$$

Reasoning inductively, having

$$u_{n-1} = F_{n-1} v_{n-1} + G_{n-1},$$

we also have

$$u_n = a_n u_{n-1} + b_n v_{n-1} + c_n,$$

$$v_n = d_n u_{n-1} + e_n v_{n-1} + f_n,$$

$$v_n = d_n (F_{n-1} v_{n-1} + G_{n-1}) + e_n v_{n-1} + f_n,$$

that is,

$$v_{n-1} - H_n v_n + L_n,$$

where

$$H_n = 1/(d_n F_{n-1} + e_n),$$

$$L_n = -H_n(d_n G_{n-1} + f_n).$$

Hence

$$u_n = a_n(F_{n-1}v_{n-1} + G_{n-1}) + b_n v_{n-1} + c_{n-1},$$

so that

$$u_n = a_n(F_{n-1}(H_n v_n + L_n) + G_{n-1}) + b_n(H_n v_n + L_n) + c_{n-1},$$

and hence

$$F_n = a_n F_{n-1} H_n + b_n H_n,$$

and

$$G_n = a_n(F_{n-1}L_n + G_{n-1}) + b_n L_n + c_{n-1}.$$

Thus,

$$u_n = F_n v_n + G_n,$$

so that, in particular,

$$u_N = F_N v_f + G_N,$$

$$v_{N-1} = H_N v_f + L_N,$$

and hence

$$u_{N-1} = F_{N-1}v_{N-1} + G_{N-1},$$

$$v_{N-2} = H_{N-1}v_{N-1} + L_{N-1},$$

and so on, up to finding,

$$u_1 = F_1 v_1 + G_1,$$

and

$$v_0 = H_0 v_1 + L_0,$$

where $H_0 = 1/e_1$ and $L_0 = -(d_1 u_0 + f_1)/e_1$.

The next step is to replace $\{\tilde{u}_n\}$ and $\{\tilde{v}_n\}$ by $\{u_n\}$ and $\{v_n\}$, respectively, and then to repeat the process up to the satisfaction of an appropriate convergence criterion.

16.7.1 An Example in Nuclear Physics

As an application of the method above exposed we develop numerical results for the system of equations relating the neutron kinetics of a nuclear reactor. Following [61], the system in question is given by

$$\begin{cases} n'(t) = \frac{(\rho(T)-\beta)}{L}n(t) + \lambda C(t) \\ C'(t) = \frac{\beta}{L}n(t) - \lambda C(t) \\ T'(t) = Hn(t), \end{cases} \tag{16.69}$$

where $n(t)$ is the neutron population, $C(t)$ is the concentration of delayed neutrons, $T(t)$ is the core temperature, $\rho(T)$ is the reactivity (which depends on the temperature T), β is the delayed neutron fraction, L is the prompt reactors generation time, λ is the average decay constant of the precursors, and H is the inverse of the reactor thermal capacity.

For our numerical examples we consider $T(0s) = 300\,\text{K}$ and $T(100\,\text{s}) = T_f = 350\,\text{K}$. Moreover we assume the relation

$$C(0) = \frac{1}{\lambda}\frac{(\beta - \rho(0))}{L}n(0),$$

where $n(0)$ is unknown (to be numerically calculated by our method such that we have $T(100\,\text{s}) = T_f$).

Also we consider

$$\rho(T) = \rho(0) - \alpha(T - T(0)).$$

The remaining values are $\beta = 0.0065$, $L = 0.0001\,\text{s}$, $\lambda = 0.00741\,\text{s}^{-1}$, $H = 0.05\,\text{K/(MWs)}$, $\alpha = 5 \cdot 10^{-5}\,\text{K}^{-1}$, and $\rho(0) = 0.2\beta$.

First we linearize the system in question about $(\tilde{n}, \tilde{T})$ obtaining (in fact it is a first approximation)

$$\begin{aligned} n'(t) = & \frac{\rho(\tilde{T}) - \beta}{L}n(t) + \frac{\rho(T) - \beta}{L}\tilde{n}(t) \\ & - \frac{\rho(\tilde{T}) - \beta}{L}\tilde{n}(t) + \lambda C(t), \end{aligned} \tag{16.70}$$

$$C'(t) = \frac{\beta}{L}n(t) - \lambda C(t),$$

$$T'(t) = Hn(t),$$

where $\rho(T) = \rho(0) - \alpha(T - T(0))$.

Discretizing such a system in finite differences, we get

$$\begin{aligned} (n_{i+1} - n_i)/d = & \frac{\rho(\tilde{T}_i) - \beta}{L}n_i + \frac{\rho(T_i) - \beta}{L}\tilde{n}_i \\ & - \frac{\rho(\tilde{T}_i) - \beta}{L}\tilde{n}_i + \lambda C_i, \end{aligned} \tag{16.71}$$

$$(C_{i+1} - C_i)/d = \frac{\beta}{L}n_i - \lambda C_i,$$

$$(T_{i+1} - T_i)/d = Hn_i,$$

where $d = 100\,\text{s}/N$, where N is the number of nodes.

Hence, we may write

$$n_{i+1} = a_i n_i + b_i T_i + d_i C_i + e_i, \tag{16.72}$$

$$C_{i+1} = f n_i + g C_i, \tag{16.73}$$

$$T_{i+1} = hT_i + mn_i, \tag{16.74}$$

where

$$a_i = 1 + \frac{\rho(\tilde{T}_i) - \beta}{L} d,$$

$$b_i = \frac{-\alpha}{L} \tilde{n}_i d,$$

$$d_i = \lambda d,$$

$$e_i = \frac{(\rho(0) + \alpha T(0) - \beta)}{L} \tilde{n}_i d - \frac{(\rho(\tilde{T}_i) - \beta)}{L} \tilde{n}_i d,$$

$$f = \frac{\beta}{L} d,$$

$$g = 1 - \lambda d,$$

$$h = 1,$$

$$m = Hd.$$

Observe that

$$C_0 = \tilde{\alpha} n_0,$$

where

$$\tilde{\alpha} = \frac{\beta - \rho(0)}{L\lambda}.$$

For $i = 0$ from (16.74) we obtain

$$n_0 = \frac{T_1 - hT_0}{m} = \alpha_1 T_1 + \beta_1, \tag{16.75}$$

where $\alpha_1 = 1/m$ and $\beta_1 = -(h/m)T_0$.

Therefore,

$$C_0 = \tilde{\alpha} n_0 = \tilde{\alpha}(\alpha_1 T_1 + \beta_1).$$

Still for $i = 0$, replacing this last relation and (16.75) into (16.72), we get

$$n_1 = a_0(\alpha_1 T_1 + \beta_1) + b_0 T_0 + d_0 \tilde{\alpha}(\alpha_1 T_1 + \beta_1) + e_0,$$

so that

$$n_1 = \tilde{\alpha}_1 T_1 + \tilde{\beta}_1, \tag{16.76}$$

where

$$\tilde{\alpha}_1 = a_0 \alpha_0 + d_0 \tilde{\alpha} \alpha_1,$$

and

$$\tilde{\beta}_1 = a_0 \beta_1 + b_0 T_0 + d_0 \tilde{\alpha} \beta_1 + e_0.$$

Finally, from (16.73),

$$\begin{aligned} C_1 &= f(\alpha_1 T_1 + \beta_1) + g\tilde{\alpha}(\alpha_1 T_1 + \beta_1) \\ &= \hat{\alpha}_1 T_1 + \hat{\beta}_1, \end{aligned} \tag{16.77}$$

where

$$\hat{\alpha}_1 = f\alpha_1 + g\tilde{\alpha}\alpha_1,$$

and

$$\hat{\beta}_1 = f\beta_1 + g\tilde{\alpha}\beta_1.$$

Reasoning inductively, having

$$n_i = \tilde{\alpha}_i T_i + \tilde{\beta}_i, \tag{16.78}$$

$$n_{i-1} = \alpha_i T_i + \beta_i, \tag{16.79}$$

$$C_i = \hat{\alpha}_i T_i + \hat{\beta}_i, \tag{16.80}$$

we are going to obtain the corresponding relations for $i+1$, $i \geq 1$. From (16.74) and (16.78) we obtain

$$T_{i+1} = hT_i + m(\tilde{\alpha}_i T_i + \tilde{\beta}_i),$$

so that

$$T_i = \eta_i T_{i+1} + \xi_i, \tag{16.81}$$

where

$$\eta_i = (h + m\tilde{\alpha}_i)^{-1},$$

and

$$\xi_i = -(m\tilde{\beta}_i)\eta_i.$$

On the other hand, from (16.72), (16.78), and (16.80), we have

$$n_{i+1} = a_i(\tilde{\alpha}_i T_i + \tilde{B}_i) + b_i T_i + d_i(\hat{\alpha}_i T_i + \hat{\beta}_i) + e_i,$$

so that from this and (16.81), we obtain

$$n_{i+1} = \tilde{\alpha}_i T_{i+1} + \tilde{\beta}_{i+1},$$

where

$$\tilde{\alpha}_{i+1} = a_i\tilde{\alpha}_i\eta_i + b_i\eta_i + d_i\hat{\alpha}_i\eta_i,$$

and

$$\tilde{\beta}_{i+1} = a_i(\tilde{\alpha}_i\xi_i + \tilde{\beta}_i) + b_i\xi_i + d_i(\hat{\alpha}_i\xi_i + \hat{\beta}_i) + e_i.$$

Also from (16.78) and (16.81) we have

$$n_i = \tilde{\alpha}_i(\eta_i T_{i+1} + \xi_i) + \tilde{\beta}_i = \alpha_{i+1} T_{i+1} + \beta_{i+1},$$

where

$$\alpha_{i+1} = \tilde{\alpha}_i\eta_i,$$

and

$$\beta_{i+1} = \tilde{\alpha}_i\xi_i + \tilde{\beta}_i.$$

Moreover,

$$\begin{aligned}
C_{i+1} &= fn_i + gC_i \\
&= f(\alpha_{i+1}T_{i+1} + \beta_{i+1}) \\
&\quad + g(\hat{\alpha}_i T_i + \hat{\beta}_i) \\
&= f(\alpha_{i+1}T_{i+1} + \beta_{i+1}) \\
&\quad + g(\hat{\alpha}_i(\eta_i T_i + \xi_i) + \hat{\beta}_i) \\
&= \hat{\alpha}_{i+1}T_{i+1} + \hat{\beta}_{i+1},
\end{aligned} \tag{16.82}$$

where

$$\hat{\alpha}_{i+1} = f\alpha_{i+1} + g\hat{\alpha}_i\eta_i,$$

and

$$\hat{\beta}_i = f\beta_{i+1} + g\hat{\alpha}_i\xi_i + g\hat{\beta}_i.$$

Summarizing, we have obtained linear functions $(F_0)_i, (F_1)_i$, and $(F_2)_i$ such that

$$T_i = (F_0)_i(T_{i+1}),$$

$$n_i = (F_1)_i(T_{i+1}),$$

$$C_i = (F_2)_i(T_{i+1}),$$

$\forall i \in \{1,\ldots,N-1\}$.

Thus, considering the known value $T_N = T_f$, we obtain

$$T_{N-1} = (F_0)_{N-1}(T_f),$$

$$n_{N-1} = (F_1)_{N-1}(T_f),$$

$$C_{N-1} = (F_2)_{N-1}(T_f),$$

and having T_{N-1}, we get

$$T_{N-2} = (F_0)_{N-2}(T_{N-1}),$$

$$n_{N-2} = (F_1)_{N-2}(T_{N-1}),$$

$$C_{N-2} = (F_2)_{N-1}(T_{N-1}),$$

and so on, up to finding

$$T_1 = (F_0)_1(T_2),$$

$$n_1 = (F_1)_1(T_2),$$

$$C_1 = (F_2)_1(T_2),$$

and $n_0 = (F_0)_1(T_1)$.

The next step is to replace $(\tilde{n}, \tilde{T})$ by the last calculated (n, T) and then to repeat the process until an appropriate convergence criterion is satisfied.

Concerning our numerical results through such a method, for the solution $n(t)$ obtained, please see Fig. 16.11. For the solution $T(t)$, see Fig. 16.12.

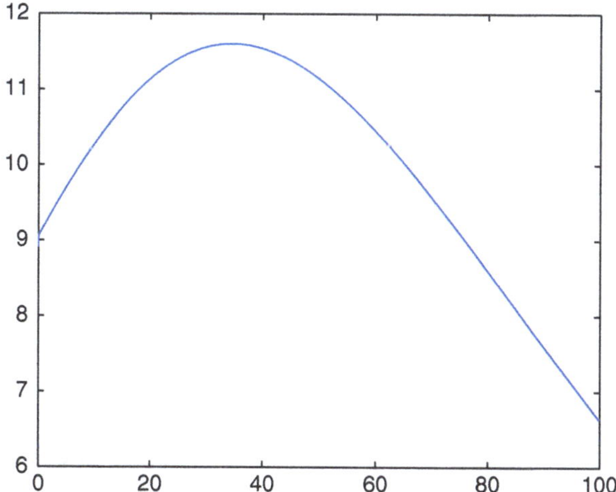

Fig. 16.11 Solution $n(t)$ for $0\,\mathrm{s} \leq t \leq 100\,\mathrm{s}$

We emphasize the numerical results here obtained are consistent with the current literature (see [61] for details).

16.8 Conclusion

In this chapter, first we have presented a duality principle for the Ginzburg–Landau system in the presence of a magnetic field. We highlight to have obtained sufficient conditions of optimality, similarly to the canonical duality procedure introduced by Gao [36]. It is worth mentioning the dual formulation is concave and amenable to numerical computation.

In a second step, we have introduced the matrix version of the generalized method of lines. Also we develop a convergent algorithm suitable for equations that present strong variational formulation and, in particular, suitable for Ginzburg–Landau-type equations. The results are rigorously proven and numerical examples are provided. We emphasize that even as the parameter ε is very small, namely, $\varepsilon = 0.0001$, the results are consistent and the convergence is very fast. Finally, in the

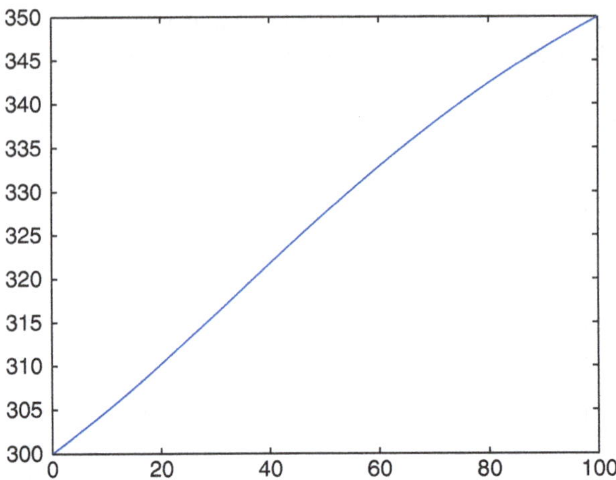

Fig. 16.12 Solution $T(t)$ for $0\,\text{s} \leq t \leq 100\,\text{s}$

last section, we develop in details Newton's method combined with the generalized method of lines main idea, with numerical results relating an example in nuclear physics.

Chapter 17
On Duality Principles for Scalar and Vectorial Multi-well Variational Problems

17.1 Introduction

Remark 17.1.1. This chapter was published in an article form by *Nonlinear Analysis-Elsevier, reference [14]*

In this chapter, our first objective is the establishment of a duality principle suitable for the variational form of some nonlinear vectorial problems in physics and engineering. The results are based on standard tools of convex analysis. As a first example we apply them to a phase transition model, which may be found in a similar format in Chenchiah and Bhattacharya [18]. It is relevant to observe that the study developed in [18] is restricted to the two-well problem, whereas our new duality principle is applicable to vectorial multi-well formulations in general, not restricted to two or three wells.

In Sect. 17.4 we discuss how the standard tools of convex analysis can be used to study the scalar case.

In Sect. 17.3 we present the main theorem in this chapter, namely Theorem 17.3.1, which corresponds, as mentioned above, to a new duality principle. It is important to emphasize that this principle stands for relaxation for a vectorial phase transition problem. In the next lines, we describe such a result.

Consider $(G\circ\Lambda) : U \to \mathbb{R}$ and $(F\circ\Lambda) : U \to \mathbb{R}$, F being a convex Gâteaux differentiable functional such that $J : U \to \mathbb{R}$ defined as

$$J(u) = (G\circ\Lambda)(u) - (F\circ\Lambda)(u) - \langle u, f\rangle_U$$

is bounded below. Here $\Lambda : U \to Y$ is a continuous linear injective operator whose the respective adjoint is denoted by $\Lambda^* : Y^* \to U^*$. Under such assumptions, we have

$$\inf_{u\in U}\{G^{**}(\Lambda u) - F(\Lambda u) - \langle u, f\rangle_U\}$$
$$\geq \sup_{v^*\in A^*}\left\{\inf_{z^*\in Y^*}\{(F\circ\Lambda)^*(\Lambda^* z^*) - G^*(v^*+z^*)\}\right\}.$$

F. Botelho, *Functional Analysis and Applied Optimization in Banach Spaces: Applications to Non-Convex Variational Models*, DOI 10.1007/978-3-319-06074-3_17,

where

$$A^* = \{v^* \in Y^* \mid \Lambda^* v^* = f\}.$$

Furthermore, under additional assumptions to be specified, we have

$$\inf_{u \in U} \{G^{**}(\Lambda u) - F(\Lambda u) - \langle u, f \rangle_U\}$$
$$= \sup_{v^* \in A^*} \left\{ \inf_{z^* \in Y^*} \{(F \circ \Lambda)^*(\Lambda^* z^*) - G^*(v^* + z^*)\} \right\}.$$

Remark 17.1.2. Henceforth by a regular boundary we mean a condition sufficient for the standard Green–Gauss theorems of integration by parts holds. Moreover, all derivatives in this text are understood in distributional sense.

Now we also present a summary of our main applied result, namely, a duality principle applied to a vectorial phase transition problem.

For an open bounded connected set $S \subset \mathbb{R}^3$ with a regular boundary denoted by ∂S, consider the functional $J : U \to \mathbb{R}$, where

$$J(u) = \frac{1}{2} \int_S \min_{k \in \{1,\ldots,N\}} \{g_k(\varepsilon(u)) + \beta_k\}\, dx - \langle u, f \rangle_U,$$

$$g_k(\varepsilon(u)) = (\varepsilon_{ij}(u) - \mathbf{e}^k_{ij}) C^k_{ijlm} (\varepsilon_{lm}(u) - \mathbf{e}^k_{lm}).$$

The operator $\varepsilon : U \to Y = Y^* = L^2(S; \mathbb{R}^9)$ is defined by

$$\varepsilon_{ij}(u) = \frac{1}{2}(u_{i,j} + u_{j,i}), \text{ for } i, j \in \{1,2,3\}.$$

Furthermore $\{C^k_{ijlm}\}$ are positive definite matrices and $\beta_k \in \mathbb{R}$ for each $k \in \{1,\ldots,N\}$, and $f \in L^2(S;\mathbb{R}^3)$ is an external load. Here $\mathbf{e}^k \in \mathbb{R}^{3\times 3}$ for $k \in \{1,\ldots,N\}$ represent the stress-free configurations or phases presented by a solid with field of displacements $u = (u_1, u_2, u_3) \in W^{1,2}(S;\mathbb{R}^3)$ (due to f). Also

$$U = \{u \in W^{1,2}(S;\mathbb{R}^3) \mid u = (0,0,0) \equiv \theta \text{ on } \partial S\} = W^{1,2}_0(S;\mathbb{R}^3).$$

Observe that we may write

$$J(u) = G(\varepsilon(u)) - F(\varepsilon(u)) - \langle u, f \rangle_U$$

where

$$G(\varepsilon(u)) = \frac{1}{2} \int_S \min_{k \in \{1,\ldots,N\}} \{g_k(\varepsilon(u)) + \beta_k\}\, dx + \frac{K}{2} \int_S (\varepsilon_{ij}(u)) H_{ijlm} (\varepsilon_{lm}(u))\, dx,$$

$$F(\varepsilon(u)) = \frac{K}{2} \int_S (\varepsilon_{ij}(u)) H_{ijlm} (\varepsilon_{lm}(u))\, dx,$$

$\{H_{ijlm}\}$ is a positive definite matrix (the identity for example) and $K > 0$ is an appropriate constant. The final duality principle is expressed by

$$\inf_{u \in U}\{G^{**}(\varepsilon(u)) - F(\varepsilon(u)) - \langle u, f\rangle_U\}$$

$$\geq \sup_{v^* \in A^*}\left\{\inf_{t \in B}\left\{\frac{1}{2K}\int_S z^*_{ij}(v^*,t)\hat{H}_{ijlm}z^*_{lm}(v^*,t)\,dx\right.\right.$$

$$\sum_{k=1}^N\left[-\int_S \frac{1}{2}t_k(v^*_{ij}+z^*_{ij}(v^*,t))D^k_{ijlm}(v^*_{lm}+z^*_{lm}(v^*,t))\,dx\right.$$

$$\left.\left.\left.+\int_S(-(v^*_{ij}+z^*_{ij}(v^*,t))t_k\hat{C}^k_{ijlm}\mathbf{e}^k_{lm}+t_k\beta_k)\,dx\right]\right\}\right\}, \tag{17.1}$$

where $z^*(v^*,t)$ is obtained through Eq. (17.44), that is,

$$\frac{1}{K}\hat{H}_{ijlm}z^*_{lm} - \sum_{k=1}^N\{t_kD^k_{ijlm}(v^*_{lm}+z^*_{lm})\} - \sum_{k=1}^N t_k\hat{C}^k_{ijlm}\mathbf{e}^k_{lm} = 0, \text{ in } S.$$

Finally,

$$A^* = \{v^* \in Y^* \mid v^*_{ij,j} + f_i = 0 \text{ in } S\},$$

and

$$B = \left\{(t_1,\ldots,t_N) \text{ measurable } \mid \right.$$

$$\left. t_k(x) \in [0,1],\ \forall k \in \{1,\ldots,N\},\ \sum_{k=1}^N t_k(x) = 1, \text{ a.e. in } S\right\}.$$

17.2 Preliminaries

We denote by U and Y Banach spaces which the topological dual spaces are identified with U^* and Y^*, respectively. Unless otherwise indicated, Y is assumed to be reflexive. The canonical duality pairing between U and U^* is denoted by $\langle\cdot,\cdot\rangle_U : U \times U^* \to \mathbb{R}$, through which the linear continuous functionals defined on U are represented.

Given $F : U \to \bar{\mathbb{R}} = \mathbb{R} \cup \{+\infty\}$ its polar $F^* : U^* \to \bar{\mathbb{R}}$ is defined as

$$F^*(u^*) = \sup_{u \in U}\{\langle u, u^*\rangle_U - F(u)\}.$$

Recall that the sub-differential $\partial F(u)$ is the subset of U^* given by

$$\partial F(u) = \{u^* \in U^*, \text{ such that } \langle v-u, u^*\rangle_U + F(u) \leq F(v),\ \forall v \in U\}.$$

Also relevant is the next definition.

Definition 17.2.1 (Adjoint Operator). Let U and Y be Banach spaces and $\Lambda : U \to Y$ a continuous linear operator. The adjoint operator related to Λ, denoted by $\Lambda^* : Y^* \to U^*$, is defined through the equation:

$$\langle u, \Lambda^* v^* \rangle_U = \langle \Lambda u, v^* \rangle_Y, \ \ \forall u \in U, \ \ v^* \in Y^*.$$

Finally, the following duality principle found in [25] will be used in this text.

Theorem 17.2.2. *Let $G : Y \to \bar{\mathbb{R}} = \mathbb{R} \cup \{+\infty\}$ and $F : U \to \mathbb{R}$ be two convex l.s.c. (lower semicontinuous) functionals so that $J : U \to \bar{\mathbb{R}}$ defined as*

$$J(u) = (G \circ \Lambda)(u) - F(u)$$

is bounded below, where $\Lambda : U \to Y$ is a continuous linear operator which the respective adjoint is denoted by $\Lambda^* : Y^* \to U^*$. Thus if there exists $\hat{u} \in U$ such that $F(\hat{u}) < +\infty$, $G(\Lambda \hat{u}) < +\infty$ being G continuous at $\Lambda \hat{u}$, we have

$$\inf_{u \in U} \{J(u)\} = \sup_{v^* \in Y^*} \{-G^*(v^*) - F^*(-\Lambda^* v^*)\}$$

and there exists at least one $v_0^* \in Y^*$ which maximizes the dual formulation. If in addition U is reflexive and

$$\lim_{\|u\| \to +\infty} J(u) = +\infty$$

then both primal and dual formulations have global extremals so that there exist $u_0 \in U$ and $v_0^* \in Y^*$ such that

$$J(u_0) = \min_{u \in U} \{J(u)\} = \max_{v^* \in Y^*} \{-G^*(v^*) - F^*(-\Lambda^* v^*)\} = -G^*(v_0^*) - F^*(-\Lambda^* v_0^*).$$

Also

$$G(\Lambda u_0) + G^*(v_0^*) = \langle \Lambda u_0, v_0^* \rangle_Y,$$
$$F(u_0) + F^*(-\Lambda^* v_0^*) = \langle u_0, -\Lambda^* v_0^* \rangle_U,$$

so that

$$G(\Lambda u_0) + F(u_0) = -G^*(v_0^*) - F^*(-\Lambda^* v_0^*).$$

Also fundamental for the construction of the main duality principle is a result found in Toland [67] (despite we have not used it directly we have followed a similar idea) which is as follows.

Theorem 17.2.3. *Consider the functionals $F, G : U \to \mathbb{R}$ through which we define $J : U \to \mathbb{R}$ as*

$$J(u) = G(u) - F(u). \tag{17.2}$$

Suppose there exists $u_0 \in U$ such that

$$J(u_0) = \inf_{u \in U} \{J(u)\} \tag{17.3}$$

and $\partial F(u_0) \neq \emptyset$.

Under such assumptions we can write

$$\inf_{u\in U}\{J(u)\} = \inf_{u^*\in U^*}\{F^*(u^*) - G^*(u^*)\} \tag{17.4}$$

and for $u_0^* \in \partial F(u_0)$ *we have*

$$F^*(u_0^*) - G^*(u_0^*) = \inf_{u^*\in U^*}\{F^*(u^*) - G^*(u^*)\}. \tag{17.5}$$

17.3 The Main Duality Principle

Now we present the main theoretical result in this chapter.

Theorem 17.3.1. *Consider the functionals* $(G\circ\Lambda): U \to \mathbb{R}$ *and* $(F\circ\Lambda): U \to \mathbb{R}$ *being* F *convex and Gâteaux differentiable and also such that* $J: U \to \mathbb{R}$ *defined as*

$$J(u) = (G\circ\Lambda)(u) - (F\circ\Lambda)(u) - \langle u, f\rangle_U$$

is bounded below. Here $\Lambda: U \to Y$ *is a continuous linear injective operator whose respective adjoint is denoted by* $\Lambda^*: Y^* \to U^*$.

Under such assumptions, we have

$$\inf_{u\in U}\{G^{**}(\Lambda u) - F(\Lambda u) - \langle u, f\rangle_U\} \geq$$

$$\sup_{v^*\in A^*}\left\{\inf_{z^*\in Y^*}\{(F\circ\Lambda)^*(\Lambda^* z^*) - G^*(v^* + z^*)\}\right\}.$$

where

$$A^* = \{v^* \in Y^* \mid \Lambda^* v^* = f\}.$$

Furthermore, assuming that $G^*: Y^* \to \mathbb{R}$ *is Lipschitz continuous, there exists* $v_0^* \in Y^*$ *such that*

$$\hat{J}^*(v_0^*) = \max_{v^*\in Y^*}\{\hat{J}^*(v^*)\},$$

where

$$J^*(v^*) = \inf_{z^*\in Y^*}\{(F\circ\Lambda)^*(\Lambda^* z^*) - G^*(v^* + z^*)\},$$

$$Ind(v^*) = \begin{cases} 0, & \text{if } \Lambda^* v^* + f = 0, \\ +\infty, & \text{otherwise,} \end{cases}$$

and

$$\hat{J}^*(v^*) = J^*(v^*) - Ind(v^*).$$

In addition we suppose that defining

$$J_1^*(v^*, z^*) = (F\circ\Lambda)^*(\Lambda^* z^*) - G^*(v^* + z^*),$$

we have that

$$J_1^*(v_0^*, z^*) \to +\infty$$

as

$$\|z^*\|_{Y^*} \to +\infty, \ \ or \ \|\Lambda^* z^*\|_{L^2} \to +\infty.$$

Furthermore, suppose that if $\{z_n^*\} \subset Y^*$ *is such that* $\|\Lambda^* z_n^*\|_{L^2} < K, \forall n \in \mathbb{N}$ *for some* $K > 0$*, then there exists* $\tilde{z}^* \in Y^*$ *such that for a not relabeled subsequence, we have*

$$\Lambda^* z_n^* \rightharpoonup \Lambda^* \tilde{z}^*, \text{ weakly in } L^2,$$

and

$$z_n^* \to \tilde{z}^*, \text{ strongly in } Y^*.$$

Under such additional assumptions, there exist $z_0^* \in Y^*$ *and* $u_0 \in U$ *such that*

$$\begin{aligned}
&\inf_{u \in U}\{G^{**}(\Lambda u) - F(\Lambda u) - \langle u, f\rangle_U\} \\
&= G^{**}(\Lambda u_0) - F(\Lambda u_0) - \langle u_0, f\rangle_U \\
&= (F \circ \Lambda)^*(\Lambda^* z_0^*) - G^*(v_0^* + z_0^*) \\
&= \sup_{v^* \in A^*}\left\{\inf_{z^* \in Y^*}\{(F \circ \Lambda)^*(\Lambda^* z^*) - G^*(v^* + z^*)\}\right\}.
\end{aligned}$$

Proof. Observe that

$$G^*(v^* + z^*) \geq \langle \Lambda u, v^*\rangle_Y + \langle \Lambda u, z^*\rangle_Y - G^{**}(\Lambda u),$$

$\forall u \in U, v^* \in Y^*$, $z^* \in Y^*$, that is,

$$\begin{aligned}
&-(F \circ \Lambda)^*(\Lambda^* z^*) + G^*(v^* + z^*) \\
&\geq \langle u, f\rangle_U - (F \circ \Lambda)^*(\Lambda^* z^*) + \langle \Lambda u, z^*\rangle_Y \\
&\quad - G^{**}(\Lambda u),
\end{aligned} \tag{17.6}$$

$\forall u \in U, v^* \in A^*, z^* \in Y^*$, and hence,

$$\begin{aligned}
&\sup_{z^* \in Y^*}\{-(F \circ \Lambda)^*(\Lambda^* z^*) + G^*(v^* + z^*)\} \\
&\geq \langle u, f\rangle_U + F(\Lambda u) - G^{**}(\Lambda u),
\end{aligned} \tag{17.7}$$

$\forall u \in U$, $v^* \in A^*$ and thus

$$\begin{aligned}
&\inf_{u \in U}\{G^{**}(\Lambda u) - F(\Lambda u) - \langle u, f\rangle_U\} \\
&\geq \sup_{v^* \in A^*}\left\{\inf_{z^* \in Y^*}\{(F \circ \Lambda)^*(\Lambda^* z^*) - G^*(v^* + z^*)\}\right\}.
\end{aligned}$$

Observe that we may write

$$\inf_{u \in U}\{G^{**}(\Lambda u) - F(\Lambda u) - \langle u, f\rangle_U\} \geq \sup_{v^* \in Y^*}\{\tilde{J}^*(v^*)\}. \tag{17.8}$$

Moreover the functional $\hat{J}^*(v^*)$ is concave, as the infimum in z^* of a family of concave functionals in v^*. On the other hand it is clear that

$$\hat{J}^*(v^*) \leq -G^*(v^*) - \text{Ind}(v^*) \; \forall v^* \in Y^*,$$

so that

$$\hat{J}^*(v^*) \to -\infty, \text{ as } \|v^*\|_{Y^*} \to +\infty \text{ or } \|\Lambda^* v^*\|_{L^2} \to +\infty.$$

Therefore if $\{v_n^*\}$ is a maximizing sequence of $\hat{J}^*(v^*)$ we have that there exists a constant $K_0 > 0$ such that

$$\|v_n^*\|_{Y^*} < K_0, \text{ and } \|\Lambda^* v^*\|_{L^2} < K_0, \forall n \in \mathbb{N}.$$

Since Y^* and L^2 are reflexive Banach spaces, there exist $v_0^* \in Y^*$ and $\tilde{v}_0^* \in L^2$, such that up to a not relabeled subsequence we have

$$v_n^* \rightharpoonup v_0^*, \text{ weakly in } Y^*,$$

and

$$\Lambda^* v_n^* \rightharpoonup \tilde{v}_0^*, \text{ weakly in } L^2.$$

Observe that given $\varphi \in C_c^\infty$ we have

$$\begin{aligned} \langle v_0^*, \Lambda \varphi \rangle_{L^2} &= \lim_{n\to\infty} \langle v_n^*, \Lambda \varphi \rangle_{L^2} \\ &= \lim_{n\to\infty} \langle \Lambda^* v_n^*, \varphi \rangle_{L^2} \\ &= \langle \tilde{v}_0^*, \varphi \rangle_{L^2}. \end{aligned} \tag{17.9}$$

Thus

$$\tilde{v}_0^* = \Lambda^* v_0^*,$$

in distributional sense, so that

$$\Lambda^* v_n^* \rightharpoonup \Lambda^* v_0^*, \text{ weakly in } L^2.$$

Hence, as $\hat{J}^*(v^*)$ is concave and strongly continuous, it is also upper semicontinuous, so that

$$\limsup_{n\to\infty} \{\hat{J}^*(v_n^*)\} \leq \hat{J}^*(v_0^*).$$

Therefore, since $\{v_n^*\}$ is a maximizing sequence, we may conclude that

$$\hat{J}^*(v_0^*) = \max_{v^* \in Y^*} \{\hat{J}^*(v^*)\}.$$

Consider now the infimum

$$\inf_{z^* \in Y^*} \{J_1(v_0^*, z^*)\},$$

where

$$J_1^*(v_0^*, z^*) = (F \circ \Lambda)^*(\Lambda^* z^*) - G^*(v_0^* + z^*).$$

From the coercivity hypothesis, if $\{z_n^*\}$ is a minimizing sequence, there exists $K_1 > 0$ such that

$$\|z_n^*\|_{Y^*} < K_1 \text{ and } \|\Lambda^* z_n^*\|_{L^2} < K_1, \ \forall n \in \mathbb{N}.$$

Also from the hypothesis, up to a not relabeled subsequence, there exist $z_0^* \in Y^*$ such that

$$\Lambda z_n^* \rightharpoonup \Lambda^* z_0^*, \ \text{ weakly in } L^2,$$

and

$$z_n^* \to z_0^*, \text{ strongly in } Y^*.$$

As $G^*(v^*)$ is strongly continuous, we obtain

$$G(v_0^* + z_n^*) \to G^*(v_0^* + z_0^*).$$

On the other hand, as $(F \circ \Lambda)^*$ is convex and strongly continuous, it is also weakly lower semicontinuous, so that

$$\liminf_{n\to\infty}\{(F \circ \Lambda)^*(\Lambda^* z_n^*)\} \geq (F \circ \Lambda)^*(\Lambda^* z_0^*).$$

Therefore

$$\liminf_{n\to\infty}\{(F \circ \Lambda)^*(\Lambda^* z_n^*) - G^*(v_0^* + z_n^*)\} \geq (F \circ \Lambda)^*(\Lambda^* z_0^*) - G^*(v_0^* + z_0^*).$$

$\{z_n^*\}$ being a minimizing sequence, we obtain

$$\inf_{z^* \in Y^*}\{J_1^*(v_0^*, z^*)\} = (F \circ \Lambda)^*(\Lambda^* z_0^*) - G^*(v_0^* + z_0^*) = \hat{J}^*(v_0^*).$$

To complete the proof observe that the extremal equation is satisfied

$$\Lambda \left[\frac{\partial (F \circ \Lambda)^*(\Lambda^* z_0^*)}{\partial \hat{z}^*}\right] \in \partial G^*(v_0^* + z_0^*), \tag{17.10}$$

where

$$\hat{z}^* = \Lambda^* z^*.$$

Defining

$$u_0 = \frac{\partial (F \circ \Lambda)^*(\Lambda^* z_0^*)}{\partial \hat{z}^*},$$

we obtain

$$\Lambda u_0 \in \partial G^*(v_0^* + z_0^*). \tag{17.11}$$

From these two last equations we obtain respectively

$$(F \circ \Lambda)^*(\Lambda^* z_0^*) = \langle u_0, \Lambda^* v_0^* \rangle_U - F(\Lambda u_0),$$

and

$$G^*(v_0^* + z_0^*) = \langle \Lambda u_0, v_0^* \rangle_Y + \langle \Lambda u_0, z_0^* \rangle_Y - G^{**}(\Lambda u_0). \tag{17.12}$$

From the fact that $v_0^* \in A^*$, we have

$$\Lambda^* v_0^* = f.$$

Note that from the last three equations, we obtain

$$\begin{aligned}(F \circ \Lambda)^*(\Lambda^* z_0^*) - G^*(v_0^* + z_0^*) \\ = G^{**}(\Lambda u_0) - F(\Lambda u_0) - \langle u_0, f \rangle_U.\end{aligned} \tag{17.13}$$

Therefore, we may conclude that

$$\begin{aligned}&\inf_{u \in U} \{G^{**}(\Lambda u) - F(\Lambda u) - \langle u, f \rangle_U\} \\ &= G^{**}(\Lambda u_0) - F(\Lambda u_0) - \langle u_0, f \rangle_U \\ &= (F \circ \Lambda)^*(\Lambda^* z_0^*) - G^*(v_0^* + z_0^*) \\ &= \sup_{\hat{v}^* \in A^*} \left\{ \inf_{z^* \in Y^*} \{(F \circ \Lambda)^*(\Lambda^* z^*) - G^*(v^* + z^*)\} \right\}.\end{aligned}$$

The proof is complete.

17.4 The Scalar Multi-well Problem

This section is dedicated to the analysis of the scalar multi-well problem via duality.

17.4.1 The Primal Variational Formulation

Consider an open bounded connected set $S \subset \mathbb{R}^n$ with a regular boundary Γ. Also consider the convex and differentiable functions $g_i : \mathbb{R}^n \to \mathbb{R}$ for each $i \in \{1, \ldots, N\}$ and $(G \circ \nabla) : U \to \mathbb{R}$ non-convex defined by

$$g(\nabla u) = \min_{i \in \{1, \ldots, N\}} \{g_i(\nabla u)\}, \tag{17.14}$$

and

$$G(\nabla u) = \int_S \min_{i \in \{1, \ldots, N\}} \{g_i(\nabla u)\} dS = \int_S g(\nabla u) dS. \tag{17.15}$$

We also assume

$$\frac{G(\nabla u)}{\|u\|_U} \to +\infty \text{ as } \|u\|_U \to \infty, \tag{17.16}$$

and

$$U = \{u \in W^{1,2}(S) \mid u = u_0 \text{ on } \Gamma\}. \quad (17.17)$$

As a preliminary result, we present Corollary 3.8, at p. 339 of Ekeland and Temam [25] (here Ω stands for S).

Theorem 17.4.1. *Let f be a Carathéodory function from $\Omega \times (\mathbb{R} \times \mathbb{R}^n)$ into $\mathbb{R}$ which satisfies*

$$a_2(x) + c_2|\xi|^\alpha \leq f(x,s,\xi) \leq a_1(x) + b|s|^\alpha + c_1|\xi|^\alpha \quad (17.18)$$

where $a_1,\ a_2 \in L^1(\Omega)$, $1 < \alpha < +\infty$, $b \geq 0$ and $c_1 \geq c_2 > 0$. Let $u_0 \in W^{1,\alpha}(\Omega)$. Under such assumptions, defining $\hat{U} = \{u \mid u - u_0 \in W_0^{1,2}(\Omega)\}$, we have

$$\inf_{u \in \hat{U}} \left\{ \int_\Omega f(x,u,\nabla u) dx \right\} = \min_{u \in \hat{U}} \left\{ \int_\Omega f^{**}(x,u;\nabla u)\, dx \right\} \quad (17.19)$$

The solutions of relaxed problem are weak cluster points in $W^{1,\alpha}(\Omega)$ of the minimizing sequences of primal problem.

Now we can enunciate the following result.

Theorem 17.4.2. *Consider the definition and assumptions about $(G \circ \nabla) : U \to \mathbb{R}$ indicated in (17.14), (17.15), and (17.16). Also assuming the hypothesis of Theorem 17.4.1, we have*

$$\inf_{u \in U} \{G(\nabla u) - \langle u, f \rangle_{L^2(S)}\} = \inf_{u \in U} \{G^{**}(\nabla u) - \langle u, f \rangle_{L^2(S)}\} \quad (17.20)$$

and there exists $u_0 \in U$ such that

$$\min_{u \in U} \{G^{**}(\nabla u) - \langle u, f \rangle_{L^2(S)}\} = G^{**}(\nabla u_0) - \langle u_0, f \rangle_{L^2(S)}. \quad (17.21)$$

The proof follows directly from Theorem 17.4.1. Our next proposition is very important to establish the subsequent results. It is simple so that we do not prove it.

Proposition 17.4.3. *Consider $g : \mathbb{R}^n \to \mathbb{R}$ defined as*

$$g(v) = \min_{i \in \{1,\ldots,N\}} \{g_i(v)\} \quad (17.22)$$

where $g_i : \mathbb{R}^n \to \mathbb{R}$ are not necessarily convex functions. Under such assumptions, we have

$$g^*(v^*) = \max_{i \in \{1,\ldots,N\}} \{g_i^*(v^*)\}. \quad (17.23)$$

Now we present the main duality principle for the scalar case.

Theorem 17.4.4. *For $(G \circ \nabla) : U \to \mathbb{R}$ defined as above, that is,*

$$G(\nabla u) = \int_S \min_{i \in \{1,\ldots,N\}} \{g_i(\nabla u)\} dS, \quad (17.24)$$

where here $g_i : \mathbb{R}^n \to \mathbb{R}$ *is convex, for all* $i \in \{1, \ldots, N\}$ *and* $F : U \to \mathbb{R}$*, defined as*

$$F(u) = \langle u, f \rangle_{L^2(S)}, \tag{17.25}$$

we have

$$\min_{u \in U} \{G^{**}(\nabla u) - F(u)\} = \sup_{v^* \in C^*} \{-G^*(v^*) + \langle u_0, v^* \cdot n \rangle_{L^2(\Gamma)}\}, \tag{17.26}$$

where

$$G^*(v^*) = \int_S \max_{i \in \{1,\ldots,N\}} \{g_i^*(v^*)\} dS \tag{17.27}$$

and

$$C^* = \{v^* \in Y^* | \, div(v^*) + f(x) = 0, \text{ in } S\}. \tag{17.28}$$

Proof. We have that

$$G^*(v^*) = G^{***}(v^*) = \sup_{v \in Y} \{\langle v, v^* \rangle_Y - G^{**}(v)\} \tag{17.29}$$

that is,

$$\begin{aligned} G^*(v^*) &\geq \langle \nabla u, v^* \rangle_Y - G^{**}(\nabla u) \\ &= \langle u, -div(v^*) \rangle_{L^2(S)} + \langle u_0, v^* \cdot n \rangle_{L^2(\Gamma)} - G^{**}(\nabla u), \end{aligned}$$

$\forall u \in U$, $v^* \in Y^*$ and thus, for $v^* \in C^*$, we can write

$$G^*(v^*) \geq \langle u_0, v^* \cdot n \rangle_{L^2(\Gamma)} + \langle u, f \rangle_{L^2(S)} - G^{**}(\nabla u), \; \forall u \in U, \tag{17.30}$$

or

$$\inf_{u \in U} \{G^{**}(\nabla u) - \langle u, f \rangle_{L^2(S)}\} \geq \sup_{v^* \in C^*} \{-G^*(v^*) + \langle u_0, v^* \cdot n \rangle_{L^2(\Gamma)}\}. \tag{17.31}$$

The equality in (17.31) follows from the hypothesis indicated in (17.16) and Theorem 17.2.2.

Observe that the dual formulation is convex but non-smooth. It is through the points of non-smoothness that the microstructure is formed, specially when the original primal formulation has no minimizers in the classical sense.

17.4.2 A Scalar Multi-well Formulation

To start this section, we present duality for the solution of a standard scalar multi-well problem. Consider an open bounded connected set $S \subset \mathbb{R}^3$, with a regular boundary Γ, and the function $(W \circ \nabla)$ defined as

$$W(\nabla u) = \min_{i \in \{1,\ldots,N\}} \left\{ \frac{1}{2}|\nabla u - a_i|^2 \right\} \tag{17.32}$$

where

$$U = \{u \in W^{1,2}(S) \mid u = u_0 \text{ on } \Gamma\} \tag{17.33}$$

a_i are known matrices, for all $i \in \{1,\ldots,N\}$, so that the energy of the system is modeled by $J : U \to \mathbb{R}$, where

$$J(u) = \int_S W(\nabla u)\, dx - \langle u, f \rangle_{L^2(S)} \tag{17.34}$$

or

$$J(u) = \frac{1}{2} \int_S \min_{i \in \{1,\ldots,N\}} \{|\nabla u - a_i|^2\}\, dx - \langle u, f \rangle_{L^2(S)}. \tag{17.35}$$

From Theorem 17.4.4 we have

$$\inf_{u \in U} \{J(u)\} = \sup_{v^* \in C^*} \left\{ -\int_S \max_{i \in \{1,\ldots,N\}} \left\{ \frac{1}{2}|v^*|^2 + v^{*T} a_i \right\} dx + \langle u_0, v^* \cdot n \rangle_{L^2(\Gamma)} \right\} \tag{17.36}$$

or

$$\inf_{u \in U} \{J(u)\}$$
$$= \sup_{v^* \in C^*} \left\{ \inf_{\lambda \in B} \left\{ -\int_S \left\{ \frac{1}{2}|v^*|^2 + \sum_{i=1}^N \lambda_i v^{*T} a_i \right\} dx \right\} + \langle u_0, v^* \cdot n \rangle_{L^2(\Gamma)} \right\}$$

where

$$B = \{\lambda = (\lambda_1,\ldots,\lambda_N) \text{ measurable} \mid \lambda_i(x) \in [0,1],\ \forall i \in \{1,\ldots,N\} \text{ and } \sum_{i=1}^N \lambda_i(x) = 1\}, \tag{17.37}$$

and

$$C^* = \{v^* \in Y^* \mid div(v^*) + f = 0, \text{ in } S\} \tag{17.38}$$

It is important to emphasize that, in general, this kind of problem does not present minimizers in the classical sense. The solution of the dual problem (which is well posed and convex) reflects the average behavior of minimizing sequences as weak cluster points (of such sequences).

17.5 Duality for a Vectorial Multi-well Model Applicable to Phase Transition Problems

In this section we consider duality for another class of multi-well problems similar as those found in [18]. However, here the format of our problem is more general, not restricted to two-well formulations. Observe that the relaxation for the case of three or more wells was so far an open question in the current literature. Our main result is summarized by the next theorem.

Theorem 17.5.1. *Consider an open bounded connected set $S \subset \mathbb{R}^3$ with a regular boundary denoted by ∂S, and the functional $J : U \to \mathbb{R}$ where*

$$J(u) = \frac{1}{2}\int_S \min_{k\in\{1,\ldots,N\}} \{g_k(\varepsilon(u)) + \beta_k\}\, dx - \langle u, f\rangle_U,$$

$$g_k(\varepsilon(u)) = (\varepsilon_{ij}(u) - \mathbf{e}^k_{ij})C^k_{ijlm}(\varepsilon_{lm}(u) - \mathbf{e}^k_{lm}).$$

The operator $\varepsilon : U \to Y = Y^* = L^2(S;\mathbb{R}^9)$ is defined by

$$\varepsilon_{ij}(u) = \frac{1}{2}(u_{i,j} + u_{j,i}),\ \textit{for } i,j \in \{1,2,3\}.$$

Furthermore $\{C^k_{ijlm}\}$ are positive definite matrices and $\beta_k \in \mathbb{R}$ for each $k \in \{1,\ldots,N\}$, and $f \in L^2(S;\mathbb{R}^3)$ is a external load. Here $\mathbf{e}^k \in \mathbb{R}^{3\times 3}$ for $k \in \{1,\ldots,N\}$ represent the stress-free configurations or phases presented by a solid with field of displacements $u = (u_1,u_2,u_3) \in W^{1,2}(S;\mathbb{R}^3)$ (due to f). Also

$$U = \{u \in W^{1,2}(S;\mathbb{R}^3) \mid u = (0,0,0) \equiv \theta \text{ on } \partial S\} = W^{1,2}_0(S;\mathbb{R}^3).$$

We may write

$$J(u) = G(\varepsilon(u)) - F(\varepsilon(u)) - \langle u, f\rangle_U$$

where

$$G(\varepsilon(u)) = \frac{1}{2}\int_S \min_{k\in\{1,\ldots,N\}} \{g_k(\varepsilon(u)) + \beta_k\}\, dx + \frac{K}{2}\int_S (\varepsilon_{ij}(u))H_{ijlm}(\varepsilon_{lm}(u))\, dx,$$

$$F(\varepsilon(u)) = \frac{K}{2}\int_S (\varepsilon_{ij}(u))H_{ijlm}(\varepsilon_{lm}(u))\, dx,$$

$\{H_{ijlm}\}$ is a positive definite matrix and $K > 0$. Under such assumptions, observing that

$$G^*(v^* + z^*) = \int_S \max_{k\in\{1,\ldots,N\}} \left\{ \frac{1}{2}(v^*_{ij} + z^*_{ij})D^k_{ijlm}(v^*_{lm} + z^*_{lm}) + (v^*_{ij} + z^*_{ij})\hat{C}^k_{ijlm}e^k_{lm} - \beta_k \right\} dx, \tag{17.39}$$

and

$$F^*(z^*) = \frac{1}{2K}\int_S z^*_{ij}\hat{H}_{ijlm}z^*_{lm}\,dx,$$

we have

$$\inf_{u\in U}\{G^{**}(\varepsilon(u)) - F(\varepsilon(u)) - \langle u,f\rangle_U\} \geq \sup_{v^*\in A^*}\left\{\inf_{z^*\in Y^*}\{F^*(z^*) - G^*(z^*+v^*)\}\right\}.$$

Furthermore, this last duality principle may be written as

$$\begin{aligned}&\inf_{u\in U}\{G^{**}(\varepsilon(u)) - F(\varepsilon(u)) - \langle u,f\rangle_U\}\\ &\quad\geq \sup_{v^*\in A^*}\left\{\inf_{t\in B}\left\{\frac{1}{2K}\int_S z^*_{ij}(v^*,t)\hat{H}_{ijlm}z^*_{lm}(v^*,t)\,dx\right.\right.\\ &\quad\sum_{k=1}^N\left[-\int_S \frac{1}{2}t_k(v^*_{ij}+z^*_{ij}(v^*,t))D^k_{ijlm}(v^*_{lm}+z^*_{lm}(v^*,t))\,dx\right.\\ &\qquad\left.\left.\left.+\int_S(-(v^*_{ij}+z^*_{ij}(v^*,t))t_k\hat{C}^k_{ijlm}\mathbf{e}^k_{lm}+t_k\beta_k)\,dx\right]\right\}\right\},\end{aligned} \tag{17.40}$$

where

$$\{\hat{H}_{ijlm}\} = \{H_{ijlm}\}^{-1},$$
$$\{D^k_{ijlm}\} = \{C^k_{ijlm}+KH_{ijlm}\}^{-1},$$

and

$$\{\hat{C}^k_{ijlm}\} = \{C^k_{ijlm}+KH_{ijlm}\}^{-1}\{C^k_{ijlm}\}.$$

Moreover, $z^*(v^*,t)$ is obtained through Eq. (17.44), that is,

$$\frac{1}{K}\hat{H}_{ijlm}z^*_{lm} - \sum_{k=1}^N\{t_kD^k_{ijlm}(v^*_{lm}+z^*_{lm})\} - \sum_{k=1}^N t_k\hat{C}^k_{ijlm}\mathbf{e}^k_{lm} = 0, \text{ in } S.$$

Also,

$$A^* = \{v^*\in Y^* \mid v^*_{ij,j}+f_i = 0 \text{ in } S\},$$

and

$$B = \left\{(t_1,\ldots,t_N) \text{ measurable } \mid t_k(x)\in[0,1],\ \forall k\in\{1,\ldots,N\},\ \sum_{k=1}^N t_k(x) = 1, \text{ a.e. in } S\right\}.$$

Finally, assuming the hypotheses of the main duality principle, there exist $u_0\in U$ and $(v^*_0,z^*_0)\in\hat{Y}^*$ such that

$$\begin{aligned}&G^{**}(\varepsilon(u_0)) - F(\varepsilon(u_0)) - \langle u_0,f\rangle_U\\ &\quad= \inf_{u\in U}\{G^{**}(\varepsilon(u)) - F(\varepsilon(u)) - \langle u,f\rangle_U\}\end{aligned}$$

$$= \sup_{v^* \in A^*} \left\{ \inf_{z^* \in Y^*} \{(F \circ \varepsilon)^*(\varepsilon^* z^*) - G^*(v^* + z^*)\} \right\}$$
$$= (F \circ \varepsilon)^*(\varepsilon^* z_0^*) - G^*(v_0^* + z_0^*). \tag{17.41}$$

Proof. Observe that

$$G^*(v^* + z^*) \geq \langle \varepsilon(u), v^* \rangle_Y + \langle \varepsilon(u), z^* \rangle_Y - G^{**}(\varepsilon(u)), \forall u \in U, v^*, z^* \in Y^*,$$

that is,

$$-F^*(z^*) + G^*(v^* + z^*) \geq \langle u, f \rangle_U + \langle \varepsilon(u), z^* \rangle_Y - F^*(z^*) - G^{**}(\varepsilon(u)),$$

$\forall u \in U, v^* \in A^*, z^* \in Y^*$.

Taking the supremum in z^* in both sides of last inequality, we obtain

$$\sup_{z^* \in Y^*} \{-F^*(z^*) + G^*(z^* + v^*)\} \geq \langle u, f \rangle_U + F(\varepsilon(u)) - G^{**}(\varepsilon(u)),$$

$\forall u \in U, v^* \in A^*$. Therefore

$$\inf_{u \in U} \{G^{**}(\varepsilon(u)) - F(\varepsilon(u)) - \langle u, f \rangle_U\} \geq \sup_{v^* \in A^*} \left\{ \inf_{z^* \in Y^*} \{F^*(z^*) - G^*(z^* + v^*)\} \right\},$$

where

$$G^*(v^* + z^*) = \int_S \max_{k \in \{1,\ldots,N\}} \left\{ \frac{1}{2}(v_{ij}^* + z_{ij}^*) D_{ijlm}^k (v_{lm}^* + z_{lm}^*) \right.$$
$$\left. + (v_{ij}^* + z_{ij}^*) \hat{C}_{ijlm}^k e_{lm}^k - \beta_k \right\} dx, \tag{17.42}$$

$$\{D_{ijlm}^k\} = \{C_{ijlm}^k + KH_{ijlm}\}^{-1},$$

and

$$\{\hat{C}_{ijlm}^k\} = \{C_{ijlm}^k + KH_{ijlm}\}^{-1}\{C_{ijlm}^k\}.$$

Hence the concerned duality principle is expressed as

$$\inf_{u \in U} \{G^{**}(\varepsilon(u)) - F(\varepsilon(u)) - \langle u, f \rangle_U\}$$
$$\geq \sup_{v^* \in A^*} \left\{ \inf_{z^* \in Y^*} \left\{ \frac{1}{2K} \int_S z_{ij}^* \hat{H}_{ijlm} z_{lm}^* \, dx \right. \right.$$
$$+ \int_S \min_{k \in \{1,\ldots,N\}} \left\{ -\frac{1}{2}(v_{ij}^* + z_{ij}^*) D_{ijlm}^k (v_{lm}^* + z_{lm}^*) \right.$$
$$\left. \left. \left. -(v_{ij}^* + z_{ij}^*) \hat{C}_{ijlm}^k \mathbf{e}_{lm}^k + \beta_k \right\} dx \right\} \right\},$$

so that

$$\inf_{u \in U} \{G^{**}(\varepsilon(u)) - F(\varepsilon(u)) - \langle u, f \rangle_U)\}$$

$$\geq \sup_{v^* \in A^*} \left\{ \inf_{t \in B} \left\{ \inf_{z^* \in Y^*} \left\{ \frac{1}{2K} \int_S z^*_{ij} \hat{H}_{ijlm} z^*_{lm} \, dx \right.\right.\right.$$
$$+ \sum_{k=1}^{N} \left[\int_S \left(-\frac{1}{2} t_k (v^*_{ij} + z^*_{ij}) D^k_{ijlm} (v^*_{lm} + z^*_{lm}) \right.\right.$$
$$\left.\left.\left.\left.\left. -(v^*_{ij} + z^*_{ij}) t_k \hat{C}^k_{ijlm} \mathbf{e}^k_{lm} + t_k \beta_k \right) dx \right] \right\} \right\} \right\},$$

where

$$B = \{(t_1, \ldots, t_N) \text{ measurable } \mid$$
$$\left. t_k(x) \in [0,1], \ \forall k \in \{1, \ldots, N\}, \ \sum_{k=1}^{N} t_k(x) = 1, \text{ a.e. in } S \right\}. \tag{17.43}$$

Observe that the infimum in z^* is attained for functions satisfying

$$\frac{1}{K} \hat{H}_{ijlm} z^*_{lm} - \sum_{k=1}^{N} \{ t_k D^k_{ijlm} (v^*_{lm} + z^*_{lm}) \} - \sum_{k=1}^{N} t_k \hat{C}^k_{ijlm} \mathbf{e}^k_{lm} = 0, \ in \ S. \tag{17.44}$$

The final format of the concerned duality principle is given by

$$\inf_{u \in U} \{ G^{**}(\varepsilon(u)) - F(\varepsilon(u)) - \langle u, f \rangle_U \}$$
$$\geq \sup_{v^* \in A^*} \left\{ \inf_{t \in B} \left\{ \frac{1}{2K} \int_S z^*_{ij}(v^*, t) \hat{H}_{ijlm} z^*_{lm}(v^*, t) \, dx \right.\right.$$
$$\sum_{k=1}^{N} \left[-\int_S \frac{1}{2} t_k (v^*_{ij} + z^*_{ij}(v^*, t)) D^k_{ijlm} (v^*_{lm} + z^*_{lm}(v^*, t)) \, dx \right.$$
$$\left.\left.\left. + \int_S (-(v^*_{ij} + z^*_{ij}(v^*, t)) t_k \hat{C}^k_{ijlm} \mathbf{e}^k_{lm} + t_k \beta_k) \, dx \right] \right\} \right\}, \tag{17.45}$$

where $z^*(v^*, t)$ is obtained through Eq. (17.44).

The remaining conclusions follow from the main duality principle.

Remark 17.5.2. In fact the only hypothesis difficult to verify, concerning the main duality principle, is the coercivity in z^*. It is worth noting that to satisfy such a hypothesis and obtain a finite value for $J^*(v^*)$, where

$$J^*(v^*) = \inf_{z^* \in Y^*} \{ (F \circ \varepsilon)(\varepsilon^* z^*) - G^*(v^* + z^*) \},$$

we may, if necessary to replace it by

$$\tilde{J}^*(v^*) = \inf_{z^* \in \tilde{Y}} \{ F^*(z^*) - G^*(v^* + z^*) \},$$

where $z^* \in \tilde{Y}$ if $z^* \in Y^*$ and

$$z^* = \frac{\partial F(\varepsilon(u))}{\partial v},$$

for some $u \in U$. We also recall that $\tilde{Y}$ may be defined through linear equations of compatibility analogously to those of linear elasticity, considering that in the present case F is a quadratic functional. We do not work these elementary details here, postponing a more extensive analysis for a future work.

Remark 17.5.3. To illustrate how the dual formulation depends on K we analyze a simple variational problem. Fix $A = 10$ and $c_1, c_2 \in \mathbb{R}$. Define $J : U \to \mathbb{R}$ by

$$J(u) = G(u') - \langle u, f \rangle_U,$$

where

$$G(u') = \int_0^1 \min\{g_1(u'), g_2(u')\}\,dx,$$

$$g_1(u') = \frac{A}{2}(u' - c_1)^2 \text{ and } g_2(u') = \frac{A}{2}(u' - c_2)^2,$$

and

$$U = W_0^{1,2}([0,1]).$$

From Theorem 17.4.4, denoting $Y^* = L^2([0,1])$, we obtain

$$\inf_{u \in u}\{J(u)\} = \sup_{v^* \in C^*}\{-G^*(v^*)\},$$

where

$$G^*(v^*) = \int_0^1 \max\{g_1^*(v^*), g_2^*(v^*)\}\,dx,$$

$$g_1^*(v^*) = \frac{1}{2A}(v^*)^2 + c_1 v^*$$

and

$$g_2^*(v^*) = \frac{1}{2A}(v^*)^2 + c_2 v^*.$$

Also

$$C^* = \{v^* \in Y^* \mid (v^*)' + f = 0 \text{ in } [0,1]\}.$$

In this case there is no duality gap between the primal and dual problems. Anyway, let us analyze the dual problem obtained as we redefine the primal formulation as indicated in the next lines.

Define

$$\hat{G}(u') = G(u') + \frac{K}{2}\int_0^1 (u')^2\,dx,$$

and

$$F(u') = \frac{K}{2}\int_0^1 (u')^2\,dx.$$

From Theorem 17.3.1 (the main duality principle) we have

$$\inf_{u \in U}\{J(u)\} \geq \sup_{v^* \in C^*}\{\inf_{z^* \in Y^*}\{F^*(z^*) - \hat{G}^*(z^*+v^*)\}\}.$$

where

$$\hat{G}^*(v^*) = \int_0^1 \max\{\hat{g}_1^*(v^*), \hat{g}_2^*(v^*)\}\,dx,$$

$$\hat{g}_1^*(v^*) = \frac{1}{2(A+K)}(v^*)^2 + \frac{Ac_1}{A+K}v^* - \frac{AKc_1^2}{2(A+K)}$$

and

$$\hat{g}_2^*(v^*) = \frac{1}{2(A+K)}(v^*)^2 + \frac{Ac_2}{A+K}v^* - \frac{AKc_2^2}{2(A+K)}.$$

Also

$$F^*(z^*) = \frac{1}{2K}\int_0^1 (z^*)^2\,dx.$$

Now defining

$$J_K^*(v^*) = \inf_{z^* \in Y^*}\{F^*(z^*) - \hat{G}^*(z^*+v^*)\},$$

we may write

$$\inf_{u \in U}\{J(u)\} \geq \sup_{v^* \in C^*}\{J_K^*(v^*)\},$$

and

$$J_K^*(v^*) = \inf_{t \in B}\{\inf_{z^* \in Y^*}\{F^*(z^*) - \int_0^1 (t\hat{g}_1^*(v^*+z^*) + (1-t)\hat{g}_2^*(v^*+z^*))\,dx\}\},$$

that is,

$$J_K^*(v^*) = \inf_{t \in B}\{\inf_{z^* \in Y^*}\{\frac{1}{2K}\int_0^1 (z^*)^2\,dx - \int_0^1 (t\hat{g}_1^*(v^*+z^*) + (1-t)\hat{g}_2^*(v^*+z^*))\,dx\}\},$$

where

$$B = \{t \text{ measurable} \mid t \in [0,1] \text{ a.e. in } [0,1]\}.$$

Evaluating the infimum in z^* we obtain the final expression for J_K^*, namely,

$$J_K^*(v^*) = \inf_{t \in B}\{\int_0^1 (a\cdot(c_1-c_2)^2 t - a\cdot(c_1-c_2)^2\cdot t^2$$
$$-tc_1v^* - (1-t)c_2v^* - \frac{1}{2\cdot 10}(v^*)^2)\,dx\}. \tag{17.46}$$

For different values of K we have different values of a, for example:

1. for $K = 5$, we have $a = 1.66667$,
2. for $K = 50$, we have $a = 4.166667$,

3. for $K = 500$, $a = 4.90196$,
4. for $K = 5{,}000$, $a = 4.99000$,
5. for $K = 50{,}000$, $a = 4.99999$.

It seems that

$$a \to 5, \text{ as } K \to \infty.$$

Also observe that

$$\begin{aligned} J_K^*(v^*) &= \inf_{t \in B}\left\{\int_0^1 (a\cdot(c_1-c_2)^2 t - a\cdot(c_1-c_2)^2\cdot t^2 \right. \\ &\quad \left. -tc_1v^* - (1-t)c_2v^* - \frac{1}{2\cdot 10}(v^*)^2)\,dx\right\} \\ &\geq \inf_{t\in B}\left\{\int_0^1 (-tc_1v^* - (1-t)c_2v^* - \frac{1}{2\cdot 10}(v^*)^2)\,dx\right\}. \end{aligned}$$

Considering the vectorial case in question, for the analogous final value of a (in this case $a = 5$), the difference observed through (17.47) will result in no duality gap between the primal and dual problems. The difference is noted for the intermediate values of t, that is, for $0 < t < 1$, and it is particularly relevant in a microstructural context.

Finally, denoting $\alpha = a\cdot(c_1-c_2)^2$, through a Lagrange multiplier λ, we may write

$$\begin{aligned} J_K^*(v^*) &= \sup_{\lambda\in C}\left\{\int_0^1 \{\frac{(\alpha-\lambda)}{4} - \frac{c_1+c_2}{2}v^*\}\,dx \right. \\ &\quad \left. -\int_0^1 \frac{(c_1-c_2)^2(v^*)^2}{4(\lambda-\alpha)}\,dx - \frac{1}{2\cdot 10}\int_0^1 (v^*)^2\,dx\right\}, \end{aligned}$$

where

$$C = \{\lambda \in Y^* \mid \lambda - \alpha > 0, \text{ a.e. in } [0,1]\}.$$

Thus the final expression of the duality principle would be

$$\begin{aligned} \inf_{u\in U}\{J(u)\} \geq \sup_{(v^*,\lambda)\in C^*\times C} &\left\{\int_0^1 \{\frac{(\alpha-\lambda)}{4} - \frac{c_1+c_2}{2}v^*\}\,dx \right. \\ &\left. -\int_0^1 \frac{(c_1-c_2)^2(v^*)^2}{4(\lambda-\alpha)}\,dx - \frac{1}{2\cdot 10}\int_0^1 (v^*)^2\,dx\right\}. \end{aligned}$$

It seems to be clear that different values of K and corresponding a may produce different optimal microstructures for the dual problem. In the next result we prove that the duality gap may become arbitrarily small as $K \to \infty$.

Our final result is concerned with the evaluation of duality gap between the primal and dual problems.

Theorem 17.5.4. *Let $\varepsilon > 0$ be a small constant. Denote $g : \mathbb{R}^9 \to \mathbb{R}$ by*

$$g(y) = \min_{k \in \{1,\ldots,N\}} \{g_k(y) + \beta_k\},$$

where g_k is as in Theorem 17.5.1. Consider U, Y, $S \subset \mathbb{R}^3$ and $\varepsilon : U \to Y$ also as in Theorem 17.5.1.

Define $J : U \to \mathbb{R}$ by

$$J(u) = G_K(\varepsilon(u)) - F_K(\varepsilon(u)) - \langle u, f\rangle_U,$$

where, for each $K \in \mathbb{N}$,

$$G_K(\varepsilon(u)) = \frac{1}{2}\int_S \min_{k \in \{1,\ldots,N\}} \{g_k(\varepsilon(u)) + \beta_k\}\, dx + \frac{K}{2}\int_S (\varepsilon_{ij}(u)) H_{ijlm}(\varepsilon_{lm}(u))\, dx,$$

$$F_K(\varepsilon(u)) = \frac{K}{2}\int_S (\varepsilon_{ij}(u)) H_{ijlm}(\varepsilon_{lm}(u))\, dx,$$

Also define $J_K^{**} : U \to \mathbb{R}$ by

$$J_K^{**}(u) = G_K^{**}(\varepsilon(u)) - F_K(\varepsilon(u)) - \langle u, f\rangle_U.$$

From the last theorem we may select a sequence $\{u_{0_K}\} \subset U$ such that

$$J_K^{**}(u_{0_K}) = \sup_{v^* \in A^*}\left\{\inf_{z^* \in Y^*}\{(F_K \circ \varepsilon)^*(\varepsilon^* z^*) - G_K^*(v^* + z^*)\}\right\} \tag{17.47}$$

$\forall K \in \mathbb{N}$.

Suppose there exists $\tilde{K}$ such that

$$\|\varepsilon(u_{0_K})\|_\infty < \tilde{K}, \forall K \in \mathbb{N}.$$

Under such assumptions, there exists $K_\varepsilon \in \mathbb{N}$ such that if $K > K_\varepsilon$ then

$$|J(u_{0_K}) - J_K^{**}(u_{0_K})| < \varepsilon,$$

and also

$$\left|J(u_{0_K}) - \sup_{v^* \in A^*}\left\{\inf_{z^* \in Y^*}\{(F_K \circ \varepsilon)^*(\varepsilon^* z^*) - G_K^*(v^* + z^*)\}\right\}\right| < \varepsilon.$$

Proof. Choose $K_1 > \tilde{K}$ sufficiently big so that defining

$$\tilde{g}(y) = \begin{cases} g(y), \text{ if } |y| < K_1, \\ +\infty, \text{ otherwise,} \end{cases}$$

we have

$$G_K(\varepsilon(u_{0K})) = (\tilde{G}_K)(\varepsilon(u_{0_K})), \forall K \in \mathbb{N}, \tag{17.48}$$

and

$$G_K^{**}(\varepsilon(u_{0K})) = (\tilde{G}_K)^{**}(\varepsilon(u_{0_K})), \forall K \in \mathbb{N}, \tag{17.49}$$

(this is possible since g is the minimum of N quadratic positive definite functions and $\|\varepsilon(u_{0_K})\|_\infty$ is uniformly bounded), where

$$(\tilde{G}_K)(\varepsilon(u)) = \frac{1}{2}\int_S \tilde{g}(\varepsilon(u))\,dx + \frac{K}{2}\int_S (\varepsilon_{ij}(u))H_{ijlm}(\varepsilon_{lm}(u))\,dx,$$

From a standard mollification, there exists $g_\delta \in C^\infty(\mathbb{R}^9)$ such that

$$|g(y) - g_\delta(y)| < \frac{\varepsilon}{|S|}, \ \forall |y| < K_1. \tag{17.50}$$

Define

$$\tilde{g}_\delta(y) = \begin{cases} g_\delta(y), \text{ if } |y| < K_1, \\ +\infty, \text{ otherwise.} \end{cases}$$

Observe that from (17.50)

$$|\tilde{G}_K(\varepsilon(u_{0K})) - (\tilde{G}_K)_\delta(\varepsilon(u_{0_K}))| < \frac{\varepsilon}{2},$$

and

$$|\tilde{G}_K^{**}(\varepsilon(u_{0K})) - (\tilde{G}_K)_\delta^{**}(\varepsilon(u_{0_K}))| < \frac{\varepsilon}{2}, \forall K \in \mathbb{N},$$

where

$$(\tilde{G}_K)_\delta(\varepsilon(u)) = \frac{1}{2}\int_S \tilde{g}_\delta(\varepsilon(u))\,dx + \frac{K}{2}\int_S (\varepsilon_{ij}(u))H_{ijlm}(\varepsilon_{lm}(u))\,dx.$$

Since $g_\delta \in C^\infty(\mathbb{R}^9)$ we have that its matrix of second derivatives is bounded in bounded sets. Thus, as

$$\|\varepsilon(u_{0_K})\|_\infty < \tilde{K}, \forall K \in \mathbb{N},$$

there exists $K_\varepsilon > 0$ such that if $K > K_\varepsilon$ then

$$(\tilde{G}_K)_\delta(\varepsilon(u_{0K})) - (\tilde{G}_K)_\delta^{**}(\varepsilon(u_{0_K})) = 0.$$

Hence, if $K > K_\varepsilon$, we obtain

$$\begin{aligned} |G_K(\varepsilon(u_{0_K})) - G_K^{**}(\varepsilon(u_{0_K}))| &\le |\tilde{G}_K(\varepsilon(u_{0_K})) - (\tilde{G}_K)_\delta(\varepsilon(u_{0_K}))| \\ &\quad + |(\tilde{G}_K)_\delta(\varepsilon(u_{0_K})) - (\tilde{G}_K)_\delta^{**}(\varepsilon(u_{0_K}))| \\ &\quad + |(\tilde{G}_K)_\delta^{**}(\varepsilon(u_{0_K})) - \tilde{G}_K^{**}(\varepsilon(u_{0_K}))| \\ &< \frac{\varepsilon}{2} + \frac{\varepsilon}{2} = \varepsilon. \end{aligned} \tag{17.51}$$

From (17.48) and (17.49) we obtain

$$|G_K(\varepsilon(u_{0_K})) - G_K^{**}(\varepsilon(u_{0_K}))| < \varepsilon,$$

so that from this last inequality and (17.47) we finally obtain

$$\left| J(u_{0_K}) - \sup_{v^* \in A^*} \left\{ \inf_{z^* \in Y^*} \{ (F_K \circ \varepsilon)^*(\varepsilon^* z^*) - G_K^*(v^* + z^*) \} \right\} \right| < \varepsilon,$$

if $K > K_\varepsilon$.

The proof is complete.

Remark 17.5.5. Through this final result we have shown that for the problem in question, the duality gap between the primal and dual formulations becomes arbitrarily small as K goes to ∞.

17.6 Conclusion

In this chapter, we have developed dual variational approaches for multi-well formulations, introducing duality as an efficient tool to tackle this kind of problem. The standard results of convex analysis can be used to clarify the understanding of mixture of phases. What is new and relevant is the duality principle for vectorial multi-well models, applicable to phase transition problems, such as those studied in the article by Chenchiah and Bhattacharya, [18]. For such problems, duality is an interesting alternative for relaxation and computation.

In our view, the importance of duality for the theoretical and numerical analysis of multi-well and related phase transition problems seems to have been clarified.

Chapter 18
More on Duality Principles for Multi-well Problems

18.1 Introduction

In this chapter we develop dual variational formulations for a more general class of non-convex multi-well variational models. Such models appear in similar form in phase transition and related problems. Please see [5, 7, 18–20, 28, 48, 64] for details. We also address problems of conductivity in composites and optimal design and control in elasticity.

18.2 The Main Duality Principle

In this section we state and prove the main result in this chapter, which is summarized by the next theorem. From now on, by a regular boundary $\partial\Omega$ of $\Omega \subset \mathbb{R}^3$, we mean regularity enough so that the standard Gauss–Green formulas of integrations by parts, the Sobolev imbedding theorem, and the trace theorem to hold. Also, $\mathbf{n}$ denotes the outward normal to $\partial\Omega$ and derivatives must be understood always in the distributional sense.

Remark 18.2.1. At some points of our analysis we refer to the problems in question after discretization. In such a case, we are referring to their approximations in a finite element or finite differences context. Finally, to simplify the notation in some parts of the text, we may denote $L^2(\Omega)$, $L^2(\Omega;\mathbb{R}^3)$ and $L^2(\Omega;\mathbb{R}^{3\times 3})$ simply by L^2.

Theorem 18.2.2. *Let $\Omega \subset \mathbb{R}^3$ be an open, bounded, connected set with a regular boundary denoted by $\partial\Omega = \Gamma$. Denote $J : U \times B \to \mathbb{R}$ by*

$$J(u,t) = G(\varepsilon(u),t) - \langle u, f\rangle_{L^2},$$

where

$$G(\varepsilon(u),t) = \frac{1}{2}\int_\Omega t_k g_k(\varepsilon(u))\,dx,$$

F. Botelho, *Functional Analysis and Applied Optimization in Banach Spaces: Applications to Non-Convex Variational Models*, DOI 10.1007/978-3-319-06074-3_18,

$$g_k(\varepsilon(u)) = H^k_{ijlm}(\varepsilon_{ij}(u) - e^k_{ij})(\varepsilon_{lm}(u) - e^k_{lm}) + \beta_k, \forall k \in \{1,\ldots,N\}.$$

Here

$$\varepsilon(u) = \{\varepsilon_{ij}(u)\} = \left\{\frac{1}{2}(u_{i,j} + u_{j,i})\right\},$$

$$B = \{t = (t_1,\ldots,t_N) \text{ measurable } |$$

$$\sum_{k=1}^N t_k(x) = 1,\ 0 \leq t_k(x) \leq 1, \text{ in } \Omega,\ \forall k \in \{1,\ldots,N\}\}, \tag{18.1}$$

and

$$U = \{u \in W^{1,2}(\Omega;\mathbb{R}^3) \mid u = u_0 \text{ on } \partial\Omega\}.$$

Also, $\beta_k \in \mathbb{R}$, $\{e^k_{ij}\} \in L^2(\Omega;\mathbb{R}^{3\times 3})$, $f \in L^2(\Omega;\mathbb{R}^3)$ *and*

$$\{H^k_{ijlm}\}$$

are fourth-order positive definite constant tensors $\forall k \in \{1,\ldots,N\}$.

Under such hypotheses,

$$\inf_{(u,t)\in U\times B}\{J(u,t)\} \geq \sup_{v^*\in A^*}\{-J^*(v^*)\},$$

where

$$J^*(v^*) = \sup_{t\in B}\{G^*(v^*,t)\} - \langle v^*_{ij}\boldsymbol{n}_j, (u_0)_i\rangle_{L^2(\Gamma)},$$

$$G^*(v^*,t) = \sup_{v\in L^2}\{\langle v, v^*\rangle_{L^2} - G(v,t)\},$$

and

$$A^* = \{v^* \in Y^* \mid v^*_{ij,j} + f_i = 0 \text{ in } \Omega\}.$$

Furthermore, there exists $v^*_0 \in A^* \subset Y^* = Y = L^2(\Omega;\mathbb{R}^{3\times 3})$ *such that*

$$-J^*(v^*_0) = \max_{v^*\in A^*}\{-J^*(v^*)\}.$$

Finally, for the discretized version of the problem, assume $t_0 \in B$ *such that*

$$G^*(v^*_0,t_0) = \sup_{t\in B}\{G^*(v^*_0,t)\},$$

is also such that the hypotheses of Corollary 11.1 are satisfied.

Under such hypotheses, for $\hat{u} \in U$ *such that*

$$\varepsilon(\hat{u}) \in \partial(J^*(v^*_0)),$$

we have

$$J(\hat{u},t_0) = \min_{(u,t)\in U\times B}\{J(u,t)\} = \max_{v^*\in A^*}\{-J^*(v^*)\} = -J^*(v^*_0).$$

Proof. Observe that

$$
\begin{aligned}
J(u,t) &= G(\varepsilon(u),t) - \langle u, f\rangle_{L^2} \\
&= -\langle \varepsilon(u), v^*\rangle_{L^2} + G(\varepsilon(u),t) \\
&\quad + \langle \varepsilon(u), v^*\rangle_{L^2} - \langle u, f\rangle_{L^2} \\
&\geq \inf_{v\in Y}\{-\langle v, v^*\rangle_{L^2} + G(v,t)\} \\
&\quad + \inf_{u\in U}\{\langle \varepsilon(u), v^*\rangle_{L^2} - \langle u, f\rangle_{L^2}\} \\
&= -G^*(v^*,t) + \langle v^*_{ij}\mathbf{n}_j, (u_0)_i\rangle_{L^2(\Gamma)} \\
&\geq \inf_{t\in B}\{-G^*(v^*,t) + \langle v^*_{ij}\mathbf{n}_j, (u_0)_i\rangle_{L^2(\Gamma)}\} \\
&= -J^*(v^*),
\end{aligned}
\tag{18.2}
$$

$\forall (u,t) \in U \times B, v^* \in A^*$.

Thus,

$$
\inf_{(u,t)\in U\times B}\{J(u,t)\} \geq \sup_{v^*\in A^*}\{-J^*(v^*)\}. \tag{18.3}
$$

The dual functional is concave and upper semicontinuous; therefore, it is weakly upper semicontinuous, so that from the direct method of calculus of variations (since it is a standard procedure we do not provide details here), there exists $v_0^* \in A^*$ such that

$$
-J^*(v_0^*) = \max_{v^*\in A^*}\{-J^*(v^*)\}.
$$

At this point and on, we consider the discretized version of the problem.

From the hypotheses, $t_0 \in B$ such that

$$
G^*(v_0^*, t_0) = \sup_{t\in B}\{G^*(v_0^*, t)\},
$$

is also such the hypotheses of Corollary 11.1 are satisfied.

From such a corollary, we may infer that for the extended functional

$$
-\langle (\hat{u})_i, v^*_{ij,j} + f_i\rangle_{L^2(\Omega)} - J^*(v^*)
$$

the optimal extremal relation

$$
\varepsilon(\hat{u}) \in \partial(J^*(v_0^*)),
$$

stands for

$$
\varepsilon(\hat{u}) = \frac{\partial G^*(v_0^*, t_0)}{\partial v^*}.
$$

Hence, since for a fixed $t_0 \in B$ $G(\varepsilon(u), t_0)$ is convex in u, we get

$$
\begin{aligned}
G^*(v_0^*, t_0) &= \langle \varepsilon(\hat{u}), v_0^*\rangle_{L^2} - G(\varepsilon(\hat{u}), t_0) \\
&= -\langle \hat{u}_i, (v_0)^*_{ij,j}\rangle_{L^2(\Omega)} + \langle (v_0)^*_{ij}\mathbf{n}_j, (u_0)_i\rangle_{L^2(\Gamma)} \\
&\quad - G(\varepsilon(\hat{u}), t_0).
\end{aligned}
\tag{18.4}
$$

Therefore,

$$G^*(v_0^*, t_0) - \langle (v_0)^*_{ij}\mathbf{n}_j, (u_0)_i \rangle_{L^2(\Gamma)}$$
$$= -G(\varepsilon(\hat{u}), t_0) + \langle \hat{u}_i, f_i \rangle_{L^2(\Omega)}. \tag{18.5}$$

From this we get

$$J^*(v_0^*) = -J(\hat{u}, t_0).$$

From this last equation and (18.3), the proof is complete.

18.3 Another Duality Principle for Phase Transition Models

In this section we state and prove another relevant result, which is summarized by the next theorem.

Theorem 18.3.1. *Let $\Omega \subset \mathbb{R}^3$ be an open, bounded, connected set with a regular boundary denoted by $\partial\Omega = \Gamma$. Denote $J : U \times B \to \mathbb{R}$ by*

$$J(u,t) = G(\varepsilon(u), t),$$

where

$$G(\varepsilon(u), t) = \frac{1}{2}\int_\Omega t_k g_k(\varepsilon(u))\, dx,$$

$$g_k(\varepsilon(u)) = H^k_{ijlm}(\varepsilon_{ij}(u) - e^k_{ij})(\varepsilon_{lm}(u) - e^k_{lm}) + \beta_k, \forall k \in \{1, \ldots, N\}.$$

Here

$$\varepsilon(u) = \{\varepsilon_{ij}(u)\} = \left\{\frac{1}{2}(u_{i,j} + u_{j,i})\right\},$$

$$B = \{t = (t_1, \ldots, t_N) \text{ measurable } |$$
$$\sum_{k=1}^N t_k(x) = 1,\ 0 \le t_k(x) \le 1, \text{ in } \Omega,$$
$$\forall k \in \{1, \ldots, N\}\}, \tag{18.6}$$

and

$$U = \{u \in W^{1,2}(\Omega; \mathbb{R}^3) \mid u = u_0 \text{ on } \partial\Omega\}.$$

Also, $\beta_k \in \mathbb{R}$, $\{e^k_{ij}\} \in L^2(\Omega; \mathbb{R}^{3\times 3}) \equiv L^2$, and as above,

$$\{H^k_{ijlm}\}$$

are fourth-order positive definite constant tensors, $\forall k \in \{1, \ldots, N\}$.

Under such hypotheses,

$$\inf_{(u,t)\in U\times B}\{J(u,t)\} = \inf_{(v^*,t)\in A^*\times B}\{J^*(v^*,t)\},$$

where

$$\begin{aligned} J^*(v^*,t) &= G_1^*(v_1^*,t) + G_2^*(v_2^*,t) \\ &\quad -\langle (v_1^*)_{ij}\boldsymbol{n}_j, (u_0)_i\rangle_{L^2(\Gamma)} - \langle (v_2^*)_{ij}\boldsymbol{n}_j, (u_0)_i\rangle_{L^2(\Gamma)}, \end{aligned} \tag{18.7}$$

$$G_1^*(v_1^*,t) = \sup_{v\in L^2}\{\langle v, v_1^*\rangle_{L^2} - G_1(v,t)\},$$

$$G_2^*(v_2^*,t) = \sup_{v\in L^2}\{\langle v, v_2^*\rangle_{L^2} - G_2(v,t)\},$$

$$-G_1(v,t) = \frac{1}{2}G(v,t) - \frac{1}{2}\int_\Omega t_k H^k_{ijlm}(v_{ij} - \boldsymbol{e}^k_{ij})(v_{lm} - \boldsymbol{e}^k_{lm})\,dx,$$

$$-G_2(v,t) = \frac{1}{2}G(-v,t) + \frac{1}{2}\int_\Omega t_k H^k_{ijlm}(v_{ij} - \boldsymbol{e}^k_{ij})(-v_{lm} - \boldsymbol{e}^k_{lm})\,dx$$

and

$$A^* = \{v^* \in Y^* \mid (v_1^*)_{ij,j} + (v_2^*)_{ij,j} = 0 \text{ in } \Omega\}.$$

Proof. Observe that

$$\begin{aligned} \inf_{(u,t)\in U\times B}\{J(u,t)\} &= \inf_{t\in B}\{\inf_{u\in U}\{G(\varepsilon(u),t)\}\} \\ &= \inf_{t\in B}\{\inf_{u\in U}\{-\frac{1}{2}\int_\Omega t_k H^k_{ijlm}(\varepsilon_{ij}(\hat{u}) - \mathbf{e}^k_{ij})(\varepsilon_{lm}(u) - \mathbf{e}^k_{lm})\,dx \\ &\quad + \frac{1}{2}G(\varepsilon(u),t) \\ &\quad + \frac{1}{2}\int_\Omega t_k H^k_{ijlm}(\varepsilon_{ij}(\hat{u}) - \mathbf{e}^k_{ij})(\varepsilon_{lm}(u) - \mathbf{e}^k_{lm})\,dx \\ &\quad + \frac{1}{2}G(\varepsilon(u),t)\}\} \\ &= \inf_{t\in B}\{\sup_{\hat{u}\in U}\{-G_1(\varepsilon(\hat{u}),t) - G_2(\varepsilon(\hat{u}),t)\}\} \\ &= \inf_{t\in B}\{\inf_{v^*\in A^*}\{G_1^*(v_1^*,t) + G_2^*(v_2^*,t) \\ &\quad -\langle (v_1^*)_{ij}\mathbf{n}_j, (u_0)_i\rangle_{L^2(\Gamma)} \\ &\quad -\langle (v_2^*)_{ij}\mathbf{n}_j, (u_0)_i\rangle_{L^2(\Gamma)}\}\} \\ &= \inf_{(v^*,t)\in A^*\times B}\{J^*(v^*,t)\}. \end{aligned} \tag{18.8}$$

The proof is complete.

18.4 Duality for a Problem on Conductivity in Composites

For the primal formulation we repeat the statements found in reference [27]. Consider a material confined into a bounded domain $\Omega \subset \mathbb{R}^N$, $N > 1$. The medium is obtained by mixing two constituents with different electric permittivity and conductivity. Let Q_0 and Q_1 denote the two $N \times N$ symmetric matrices of electric permittivity corresponding to each phase. For each phase, we also denote by L_j, $j = 0,1$ the anisotropic $N \times N$ symmetric matrix of conductivity. Let $0 < t_1 < 1$ be the proportion of the constituent 1 into the mixture. Constituent 1 occupies a space in the physical domain Ω which we denote by $E \subset \Omega$. Regarding the set E as our design variable, we introduce the characteristic function $\chi : \Omega \to \{0,1\}$:

$$\chi(x) = \begin{cases} 1, & \text{if } x \in E, \\ 0, & \text{otherwise}, \end{cases} \tag{18.9}$$

Thus,

$$\int_E dx = \int_\Omega \chi(x)dx = t_1 \int_\Omega dx = t_1|\Omega|. \tag{18.10}$$

The matrix of conductivity corresponding to the material as a whole is $L = \chi L_1 + (1-\chi)L_0$.

Finally, the electrostatic potential, denoted by $u : \Omega \to \mathbb{R}$, is supposed to satisfy the equation

$$div[\chi L_1 \nabla u + (1-\chi)L_0 \nabla u] + f(x) = 0, \text{ in } \Omega, \tag{18.11}$$

with the boundary conditions

$$u = u_0, \text{ on } \partial\Omega \tag{18.12}$$

where $f : \Omega \to \mathbb{R}$ is a given source or sink of current (we assume $f \in L^2(\Omega)$). From now on we assume $N = 3$. Consider the slightly different problem of minimizing the cost functional

$$I(\chi,u) = \int_\Omega \left(\frac{\chi}{2}(\nabla u)^T Q_1 \nabla u + \frac{(1-\chi)}{2}(\nabla u)^T Q_0 \nabla u \right) dx \tag{18.13}$$

subject to

$$div[\chi L_1 \nabla u + (1-\chi)L_0 \nabla u] + f(x) = 0 \tag{18.14}$$

and

$$\int_\Omega \chi \, dx \leq t_1|\Omega|,$$

where $u \in U$, here $U = \{u \in W^{1,2}(\Omega) \mid u = u_0 \text{ on } \partial\Omega\}$.

Our main duality principle for such a non-convex optimization problem is summarized by the next theorem.

Theorem 18.4.1. *Let $\Omega \subset \mathbb{R}^3$ be an open, bounded, connected set with a regular boundary denoted by $\partial\Omega = \Gamma$. Redefine without relabeling it, $J : U \times B \to \overline{\mathbb{R}} = \mathbb{R} \cup \{+\infty\}$ by*

$$J(u,t) = G(\nabla u, t) + Ind(u,t),$$

where

$$U = \{u \in W^{1,2}(\Omega) \mid u = u_0 \text{ on } \partial\Omega\},$$
$$Y = Y^* = L^2(\Omega;\mathbb{R}^3) \equiv L^2,$$

$$B = \{t \text{ measurable } \mid t \in \{0,1\}, \text{ a.e. in } \Omega \text{ and } \int_\Omega t\, dx \leq t_1 |\Omega|\}, \tag{18.15}$$

and

$$0 < t_1 < 1.$$

Also,

$$G(\nabla u, t) = \frac{1}{2} \int_\Omega \{t(\nabla u^T Q_1 \nabla u) + (1-t)(\nabla u^T Q_0 \nabla u)\}\, dx,$$

$$Ind(u,t) = \begin{cases} 0 & \text{if } (u,t) \in A \\ +\infty & \text{otherwise,} \end{cases} \tag{18.16}$$

where

$$A = \{(u,t) \in U \times B \mid div((tL_1 + (1-t)L_0)\nabla u) + f = 0 \text{ in } \Omega\}. \tag{18.17}$$

Under such hypotheses,

$$\inf_{(u,t) \in U \times B} \{J(u,t)\} \geq \sup_{(v^*,\lambda) \in A^* \times U_1} \{-J^*(v^*,\lambda)\}.$$

Here,

$$J^*(v^*,\lambda) = \sup_{t \in B} \{G^*(v^* - (tL_1 + (1-t)L_0)\nabla \lambda)\} + \langle \lambda, f \rangle_{L^2(\Omega)} - \int_\Gamma (v^* \cdot \mathbf{n}) u_0\, d\Gamma, \tag{18.18}$$

$$G^*(v^* - (tL_1 + (1-t)L_0)\nabla \lambda) = \sup_{v \in Y} \{\langle v, v^* \rangle_{L^2} - G(v,t) - \langle (tL_1 + (1-t)L_0)\nabla \lambda, v \rangle_{L^2}\},$$

and

$$A^* = \{v^* \in Y^* \mid div(v^*) = 0 \text{ in } \Omega\}.$$

Moreover, there exists $(v_0^*, \lambda_0) \in A^* \times U_1$ *such that*

$$J^*(v_0^*, \lambda_0) = \max_{(v^*,\lambda) \in A^* \times U_1} \{-J^*(v^*,\lambda)\},$$

where $U_1 = W_0^{1,2}(\Omega)$.

Assume, after discretization, that $t_0 \in B$ *such that*

$$G^*(v_0^* - (t_0 L_1 + (1-t_0)L_0)\nabla \lambda_0) = \sup_{t \in B} \{G^*(v_0^* - (tL_1 + (1-t)L_0)\nabla \lambda_0)\},$$

is also such that for an appropriate $\hat{u} \in U$ the optimal inclusions

$$\nabla \hat{u} \in \partial_{v^*} J(v_0^*, \lambda_0),$$

and

$$\theta \in \partial_\lambda J^*(v_0^*, \lambda_0),$$

stand for

$$\nabla \hat{u} = \frac{\partial G^*(v_0^* - (t_0 L_1 + (1-t_0)L_0)\nabla \lambda_0)}{\partial v^*},$$

and

$$\theta = div\left((t_0 L_1 + (1-t_0)L_0)\frac{\partial G^*(v_0^* - (t_0 L_1 + (1-t_0)L_0)\nabla \lambda_0)}{\partial v^*}\right) + f.$$

Under such additional hypotheses we have

$$\begin{aligned} J(\hat{u}, t_0) &= \inf_{(u,t) \in U \times B} \{J(u,t)\} \\ &= \max_{(v^*,\lambda) \in A^* \times U_1} \{-J^*(v^*, \lambda)\} \\ &= -J^*(v_0^*, \lambda_0). \end{aligned} \quad (18.19)$$

Proof. Observe that

$$\begin{aligned} J(u,t) &= G(\nabla u, t) + Ind(u,t) \\ &\geq G(\nabla u, t) - \langle \lambda, \text{div}((tL_1 + (1-t)L_0)\nabla u + f\rangle_{L^2(\Omega)} \\ &= G(\nabla u, t) + \langle (tL_1 + (1-t)L_0)\nabla \lambda, \nabla u\rangle_{L^2} - \langle \lambda, f\rangle_{L^2} \\ &= -\langle \nabla u, v^*\rangle_{L^2} + G(\nabla u, t) + \langle (tL_1 + (1-t)L_0)\nabla \lambda, \nabla u\rangle_{L^2} \\ &\quad -\langle \lambda, f\rangle_{L^2} + \langle \nabla u, v^*\rangle_{L^2} \\ &\geq \inf_{v \in Y}\{-\langle v, v^*\rangle_{L^2} + G(v,t) + \langle (tL_1 + (1-t)L_0)\nabla \lambda, v\rangle_{L^2} \\ &\quad + \inf_{u \in U}\{-\langle \lambda, f\rangle_{L^2} + \langle \nabla u, v^*\rangle_{L^2}\} \\ &= -G^*(v^* - (tL_1 + (1-t)L_0)\nabla \lambda) - \langle \lambda, f\rangle_{L^2} \\ &\quad + \int_\Gamma (v^* \cdot \mathbf{n}) u_0 \, d\Gamma \\ &\geq \inf_{t \in B}\{-G^*(v^* - (tL_1 + (1-t)L_0)\nabla \lambda)\} - \langle \lambda, f\rangle_{L^2} \\ &\quad + \int_\Gamma (v^* \cdot \mathbf{n}) u_0 \, d\Gamma \\ &= -J^*(v, \lambda), \forall (u,t) \in U \times B, \; (v^*, \lambda) \in A^* \times U_1. \end{aligned} \quad (18.20)$$

Thus,

$$\inf_{(u,t) \in U \times B} \{J(u,t)\} \geq \sup_{(v^*,\lambda) \in A^* \times U_1} \{-J^*(v^*, \lambda)\}. \quad (18.21)$$

From the remaining hypotheses, after discretization, we have that $t_0 \in B$ such that

$$G^*(v_0^* - (t_0 L_1 + (1-t_0)L_0)\nabla\lambda_0) = \sup_{t \in B}\{G^*(v_0^* - (tL_1 + (1-t)L_0)\nabla\lambda_0),$$

is also such that for an appropriate $\hat{u} \in U$ the optimal inclusions

$$\nabla\hat{u} \in \partial_{v^*} J(v_0^*, \lambda_0),$$

and

$$\theta \in \partial_\lambda J^*(v_0^*, \lambda_0),$$

stand for

$$\nabla\hat{u} = \frac{\partial G^*(v_0^* - (t_0 L_1 + (1-t_0)L_0)\nabla\lambda_0)}{\partial v^*},$$

and

$$\theta = \operatorname{div}\left((t_0 L_1 + (1-t_0)L_0)\frac{\partial G^*(v_0^* - (t_0 L_1 + (1-t_0)L_0)\nabla\lambda_0)}{\partial v^*}\right) + f.$$

Hence

$$\operatorname{div}((t_0 L_1 + (1-t_0)L_0)\nabla\hat{u}) + f = \theta, \text{ in } \Omega,$$

and

$$\begin{aligned} G^*(v_0^* - (t_0 L_1 + (1-t_0)L_0)\nabla\lambda_0) &= \langle \nabla\hat{u}, v_0^* \rangle_{L^2} - G(\nabla\hat{u}, t_0) \\ &\quad - \langle (t_0 L_1 + (1-t_0)L_0)\nabla\lambda_0, \nabla\hat{u} \rangle_{L^2} \\ &= \int_\Gamma (v_0^* \cdot \mathbf{n}) u_0 \, d\Gamma - G(\nabla\hat{u}, t_0) \\ &\quad - \langle (t_0 L_1 + (1-t_0)L_0)\nabla\lambda_0, \nabla\hat{u} \rangle_{L^2} \\ &= \int_\Gamma (v_0^* \cdot \mathbf{n}) u_0 \, d\Gamma - G(\nabla\hat{u}, t_0) \\ &\quad - \langle \lambda_0, f \rangle_{L^2}. \end{aligned} \tag{18.22}$$

Therefore,

$$\begin{aligned} G(\nabla\hat{u}, t_0) + Ind(\hat{u}, t_0) &= -G^*(v_0^* - (t_0 L_1 + (1-t_0)L_0)\nabla\lambda_0) \\ &\quad - \langle \lambda_0, f \rangle_{L^2} + \int_\Gamma (v_0^* \cdot \mathbf{n}) u_0 \, d\Gamma, \end{aligned} \tag{18.23}$$

so that

$$J(\hat{u}, t_0) = -J^*(v_0^*, \lambda_0).$$

From this and (18.21), the proof is complete.

18.5 Optimal Design and Control for a Plate Model

In this section, we develop duality for the optimal control of a two-phase plate model. The control variable is t_1 and is related to the elastic constant distribution $K(t_1)$ given by

$$K(t_1) = t_1 K_1 + (1 - t_1) K_2$$

where $K_1 \gg K_2 > 0$. Moreover, the plate stiffness is given by the tensor $H_{\alpha\beta\lambda\mu}(t)$ as specified in the next theorem. We are concerned with the calculation of the optimal t, t_1 which minimizes the plate compliance (or, equivalently, its inner work), under appropriate constraints. The plate model in question is the Kirchhoff one, where $\Omega \subset \mathbb{R}^2$ denotes the plate mid-surface. Moreover, $w \in W^{1,2}(\Omega)$ denotes the field of displacements resulting from a vertical load $f \in L^2(\Omega)$ action. Please see the next theorem for details.

18.5.1 The Duality Principle for the Plate Model

We start this subsection with the following theorem.

Theorem 18.5.1. *Let $\Omega \subset \mathbb{R}^2$ be an open bounded connected set with a regular boundary denoted by $\partial\Omega = \Gamma$. Define $J : U \times B \times B_1 \to \overline{\mathbb{R}} = \mathbb{R} \cup \{+\infty\}$, by*

$$J(w,t,t_1) = G_1(\Lambda w, t) + G_2(w, t_1) + Ind(w,t,t_1),$$

where $U = W_0^{2,2}(\Omega)$,

$$\begin{aligned} B = \{ & t \text{ measurable } \mid t \in \{0,1\}, \text{ a.e. in } \Omega \\ & \text{and } \int_\Omega t \, dx \le \tilde{t}|\Omega|\}, \end{aligned} \tag{18.24}$$

$$0 < \tilde{t} < 1,$$

$$\begin{aligned} B_1 = \{ & t_1 \text{ measurable } \mid 0 \le t_1 \le 1, \text{ a.e. in } \Omega \\ & \text{and } \int_\Omega t_1 \, dx \le \tilde{t}_1 |\Omega|\}, \end{aligned} \tag{18.25}$$

and

$$0 < \tilde{t}_1 < 1.$$

Also, $\Lambda :\to Y = Y^* = L^2(\Omega; \mathbb{R}^{2\times 2}) \equiv L^2$ is given by

$$\Lambda w = \{w_{,\alpha\beta}\},$$

$$G_1(\Lambda w, t) = \frac{1}{2} \int_\Omega H_{\alpha\beta\lambda\mu}(t) w_{,\alpha\beta} w_{,\lambda\mu} \, dx,$$

$$G_2(w,t_1) = \frac{1}{2}\int_\Omega K(t_1)w^2\,dx,$$

$$H_{\alpha\beta\lambda\mu}(t) = t(H_0)_{\alpha\beta\lambda\mu} + (1-t)(H_1)_{\alpha\beta\lambda\mu},$$

where $\{(H_0)_{\alpha\beta\lambda\mu}\}$ *and* $\{(H_1)_{\alpha\beta\lambda\mu}\}$ *are fourth-order positive definite constant tensors. Also,*

$$K(t_1) = t_1K_1 + (1-t_1)K_2,$$

where

$$K_1 \gg K_2 > 0.$$

Moreover,

$$Ind(w,t,t_1) = \begin{cases} 0 & \text{if } (w,t,t_1) \in A \\ +\infty & \text{otherwise,} \end{cases} \tag{18.26}$$

where

$$A = \{(w,t,t_1) \in U \times B \times B_1 \mid (H_{\alpha\beta\lambda\mu}(t)w_{,\lambda\mu})_{,\alpha\beta} + K(t_1)w - f = 0 \text{ in } \Omega\}. \tag{18.27}$$

Under such hypotheses,

$$\inf_{(w,t,t_1)\in U\times B\times B_1}\{J(w,t,t_1)\} = \inf_{(v^*,t,t_1)\in Y^*\times B\times B_1}\{J^*(v^*,t,t_1)\}.$$

Here,

$$J^*(v^*,t,t_1) = G_1^*(v^*,t) + G_2^*(\Lambda^*v^* - f, t_1),$$

where

$$\begin{aligned} G_1^*(v^*,t) &= \sup_{v\in Y}\{\langle v,v^*\rangle_{L^2} - G_1(v,t)\} \\ &= \frac{1}{2}\int_\Omega \overline{H}_{\alpha\beta\lambda\mu}(t)v^*_{\alpha\beta}v^*_{\lambda\mu}\,dx, \end{aligned} \tag{18.28}$$

$$\overline{H}_{\alpha\beta\lambda\mu}(t) = \{H_{\alpha\beta\lambda\mu}(t)\}^{-1},$$

$$\begin{aligned} G_2^*(v^*,t_1) &= \sup_{w\in U}\{-\langle \Lambda w, v^*\rangle_{L^2} - G_2(w,t_1) + \langle w,f\rangle_{L^2(\Omega)}\} \\ &= \frac{1}{2}\int_\Omega \overline{K}(t_1)((v^*_{\alpha\beta})_{,\alpha\beta} - f)^2;dx, \end{aligned} \tag{18.29}$$

$$\overline{K}(t) = \{K(t)\}^{-1}.$$

Furthermore, also under the mentioned hypotheses, we have

$$\inf_{(w,t,t_1)\in U\times B\times B_1}\{J(w,t,t_1)\} \geq \sup_{\hat{w}\in U}\{-\hat{J}(\hat{w})\},$$

where

$$\hat{J}(\hat{w}) = \sup_{(t,t_1)\in B\times B_1} \{G_1(\Lambda\hat{w},t)+G_2(\hat{w},t_1)-\langle \hat{w},f\rangle_{L^2}\}.$$

Finally, there exists $\hat{w}_0 \in U$ *such that*

$$-\hat{J}(\hat{w}_0) = \max_{\hat{w}\in U}\{-\hat{J}(\hat{w})\}.$$

Suppose, after discretization, that $(t_0,(t_1)_0) \in B\times B_1$ *such that*

$$\begin{aligned}\hat{J}(\hat{w}_0) = & \sup_{(t,t_1)\in B\times B_1} \{G_1(\Lambda\hat{w}_0,t)+G_2(\hat{w}_0,t_1) \\ & -\langle \hat{w}_0,f\rangle_{L^2}\} \\ = & G_1(\Lambda\hat{w}_0,t_0)+G_2(\hat{w}_0,(t_1)_0) \\ & -\langle \hat{w}_0,f\rangle_{L^2}.\end{aligned} \tag{18.30}$$

is also such that the optimal inclusion

$$\theta \in \partial\hat{J}(\hat{w}_0)$$

stands for

$$\delta_{\hat{w}}\{G_1(\Lambda\hat{w}_0,t_0)+G_2(\hat{w}_0,(t_1)_0)-\langle\hat{w}_0,f\rangle_{L^2}\}=\theta.$$

Under such hypotheses

$$\inf_{(w,t,t_1)\in U\times B\times B_1}\{J(w,t,t_1)\} = J(\hat{w}_0,t_0,(t_1)_0) = -\hat{J}(\hat{w}_0) = \max_{\hat{w}\in U}\{-\hat{J}(\hat{w})\}.$$

Proof. Observe that

$$\begin{aligned}\inf_{(w,t,t_1)\in U\times B\times B_1}\{J(w,t,t_1)\} = & \inf_{(t,t_1)\in B\times B_1}\{\inf_{w\in U}\{J(w,t,t_1)\}\} \\ = & \inf_{(t,t_1)\in B\times B_1}\{\sup_{\hat{w}\in U}\{\inf_{w\in U}\{G_1(\Lambda w,t)+G_2(w,t_1) \\ & -\langle \hat{w},(H_{\alpha\beta\lambda\mu}(t)w_{,\lambda\mu})_{,\alpha\beta}+K(t_1)w-f\rangle_{L^2}\}\} \\ = & \inf_{(t,t_1)\in B\times B_1}\{\sup_{\hat{w}\in U}\{-G_2(\Lambda\hat{w},t)-G_2(\hat{w},t_1) \\ & +\langle\hat{w},f\rangle_{L^2}\}\} \\ = & \inf_{(t,t_1)\in B\times B_1}\{\inf_{v^*\in Y^*}\{G_1^*(v^*,t) \\ & +G_2^*(\Lambda^*v^*-f,t_1)\}\} \\ = & \inf_{(v^*,t,t_1)\in Y^*\times B\times B_1}\{J^*(v^*,t,t_1)\}.\end{aligned} \tag{18.31}$$

Moreover,

$$\begin{aligned}J(w,t,t_1) \geq & \inf_{w\in U}\{J(w,t,t_1)\} \\ \geq & \inf_{w\in U}\{G_1(\Lambda w,t)+G_2(w,t_1)\end{aligned}$$

$$\begin{aligned}
&-\langle \hat{w}, (H_{\alpha\beta\lambda\mu}(t)w_{,\lambda\mu})_{,\alpha\beta} + K(t_1)w - f\rangle_{L^2(\Omega)}\} \\
&= -G_1(\Lambda\hat{w}, t) - G_2(\hat{w}, t_1) + \langle \hat{w}, f\rangle_{L^2} \\
&\geq \inf_{(t,t_1)\in B\times B_1} \{-G_1(\Lambda\hat{w}, t) - G_2(\hat{w}, t_1) + \langle \hat{w}, f\rangle_{L^2}\} \\
&= -\hat{J}(\hat{w}), \forall (w,t,t_1) \in U \times B \times B_1,\ \hat{w} \in U. \qquad (18.32)
\end{aligned}$$

Hence,

$$\inf_{(w,t,t_1)\in U\times B\times B_1} \{J(w,t,t_1)\} \geq \sup_{\hat{w}\in U}\{-\hat{J}(\hat{w})\}. \qquad (18.33)$$

Also, since $-\hat{J} : U \to \mathbb{R}$ is concave and continuous, it is weakly upper semicontinuous, so that by the direct method of calculus of variations (since it is a standard procedure, we do not provide more detail here), there exists $\hat{w}_0 \in U$ such that

$$-\hat{J}(\hat{w}_0) = \max_{\hat{w}\in U}\{-\hat{J}(\hat{w})\}.$$

Finally, from the hypotheses, after discretization, we have that $(t_0, (t_1)_0) \in B \times B_1$ such that

$$\begin{aligned}
\hat{J}(\hat{w}) = &\sup_{(t,t_1)\in B\times B_1} \{G_1(\Lambda\hat{w}_0, t) + G_2(\hat{w}_0, t_1) \\
&-\langle \hat{w}_0, f\rangle_{L^2}\} \\
= &\ G_1(\Lambda\hat{w}_0, t_0) + G_2(\hat{w}_0, (t_1)_0) \\
&-\langle \hat{w}_0, f\rangle_{L^2}. \qquad (18.34)
\end{aligned}$$

is also such that the optimal inclusion

$$\theta \in \partial\hat{J}(\hat{w}_0)$$

stands for

$$\delta_{\hat{w}}\{G_1(\Lambda\hat{w}_0, t_0) + G_2(\hat{w}_0, (t_1)_0) - \langle \hat{w}_0, f\rangle_{L^2}\} = \theta.$$

Thus,

$$\begin{aligned}
-\hat{J}(\hat{w}_0) &= -G_1(\Lambda\hat{w}_0, t_0) - G_2(\hat{w}_0, (t_1)_0) \\
&\quad + \langle \hat{w}_0, f\rangle_{L^2} \\
&= G_1(\Lambda\hat{w}_0, t_0) + G_2(\hat{w}_0, (t_1)_0) + Ind(\hat{w}_0, t_0, (t_1)_0) \\
&= J(\hat{w}_0, t_0, (t_1)_0). \qquad (18.35)
\end{aligned}$$

From this (18.33), the proof is complete.

18.6 A Numerical Example

In this section we develop a numerical example. In fact we address the problem of establishing the optimal distribution of springs on a beam similarly as above specified, in order to minimize its compliance. Here the control variable is t, where the function relating to the constant spring distribution is given by $K(t) = t(x)K_1 + (1-t(x))K_2$, where $K_1 \gg K_2 > 0$ are the constants related to a strong and a weak spring, respectively. Our main result is summarized by the following theorem, which may be proven similarly as the last one (we do not provide a proof here).

Theorem 18.6.1. *Consider a straight beam with rectangular cross section in which the axis corresponds to the set* $\Omega = [0,l]$. *Define* $\tilde{J} : U \times B \to \overline{\mathbb{R}} = \mathbb{R} \cup \{+\infty\}$ *by*

$$\tilde{J}(w,t) = \frac{1}{2}\int_0^l EIw_{,xx}^2\,dx + \frac{1}{2}\int_0^l K(t)w^2\,dx,$$

subject to

$$EIw_{xxxx} + K(t)w - f = 0, \text{ in } [0,l], \tag{18.36}$$

where

$$U = \{w \in W^{2,2}(\Omega) \mid w(0) = w(l) = 0\},$$

$$B = \{t \text{ measurable } \mid 0 \le t \le 1, a.e. \text{ in } \Omega \text{ and } \int_0^1 t\,dx \le c_0\Omega\},$$

and $0 < c_0 < 1, K_1 \gg K_2 > 0$.
Now, define $J : U \times B \to \overline{\mathbb{R}} = \mathbb{R} \cup \{+\infty\}$ *by*

$$J(w,t) = G(\Lambda w) + F(w,t) + Ind(w,t)$$

where

$$\Lambda : U \to Y = Y^* = L^2(\Omega) \equiv L^2,$$

is expressed by

$$\Lambda w = w_{,xx},$$

$$G(\Lambda w) = \frac{1}{2}\int_0^l EIw_{,xx}^2\,dx,$$

$$F(w,t) = \frac{1}{2}\int_0^l K(t)w^2\,dx,$$

$$Ind(w,t) = \begin{cases} 0, & \text{if } (w,t) \in B_0 \\ +\infty, & \text{otherwise,} \end{cases} \tag{18.37}$$

and

$$B_0 = \{(w,t) \in U \times B \text{ such that } (18.36) \text{ is satisfied }\}.$$

Under such hypotheses, we have

$$\inf_{(w,t)\in U\times B}\{J(w,t)\} = \inf_{(v^*,t)\in A^*\times B}\{J^*(v^*,t)\},$$

where $A^* = U$,

$$J^*(v^*,t) = G^*(v^*) + F^*(\Lambda^* v^* - f, t),$$

$$\begin{aligned} G^*(v^*) &= \sup_{v \in Y}\{\langle v, v^*\rangle_{L^2} - G(v)\} \\ &= \frac{1}{2EI}\int_0^l (v^*)^2\,dx, \end{aligned} \tag{18.38}$$

$$\begin{aligned} F^*(\Lambda^* v^*, t) &= \sup_{w \in U}\{-\langle w_{xx}, v^*\rangle_{L^2} - F(w,t) \\ &\quad + \langle w, f\rangle_{L^2}\} \\ &= \frac{1}{2}\int_0^l \frac{((v^*)_{xx} - f)^2}{K(t)}\,dx. \end{aligned} \tag{18.39}$$

Furthermore, also under the mentioned hypotheses, we have

$$\inf_{(w,t) \in U \times B}\{J(w,t)\} \geq \sup_{\hat{w} \in U}\{-\hat{J}(\hat{w})\},$$

where

$$\hat{J}(\hat{w}) = G(\Lambda \hat{w}) + \hat{F}(\hat{w}) - \langle \hat{w}, f\rangle_{L^2},$$

and

$$\hat{F}(\hat{w}) = \sup_{t \in B}\{F(\hat{w}, t)\}.$$

Finally, there exists $\hat{w}_0 \in U$ *such that*

$$-\hat{J}(\hat{w}_0) = \max_{\hat{w} \in U}\{-\hat{J}(\hat{w})\}.$$

Suppose, after discretization, that $t_0 \in B$ *such that*

$$\hat{F}(\hat{w}_0) = F(\hat{w}_0, t_0),$$

is also such that the optimal inclusion

$$\theta \in \partial \hat{J}(\hat{w}_0)$$

stands for

$$\delta_{\hat{w}}\{G(\Lambda \hat{w}_0) + F(\hat{w}_0, t_0) - \langle \hat{w}_0, f\rangle_{L^2}\} = \theta.$$

Under such hypotheses

$$\inf_{(w,t) \in U \times B}\{J(w,t)\} = J(\hat{w}_0, t_0) = -\hat{J}(\hat{w}_0) = \max_{\hat{w} \in U}\{-\hat{J}(\hat{w})\}.$$

We have computed, through the dual formulations, the solution for the numerical values $EI = 10^5$, $K_1 = 990{,}000$, $K_2 = 10$, and $P = 1{,}000$, $t_1 = 0.5$, $l = 1.0$ with units relating the international system. We consider two cases: first, for t, the constraint $0 \leq t \leq 1$ and, in a second step, the constraint $t \in \{0,1\}$.

For the results, please see Figs. 18.1, 18.2, 18.3, and 18.4, below indicated.

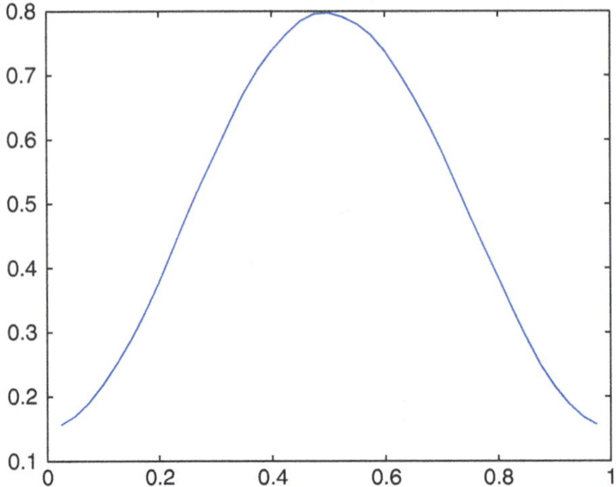

Fig. 18.1 Parameter $0 \leq t \leq 1$ relating the distribution of spring constants

18.7 Conclusion

At some point of our analysis, for almost all results, we have considered the discretized problem version. The reason is that we may not guarantee the attainability of a measurable $t_0 \in B$ such that

$$G^*(v_0^*, t_0) = \sup_{t \in B}\{G^*(v_0^*, t)\},$$

before discretization. It seems to be clear that after discretization, we are more likely to satisfy the hypotheses of Corollary 11.1. We emphasize again that if such hypotheses are satisfied by $(v_0^*, t_0) \in A^* \times B$, then the duality gap between the primal and dual formulations is zero for the different problems addressed. For the second result we obtain directly a duality principle with no duality gap between the primal and dual problems and such that the dual formulation computation is relatively easy. We also highlight again that for some of the results developed, the dual formulations are concave and also useful for relaxation and to obtain numerical results. In the last section we present a numerical example. We also emphasize the results obtained are consistent with the problem physics. Finally, for this same numerical example, it is worth mentioning that through the theoretical results developed, we may assert that

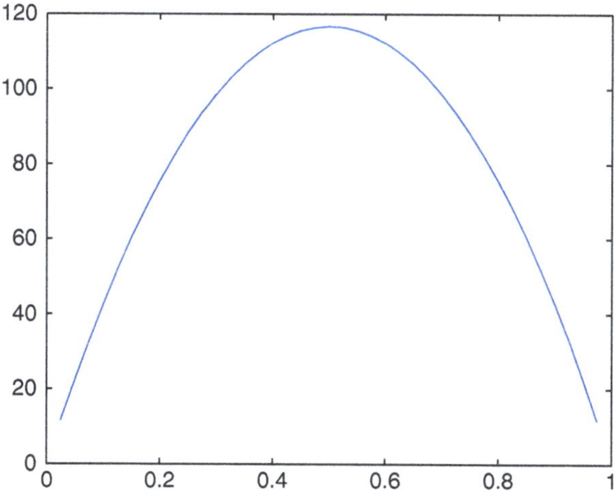

Fig. 18.2 Moments $v^*(x)$ relating the case $0 \leq t \leq 1$

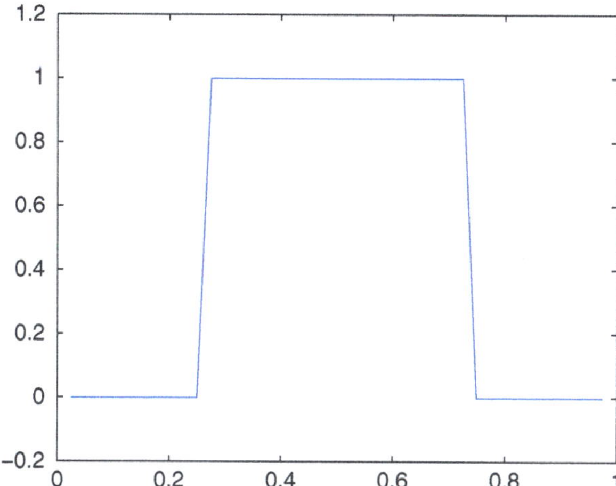

Fig. 18.3 Parameter $t \in \{0, 1\}$ relating the distribution of spring constants

the solution related to Figs. 18.3 and 18.4 is the global optimal one. On the other hand, the solution related to Figs. 18.1 and 18.2 is just a critical point (anyway, a possible candidate to global optimal solution).

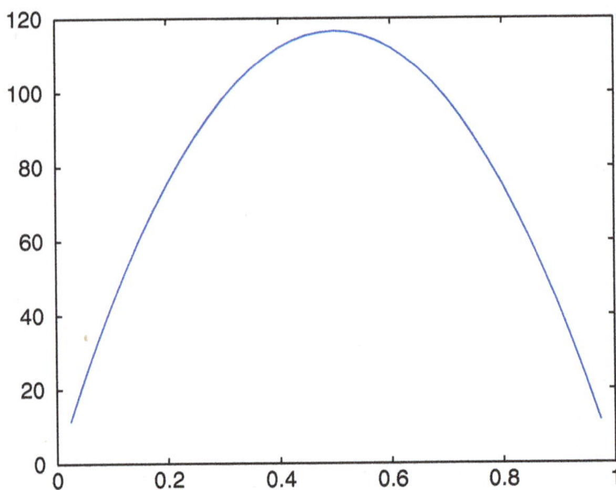

Fig. 18.4 Moments $v^*(x)$ relating the case $t \in \{0, 1\}$

Chapter 19
Duality and Computation for Quantum Mechanics Models

Fabio Botelho and Anderson Ferreira

19.1 Introduction

Our first objective is to obtain a duality principle for a class of nonlinear eigenvalue problems. The results are closely related to the canonical duality procedure; for details see Gao [36]. In a second step we apply the method to compute examples of nonlinear Schrödinger equation. We highlight the nonlinear Hamiltonian part refers to a kind of wave function self-interacting term, which models a great variety of physical phenomena. Among others, we would mention the nonlinear dynamics of superfluids, the Ginzburg–Landau theory of phase transitions, and the propagation of electromagnetic waves in plasmas.

At this point we start to describe the equation in question and respective variational formulation.

Let $\Omega \subset \mathbb{R}^3$ be an open, bounded, connected set with a regular boundary denoted by $\partial\Omega$. By a regular boundary we mean regularity enough so that the Sobolev imbedding theorem, the trace theorem, and the standard Gauss–Green formulas of integration by parts hold.

Moreover, we define

$$U = \{\varphi \in W^{1,2}(\Omega) \mid \varphi = 0 \text{ on } \partial\Omega\} = W_0^{1,2}(\Omega).$$

We emphasize the derivatives must be understood in the distributional sense, whereas the boundary conditions are in the sense of traces.

F. Botelho (✉)
Department of Mathematics and Statistics, Federal University of Pelotas, Pelotas, RS-Brazil
e-mail: fabio.silva.botelho@gmail.com

A. Ferreira
Department of Physics, Federal University of Pelotas, Brazil
e-mail: anderson.augusto@ufpel.edu.br

F. Botelho, *Functional Analysis and Applied Optimization in Banach Spaces: Applications to Non-Convex Variational Models*, DOI 10.1007/978-3-319-06074-3_19,

Consider the eigenvalue problem given by the solution in U of equation

$$-\frac{\hbar^2}{2m}\nabla^2\varphi+2\alpha|\varphi|^2\varphi+V(x)\varphi-\mu\varphi=0 \text{ in } \Omega, \tag{19.1}$$

where $\mu\in\mathbb{R}$ is a suitable Lagrange multiplier to be obtained such that the following constraint is satisfied:

$$\int_\Omega |\varphi|^2\,dx=c. \tag{19.2}$$

Here $c>0$ is appropriate constant to be specified.

Furthermore, α, $\hbar$, m are positive constants and $V\in L^2(\Omega)$.

In case $\varphi\in U$ satisfies (19.1) and (19.2), the function $\psi(x,t)$, given by

$$\psi(x,t)=e^{-\frac{i\mu t}{\hbar}}\varphi(x),$$

solves the well-known nonlinear Schrödinger equation given by

$$i\hbar\frac{\partial\psi(x,t)}{\partial t}=-\frac{\hbar^2}{2m}\nabla^2\psi(x,t)+2\alpha|\psi(x,t)|^2\psi(x,t)+V(x)\psi(x,t) \text{ in } \Omega\times[0,\infty),$$

with the boundary condition $\psi=0$ on $\partial\Omega\times[0,\infty)$, so that

$$\int_\Omega |\psi(x,t)|^2\,dx=c, \forall t\in[0,\infty).$$

Remark 19.1.1. About the references, we highlight that details on the Sobolev spaces involved may be found in [1, 26]. Duality principles for related problems are addressed in [13]. Also, an extensive study on Lagrange multipliers may be found in [40, 47]. For the numerical results, details on finite difference schemes are presented in [63]. Finally, details on related physics problems are developed in [4, 45, 46].

19.2 The Duality Principle

The corresponding primal variational formulation of the system above described is given by the functional $J:U\to\overline{\mathbb{R}}=\mathbb{R}\cup\{+\infty\}$, where unless otherwise indicated, we denote $x=(x_1,x_2,x_3)\in\mathbb{R}^3$, $dx=dx_1\,dx_2\,dx_3$ and

$$\begin{aligned} J(\varphi) = {} & \frac{\hbar^2}{4m}\int_\Omega \nabla\varphi\cdot\nabla\varphi\,dx+\frac{\alpha}{2}\int_\Omega|\varphi|^4\,dx \\ & +\frac{1}{2}\int_\Omega V(x)|\varphi|^2\,dx+Ind(\varphi), \end{aligned} \tag{19.3}$$

where

$$Ind(\varphi)=\begin{cases} 0, & \text{if } \int_\Omega|\varphi|^2\,dx=c, \\ +\infty, & \text{otherwise.}\end{cases} \tag{19.4}$$

For such a nonlinear optimization problem, we have the following duality principle.

Theorem 19.2.1. *Redefine $J : U \to \overline{\mathbb{R}}$ by*

$$J(\varphi) = G_1(\nabla\varphi) + G_2(|\varphi|^2) + G_3(\varphi) + Ind(\varphi),$$

where

$$G_1(\nabla\varphi) = \frac{\hbar^2}{4m}\int_\Omega \nabla\varphi\cdot\nabla\varphi\,dx,$$
$$G_2(|\varphi|^2) = \frac{\alpha}{2}\int_\Omega |\varphi|^4\,dx,$$
$$G_3(\varphi) = \frac{1}{2}\int_\Omega V(x)|\varphi|^2\,dx,$$

and, as above indicated,

$$Ind(\varphi) = \begin{cases} 0, & \text{if } \int_\Omega |\varphi|^2\,dx = c, \\ +\infty, & \text{otherwise.}\end{cases} \tag{19.5}$$

Under such hypotheses, we have

$$\inf_{\varphi\in U}\{J(\varphi)\} \geq \sup_{v^*\in Y^*}\{-J^*(v^*)\},$$

where $Y = Y^* = L^2(\Omega)$,

$$J^*(v^*) = G_2^*(v^*) + \tilde{F}^*(v^*),$$

$$\begin{aligned} G_2^*(v^*) &= \sup_{v\in Y}\{\langle v, v^*\rangle_{L^2(\Omega)} - G_2(v)\} \\ &= \frac{1}{2\alpha}\int_\Omega |v^*|^2\,dx. \end{aligned} \tag{19.6}$$

Moreover,

$$\begin{aligned} \tilde{F}^*(v^*) = \sup_{\varphi\in U}\{&-\langle|\varphi|^2, v^*\rangle_{L^2(\Omega)} \\ &- G_1(\nabla\varphi) - G_3(\varphi) - Ind(\varphi)\}. \end{aligned} \tag{19.7}$$

Finally, assume there exists $(\varphi_0, \mu_0) \in U \times \mathbb{R}$, such that

$$\delta_\varphi\{J_{\mu_0}(\varphi_0)\} = \theta,$$

where

$$\begin{aligned} J_{\mu_0}(\varphi) = {} & G_1(\nabla\varphi) + G_2(|\varphi|^2) + G_3(\varphi) \\ & -\frac{\mu_0}{2}\left(\int_\Omega |\varphi|^2\,dx - c\right), \end{aligned} \tag{19.8}$$

and so that

$$\int_\Omega |\varphi_0|^2\, dx - c = 0.$$

Also, suppose that

$$\tilde{J}_{\mu_0}(\varphi) \geq 0, \forall \varphi \in U,$$

where

$$\begin{aligned}\tilde{J}_{\mu_0}(\varphi) &= \langle |\varphi|^2, v_0^* \rangle_{L^2(\Omega)} + G_1(\nabla \varphi) \\ &\quad + G_3(\varphi) - \frac{\mu_0}{2} \int_\Omega |\varphi|^2\, dx, \end{aligned} \tag{19.9}$$

and

$$\begin{aligned} v_0^* &= \frac{\partial G_2(|\varphi_0|^2)}{\partial v} \\ &= \alpha |\varphi_0|^2. \end{aligned} \tag{19.10}$$

Under such hypotheses, we have

$$\min_{\varphi \in U}\{J(\varphi)\} = J(\varphi_0) = -J^*(v_0^*) = \max_{v^* \in Y^*}\{-J^*(v^*)\}.$$

Proof. Observe that

$$\begin{aligned} J(\varphi) &= -\langle |\varphi|^2, v^* \rangle_{L^2(\Omega)} + G_2(|\varphi|^2) \\ &\quad + \langle |\varphi|^2, v^* \rangle_{L^2(\Omega)} + G_1(\nabla \varphi) \\ &\quad + G_3(\varphi) + Ind(\varphi) \\ &\geq \inf_{v \in Y}\{-\langle v, v^* \rangle_{L^2(\Omega)} + G_2(v)\} \\ &\quad + \inf_{\varphi \in U}\{\langle |\varphi|^2, v^* \rangle_{L^2(\Omega)} + G_1(\nabla \varphi) \\ &\quad + G_3(\varphi) + Ind(\varphi)\} \\ &= -G_2^*(v^*) - \tilde{F}^*(v^*) \\ &= -J^*(v^*), \end{aligned} \tag{19.11}$$

$\forall \varphi \in U$, $v^* \in Y^*$. Thus,

$$\inf_{\varphi \in U}\{J(\varphi)\} \geq \sup_{v^* \in Y^*}\{-J^*(v^*)\}. \tag{19.12}$$

Finally, from the additional hypotheses, we may infer that

$$\begin{aligned} -\tilde{F}^*(v_0^*) &= \langle |\varphi_0|^2, v_0^* \rangle_{L^2(\Omega)} + G_1(\nabla \varphi_0) \\ &\quad + G_3(\varphi_0) - \frac{\mu_0}{2}\left(\int_\Omega |\varphi_0|^2\, dx - c\right), \end{aligned} \tag{19.13}$$

that is,

$$-\tilde{F}^*(v_0^*) = \langle |\varphi_0|^2, v_0^* \rangle_{L^2(\Omega)} + G_1(\nabla \varphi_0) + G_3(\varphi_0) + Ind(\varphi_0). \quad (19.14)$$

From the definition of v_0^* we get

$$G_2^*(v_0^*) = \langle |\varphi_0|^2, v_0^* \rangle_{L^2(\Omega)} - G_2(|\varphi_0|^2),$$

so that from these last two equations, we obtain

$$-G_2^*(v_0^*) - \tilde{F}^*(v_0^*) = G_1(\nabla \varphi_0) + G_2(|\varphi_0|^2) + G_3(\varphi_0) + Ind(\varphi_0).$$

Hence,

$$-J^*(v_0^*) = J(\varphi_0).$$

From this and (19.12) the proof is complete.

19.3 Numerical Examples

Before presenting the numerical examples, we introduce the following remark.

Remark 19.3.1. Consider the function $f : \mathbb{R}^2 \setminus \{(0,0)\} \to \mathbb{R}^2$, where

$$f(x,y) = \left(\frac{x}{\sqrt{x^2+y^2}}, \frac{y}{\sqrt{x^2+y^2}} \right).$$

We recall that, for the Euclidean norm in question,

$$\nabla f(x,y) = \left(\nabla \left(\frac{x}{\sqrt{x^2+y^2}} \right), \nabla \left(\frac{y}{\sqrt{x^2+y^2}} \right) \right),$$

so that

$$\|\nabla f(x,y)\| = \sqrt{\left\| \nabla \left(\frac{x}{\sqrt{x^2+y^2}} \right) \right\|^2 + \left\| \nabla \left(\frac{y}{\sqrt{x^2+y^2}} \right) \right\|^2}.$$

We may compute

$$\|\nabla f(x,y)\| = \frac{1}{\sqrt{x^2+y^2}}. \quad (19.15)$$

A similar result is valid for a $N > 2$-dimensional vector representing the discretized version of a function.

Thus denoting $\phi = (x_1,\ldots,x_N) \in \mathbb{R}^N$ we may symbolically write

$$\left\| \delta\left(\frac{\phi}{\|\phi\|}\right) \right\| = \frac{1}{\|\phi\|}. \tag{19.16}$$

We present numerical results, first, for the following one-dimensional closely related eigenvalue problem with $\alpha = 1/2$

$$-\frac{1}{\lambda^2}\frac{d^2\varphi(x)}{dx^2} + \varphi^3(x) + V(x)\varphi(x) - \mu\varphi(x) = 0, \text{ in } \Omega = [0,1],$$

with the boundary conditions

$$\varphi(0) = \varphi(1) = 0.$$

Here $\mu \in \mathbb{R}$ is such that

$$\int_\Omega \varphi^2(x)\,dx = 1.$$

Moreover, the potential $V(x)$ is given by

$$V(x) = \begin{cases} 0, & \text{if } x \in (0,1), \\ +\infty, & \text{otherwise.} \end{cases} \tag{19.17}$$

19.3.1 The Algorithm

We denote by $I(\varphi)$ the diagonal matrix which the diagonal is the vector φ. Furthermore, if $\varphi \in \mathbb{R}^N$ is such that

$$\varphi = [\varphi_1,\ldots,\varphi_N],$$

we denote

$$\varphi^p = [\varphi_1^p,\ldots,\varphi_N^p],\ \forall p \in \mathbb{N}.$$

The above equation, after discretization, in finite differences may be expressed by

$$-\frac{1}{\lambda^2 d^2} M_2\varphi + I(\varphi^2)\varphi - \mu\varphi = 0,$$

where

$$M_2 = \begin{bmatrix} -2 & 1 & 0 & 0 & \ldots & 0 \\ 1 & -2 & 1 & 0 & \ldots & 0 \\ 0 & 1 & -2 & 1 & 0.. & 0 \\ \ldots & \ldots & \ldots & \ldots & \ldots & \ldots \\ 0 & 0 & \ldots & 1 & -2 & 1 \\ 0 & 0 & \ldots & \ldots & 1 & -2 \end{bmatrix}. \tag{19.18}$$

Here M_2 is a $N \times N$ matrix, where N is the number of nodes and $d = 1/N$.

In the next lines we describe the algorithm:

1. Set $n = 1$.
2. Choose $\tilde{\varphi}_1 \in W^{1,2}(\Omega)$, so that $\tilde{\varphi}_1 \neq 0, a.e.$ in Ω.
3. Define
$$v_n^* = \alpha\left(\frac{|\tilde{\varphi}_n|^2}{\|\tilde{\varphi}_n\|^2}\right).$$
4. Obtain $\tilde{\varphi}_{n+1} \in U$ as the solution of the linear equation
$$-\frac{1}{\lambda^2 d^2} M_2 \tilde{\varphi}_{n+1} + 2I(v_n^*)\tilde{\varphi}_{n+1} - \frac{\tilde{\varphi}_n}{\|\tilde{\varphi}_n\|} = 0.$$
5. Set $n \to n+1$, and go to step 3, up to the satisfaction of an appropriate convergence criterion.

Here we present a rather informal discussion about the algorithm convergence.

First, observe that the equation in question may be written as

$$-M_2\tilde{\varphi}_{n+1} + I\left[\left(\frac{\tilde{\varphi}_n}{\|\tilde{\varphi}_n\|}\right)^2\right]\tilde{\varphi}_{n+1} d^2\lambda^2 - \frac{\tilde{\varphi}_n}{\|\tilde{\varphi}_n\|} d^2\lambda^2 = 0,$$

so that

$$\left\|-M_2 + I\left[\left(\frac{\tilde{\varphi}_n}{\|\tilde{\varphi}_n\|}\right)^2\right] d^2\lambda^2\right\| \|\tilde{\varphi}_{n+1}\| \geq d^2\lambda^2,$$

that is,

$$\frac{1}{\|\tilde{\varphi}_{n+1}\|} \leq \frac{1}{d^2\lambda^2}\left\|-M_2 + I\left[\left(\frac{\tilde{\varphi}_n}{\|\tilde{\varphi}_n\|}\right)^2\right] d^2\lambda^2\right\|, \forall n \in \mathbb{N}. \tag{19.19}$$

Also, we may denote

$$\tilde{\varphi}_{n+1} = G(\tilde{\varphi}_n),$$

where

$$G(\tilde{\varphi}_n) = \left(-M_2 + I\left[\left(\frac{\tilde{\varphi}_n}{\|\tilde{\varphi}_n\|}\right)^2\right] d^2\lambda^2\right)^{-1} \cdot \left(\frac{\tilde{\varphi}_n}{\|\tilde{\varphi}_n\|}\right) d^2\lambda^2.$$

Therefore

$$\tilde{\varphi}_{n+1} = G(\tilde{\varphi}_n)$$

and

$$\tilde{\varphi}_{n+2} = G(\tilde{\varphi}_{n+1}),$$

so that, denoting for $t \in [0,1]$,

$$(\tilde{\varphi}_t)_n = t\varphi_n + (1-t)\varphi_{n+1}$$

from the generalized mean value inequality, we obtain

$$\|\tilde{\varphi}_{n+2}-\tilde{\varphi}_{n+1}\| \leq \sup_{t\in[0,1]} \|G'((\tilde{\varphi}_t)_n)\| \|\tilde{\varphi}_{n+1}-\tilde{\varphi}_n\|, \tag{19.20}$$

Computing the derivative and from (19.19), we get the estimate

$$\begin{aligned}
\|G'(\tilde{\varphi}_n)\| \leq & \left\| \left(-M_2 - I\left[\left(\frac{\tilde{\varphi}_n}{\|\tilde{\varphi}_n\|}\right)^2\right] d^2\lambda^2\right) \cdot \left(-M_2 + I\left[\left(\frac{\tilde{\varphi}_n}{\|\tilde{\varphi}_n\|}\right)^2\right] d^2\lambda^2\right)^{-2} \right\| \\
& \times \left\| \delta\left(\frac{\tilde{\varphi}_n}{\|\tilde{\varphi}_n\|}\right)\right\| d^2\lambda^2 \\
\leq & \left\| \left(-M_2 - I\left[\left(\frac{\tilde{\varphi}_n}{\|\tilde{\varphi}_n\|}\right)^2\right] d^2\lambda^2\right)\right\| \\
& \times \left\| \left(-M_2 + I\left[\left(\frac{\tilde{\varphi}_n}{\|\tilde{\varphi}_n\|}\right)^2\right] d^2\lambda^2\right)^{-2} \right\| \\
& \times \left(\frac{1}{\|\tilde{\varphi}_n\|}\right) d^2\lambda^2 \\
\leq & \left\| \left(-M_2 - I\left[\left(\frac{\tilde{\varphi}_n}{\|\tilde{\varphi}_n\|}\right)^2\right] d^2\lambda^2\right)\right\| \\
& \times \left\| \left(-M_2 + I\left[\left(\frac{\tilde{\varphi}_n}{\|\tilde{\varphi}_n\|}\right)^2\right] d^2\lambda^2\right)^{-2} \right\| \\
& \times \left\| \left(-M_2 + I\left[\left(\frac{\tilde{\varphi}_n}{\|\tilde{\varphi}_n\|}\right)^2\right] d^2\lambda^2\right) \right\|,
\end{aligned} \tag{19.21}$$

where we have used equality (19.16), as it is indicated at Remark 19.3.1, that is, for a vector $\tilde{\varphi}_n \in \mathbb{R}^N$,

$$\left\| \delta\left(\frac{\tilde{\varphi}_n}{\|\tilde{\varphi}_n\|}\right)\right\| = \frac{1}{\|\tilde{\varphi}_n\|}.$$

Recalling that $-M_2$ is positive definite, at this point, we assume that for an appropriate choice of d, there exist $n_0 \in \mathbb{N}$ and $0<\beta<1$ such that

$$\begin{aligned}
& \left\| \left(-M_2 - I\left[\left(\frac{\tilde{\varphi}_n}{\|\tilde{\varphi}_n\|}\right)^2\right] d^2\lambda^2\right)\right\| \\
& \times \left\| \left(-M_2 + I\left[\left(\frac{\tilde{\varphi}_n}{\|\tilde{\varphi}_n\|}\right)^2\right] d^2\lambda^2\right)^{-2} \right\| \\
& \times \left\| \left(-M_2 + I\left[\left(\frac{\tilde{\varphi}_n}{\|\tilde{\varphi}_n\|}\right)^2\right] d^2\lambda^2\right) \right\| \leq \beta < 1, \ \forall n > n_0.
\end{aligned}$$

We have thus obtained

$$\|G'(\tilde{\varphi}_n)\| \leq \beta, \forall n > n_0, \tag{19.22}$$

where $0 < \beta < 1$.

As we have mentioned above, this is just an informal discussion.

Suppose that from (19.22) we may obtain that (in fact we do not provide details here)

$$\|G'((\tilde{\varphi}_t)_n)\| \leq \beta, \forall t \in [0,1],\ n > n_0. \tag{19.23}$$

From this and (19.20), we have

$$\|\tilde{\varphi}_{n+2} - \tilde{\varphi}_{n+1}\| \leq \beta \|\tilde{\varphi}_{n+1} - \tilde{\varphi}_n\|, \forall n > n_0.$$

So, we may infer that

$$\{\tilde{\varphi}_n\}$$

is a Cauchy sequence, that is, it is strongly converging to some $\tilde{\varphi}_0$.

Thus, we may write

$$-M_2\tilde{\varphi}_0 + I\left[\left(\frac{\tilde{\varphi}_0}{\|\tilde{\varphi}_0\|}\right)^2\right]\tilde{\varphi}_0 d^2\lambda^2 - \frac{\tilde{\varphi}_0}{\|\tilde{\varphi}_0\|}d^2\lambda^2 = 0,$$

so that denoting

$$\varphi_0 = \frac{\tilde{\varphi}_0}{\|\tilde{\varphi}_0\|},$$

we get

$$-M_2\varphi_0 + I(\varphi_0^2)\varphi_0 d^2\lambda^2 - \frac{\varphi_0}{\|\tilde{\varphi}_0\|}d^2\lambda^2 = 0.$$

That is,

$$-\frac{1}{\lambda^2 d^2}M_2\varphi_0 + I(\varphi_0^2)\varphi_0 - \mu_0\varphi_0 = 0,$$

where

$$\mu_0 = \frac{1}{\|\tilde{\varphi}_0\|}.$$

Clearly we have

$$\int_\Omega |\varphi_0|^2\,dx = 1.$$

Remark 19.3.2. Observe that we have not formally proven that the algorithm converges. However, from the analysis developed, we have a strong indication that such a convergence, under mild hypotheses, holds. Indeed, all numerical examples so far worked have converged very easily.

An analogous analysis is valid for the two-dimensional case.

About the numerical results for this one-dimensional example, the ground state is given in Fig. 19.1. Other solution with a greater eigenvalue is plotted in Fig. 19.2.

We highlight the numerical results obtained perfectly agree with the well-known analytic solutions for this one-dimensional model. See [17] for details.

Finally, we present an analogous two-dimensional example, that is, we develop results for the eigenvalue problem

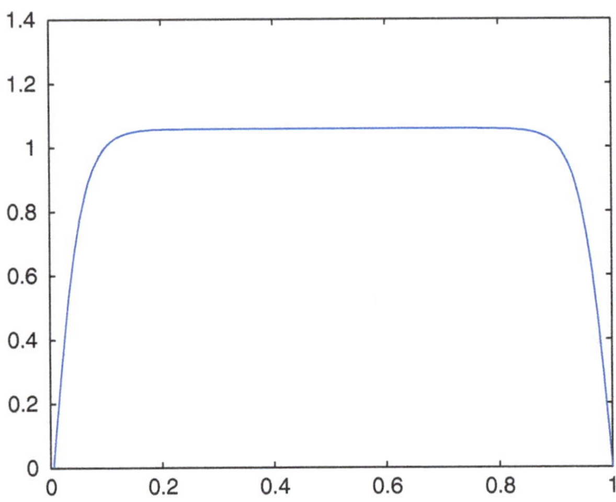

Fig. 19.1 Ground state wave function $\varphi(x)$ for $\lambda = 25$

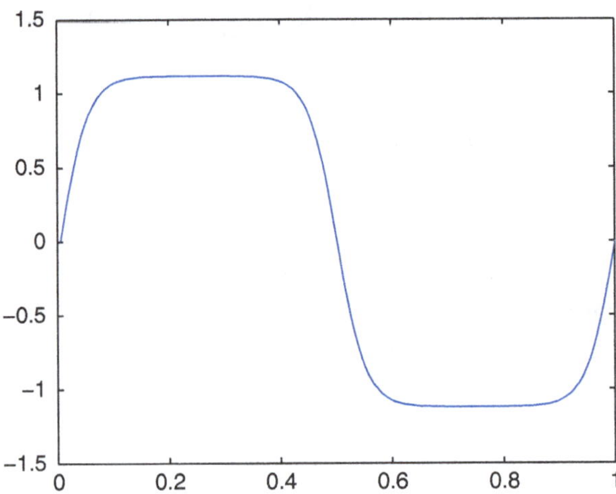

Fig. 19.2 Other wave function $\varphi(x)$ for $\lambda = 25$

$$-\frac{1}{\lambda^2}\nabla^2\varphi(x,y)+\varphi^3(x,y)+V(x,y)\varphi(x,y)-\mu\varphi(x,y)=0, \text{ in } \Omega=[0,1]\times[0,1],$$

with the boundary condition

$$\varphi = 0 \text{ on } \partial\Omega.$$

Here $\mu \in \mathbb{R}$ is such that

$$\int_\Omega \varphi^2(x,y)\, dxdy = 1.$$

Moreover, the potential $V(x,y)$ is given by

$$V(x,y) = \begin{cases} 0, & \text{if } (x,y) \in (0,1)\times(0,1), \\ +\infty, & \text{otherwise.} \end{cases} \tag{19.24}$$

For such a two-dimensional case, the ground state for $\lambda = 1$ is given in Fig. 19.3. The ground state, for $\lambda = 25$, is given in Fig. 19.4.

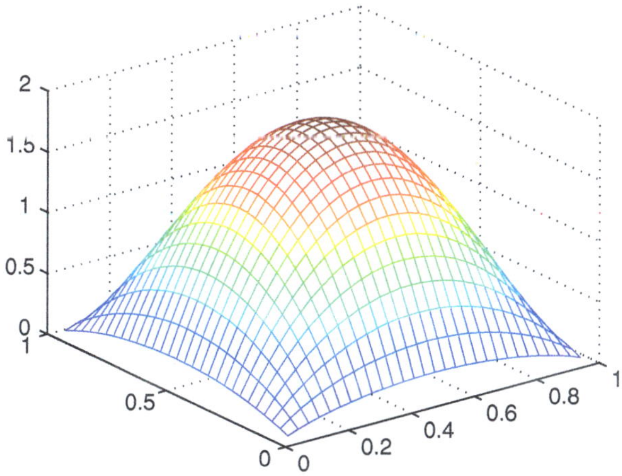

Fig. 19.3 Ground state wave function $\varphi(x,y)$ for $\lambda = 1$

Other solution, also for $\lambda = 25$, is plotted in Fig. 19.5.

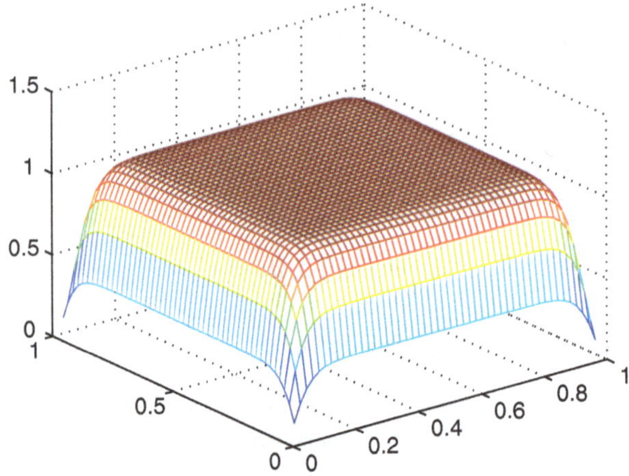

Fig. 19.4 Ground state wave function $\varphi(x,y)$ for $\lambda = 25$

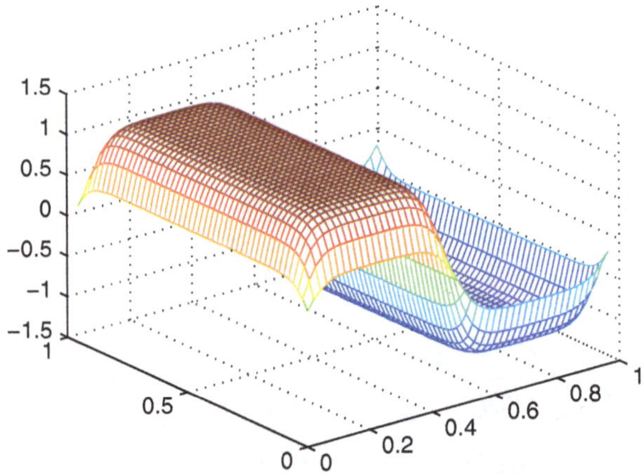

Fig. 19.5 Other wave function $\varphi(x,y)$ for $\lambda = 25$

19.4 Conclusion

In this chapter we develop a concave dual variational formulation for a nonlinear model in quantum mechanics. In practice, the results may be applied to obtain the ground state for the Schrödinger equation, which may be verified through the optimality conditions obtained. Finally, we emphasize the approach here developed may be applied to many other situations, such as for problems in quantum mechanics involving a large number of variables.

Chapter 20
Duality Applied to the Optimal Design in Elasticity

Fabio Botelho and Alexandre Molter

20.1 Introduction

In this chapter we develop duality for an optimal design problem in elasticity. We start by describing the primal formulation.

Consider $\Omega \subset \mathbb{R}^3$, an open, bounded, and connected set with a regular boundary denoted by $\partial\Omega = \Gamma_0 \cup \Gamma_1$, where $\Gamma_0 \cap \Gamma_1 = \emptyset$. By a regular boundary $\partial\Omega$ we mean regularity enough so that the Sobolev imbedding theorem and relating results, the trace theorem, and the standard Gauss–Green formulas of integration by parts hold.

Here Ω stands for the volume of an elastic solid under the action of a load $P \in L^2(\Omega;\mathbb{R}^3)$. We assume $|\Gamma_0| > 0$, where $|\Gamma_0|$ denotes the Lebesgue measure of Γ_0. Also, we denote by $\mathbf{n}$ the outward normal to the solid surface. The field of displacements is denoted by $u = (u_1, u_2, u_3) \in U$, where

$$U = \{u \in W^{1,2}(\Omega;\mathbb{R}^3) \mid u = (0,0,0) \text{ on } \Gamma_0\}. \tag{20.1}$$

The strain tensor, given by $e = \{e_{ij}\}$, is defined by

$$e_{ij}(u) = \frac{1}{2}(u_{i,j} + u_{j,i}). \tag{20.2}$$

Denoting by H^1_{ijkl} and H^0_{ijkl} two symmetric positive definite fourth-order constant tensors, first we define the optimization problem of minimizing $J(u,t)$ where

$$J(u,t) = \frac{1}{2}\int_\Omega \left(tH^1_{ijkl}e_{ij}(u)e_{kl}(u) + (1-t)H^0_{ijkl}e_{ij}(u)e_{kl}(u)\right)\,dx,$$

F. Botelho (✉)
Department of Mathematics and Statistics, Federal University of Pelotas, Pelotas, RS-Brazil
e-mail: fabio.silva.botelho@gmail.com

A. Molter
Department of Mathematics and Statistics, Federal University of Pelotas, Brazil
e-mail: alexandre.molter@ufpel.edu.br

F. Botelho, *Functional Analysis and Applied Optimization in Banach Spaces: Applications to Non-Convex Variational Models*, DOI 10.1007/978-3-319-06074-3_20,

subject to

$$\left(tH^1_{ijkl}e_{kl}(u)+(1-t)H^0_{ijkl}e_{kl}(u)\right)_{,j}+P_i=0, \text{ in } \Omega, \tag{20.3}$$

$$\left(tH^1_{ijkl}e_{kl}(u)+(1-t)H^0_{ijkl}e_{kl}(u)\right)\mathbf{n}_j=0, \text{ on } \Gamma_1, \tag{20.4}$$

$\forall i \in \{1,2,3\}$, $u \in U$, $t \in \{0,1\}$, a.e. in Ω, and

$$\int_\Omega t\,dx \leq t_1|\Omega|. \tag{20.5}$$

Here $0 < t_1 < 1$ and $|\Omega|$ denote the Lebesgue measure of Ω.

We relax such an original problem now allowing the parameter t to assume values in $[0,1]$ on Ω. Also, a penalization constant $p \geq 1$ is introduced in order to replace t by t^p in the energy functional, with the objective of approximating the resulting design variable to the set $\{0,1\}$. A standard value for p would be $p=3$, for example. Thus, we rewrite $J: U \times B \to \overline{\mathbb{R}} = \mathbb{R}\cup\{+\infty\}$ as

$$J(u,t)=\int_\Omega \frac{H_{ijkl}(t^p)}{2}e_{ij}(u)e_{kl}(u)\,dx+Ind(u,t), \tag{20.6}$$

where

$$H_{ijkl}(t^p)=t^pH^1_{ijkl}+(1-t^p)H^0_{ijkl},$$

$$Ind(u,t)=Ind_1(u,t)+Ind_2(u,t),$$

$$Ind_1(u,t)=\begin{cases}0, & \text{if } (H_{ijkl}(t^p)e_{kl}(u))_{,j}+P_i=0, \text{ in } \Omega, \forall i\in\{1,2,3\},\\ +\infty, & \text{otherwise,}\end{cases}$$

$$Ind_2(u,t)=\begin{cases}0, & \text{if } (H_{ijkl}(t^p)e_{kl}(u))\mathbf{n}_j=0, \text{ on } \Gamma_1, \ \forall i\in\{1,2,3\}\\ +\infty, & \text{otherwise,}\end{cases}$$

and

$$B=\left\{t \text{ measurable} \mid 0\leq t(x)\leq 1, \text{ a.e. in } \Omega, \int_\Omega t(x)\,dx\leq t_1|\Omega|\right\}.$$

Also $Y=Y^*=L^2(\Omega;\mathbb{R}^{3\times3})$ and from now on we denote

$$\{\overline{H}_{ijkl}(t^p)\}=\{H_{ijkl}(t^p)\}^{-1}.$$

20.2 On the Duality Principle

In this section we develop a duality principle for the problem in question. Similar problems are addressed in [3, 8, 13, 14, 29]. Details on general Sobolev spaces theory may be found in [1, 26]. We start with the next theorem. It is worth emphasizing the result indicated in (20.8) is well known, whereas the chain of equalities indicated in (20.15) and (20.17) we believe they are both new for the relaxed problem, that is, for the case $p > 1$ where the min–max theorem does not apply.

Theorem 20.2.1. *Considering the above expressed function $Ind(u,t)$, defining*

$$G(e(u),t) = \int_\Omega \frac{H_{ijkl}(t^p)}{2} e_{ij}(u) e_{kl}(u)\, dx, \tag{20.7}$$

and

$$J(u,t) = G(e(u),t) + Ind(u,t), \forall (u,t) \in U \times B,$$

we have that

$$\begin{aligned} \inf_{(u,t)\in U\times B} \{J(u,t)\} &= \inf_{(\sigma,t)\in A^*\times B} \{\tilde{G}^*(\sigma,t)\} \\ &\geq \sup_{\hat{u}\in U} \{-\tilde{J}^*(\hat{u})\}, \end{aligned} \tag{20.8}$$

where

$$-\tilde{J}^*(\hat{u}) = \inf_{t\in B} \{-G(e(\hat{u}),t) + \langle \hat{u}, P\rangle_{L^2}\},\ \forall \hat{u} \in U. \tag{20.9}$$

Under such definitions, there exists $\hat{u}_0 \in U$ such that

$$-\tilde{J}^*(\hat{u}_0) = \max_{u\in U} \{-\tilde{J}^*(\hat{u})\}. \tag{20.10}$$

Moreover,

$$\begin{aligned} \tilde{G}^*(\sigma,t) &= \sup_{v\in Y} \{\langle v,\sigma\rangle_{L^2} - G(v,t)\} \\ &= \frac{1}{2}\int_\Omega \overline{H}_{ijkl}(t^p)\sigma_{ij}\sigma_{kl}\, dx, \end{aligned} \tag{20.11}$$

and

$$A^* = \{\sigma \in Y^* \mid \sigma_{ij,j} + P_i = 0, \text{ in } \Omega,\ \sigma_{ij}\mathbf{n}_j = 0 \text{ on } \Gamma_1\}.$$

Also, the following representation holds:

$$\begin{aligned} G(e(\hat{u}),t) = \sup_{v\in Y} &\left\{ -\frac{1}{2}\int_\Omega H_{ijkl}(t^p) v_{ij} v_{kl}\, dx + \langle e_{ij}(\hat{u}), H_{ijkl}(t^p) v_{kl}\rangle_{L^2} \right\} \\ &- \frac{1}{2}\int_\Omega H_{ijkl}(t^p) e_{ij}(\hat{u}) e_{kl}(\hat{u})\, dx. \end{aligned} \tag{20.12}$$

Assume there exists $(\sigma_0, \tilde{t}_0, \tilde{u}_0) \in A^* \times B \times U$ such that

$$\delta\{J^*_\lambda(\sigma_0, \tilde{t}_0, \tilde{u}_0)\} = \theta,$$

where

$$J^*_\lambda(\sigma,t,\hat{u}) = \tilde{G}^*(\sigma,t) + \langle \hat{u}_i, -\sigma_{ij,j} - P_i\rangle_{L^2} + \int_{\Gamma_1} \hat{u}_i\sigma_{ij}\mathbf{n}_j \, d\Gamma$$
$$+ \int_\Omega \lambda_1(t^2 - t)\, dx + \lambda_2 \left(\int_\Omega t\, dx - t_1|\Omega| \right). \tag{20.13}$$

Here $(\lambda_1, \lambda_2) = \lambda$ *are appropriate Lagrange multipliers.*

Furthermore, suppose

$$-\tilde{J}^*(\tilde{u}_0) = \inf_{t\in B}\{-G(e(\tilde{u}_0),t) + \langle \tilde{u}_0, P\rangle_{L^2}\}$$
$$= -G(e(\tilde{u}_0),\tilde{t}_0) + \langle \tilde{u}_0, P\rangle_{L^2}. \tag{20.14}$$

Under such hypotheses we have

$$\min_{(u,t)\in U\times B}\{J(u,t)\} = J(\tilde{u}_0,\tilde{t}_0)$$
$$= \tilde{G}^*(\sigma_0,\tilde{t}_0)$$
$$= \min_{(\sigma,t)\in A^*\times B}\{\tilde{G}^*(\sigma,t)\}$$
$$= \max_{\hat{u}\in U}\{-\tilde{J}^*(\hat{u})\}$$
$$= -\tilde{J}^*(\tilde{u}_0), \tag{20.15}$$

where

$$\sigma_0 = \{(\sigma_0)_{ij}\} = \{H_{ijkl}(\tilde{t}_0^p)e_{kl}(\tilde{u}_0)\}.$$

Finally, considering the same problem after discretization, for the optimal $\hat{u}_0 \in U$ *satisfying (20.10), assume* $t_0 \in B$ *defined by*

$$-\tilde{J}^*(\hat{u}_0) = \inf_{t\in B}\{-G(e(\hat{u}_0),t) + \langle \hat{u}_0, P\rangle_{L^2}\}$$
$$= -G(e(\hat{u}_0),t_0) + \langle \hat{u}_0, P\rangle_{L^2}, \tag{20.16}$$

is such that $(\hat{u}_0, t_0)$ *are also such that the hypotheses of Corollary 11.1 to hold.*

Under such hypotheses, we have

$$\min_{(u,t)\in U\times B}\{J(u,t)\} = J(\hat{u}_0,t_0) = -\tilde{J}^*(\hat{u}_0) = \max_{\hat{u}\in U}\{-\tilde{J}^*(\hat{u})\} \tag{20.17}$$

We emphasize to have denoted $L^2(\Omega)$, $L^2(\Omega;\mathbb{R}^3)$, *or* $L^2(\Omega;\mathbb{R}^{3\times 3})$ *simply by* L^2, *as their meaning is clear.*

Proof. First observe that

$$\inf_{(u,t)\in U\times B}\{J(u,t)\}$$
$$= \inf_{t\in B}\{\inf_{u\in U}\{J(u,t)\}\}$$
$$= \inf_{t\in B}\{\inf_{u\in U}\{G(e(u),t) + Ind(u,t)\}\}$$

$$= \inf_{t \in B} \left\{ \sup_{\hat{u} \in U} \left\{ \inf_{u \in U} \{ G(e(u),t) + \langle \hat{u}_i, (H_{ijkl}(t^P) e_{kl}(u))_{,j} + P_i \rangle_{L^2} \right.\right.$$
$$\left.\left. - \int_{\Gamma_1} \hat{u}_i H_{ijkl}(t^P) e_{kl}(u) \mathbf{n}_j \, d\Gamma \} \right\} \right\},$$

so that

$$\begin{aligned}
&\inf_{(u,t) \in U \times B} \{J(u,t)\} \\
&= \inf_{t \in B} \left\{ \sup_{\hat{u} \in U} \left\{ \inf_{u \in U} \{ G(e(u),t) - \langle e_{ij}(\hat{u}), H_{ijkl}(t^P) e_{kl}(u) \rangle_{L^2} + \langle \hat{u}_i, P_i \rangle_{L^2} \} \right\} \right\} \\
&= \inf_{t \in B} \left\{ \sup_{\hat{u} \in U} \{ -G(e(\hat{u}),t) + \langle \hat{u}_i, P_i \rangle_{L^2} \} \right\} \\
&= \inf_{t \in B} \left\{ \inf_{\sigma \in A^*} \{ \tilde{G}^*(\sigma,t) \} \right\} \\
&= \inf_{(\sigma,t) \in A^* \times B} \{ \tilde{G}^*(\sigma,t) \}.
\end{aligned} \tag{20.18}$$

On the other hand,

$$\begin{aligned}
J(u,t) &= G(e(u),t) + Ind(u,t) \\
&\geq G(e(u),t) + \langle \hat{u}_i, (H_{ijkl}(t^P) e_{kl}(u))_{,j} + P_i \rangle_{L^2} \\
&\quad - \int_{\Gamma_1} \hat{u}_i H_{ijkl}(t^P) e_{kl}(u) \mathbf{n}_j \, d\Gamma \\
&= G(e(u),t) - \langle e_{ij}(\hat{u}), H_{ijkl}(t^P) e_{kl}(u) \rangle_{L^2} + \langle \hat{u}_i, P_i \rangle_{L^2} \\
&\geq \inf_{v \in Y} \{ G(v,t) - \langle e_{ij}(\hat{u}), H_{ijkl}(t^P) v_{kl} \rangle_{L^2} + \langle \hat{u}_i, P_i \rangle_{L^2} \} \\
&= -G(e(\hat{u}),t) + \langle \hat{u}_i, P_i \rangle_{L^2} \\
&\geq \inf_{t \in B} \{ -G(e(\hat{u}),t) + \langle \hat{u}_i, P_i \rangle_{L^2} \} \\
&= -\tilde{J}^*(\hat{u}),
\end{aligned} \tag{20.19}$$

$\forall u, \hat{u} \in U$.

Therefore,

$$\inf_{(u,t) \in U \times B} \{J(u,t)\} \geq \sup_{\hat{u} \in U} \{ -\tilde{J}^*(\hat{u}) \}, \tag{20.20}$$

so that from (20.18) and (20.19) we obtain

$$\begin{aligned}
\inf_{(u,t) \in U \times B} \{J(u,t)\} &= \inf_{(\sigma,t) \in A^* \times B} \{ \tilde{G}^*(\sigma,t) \} \\
&\geq \sup_{\hat{u} \in U} \{ -\tilde{J}^*(\hat{u}) \},
\end{aligned} \tag{20.21}$$

From the hypotheses, there exists $(\sigma_0, \tilde{t}_0, \tilde{u}_0) \in A^* \times B \times U$ such that

$$\delta\{J^*_\lambda(\sigma_0, \tilde{t}_0, \tilde{u}_0)\} = \theta,$$

where

$$J^*_\lambda(\sigma, t, \hat{u}) = \tilde{G}^*(\sigma, t) + \langle \hat{u}_i, -\sigma_{ij,j} - P_i \rangle_{L^2} + \int_{\Gamma_1} \hat{u}_i \sigma_{ij} \mathbf{n}_j \, d\Gamma$$
$$+ \int_\Omega \lambda_1 (t^2 - t) \, dx + \lambda_2 \left(\int_\Omega t \, dx - t_1 |\Omega| \right), \qquad (20.22)$$

where $(\lambda_1, \lambda_2) = \lambda$ are appropriate Lagrange multipliers and so that

$$\begin{aligned} -\tilde{J}^*(\tilde{u}_0) &= \inf_{t \in B} \{-G(e(\tilde{u}_0), t) + \langle \tilde{u}_0, P \rangle_{L^2}\} \\ &= -G(e(\tilde{u}_0), \tilde{t}_0) + \langle \tilde{u}_0, P \rangle_{L^2}. \end{aligned} \qquad (20.23)$$

By this last equation, we get

$$\begin{aligned} -\tilde{J}^*(\tilde{u}_0) &= -G(e(\tilde{u}_0), \tilde{t}_0) + \langle \tilde{u}_0, P \rangle_{L^2} \\ &= -\frac{1}{2} \int_\Omega H_{ijkl}(\tilde{t}_0^p) e_{ij}(\tilde{u}_0) e_{kl}(\tilde{u}_0) \, dx + \langle \tilde{u}_0, P \rangle_{L^2} \\ &= \frac{1}{2} \int_\Omega H_{ijkl}(\tilde{t}_0^p) e_{ij}(\tilde{u}_0) e_{kl}(\tilde{u}_0) \, dx \\ &= G(e(\tilde{u}_0), \tilde{t}_0) + Ind(\tilde{u}_0, \tilde{t}_0) \\ &= J(\tilde{u}_0, \tilde{t}_0) \\ &= -G(e(\tilde{u}_0), \tilde{t}_0) + \langle \tilde{u}_0, P \rangle_{L^2} \\ &= -G(e(\tilde{u}_0), \tilde{t}_0) + \langle (\tilde{u}_0)_i, -(\sigma_0)_{ij,j} \rangle_{L^2} \\ &= -G(e(\tilde{u}_0), \tilde{t}_0) + \langle e_{ij}(\tilde{u}_0), (\sigma_0)_{ij} \rangle_{L^2} \\ &= \tilde{G}^*(\sigma_0, \tilde{t}_0). \end{aligned} \qquad (20.24)$$

From this and (20.21), we have proven (20.15).

For proving (20.10) and (20.17), observe that from Korn's inequality (in fact it may be shown that $\|u\|_U$ and $|e(u)|_0 = (\int_\Omega e_{ij}(u) e_{ij}(u) \, dx)^{1/2}$ are equivalent norms; for details see [21]) we have

$$-\tilde{J}^*(\hat{u}) \to -\infty, \text{ as } \|\hat{u}\|_U \to \infty.$$

Furthermore, $-\tilde{J}^*(\hat{u})$ is concave since it is the infimum of a family of concave functions. From the coerciveness above verified, if $\{\hat{u}_n\}$ is a maximizing sequence, there exists $K_1 > 0$ such that

$$\|\hat{u}_n\|_U \leq K_1, \forall n \in \mathbb{N}.$$

Hence, there exists $\hat{u}_0 \in U$, such that up to a subsequence not relabeled, we have

$$e(\hat{u}_n) \rightharpoonup e(\hat{u}_0), \text{ weakly in } L^2.$$

$$\hat{u}_n \to \hat{u}_0, \text{ strongly in } L^2,$$

as $n \to \infty$.

From the concavity and weak upper semicontinuity of $-\tilde{J}^*(\hat{u})$ we get

$$\sup_{\hat{u} \in U}\{-\tilde{J}^*(\hat{u})\} = \limsup_{n \to \infty}\{-\tilde{J}^*(\hat{u}_n)\} \leq -\tilde{J}^*(\hat{u}_0).$$

Thus,

$$\max_{\hat{u} \in U}\{-\tilde{J}^*(\hat{u})\} = -\tilde{J}^*(\hat{u}_0).$$

At this point and on we consider the problem in question after discretization. Recall we have assumed that $t_0 \in B$ such that

$$\begin{aligned} -\tilde{J}^*(\hat{u}_0) &= \inf_{t \in B}\{-G(e(\hat{u}_0),t) + \langle \hat{u}_0, P\rangle_{L^2}\} \\ &= -G(e(\hat{u}_0),t_0) + \langle \hat{u}_0, P\rangle_{L^2}, \end{aligned} \tag{20.25}$$

is such that $(\hat{u}_0, t_0)$ are also such that the hypotheses of Corollary 11.1 are satisfied.

From such a corollary, the optimal equation

$$\delta \tilde{J}^*(\hat{u}_0) = \theta$$

stands for

$$\frac{\partial\{G(e(\hat{u}_0),t_0) - \langle \hat{u}_0, P\rangle_{L^2}\}}{\partial u} = \theta.$$

Hence,

$$(H_{ijkl}(t_0^p)e_{kl}(\hat{u}_0))_{,j} + P_i = 0 \text{ in } \Omega, \tag{20.26}$$

and

$$(H_{ijkl}(t_0^p)e_{kl}(\hat{u}_0))\mathbf{n}_j = 0 \text{ on } \Gamma_1, \tag{20.27}$$

$\forall i \in \{1,2,3\}$.

By (20.26) and (20.27), we obtain

$$\begin{aligned} -\tilde{J}^*(\hat{u}_0) &= -\frac{1}{2}\int_\Omega H_{ijkl}(t_0^p)e_{ij}(\hat{u}_0)e_{kl}(\hat{u}_0)\,dx + \langle \hat{u}_0, P\rangle_{L^2} \\ &= \frac{1}{2}\int_\Omega H_{ijkl}(t_0^p)e_{ij}(\hat{u}_0)e_{kl}(\hat{u}_0)\,dx + Ind(\hat{u}_0,t_0) \\ &\quad - G(e(\hat{u}_0),t_0) + Ind(\hat{u}_0,t_0). \end{aligned} \tag{20.28}$$

Observe that

$$\begin{aligned} G(e(\hat{u}_0),t_0) + Ind(\hat{u}_0,t_0) &\geq \inf_{(u,t) \in U \times B}\{G(e(u),t) + Ind(u,t)\} \\ &\geq -\tilde{J}^*(\hat{u}_0). \end{aligned} \tag{20.29}$$

By (20.28) and (20.29), we obtain

$$\begin{aligned} G(e(\hat{u}_0),t_0)+Ind(\hat{u}_0,t_0) &= \inf_{(u,t)\in U\times B}\{G(e(u),t)+Ind(u,t)\} \\ &= -\tilde{J}^*(\hat{u}_0), \end{aligned} \tag{20.30}$$

so that from this and (20.20) we finally have

$$\begin{aligned} \min_{(u,t)\in U\times B}\{J(u,t)\} &= J(\hat{u}_0,t_0) \\ &= G(e(\hat{u}_0),t_0)+Ind(\hat{u}_0,t_0) \\ &= -\tilde{J}^*(\hat{u}_0) \\ &= \max_{\hat{u}\in U}\{-\tilde{J}^*(\hat{u})\}. \end{aligned} \tag{20.31}$$

The proof is complete.

Remark 20.2.2. After discretization, for the case $p=1$, the chain of equalities indicated in (20.31) may be easily obtained from the min–max theorem. However, to improve the numerical results in practical situations, it is desirable to use $p>1$ and in such a case the min–max theorem does not apply ($p=3$, e.g., is a standard choice).

20.3 A Numerical Example

As an example, we present the problem of finding the optimal plate thickness distribution relating the structural inner work minimization. For, consider a plate which the middle surface is denoted by $\Omega\subset\mathbb{R}^2$, where Ω is an open, bounded, connected set with a sufficiently regular boundary denoted by $\partial\Omega$. As mentioned above the design variable, the plate thickness $h(x)$, is such that $h_0\leq h(x)\leq h_1$, where $x=(x_1,x_2)\in\Omega\subset\mathbb{R}^2$. The field of normal displacements to Ω, due to a external load $P\in L^2(\Omega)$, is denoted by $w:\Omega\to\mathbb{R}$.

Such an optimization problem is represented by the minimization of $J:U\times B\to\overline{\mathbb{R}}=\mathbb{R}\cup\{+\infty\}$, where

$$J(w,t)=\int_\Omega \frac{H_{\alpha\beta\lambda\mu}(t)}{2}w_{,\alpha\beta}w_{,\lambda\mu}\,dx+Ind(w,t), \tag{20.32}$$

$$Ind(w,t)=Ind_1(w,t)+Ind_2(w,t),$$

$$Ind_1(w,t)=\begin{cases}0, & \text{if } (H_{\alpha\beta\lambda\mu}(t)w_{,\lambda\mu})_{,\alpha\beta}-P=0, \text{ in } \Omega,\\ +\infty, & \text{otherwise,}\end{cases}$$

$$Ind_2(w,t)=\begin{cases}0, & \text{if } H_{\alpha\beta\lambda\mu}(t)w_{,\lambda\mu}\mathbf{n}_\alpha\mathbf{n}_\beta=0, \text{ on } \partial\Omega,\\ +\infty, & \text{otherwise,}\end{cases}$$

and

$$B = \{t \text{ measurable} \mid t(x) \in [0,1], \text{ a.e. in } \Omega \\ \text{and } \int_\Omega (th_1^3 + (1-t)h_0^3)\, dx \leq t_1 h_1^3 |\Omega|\}. \tag{20.33}$$

Moreover, $|\Omega|$ denotes the Lebesgue measure of Ω and

$$U = \left\{ w \in W^{2,2}(\Omega) \mid w = 0 \text{ on } \partial\Omega \right\}. \tag{20.34}$$

We develop numerical results for the particular case, where $t_1 = 0.52$,

$$H_{\alpha\beta\lambda\mu}(t) = H(t) = \hat{h}(t)E, \tag{20.35}$$

$\hat{h}(t) = th_1^3 + (1-t)h_0^3$, $h_1 = 0.1$, $h_0 = 10^{-4}$, and $E = 10^7$, with the units related to the international system (we emphasize to denote $x = (x_1, x_2)$). Observe that $0 \leq t(x) \leq 1$, a.e. in Ω. Similarly, as in the last section (see [13] for details), we may obtain the following duality principle (here in a slightly different version):

$$\inf_{(w,t) \in U \times B} \{J(w,t)\} - \inf_{(t,\{M_{\alpha\beta}\}) \in B \times D^*} \left\{ \frac{1}{2} \int_\Omega \overline{H}_{\alpha\beta\lambda\mu}(t) M_{\alpha\beta} M_{\lambda\mu}\, dx \right\},$$

where

$$\{\overline{H}_{\alpha\beta\lambda\mu}(t)\} = \{H_{\alpha\beta\lambda\mu}(t)\}^{-1},$$

and

$$D^* = \{\{M_{\alpha\beta}\} \in Y^* \mid M_{\alpha\beta,\alpha\beta} + P = 0, \text{ in } \Omega,\ M_{\alpha\beta}\mathbf{n}_\alpha \mathbf{n}_\beta = 0 \text{ on } \partial\Omega\}.$$

We have computed the dual problem for $\Omega = [0,1] \times [0,1]$ and a vertical load acting on the plate given by $P(x) = 10{,}000$, obtaining the results indicated in the respective Figs. 20.1 and 20.2, for $t_0(x)$ and $w_0(x)$. We emphasize they are critical points, that is, just candidates to optimal points. Observe that for the concerned critical point

$$\begin{aligned} J(w_0,t_0) &= \int_\Omega \frac{H_{\alpha\beta\lambda\mu}(t_0)}{2} (w_0)_{,\alpha\beta} (w_0)_{,\lambda\mu}\, dx + Ind(w_0,t_0) \\ &= \int_\Omega \frac{H_{\alpha\beta\lambda\mu}(t_0)}{2} (w_0)_{,\alpha\beta} (w_0)_{,\lambda\mu}\, dx \\ &= \frac{1}{2} \int_\Omega \overline{H}_{\alpha\beta\lambda\mu}(t_0) (M_0)_{\alpha\beta} (M_0)_{\lambda\mu}\, dx \\ &= \tilde{G}^*(\sigma_0, t_0), \end{aligned} \tag{20.36}$$

where the moments $\{(M_0)_{\alpha\beta}\}$ are given by

$$\{(M_0)_{\alpha\beta}\} = \{-H_{\alpha\beta\lambda\mu}(t_0)(w_0)_{,\lambda\mu}\}.$$

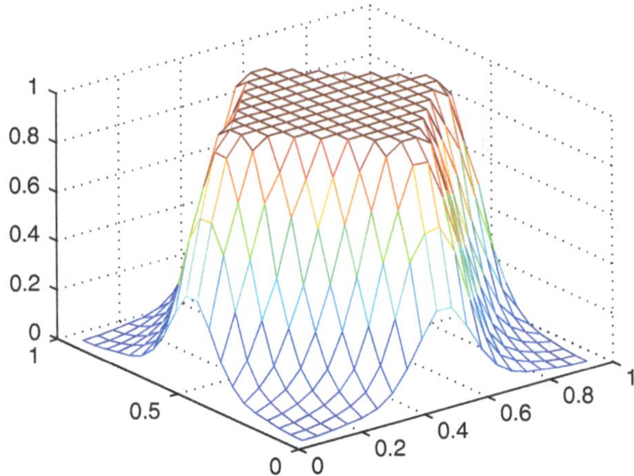

Fig. 20.1 $t_0(x)$-function relating the plate thickness distribution

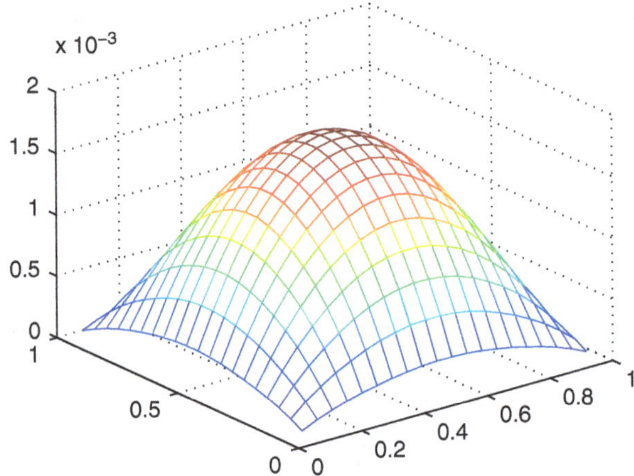

Fig. 20.2 $w_0(x)$-field of displacements

20.4 Another Numerical Example

In the paper [60] Sigmund presents an algorithm for shape optimization. Such an algorithm refers to find critical points of the Lagrangian functional L_λ (slightly changed considering the development established in this text) given by

$$\begin{aligned} L_\lambda(u,\hat{u},t) = {} & \langle u,P\rangle_{L^2} - \int_\Omega H_{ijkl}(t^p)e_{ij}(u)e_{kl}(\hat{u})\,dx + \langle \hat{u},P\rangle_{L^2} \\ & + \int_\Omega \lambda_1^+(t(x)-1)\,dx + \int_\Omega \lambda_1^-(t_{min}-t(x))\,dx \\ & + \lambda_2\left(\int_\Omega t(x)\,dx - t_1|\Omega|\right). \end{aligned} \quad (20.37)$$

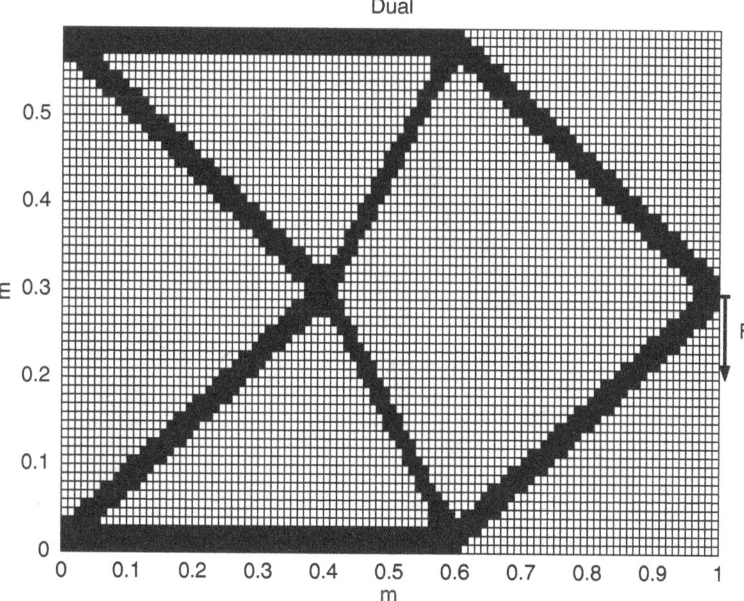

Fig. 20.3 The cantilever beam density through the dual formulation, $t_1 = 0.50$

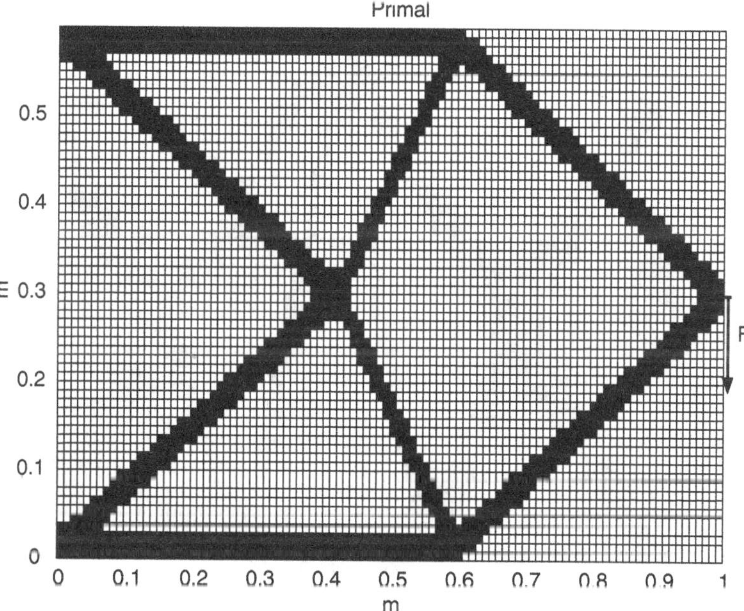

Fig. 20.4 The cantilever beam density through the primal formulation, $t_1 = 0.50$

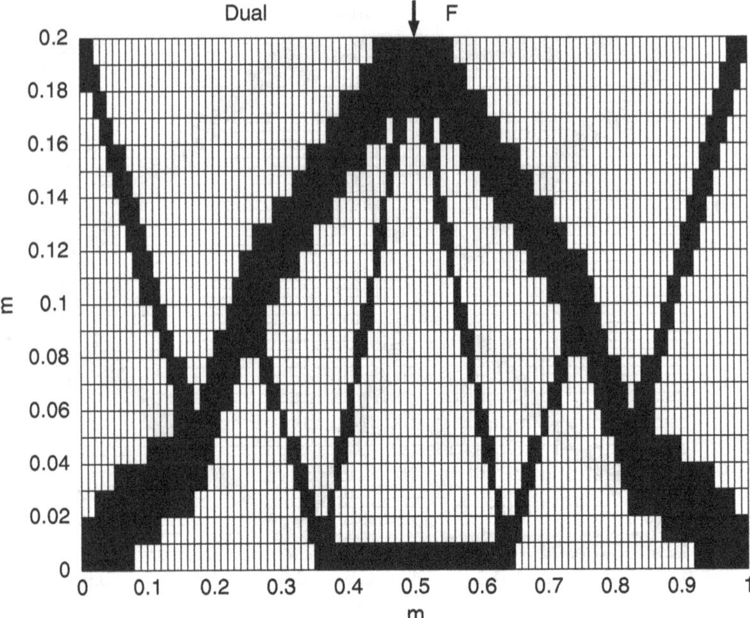

Fig. 20.5 Clamped beam density through the dual formulation, $t_1 = 0.50$

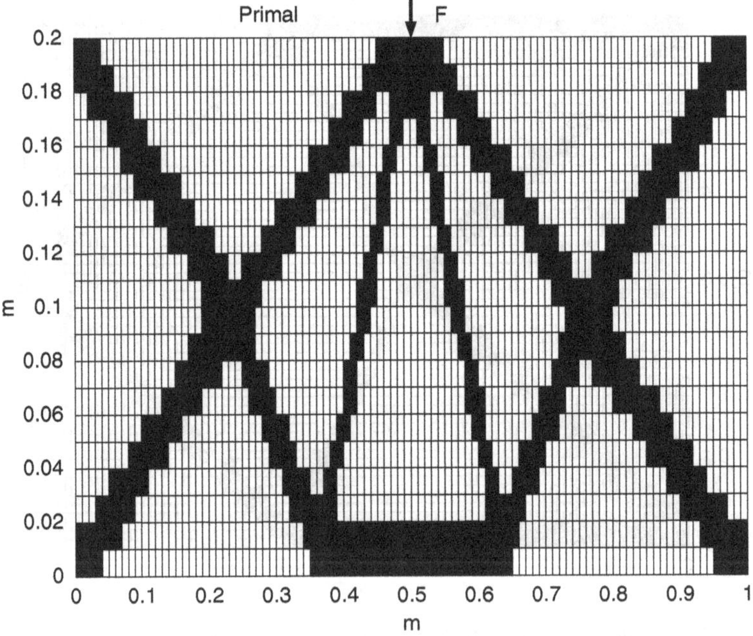

Fig. 20.6 Clamped beam density through the primal formulation, $t_1 = 0.50$

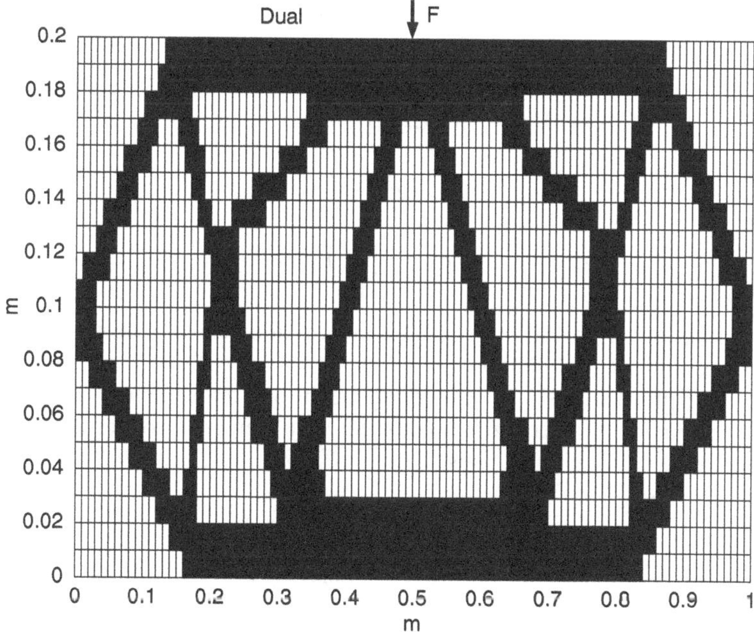

Fig. 20.7 Simply supported beam density through the dual formulation, $t_1 = 0.50$

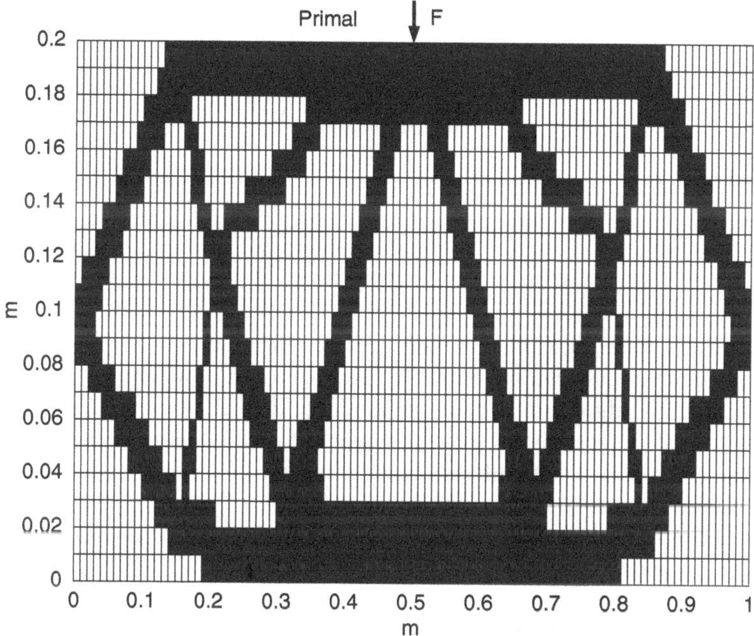

Fig. 20.8 Simply supported beam density through the primal formulation, $t_1 = 0.50$

We made a change in the 99-line Sigmund algorithm, now updating the density t through a critical point of the dual functional. The optimal equation would be

$$\delta_t \left\{ \int_\Omega \overline{H}_{ijkl}(t^p)\sigma_{ij}\sigma_{kl}\,dx \right.$$
$$+ \int_\Omega \lambda_1^+(t(x)-1)\,dx + \int_\Omega \lambda_1^-(t_{min}-t(x))\,dx$$
$$\left. +\lambda_2\left(\int_\Omega t(x)\,dx - t_1|\Omega|\right)\right\} = \theta. \tag{20.38}$$

For a fixed σ we update t through this last equation and update σ, for a fixed t, through the equilibrium equation.

In a more practical fashion, concerning the 99-line finite element algorithm and respective notation, the heuristic equation for each element through which t_e is updated up to the OC function, that is,

$$t_e^{new} = t_e \left(\frac{p(t_e)^{p-1} u_e^T K_0 u_e}{\lambda_2 \frac{\partial V}{\partial t_e}} \right)^\eta, \tag{20.39}$$

is replaced by

$$\frac{\partial\left((t_e^p K_0 u_e)^T (t_e^{new})^{-p} K_0^{-1} (t_e^p K_0 u_e)\right)}{\partial t_e^{new}} + \lambda_2 = 0, \tag{20.40}$$

so that

$$t_e^{new} = \left(\frac{p(t_e^{2p}) u_e^T K_0 u_e}{\lambda_2} \right)^{\frac{1}{1+p}}.$$

Hence,

$$t_e^{new} = t_e \left(\frac{p(t_e^{p-1}) u_e^T K_0 u_e}{\lambda_2} \right)^{\frac{1}{1+p}}, \tag{20.41}$$

It is worth emphasizing in such a procedure we must have $0 < t_e, t_e^{new} < 1$. To include the general case we have used an analogous OC function as found in the original Sigmund algorithm. We may observe that Eqs. (20.41) and (20.39) are very similar, so that in fact this last heuristic equation (20.39) almost corresponds, for different values of η, to the Euler–Lagrange equations for the dual problem, which, to some extent, justify the algorithm convergence. Based on the 99-line O. Sigmund software with inclusion of the mentioned change (which corresponds, as above indicated, to optimization through the dual formulation), the second author of this chapter (Alexandre Molter) designed a code suitable for the different situations here addressed, through which we have developed numerical examples.

We develop examples concerning a two-dimensional elastic beam with different types of boundary conditions and in all cases we have initially obtained a good approximation for the restriction $t \in \{0,1\}$, *a.e.*. However, it is worth mentioning, for the figures in question, in the gray scale obtained in $[0,1]$ we have post-processed such figures qualitatively by setting $t \equiv 0$ if for the original result $0 \leq t < 0.5$ and setting $t \equiv 1$ if $0.5 \leq t \leq 1$.

The numerical values for such examples are $E_1 = 210 * 10^9$ (Young modulus), $E_0 \ll E_1$ (for simulating absence of material), $\nu = 0.3$ (Poisson coefficient), and $F = 10{,}000$, where the position of applied F is indicated in each figure, the volume fraction is 0.5, and units refer to the international system. Please see Figs. 20.3, 20.5, and 20.7 for the solutions through the dual formulation and Figs. 20.4, 20.6, and 20.8 for solutions through the primal one.

We highlight the beam dimensions are:

1 m × 0.6 m for the cantilever ones indicated in Figs. 20.3 and 20.4.

1 m × 0.4 m for the beams clamped at both extremals indicated in Figs. 20.5 and 20.6.

1 m × 0.2 m for the beams simply supported at both extremals (in fact boundary conditions refer just to zero vertical displacement at $x = 0$ and $x = 1.0$) indicated in Figs. 20.7 and 20.8. Finally, we may observe the results are qualitatively similar through the dual and primal formulations. The main differences are found just in Figs. 20.5 and 20.6. For related results, see [8, 49].

20.5 Conclusion

In this chapter we have developed a dual variational formulation for an optimal design problem in elasticity. The infimum in t indicated in the dual formulation represents the structure search for stiffness in the optimization process, which implies the internal work minimization. In some cases the primal problem, before discretization, may not have solutions, so that the solution of dual problem is a weak cluster point of minimizing sequences for the primal one. After discretization, it has been established conditions for the duality gap between the primal and dual problems be zero. We expect the results obtained can be used as engineering project tools.

Chapter 21
Duality Applied to Micro-Magnetism

21.1 Introduction

In this chapter we develop dual variational formulations for models in micro-magnetism. For the primal formulation we refer to references [10, 43, 44, 51] for details. In particular we refer to the original results presented in [10], emphasizing that the present work is their natural continuation and extension.

At this point we start to describe the primal formulation.

Let $\Omega \subset \mathbb{R}^3$ be an open bounded set with a finite Lebesgue measure and a regular boundary denoted by $\partial\Omega$. By a regular boundary $\partial\Omega$ we mean regularity enough so that the Sobolev imbedding theorem and relating results, the trace theorem and the standard Gauss–Green formulas of integration by parts to hold. The corresponding outward normal is denoted by $\mathbf{n}$. Also, we denote by θ either the zero vector in $\mathbb{R}^3$ or the zero in an appropriate function space.

Under such assumptions and notations, consider problem of finding the magnetization $m : \Omega \to \mathbb{R}^3$, which minimizes the functional

$$J(m,f) = \frac{\alpha}{2}\int_\Omega |\nabla m|_2^2\,dx + \int_\Omega \varphi(m(x))\,dx - \int_\Omega H(x)\cdot m\,dx \\ +\frac{1}{2}\int_{\mathbb{R}^3} |f(x)|_2^2\,dx, \tag{21.1}$$

where

$$m = (m_1, m_2, m_3) \in W^{1,2}(\Omega;\mathbb{R}^3) \equiv Y_1,\ |m(x)|_2 = 1, \text{ a.e. in } \Omega \tag{21.2}$$

and $f \in L^2(\mathbb{R}^3;\mathbb{R}^3) = Y_2$ is the unique field determined by the simplified Maxwell's equations

$$Curl(f) = \theta,\ div(-f + m\chi_\Omega) = 0, \text{ a.e. in } \mathbb{R}^3. \tag{21.3}$$

Here $H \in L^2(\Omega;\mathbb{R}^3)$ is a known external field and χ_Ω is a function defined by

F. Botelho, *Functional Analysis and Applied Optimization in Banach Spaces: Applications to Non-Convex Variational Models*, DOI 10.1007/978-3-319-06074-3_21,

$$\chi_\Omega(x) = \begin{cases} 1, & \text{if } x \in \Omega, \\ 0, & \text{otherwise.} \end{cases} \tag{21.4}$$

The term

$$\frac{\alpha}{2} \int_\Omega |\nabla m|_2^2 \, dx$$

is called the exchange energy, where

$$|m|_2 = \sqrt{\sum_{k=1}^{3} m_k^2}$$

and

$$|\nabla m|_2^2 = \sum_{k=1}^{3} |\nabla m_k|_2^2.$$

Finally, $\varphi(m)$ represents the anisotropic contribution and is given by a multi-well functional whose minima establish the preferred directions of magnetization.

Remark 21.1.1. Here are some brief comments on the references. Relating and similar problems are addressed in [7, 11, 13, 14, 55, 56]. The basic results on convex and variational analysis used in this text may found in [13, 24, 25, 40, 47]. Finally, an extensive study on Sobolev spaces may be found in [1].

Remark 21.1.2. At some points of our analysis we refer to the problems in question after discretization. In such a case we refer to their approximations in a finite element or finite differences context.

21.2 Summary of Results for the Hard Uniaxial Case

We consider first the case of a uniaxial material with no exchange energy. That is, $\alpha = 0$ and $\varphi(m) = \beta(1 - |m \cdot e|)$.

Observe that

$$\varphi(m) = \min\{\beta(1 + m \cdot e), \beta(1 - m \cdot e)\}$$

where $\beta > 0$ and $e \in \mathbb{R}^3$ is a unit vector.

In the next lines we present the primal formulation and related duality principle.

Define $J : Y_1 \times Y_2 \times B \to \overline{\mathbb{R}}$ by

$$J(m, f, t) = G_1(m, f, t) + G_2(f),$$

$$\begin{aligned} G_1(m, f, t) = & \int_\Omega (t g_1(m) + (1 - t) g_2(m)) \, dx \\ & + Ind_0(m) + Ind_1(f) + Ind_2(m, f) \\ & - \int_\Omega H(x) \cdot m \, dx, \end{aligned} \tag{21.5}$$

and

$$G_2(f) = \frac{1}{2}\int_{\mathbb{R}^3} |f(x)|_2^2\, dx.$$

Also,

$$g_1(m) = \beta(1 + m \cdot e),$$

$$g_2(m) = \beta(1 - m \cdot e),$$

$$Ind_0(m) = \begin{cases} 0, & \text{if } |m(x)|_2 = 1 \text{ a.e. in } \Omega, \\ +\infty, & \text{otherwise,} \end{cases}$$

$$Ind_1(m,f) = \begin{cases} 0, & \text{if } div(-f + m\chi_\Omega) = 0 \text{ a.e. in } \mathbb{R}^3, \\ +\infty, & \text{otherwise,} \end{cases}$$

and

$$Ind_2(f) = \begin{cases} 0, & \text{if } Curl(f) = \theta, \text{ a.e. in } \mathbb{R}^3, \\ +\infty, & \text{otherwise.} \end{cases}$$

Observe that as abovementioned,

$$\begin{aligned} \int_\Omega \varphi(m)\, dx &= \int_\Omega \beta(1 - |m \cdot e|)\, dx \\ &= \min_{t \in B}\left\{\int_\Omega (tg_1(m) + (1-t)g_2(m))\, dx\right\}. \end{aligned} \tag{21.6}$$

Under additional assumptions to be specified, we have

$$\inf_{(m,f,t) \in Y_1 \times Y_2 \times B}\{J(m,f,t)\} = \sup_{\lambda \in \hat{Y}^*}\{-J^*(\lambda)\} \tag{21.7}$$

where

$$J^*(\lambda) = \tilde{G}_1^*(\lambda) + G_2^*(\lambda) - \int_\Omega \beta\, dx,$$

$$\begin{aligned} \tilde{G}_1^*(\lambda) &= \sup_{t \in B}\left\{\int_\Omega \left(\sum_{i=1}^3 \left(\frac{\partial \lambda_2}{\partial x_i} + H_i + \beta(1-2t)e_i\right)^2\right)^{1/2} dx\right\} \\ &= \sup_{t \in B}\{\hat{G}_1^*(\lambda, t)\}, \end{aligned} \tag{21.8}$$

where

$$\hat{G}_1^*(\lambda, t) = \int_\Omega \left(\sum_{i=1}^3 \left(\frac{\partial \lambda_2}{\partial x_i} + H_i + \beta(1-2t)e_i\right)^2\right)^{1/2} dx,$$

and

$$G_2^*(\lambda) = \frac{1}{2}\int_{\mathbb{R}^3} |Curl^*\lambda_1 - \nabla\lambda_2|_2^2\, dx.$$

Finally,

$$B = \{t \text{ measurable} \mid t(x) \in [0,1], \text{ a.e. in } \Omega\},$$

and

$$\hat{Y}^* = \{\lambda = (\lambda_1, \lambda_2) \in W^{1,2}(\mathbb{R}^3; \mathbb{R}^3) \times W^{1,2}(\mathbb{R}^3) \mid \lambda_2 = 0 \text{ on } \partial\Omega\}.$$

21.3 The Duality Principle for the Hard Case

In the next lines we present one of our main results, which is summarized by the following theorem.

Theorem 21.3.1. *Define $J : Y_1 \times Y_2 \times B \to \overline{\mathbb{R}}$ by*

$$J(m, f, t) = G_1(m, f, t) + G_2(f),$$

$$\begin{aligned} G_1(m, f, t) = & \int_\Omega (tg_1(m) + (1-t)g_2(m))\,dx \\ & + Ind_0(m) + Ind_1(f) + Ind_2(m, f) \\ & - \int_\Omega H(x) \cdot m\,dx, \end{aligned} \tag{21.9}$$

and

$$G_2(f) = \frac{1}{2} \int_{\mathbb{R}^3} |f(x)|_2^2\,dx.$$

Also,

$$g_1(m) = \beta(1 + m \cdot e),$$

$$g_2(m) = \beta(1 - m \cdot e),$$

$$Ind_0(m) = \begin{cases} 0, & \text{if } |m(x)|_2 = 1 \text{ a.e. in } \Omega, \\ +\infty, & \text{otherwise,} \end{cases}$$

$$Ind_1(m, f) = \begin{cases} 0, & \text{if } div(-f + m\chi_\Omega) = 0 \text{ a.e. in } \mathbb{R}^3, \\ +\infty, & \text{otherwise,} \end{cases}$$

and

$$Ind_2(f) = \begin{cases} 0, & \text{if } Curl(f) = \theta, \text{ a.e. in } \mathbb{R}^3, \\ +\infty, & \text{otherwise.} \end{cases}$$

This case refers to a uniaxial material with no exchange energy, that is, $\alpha = 0$. Observe that

$$\begin{aligned} \int_\Omega \varphi(m)\,dx &= \int_\Omega \beta(1 - |m \cdot e|)\,dx \\ &= \min_{t \in B} \left\{ \int_\Omega (tg_1(m) + (1-t)g_2(m))\,dx \right\}. \end{aligned} \tag{21.10}$$

Under such assumptions, we have

$$\inf_{(m,f,t)\in Y_1\times Y_2\times B}\{J(m,f,t)\} \geq \sup_{\lambda\in\hat{Y}^*}\{-J^*(\lambda)\} \tag{21.11}$$

where

$$J^*(\lambda) = \tilde{G}_1^*(\lambda) + G_2^*(\lambda) - \int_\Omega \beta\, dx,$$

$$\begin{aligned}\tilde{G}_1^*(\lambda) = & \sup_{t\in B}\left\{\int_\Omega \left(\sum_{i=1}^3 \left(\frac{\partial\lambda_2}{\partial x_i} + H_i + \beta(1-2t)e_i\right)^2\right)^{1/2} dx\right\} \\ & - \sup_{t\in B}\{\hat{G}_1^*(\lambda,t)\},\end{aligned} \tag{21.12}$$

where

$$\hat{G}_1^*(\lambda,t) = \int_\Omega \left(\sum_{i=1}^3 \left(\frac{\partial\lambda_2}{\partial x_i} + H_i + \beta(1-2t)e_i\right)^2\right)^{1/2} dx,$$

$$G_2^*(\lambda) = \frac{1}{2}\int_{\mathbb{R}^3} |Curl^*\lambda_1 - \nabla\lambda_2|_2^2\, dx,$$

$$B = \{t \text{ measurable} \mid t(x) \in [0,1], \text{ a.e. in } \Omega\},$$

and

$$\hat{Y}^* = \{\lambda = (\lambda_1,\lambda_2) \in W^{1,2}(\mathbb{R}^3;\mathbb{R}^3)\times W^{1,2}(\mathbb{R}^3) \mid \lambda_2 = 0 \text{ on } \partial\Omega\}.$$

Furthermore, under these last assumptions, there exists $\lambda_0 \in \hat{Y}^*$ *such that*

$$-J^*(\lambda_0) = \max_{\lambda\in\hat{Y}^*}\{-J^*(\lambda)\}.$$

Moreover, after discretization, suppose that $t_0 \in B$ *such that*

$$\tilde{G}_1^*(\lambda_0) = \hat{G}_1^*(\lambda_0,t_0),$$

is also such that $\hat{G}_1^*(\lambda,t)$ *is locally Lipschitz continuous in a neighborhood of* (λ_0,t_0).

Also assume (λ_0,t_0) *is such that the hypotheses of Corollary 11.1 are satisfied. Under such hypotheses, defining*

$$(m_0)_i = \frac{\frac{\partial(\lambda_0)_2}{\partial x_i} + H_i + \beta(1-2t_0)e_i}{\sqrt{\sum_{i=1}^3\left(\frac{\partial(\lambda_0)_2}{\partial x_i} + H_i + \beta(1-2t_0)e_i\right)^2}}, \quad \forall i \subset \{1,2,3\}$$

and

$$f_0 = Curl^*(\lambda_0)_1 - \nabla(\lambda_0)_2,$$

we have that

$$\begin{aligned} J(m_0,f_0,t_0) &= \min_{(m,f,t)\in\tilde{Y}}\{J(m,f,t)\} \\ &= \max_{\lambda\in\hat{Y}^*}\{-J^*(\lambda)\} \\ &= -J^*(\lambda_0). \end{aligned} \tag{21.13}$$

Proof. Observe that denoting $G_0 : Y_1 \times B \to \mathbb{R}$ by

$$G_0(m,t) = \int_\Omega (tg_1(m)+(1-t)g_2(m))dx - \int_\Omega H\cdot m\,dx,$$

we have that

$$\begin{aligned} J(m,f,t) &= G_1(m,f,t)+G_2(f) \\ &\geq G_0(m,t)+G_2(f) \\ &\quad +\int_\Omega \frac{\lambda_3}{2}\left(\sum_{i=1}^3 m_i^2-1\right)dx+\langle Curl(f),\lambda_1\rangle_{L^2(\mathbb{R}^3,\mathbb{R}^3)} \\ &\quad +\langle div(-f+m\chi_\Omega),\lambda_2\rangle_{L^2(\mathbb{R}^3)} \\ &\geq \inf_{(m,f)\in Y_1\times Y_2}\{G_0(m,t)+G_2(f) \\ &\quad +\int_\Omega \frac{\lambda_3}{2}\left(\sum_{i=1}^3 m_i^2-1\right)dx+\langle Curl(f),\lambda_1\rangle_{L^2(\mathbb{R}^3,\mathbb{R}^3)} \\ &\quad +\langle div(-f+m\chi_\Omega),\lambda_2\rangle_{L^2(\mathbb{R}^3)}\} \\ &= \inf_{(m,f)\in Y_1\times Y_2}\{\int_\Omega (tg_1(m)+(1-t)g_2(m))\,dx \\ &\quad -\int_\Omega H(x)\cdot m\,dx+\frac{1}{2}\int_{\mathbb{R}^3}|f(x)|_2^2\,dx \\ &\quad +\int_\Omega \frac{\lambda_3}{2}\left(\sum_{i=1}^3 m_i^2-1\right)dx+\langle Curl(f),\lambda_1\rangle_{L^2(\mathbb{R}^3,\mathbb{R}^3)} \\ &\quad +\langle div(-f+m\chi_\Omega),\lambda_2\rangle_{L^2(\mathbb{R}^3)}\}. \end{aligned} \tag{21.14}$$

This last infimum indicated is attained for functions satisfying the equations

$$H_i+\beta(1-2t)e_i-\lambda_3 m_i+\frac{\partial\lambda_2}{\partial x_i}=0,$$

if $\lambda_2=0$ on $\partial\Omega$.

That is,

$$m_i=\frac{H_i+\beta(1-2t)e_i+\frac{\partial\lambda_2}{\partial x_i}}{\lambda_3}$$

and thus from the constraint

$$\sum_{i=1}^{3} m_i^2 - 1 = 0$$

we obtain

$$\lambda_3 = \left(\sum_{i=1}^{3} \left(H_i + \beta(1-2t)e_i + \frac{\partial \lambda_2}{\partial x_i} \right)^2 \right)^{1/2}.$$

Also, the infimum in f is attained for functions satisfying

$$-f + Curl^* \lambda_1 - \nabla \lambda_2 = \theta.$$

Through such results we get

$$\begin{aligned}
&\inf_{(m,f)\in Y_1\times Y_2} \{G_0(m,t) + G_2(f) \\
&\quad + \int_\Omega \frac{\lambda_3}{2}\left(\sum_{i=1}^{3} m_i^2 - 1\right) dx + \langle Curl(f), \lambda_1\rangle_{L^2(\mathbb{R}^3,\mathbb{R}^3)} \\
&\quad + \langle div(-f + m\chi_\Omega), \lambda_2\rangle_{L^2(\mathbb{R}^3)}\} \\
&= -\int_\Omega \left(\sum_{i=1}^{3} \left(H_i + \beta(1-2t)e_i + \frac{\partial \lambda_2}{\partial x_i} \right)^2 \right)^{1/2} dx \\
&\quad - \frac{1}{2}\int_{\mathbb{R}^3} |Curl^*\lambda_1 - \nabla\lambda_2|_2^2 \, dx + \int_\Omega \beta dx \\
&= -\hat{G}^*(\lambda,t) - G_2^*(\lambda) + \int_\Omega \beta \, dx.
\end{aligned} \tag{21.15}$$

From this and (21.14) we obtain

$$\begin{aligned}
J(m,f,t) &\geq -\hat{G}^*(\lambda,t) - G_2^*(\lambda) + \int_\Omega \beta \, dx \\
&\geq \inf_{t\in B}\{-\hat{G}^*(\lambda,t)\} - G_2^*(\lambda) + \int_\Omega \beta \, dx \\
&= -\tilde{G}_1^*(\lambda) - G_2^*(\lambda) + \int_\Omega \beta \, dx \\
&= -J^*(\lambda),
\end{aligned} \tag{21.16}$$

$\forall (m,f,t) \in \tilde{Y} = Y_1 \times Y_2 \times B$, $\lambda \in \hat{Y}^*$.

Therefore,

$$\inf_{(m,f,t)\in Y_1\times Y_2\times B} \{J(m,f,t)\} \geq \sup_{\lambda\in\hat{Y}^*} \{-J^*(\lambda)\} \tag{21.17}$$

Finally, from the concavity, continuity, and coerciveness of $-J^* : \hat{Y}^* \to \mathbb{R}$, by an application of the direct method of calculus of variations (since it is a standard procedure we do not give more details here), we have that there exists $\lambda_0 \in \hat{Y}^*$ such that

$$-J^*(\lambda_0) = \max_{\lambda\in\hat{Y}^*}\{-J^*(\lambda)\}.$$

Observe that after discretization, we have

$$\begin{aligned} -\tilde{G}_1^*(\lambda_0) &= \inf_{t \in B}\{-\hat{G}_1^*(\lambda_0,t)\} \\ &= -\hat{G}_1^*(\lambda_0,t_0). \end{aligned} \tag{21.18}$$

Also after discretization, from the hypotheses and Corollary 11.1, we have

$$\delta\{\tilde{G}_1^*(\lambda_0)\} = \frac{\partial \hat{G}_1^*(\lambda_0,t_0)}{\partial \lambda}. \tag{21.19}$$

Thus, the extremal equation

$$\delta\{-J^*(\lambda_0)\} = \theta,$$

stands for

$$-\frac{\partial \hat{G}_1^*(\lambda_0,t_0)}{\partial \lambda_2} - \frac{\partial G_2^*(\lambda_0)}{\partial \lambda_2} = \theta,$$

and

$$-\frac{\partial G_2^*(\lambda_0)}{\partial \lambda_1} = \theta,$$

that is,

$$\begin{aligned} &\sum_{i=1}^{3} \frac{\partial}{\partial x_i}\left(\frac{\frac{\partial(\lambda_0)_2}{\partial x_i} + H_i + \beta(1-2t_0)e_i}{\sqrt{\sum_{i=1}^{3}\left(\frac{\partial(\lambda_0)_2}{\partial x_i} + H_i + \beta(1-2t_0)e_i\right)^2}}\chi_\Omega\right) \\ &-div(Curl^*(\lambda_0)_1 - \nabla(\lambda_0)_2) = 0, \end{aligned} \tag{21.20}$$

a.e. in $\mathbb{R}^3$, and

$$Curl(Curl^*(\lambda_0)_1 - \nabla(\lambda_0)_2) = \theta, \text{ a.e. in } \mathbb{R}^3.$$

Hence

$$div(m_0\chi_\Omega - f_0) = 0, \text{ a.e. in } \mathbb{R}^3,$$

and

$$Curl(f_0) = \theta, \text{ a.e. in } \mathbb{R}^3.$$

Now observe that from the definition of m_0 we get

$$(m_0)_i = \frac{\partial \hat{G}_1^*(\lambda_0,t_0)}{\partial v_i}, \ \forall i \in \{1,2,3\}$$

where $v_i = \frac{\partial \lambda_2}{\partial x_i}$, so that, from a well-known property of Legendre transform, we obtain

$$\hat{G}_1^*(\lambda_0,t_0) - \int_\Omega \beta \, dx$$

$$= \left\langle (m_0)_i, \frac{\partial(\lambda_0)_2}{\partial x_i} \right\rangle_{L^2(\Omega)} - G_0(m_0,t_0)$$
$$- \int_\Omega \frac{(\lambda_0)_3}{2} \left(\sum_{i=1}^3 (m_0)_i^2 - 1 \right) dx. \tag{21.21}$$

On the other hand, from the definition of f_0, we get

$$f_0 = \frac{\partial G_2^*(\lambda_0)}{\partial v_1},$$

where

$$v_1 = Curl^* \lambda_1 - \nabla \lambda_2,$$

so that

$$\begin{aligned} G_2^*(\lambda_0) &= \langle f_0, Curl^*(\lambda_0)_1 - \nabla(\lambda_0)_2 \rangle_{L^2(\Omega;\mathbb{R}^3)} - \int_{\mathbb{R}^3} |f_0(x)|^2 \, dx \\ &= \langle f_0, Curl^*(\lambda_0)_1 - \nabla(\lambda_0)_2 \rangle_{L^2(\Omega;\mathbb{R}^3)} - G_2(f_0). \end{aligned} \tag{21.22}$$

From (21.21) and (21.22) we obtain

$$\begin{aligned} J(m_0,f_0,t_0) &= G_1(m_0,f_0,t_0) + G_2(f_0) \\ &\quad - G_0(m_0,t_0) + G_2(f_0) \\ &\quad + \int_\Omega \frac{(\lambda_0)_3}{2} \left(\sum_{i=1}^3 (m_0)_i^2 - 1 \right) dx \\ &\quad + \langle Curl(f_0), (\lambda_0)_1 \rangle_{L^2(\mathbb{R}^3,\mathbb{R}^3)} \\ &\quad + \langle div(-f_0 + m_0 \chi_\Omega), (\lambda_0)_2 \rangle_{L^2(\mathbb{R}^3)} \\ &= -\hat{G}_1^*(\lambda_0,t_0) - G_2^*(\lambda_0) + \int_\Omega \beta \, dx \\ &= -\tilde{G}_1^*(\lambda_0) - G_2^*(\lambda_0) + \int_\Omega \beta \, dx \\ &= -J^*(\lambda_0). \end{aligned} \tag{21.23}$$

From (21.17) we have

$$\inf_{(m,f,t) \in \tilde{Y}} \{J(m,f,t)\} \geq \sup_{\lambda \in \hat{Y}^*} \{-J^*(\lambda)\}.$$

From this and (21.23) we may infer that

$$\begin{aligned} J(m_0,f_0,t_0) &= \min_{(m,f,t) \in \tilde{Y}} \{J(m,f,t)\} \\ &= \max_{\lambda \in \hat{Y}^*} \{-J^*(\lambda)\} \\ &= -J^*(\lambda_0). \end{aligned} \tag{21.24}$$

This completes the proof.

Remark 21.3.2. The reason we consider the problem just after discretization at some point refers to the fact that we cannot guarantee the equation $\sup_{t\in B}\{\hat{G}_1^*(\lambda_0,t)\} = \hat{G}_1^*(\lambda_0,t_0)$ is satisfied, for the infinite dimensional original problem, by a measurable $t_0 \in B$. In fact, the global optimal point for the primal formulation may not be attained for the infinite dimensional problem, but surely it is attained for its finite dimensional approximation. If the last theorem hypotheses are satisfied, the solution for primal finite dimensional problem may be obtained by the corresponding dual one with no duality gap.

21.4 The Semi-Linear Case

In this section we present another relevant result, which is summarized by the following theorem.

Theorem 21.4.1. *Define* $J : Y_1 \times Y_2 \times B \to \overline{\mathbb{R}}$ *by*

$$J(m,f,t) = G_0(m) + G_1(m,f,t) + G_2(f),$$

$$G_0(m) = \frac{\alpha}{2}\int_\Omega |\nabla m|_2^2\,dx,$$

$$\begin{aligned} G_1(m,f,t) = &\int_\Omega (tg_1(m) + (1-t)g_2(m))\,dx \\ &+ Ind_0(m) + Ind_1(f) + Ind_2(m,f) \\ &- \int_\Omega H(x)\cdot m\,dx, \end{aligned} \tag{21.25}$$

and

$$G_2(f) = \frac{1}{2}\int_{\mathbb{R}^3} |f(x)|_2^2\,dx.$$

Also,

$$g_1(m) = \beta(1 + m\cdot e),$$

$$g_2(m) = \beta(1 - m\cdot e),$$

$$Ind_0(m) = \begin{cases} 0, & \text{if } |m(x)|_2 = 1 \text{ a.e. in } \Omega, \\ +\infty, & \text{otherwise,} \end{cases}$$

$$Ind_1(m,f) = \begin{cases} 0, & \text{if } div(-f + m\chi_\Omega) = 0 \text{ a.e. in } \mathbb{R}^3, \\ +\infty, & \text{otherwise,} \end{cases}$$

and

$$Ind_2(f) = \begin{cases} 0, & \text{if } Curl(f) = \theta, \text{ a.e. in } \mathbb{R}^3, \\ +\infty, & \text{otherwise.} \end{cases}$$

We recall the present case refers to a uniaxial material with exchange energy. That is, $\alpha > 0$ *and* $\varphi(m) = \beta(1 - |m\cdot e|)$.

Under such assumptions, we have

$$\inf_{(m,f,t)\in Y_1\times Y_2\times B}\{J(m,f,t)\} \geq \sup_{(m^*,\lambda,\tilde{m})\in A^*}\{-J^*(m^*,\lambda,\tilde{m})\} \tag{21.26}$$

where

$$J^*(m^*,\lambda,\tilde{m}) = \hat{G}_0^*(m^*,\lambda)+\tilde{G}_1^*(m^*,\lambda) + G_2^*(\lambda)+\frac{1}{2}\int_\Omega \lambda_3\,dx-\int_\Omega \beta\,dx, \tag{21.27}$$

$$\hat{G}_0^*(m^*,\lambda) = \sup_{m\in Y_1}\{-G_0(m)+\frac{1}{2}\langle m_i^2,m_i^*\rangle_{L^2(\Omega)}-\langle m_i\boldsymbol{n}_i,\lambda_2\rangle_{L^2(\partial\Omega)}\},$$

$$\tilde{G}_1^*(m^*,\lambda) = \sup_{t\in B}\left\{\frac{1}{2}\int_\Omega\left(\sum_{i=1}^3\frac{\left(\frac{\partial\lambda_2}{\partial x_i}+H_i+\beta(1-2t)e_i\right)^2}{m_i^*+\lambda_3}\right)dx\right\} = \sup_{t\in B}\{\hat{G}_1^*(m^*,\lambda,t)\}, \tag{21.28}$$

where

$$\hat{G}_1^*(m^*,\lambda,t) = \frac{1}{2}\int_\Omega\left(\sum_{i=1}^3\frac{\left(\frac{\partial\lambda_2}{\partial x_i}+H_i+\beta(1-2t)e_i\right)^2}{m_i^*+\lambda_3}\right)dx,$$

$$G_2^*(\lambda) = \frac{1}{2}\int_{\mathbb{R}^3}|Curl^*\lambda_1-\nabla\lambda_2|_2^2\,dx,$$

and

$$B = \{t \text{ measurable} \mid t(x)\in[0,1],\ a.e.\ in\ \Omega\}.$$

Also,

$$A^* = A_1\cap A_2\cap A_3\cap A_4,$$

where

$$Y_3 = L^2(\Omega;\mathbb{R}^3)\times W^{1,2}(\mathbb{R}^3;\mathbb{R}^3)\times W^{1,2}(\mathbb{R}^3)\times L^2(\Omega),$$

$$A_1 = \{(m^*,\lambda)\in Y_3 \mid m_i^*+\lambda_3>0 \text{ in } \Omega, \forall i\in\{1,2,3\}\}.$$

$$A_2 = \{(m^*,\lambda,m)\in Y_4 = Y_3\times W^{1,2}(\Omega;\mathbb{R}^3) \mid \lambda_2\boldsymbol{n}_i+\frac{\partial\tilde{m}_l}{\partial\boldsymbol{n}}=0 \text{ on } \partial\Omega, \forall i\in\{1,2,3\}\}, \tag{21.29}$$

$$A_3 = \{(m^*,\lambda)\in Y_3 \mid \tilde{J}(m)>0, \forall m\in Y_1 \text{ such that } m\neq\theta\}.$$

Here,

$$\lambda = (\lambda_1, \lambda_2, \lambda_3)$$

and

$$\begin{aligned}\tilde{J}(m) &= G_0(m) - \frac{1}{2}\langle m_i^2, m_i^* \rangle_{L^2(\Omega)} \\ &= \frac{\alpha}{2}\int_\Omega |\nabla m|_2^2\,dx - \frac{1}{2}\langle m_i^2, m_i^* \rangle_{L^2(\Omega)}. \end{aligned} \tag{21.30}$$

And also,

$$A_4 = \{(m^*, \lambda, \tilde{m}) \in Y_4 = Y_3 \times W^{1,2}(\Omega; \mathbb{R}^3) \mid \alpha \nabla^2 \tilde{m}_i + m_i^* \tilde{m}_i = 0, \text{ in } \Omega\}.$$

Suppose there exists $(m_0^*, \lambda_0, \tilde{m}_0, m_0) \in A^* \times Y_1$ *such that*

$$\delta\left\{-J^*(m_0^*, \lambda_0, \tilde{m}_0) - \frac{1}{2}\langle (m_0)_i, (\tilde{m}_0)_i (m_0)_i^* + \alpha \nabla^2 (\tilde{m}_0)_i \rangle_{L^2(\Omega)}\right\} = \theta.$$

Moreover, considering the problem in question after discretization, assume $t_0 \in B$ *such that*

$$\tilde{G}_1^*(m_0^*, \lambda_0) = \tilde{G}_1^*(m_0^*, \lambda_0, t_0),$$

is also such that $\hat{G}_1^*(m^*, \lambda, t)$ *is locally Lipschitz continuous in a neighborhood of* (m_0^*, λ_0, t_0).

Also assume (m_0^*, λ_0, t_0) *is such that the hypotheses of Corollary 11.1 are satisfied. Under such hypotheses, we have*

$$(m_0)_i = \frac{\frac{\partial (\lambda_0)_2}{\partial x_i} + H_i + \beta(1 - 2t_0)e_i}{(m_0)_i^* + (\lambda_0)_3}, \quad \forall i \in \{1, 2, 3\}$$

and defining

$$f_0 = Curl^*(\lambda_0)_1 - \nabla(\lambda_0)_2,$$

we have also that

$$\begin{aligned} J(m_0, f_0, t_0) &= \min_{(m,f,t) \in \tilde{Y}} \{J(m, f, t)\} \\ &= \max_{(m^*, \lambda, \tilde{m}) \in A^*} \{-J^*(m^*, \lambda, \tilde{m})\} \\ &= -J^*(m_0^*, \lambda_0, m_0). \end{aligned} \tag{21.31}$$

Proof. Observe that defining $\tilde{G}_0 : Y_1 \times Y_2 \times B \to \mathbb{R}$ by

$$\tilde{G}_0(m, f, t) = \int_\Omega (t g_1(m) + (1 - t) g_2(m))dx - \int_\Omega H \cdot m\,dx,$$

we have that

$$\begin{aligned}
J(m,f,t) &= G_0(m)+G_1(m,f,t)+G_2(f) \\
&\geq G_0(m)+\tilde{G}_0(m,f,t)+G_2(f) \\
&\quad +\int_\Omega \frac{\lambda_3}{2}\left(\sum_{i=1}^3 m_i^2-1\right)dx+\langle Curl(f),\lambda_1\rangle_{L^2(\mathbb{R}^3,\mathbb{R}^3)} \\
&\quad +\langle div(-f+m\chi_\Omega),\lambda_2\rangle_{L^2(\mathbb{R}^3)} \\
&= \left\{G_0(m)-\frac{1}{2}\langle m_i^2,m_i^*\rangle_{L^2(\Omega)}\right\}+\tilde{G}_0(m,f,t)+G_2(f) \\
&\quad +\int_\Omega \frac{\lambda_3}{2}\left(\sum_{i=1}^3 m_i^2-1\right)dx+\langle Curl(f),\lambda_1\rangle_{L^2(\mathbb{R}^3,\mathbb{R}^3)} \\
&\quad +\langle div(-f+m\chi_\Omega),\lambda_2\rangle_{L^2(\mathbb{R}^3)}+\frac{1}{2}\langle m_i^2,m_i^*\rangle_{L^2(\Omega)}.
\end{aligned}$$

Thus,

$$\begin{aligned}
J(m,f,t) &\geq \inf_{m\in Y_1}\left\{G_0(m)-\frac{1}{2}\langle m_i^2,m_i^*\rangle_{L^2(\Omega)}+\langle m_i\mathbf{n}_i,\lambda_2\rangle_{L^2(\partial\Omega)}\right\} \\
&\quad +\inf_{(m,f)\in Y_1\times Y_2}\{\tilde{G}_0(m,f,t)+G_2(f) \\
&\quad +\int_\Omega \frac{\lambda_3}{2}\left(\sum_{i=1}^3 m_i^2-1\right)\,dx+\langle Curl(f),\lambda_1\rangle_{L^2(\mathbb{R}^3,\mathbb{R}^3)} \\
&\quad -\langle(-f+m\chi_\Omega),\nabla\lambda_2\rangle_{L^2(\mathbb{R}^3;\mathbb{R}^3)}+\frac{1}{2}\langle m_i^2,m_i^*\rangle_{L^2(\Omega)}\} \\
&= -\hat{G}_0^*(m^*,\lambda)+\inf_{(m,f)\in Y_1\times Y_2}\{\int_\Omega(tg_1(m)+(1-t)g_2(m))\,dx \\
&\quad -\int_\Omega H(x)\cdot m\,dx+\frac{1}{2}\int_{\mathbb{R}^3}|f(x)|_2^2\,dx \\
&\quad +\int_\Omega \frac{\lambda_3}{2}\left(\sum_{i=1}^3 m_i^2-1\right)\,dx+\langle Curl(f),\lambda_1\rangle_{L^2(\mathbb{R}^3,\mathbb{R}^3)} \\
&\quad -\langle(-f+m\chi_\Omega),\nabla\lambda_2\rangle_{L^2(\mathbb{R}^3;\mathbb{R}^3)}+\frac{1}{2}\langle m_i^2,m_i^*\rangle_{L^2(\Omega)}\}. \qquad (21.32)
\end{aligned}$$

This last infimum in m indicated is attained for functions satisfying the equations

$$H_i+\beta(1-2t)e_i-(m_i^*+\lambda_3)m_i+\frac{\partial\lambda_2}{\partial x_i}=0.$$

That is,

$$m_i=\frac{H_i+\beta(1-2t)e_i+\frac{\partial\lambda_2}{\partial x_i}}{m_i^*+\lambda_3}.$$

Also, the infimum in f is attained for functions satisfying

$$f = Curl^*\lambda_1 - \nabla\lambda_2.$$

Through such results we get

$$\begin{aligned}
&\tilde{G}_0(m,f,t) + G_2(f) \\
&\quad + \int_\Omega \frac{\lambda_3}{2}\left(\sum_{i=1}^3 m_i^2 - 1\right) dx + \langle Curl(f), \lambda_1\rangle_{L^2(\mathbb{R}^3,\mathbb{R}^3)} \\
&\quad - \langle(-f + m\chi_\Omega), \nabla\lambda_2\rangle_{L^2(\mathbb{R}^3;\mathbb{R}^3)} + \frac{1}{2}\langle m_i^2, m_i^*\rangle_{L^2(\Omega)} \\
&\geq -\frac{1}{2}\int_\Omega \left(\sum_{i=1}^3 \frac{\left(\frac{\partial\lambda_2}{\partial x_i} + H_i + \beta(1-2t)e_i\right)^2}{m_i^* + \lambda_3}\right) dx \\
&\quad - \frac{1}{2}\int_{\mathbb{R}^3} |Curl^*\lambda_1 - \nabla\lambda_2|_2^2\, dx - \frac{1}{2}\int_\Omega \lambda_3\, dx + \int_\Omega \beta dx \\
&= -\hat{G}^*(m^*,\lambda,t) - G_2^*(\lambda) - \frac{1}{2}\int_\Omega \lambda_3\, dx + \int_\Omega \beta dx,
\end{aligned} \tag{21.33}$$

if

$$\lambda_2 \mathbf{n}_i + \frac{\partial \tilde{m}_i}{\partial \mathbf{n}} = 0 \text{ on } \partial\Omega,\ \forall i \in \{1,2,3\}.$$

From this and (21.32) we obtain

$$\begin{aligned}
J(m,f,t) &\geq -\hat{G}_0^*(m^*,\lambda) - \hat{G}^*(m^*,\lambda,t) - G_2^*(\lambda) \\
&\quad -\frac{1}{2}\int_\Omega \lambda_3\, dx + \int_\Omega \beta\, dx \\
&\geq -\hat{G}_0^*(m^*,\lambda) + \inf_{t\in B}\{-\hat{G}^*(m^*,\lambda,t)\} - G_2^*(\lambda) \\
&\quad -\frac{1}{2}\int_\Omega \lambda_3\, dx + \int_\Omega \beta\, dx \\
&= -\hat{G}_0^*(m^*,\lambda) - \tilde{G}_1^*(m^*,\lambda) - G_2^*(\lambda) \\
&\quad -\frac{1}{2}\int_\Omega \lambda_3\, dx + \int_\Omega \beta\, dx \\
&= -J^*(m^*,\lambda,\tilde{m}),
\end{aligned} \tag{21.34}$$

$\forall (m,f,t) \in \tilde{Y},\ (m^*,\lambda,\tilde{m}) \in A^*$.

Therefore,

$$\inf_{(m,f,t)\in Y_1\times Y_2\times B}\{J(m,f,t)\} \geq \sup_{(m^*,\lambda,\tilde{m})\in A^*}\{-J^*(m^*,\lambda,\tilde{m})\} \tag{21.35}$$

Finally, from now on considering the problem after discretization, we have

$$\begin{aligned} -\tilde{G}_1^*(m_0^*,\lambda_0) &= \inf_{t\in B}\{-\hat{G}_1^*(m_0^*,\lambda_0,t)\} \\ &= -\hat{G}_1^*(m_0^*,\lambda_0,t_0). \end{aligned} \tag{21.36}$$

From the hypotheses and Corollary 11.1 we have that

$$\delta\{\tilde{G}_1^*(m_0^*,\lambda_0)\} = \left\{\frac{\partial \hat{G}_1^*(m_0^*,\lambda_0,t_0)}{\partial m^*}, \frac{\partial \hat{G}_1^*(m_0^*,\lambda_0,t_0)}{\partial \lambda}\right\}. \tag{21.37}$$

Thus, considering that the following representation holds,

$$\hat{G}_0^*(m_0^*,\lambda_0) = -\langle (\tilde{m}_0)_i \mathbf{n}_i, (\lambda_0)_2\rangle_{L^2(\partial\Omega)}/2,$$

from these last results and hypotheses, the extremal equation

$$\delta\left\{-J^*(m_0^*,\lambda_0,\tilde{m}_0) - \frac{1}{2}\langle (m_0)_i, (\tilde{m}_0)_i (m_0)_i^* + \alpha\nabla^2(\tilde{m}_0)_i\rangle_{L^2(\Omega)}\right\} = \theta,$$

stands for

$$\frac{\partial\left(-\hat{G}_1^*(m_0^*,\lambda_0,t_0) - \frac{1}{2}\langle (m_0)_i, (\tilde{m}_0)_i (m_0)_i^* + \alpha\nabla^2(\tilde{m}_0)_i\rangle_{L^2(\Omega)}\right)}{\partial m_i^*} = \theta, \tag{21.38}$$

$$-\frac{\partial\left(\frac{1}{2}\langle (m_0)_i, (\tilde{m}_0)_i (m_0)_i^* + \alpha\nabla^2(\tilde{m}_0)_i\rangle_{L^2(\Omega)}\right)}{\partial \tilde{m}_i} = \theta, \tag{21.39}$$

$$\frac{\partial\left(\frac{1}{2}\langle (m_0)_i, (\tilde{m}_0)_i (m_0)_i^* + \alpha\nabla^2(\tilde{m}_0)_i\rangle_{L^2(\Omega)}\right)}{\partial m_i} = \theta, \tag{21.40}$$

$$-\frac{\partial \hat{G}_1^*(m_0^*,\lambda_0,t_0)}{\partial \lambda_3} - \frac{1}{2} = \theta, \tag{21.41}$$

that is,

$$\frac{1}{2}\sum_{i=1}^{3}\left(\frac{\frac{\partial(\lambda_0)_2}{\partial x_i} + H_i + \beta(1-2t_0)e_i}{(m_0^*)_i + (\lambda_0)_3}\right)^2 - \frac{1}{2} = 0, \text{ a.e. in } \Omega, \tag{21.42}$$

Also,

$$-\frac{\partial \hat{G}_0^*(m_0^*,\lambda_0)}{\partial \lambda_2} - \frac{\partial \hat{G}_1^*(m_0^*,\lambda_0,t_0)}{\partial \lambda_2} - \frac{\partial G_2^*(\lambda_0)}{\partial \lambda_2} = \theta, \tag{21.43}$$

and

$$-\frac{\partial G_2^*(\lambda_0)}{\partial \lambda_1} = \theta. \tag{21.44}$$

That is, from (21.38),

$$-\frac{1}{2}\left(\frac{\frac{\partial(\lambda_0)_2}{\partial x_i}+H_i+\beta(1-2t_0)e_i}{(m_0^*)_i+(\lambda_0)_3}\right)^2+\frac{(m_0)_i(\tilde{m}_0)_i}{2}=0, \text{ a.e. in } \Omega, \tag{21.45}$$

and by (21.39) and (21.40)

$$(m_0)_i(m_0)_i^*+\alpha\nabla^2(m_0)_i=0, \text{ a.e. in } \Omega,$$

$$(\tilde{m}_0)_i(m_0)_i^*+\alpha\nabla^2(\tilde{m}_0)_i=0, \text{ a.e. in } \Omega,$$

so that

$$m_0=\tilde{m}_0, \text{ a.e. in } \Omega, \tag{21.46}$$

and hence,

$$\hat{G}_0^*(m_0^*,\lambda_0)=-G_0(m_0)+\frac{1}{2}\langle(m_0)_i^2,(m_0)_i^*\rangle_{L^2(\Omega)}-\langle(m_0)_i\mathbf{n}_i,(\lambda_0)_2\rangle_{L^2(\partial\Omega)}. \tag{21.47}$$

Moreover, by (21.43),

$$\sum_{i=1}^{3}\frac{\partial}{\partial x_i}\left(\frac{\frac{\partial(\lambda_0)_2}{\partial x_i}+H_i+\beta(1-2t_0)e_i}{(m_0^*)_i+(\lambda_0)_3}\chi_\Omega\right)$$
$$-div(Curl^*(\lambda_0)_1-\nabla(\lambda_0)_2)=0, \tag{21.48}$$

a.e. in $\mathbb{R}^3$, and

$$\left(\frac{\frac{\partial(\lambda_0)_2}{\partial x_i}+H_i+\beta(1-2t_0)e_i}{(m_0^*)_i+(\lambda_0)_3}\right)\mathbf{n}_i-(\tilde{m}_0)_i\mathbf{n}_i=0, \text{ on } \partial\Omega.$$

Thus, from this, (21.45), (21.46), and (21.42), we obtain

$$(m_0)_i=\frac{\frac{\partial(\lambda_0)_2}{\partial x_i}+H_i+\beta(1-2t_0)e_i}{(m_0^*)_i+(\lambda_0)_3},$$

and

$$\sum_{i=1}^{3}(m_0)_i^2=1, \text{ a.e. in } \Omega.$$

From (21.44),

$$Curl(Curl^*(\lambda_0)_1-\nabla(\lambda_0)_2)=\theta, \text{ a.e. in } \mathbb{R}^3.$$

Hence

$$div(m_0\chi_\Omega-f_0)=0, \text{ a.e. in } \mathbb{R}^3,$$

and

$$Curl(f_0) = \theta, \text{ a.e. in } \mathbb{R}^3.$$

Now observe that from the expression of m_0 we get

$$(m_0)_i = \frac{\partial \hat{G}_1^*(m_0^*, \lambda_0, t_0)}{\partial v_i}, \ \forall i \in \{1,2,3\}$$

where $v_i = \frac{\partial \lambda_2}{\partial x_i}$, so that from a well-known property of Legendre transform, we obtain

$$\begin{aligned}
&\hat{G}_1^*(m_0^*, \lambda_0, t_0) + \frac{1}{2}\int_\Omega (\lambda_0)_3 \, dx - \int_\Omega \beta \, dx \\
&= \left\langle (m_0)_i, \frac{\partial (\lambda_0)_2}{\partial x_i} \right\rangle_{L^2(\Omega)} - \tilde{G}_0(m_0, f_0, t_0) \\
&- \int_\Omega \frac{(\lambda_0)_3}{2} \left(\sum_{i=1}^3 (m_0)_i^2 - 1 \right) dx \\
&- \frac{1}{2} \langle (m_0)_i^2, (m_0)_i^* \rangle_{L^2(\Omega)}.
\end{aligned} \tag{21.49}$$

On the other hand, from the definition of f_0, we get

$$f_0 = \frac{\partial G_2^*(\lambda_0)}{\partial v_1},$$

where

$$v_1 = Curl^* \lambda_1 - \nabla \lambda_2,$$

so that

$$\begin{aligned}
G_2^*(\lambda_0) &= \langle f_0, Curl^*(\lambda_0)_1 - \nabla(\lambda_0)_2 \rangle_{L^2(\Omega;\mathbb{R}^3)} - \int_{\mathbb{R}^3} |f_0(x)|^2 \, dx \\
&= \langle f_0, Curl^*(\lambda_0)_1 - \nabla(\lambda_0)_2 \rangle_{L^2(\Omega;\mathbb{R}^3)} - G_2(f_0).
\end{aligned} \tag{21.50}$$

From (21.47), (21.49), and (21.50) we obtain

$$\begin{aligned}
J(m_0, f_0, t_0) &= G_0(m_0) + G_1(m_0, f_0, t_0) + G_2(f_0) \\
&= G(m_0) - \frac{1}{2} \langle (m_0)_i^2, (m_0^*)_i \rangle_{L^2(\Omega)} + \tilde{G}_0(m_0, f_0, t_0) + G_2(f_0) \\
&\quad + \int_\Omega \frac{(\lambda_0)_3}{2} \left(\sum_{i=1}^3 (m_0)_i^2 - 1 \right) dx + \langle Curl(f_0), (\lambda_0)_1 \rangle_{L^2(\mathbb{R}^3, \mathbb{R}^3)} \\
&\quad + \langle div(-f_0 + m_0 \chi_\Omega), (\lambda_0)_2 \rangle_{L^2(\mathbb{R}^3)} + \frac{1}{2} \langle (m_0)_i^2, (m_0)_i^* \rangle_{L^2(\Omega)} \\
&= -\hat{G}_0^*(m_0^*, \lambda_0) - \hat{G}_1^*(m_0^*, \lambda_0, t_0) - G_2^*(\lambda_0) \\
&\quad - \frac{1}{2} \int_\Omega (\lambda_0)_3 \, dx + \int_\Omega \beta \, dx
\end{aligned}$$

$$\begin{aligned}&= -\hat{G}_0^*(m_0^*,\lambda_0) - \tilde{G}_1^*(m_0^*,\lambda_0) - G_2^*(\lambda_0) \\ &\quad -\frac{1}{2}\int_\Omega (\lambda_0)_3\,dx + \int_\Omega \beta\,dx \\ &= -J^*(m_0^*,\lambda_0,m_0). \end{aligned} \tag{21.51}$$

From (21.35) we have

$$\inf_{(m,f,t)\in\tilde{Y}}\{J(m,f,t)\} \geq \sup_{(m^*,\lambda,\tilde{m})\in A^*}\{-J^*(m^*,\lambda,\tilde{m})\}.$$

From this and (21.51) we may infer that

$$\begin{aligned} J(m_0,f_0,t_0) &= \min_{(m,f,t)\in\tilde{Y}}\{J(m,f,t)\} \\ &= \max_{(m^*,\lambda,\tilde{m})\in A^*}\{-J^*(m^*,\lambda,\tilde{m})\} \\ &= -J^*(m_0^*,\lambda_0,m_0). \end{aligned} \tag{21.52}$$

The proof is complete.

21.5 Numerical Examples

In this section we present a numerical two-dimensional example concerning the hard case. Consider $\Omega = [0,1]\times[0,1] \subset \mathbb{R}^2$, the region corresponding to a micro-magnetic sample. We develop numerical results for the minimization of the simplified dual functional

$$\begin{aligned} J^*(\lambda_2) &= \sup_{t\in B}\left\{\int_\Omega\left(\sum_{i=1}^2\left(\frac{\partial\lambda_2}{\partial x_i} + H_i + \beta(1-2t)e_i\right)^2\right)^{1/2} dx\right\} \\ &\quad + \frac{1}{2}\int_\Omega |\nabla\lambda_2|_2^2\,dx, \end{aligned} \tag{21.53}$$

with the boundary condition

$$\lambda_2 = 0 \text{ on } \partial\Omega.$$

In such a case we have neglected the external induced magnetic field. Anyway, observe that

$$Curl(f_0) = \theta$$

(in fact it is an appropriate version for the two-dimensional case) stands for the obviously satisfied equation

$$Curl(\nabla\lambda_2) = \theta.$$

Finally, units are related to the international system.

21.5.1 First Example

For such an example, for a fixed $\beta > 0$, we consider the cases $H_0/\beta = 1.0, H_0/\beta = 10$, and $H_0/\beta = 100$, where the magnetic field $\mathbf{H}$ is given by

$$\mathbf{H} = H_0\mathbf{i} + 0\mathbf{j},$$

where

$$\mathbf{i} = (1,0) \text{ and } \mathbf{j} = (0,1).$$

Moreover, $e_1 = \sqrt{2}/2$ and $e_2 = \sqrt{2}/2$, where (e_1,e_2) is the preferred direction of magnetization. We have plotted the stream lines for the vector field $m = (m_1,m_2)$ for these three cases. Please see Figs. 21.1, 21.2, and 21.3. We observe that as H_0 increases, the magnetization m direction approaches the magnetic field $\mathbf{H}$ one, which in such an example is given by $\mathbf{i} = (1,0)$.

Remark 21.5.1. It is worth mentioning that as H_0/β is smaller the magnetization m is closer to (e_1,e_2). However its direction approaches the $\mathbf{H}$ one, as H_0 increases. Such a result is consistent with the concerned problem physics.

21.5.2 Second Example

For such an example, for a fixed $\beta > 0$, we consider the cases $H_0/\beta = 0.5, H_0/\beta = 5.0$, and $H_0/\beta = 50$, where the magnetic field $\mathbf{H}$ is given by

$$\mathbf{H} = H_0\mathbf{H}_a(x,y),$$

where

$$\mathbf{H}_a = x(0.5-y)\mathbf{i} - y(0.5-x)\mathbf{j}.$$

Also, again $e_1 = \sqrt{2}/2$ and $e_2 = \sqrt{2}/2$, where (e_1,e_2) is the preferred direction of magnetization. For the stream lines of $\mathbf{H}_a$, please see Fig. 21.4. For the magnetization m for these three different cases see Figs. 21.5, 21.6, and 21.7. For the parameter t related to the case $H_0/\beta = 0.5$, see Fig. 21.8.

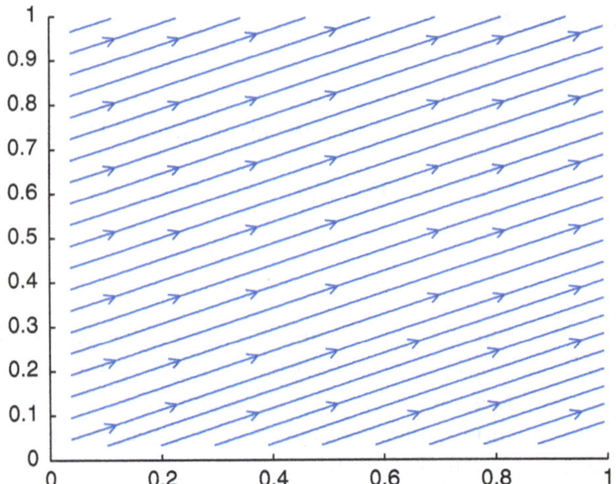

Fig. 21.1 First example—stream lines for the magnetization m for $H_0/\beta = 1.0$

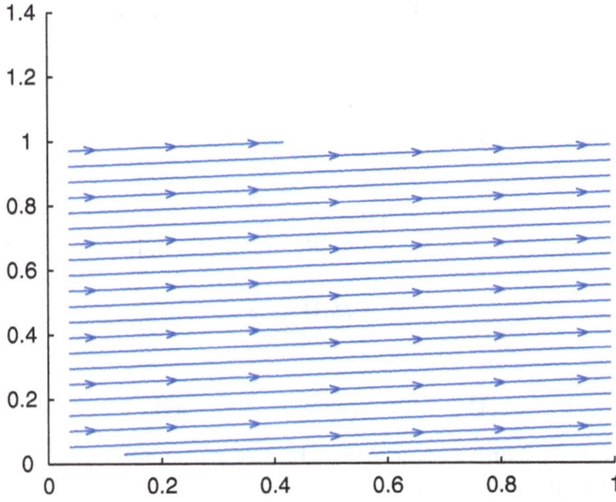

Fig. 21.2 First example—stream lines for the magnetization m for $H_0/\beta = 10.0$

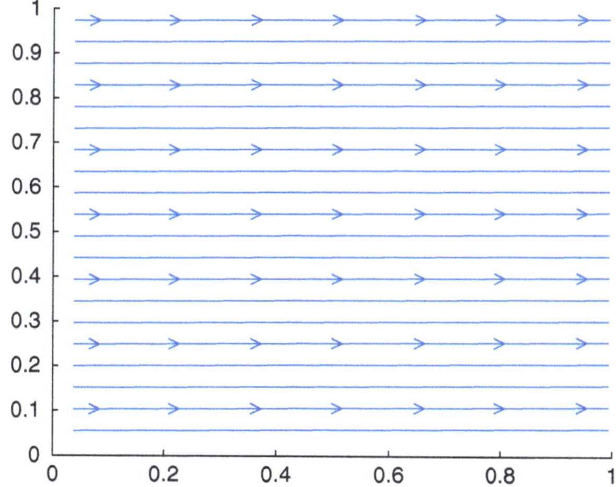

Fig. 21.3 First example—stream lines for the magnetization m for $H_0/\beta = 100.0$

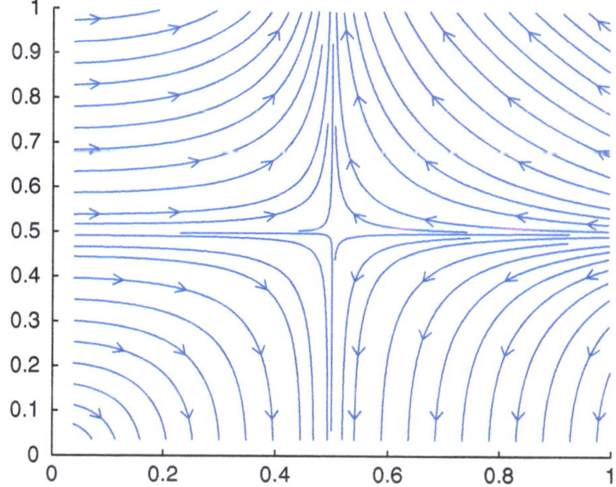

Fig. 21.4 Second example—stream lines for external magnetic field $\mathbf{H}_a$

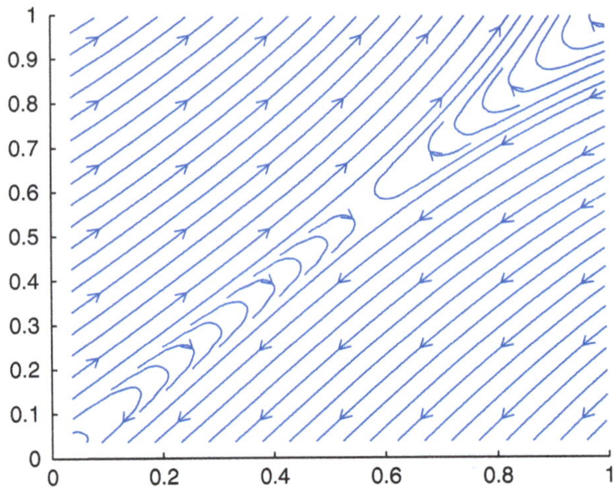

Fig. 21.5 Second example—stream lines for the magnetization m for $H_0/\beta = 0.5$

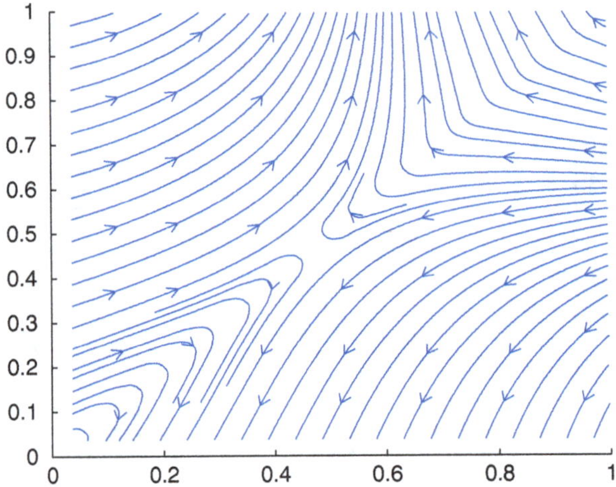

Fig. 21.6 Second example—stream lines for the magnetization m for $H_0/\beta = 5.0$

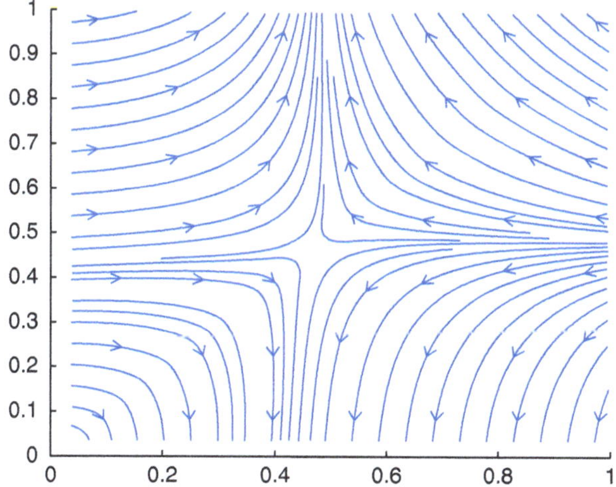

Fig. 21.7 Second example—stream lines for the magnetization m for $H_0/\beta = 50.0$

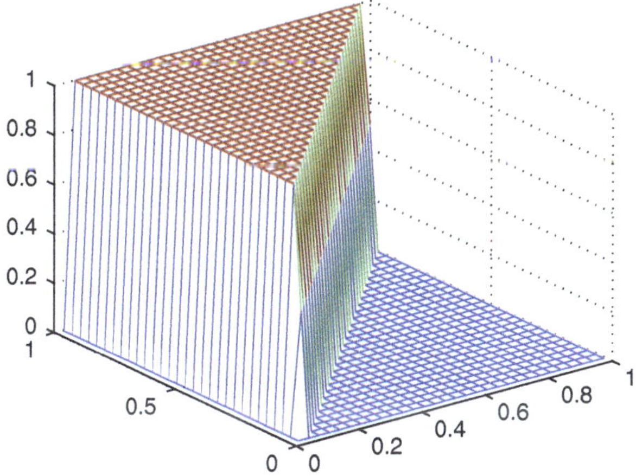

Fig. 21.8 Second example—parameter $t(x)$ for $H_0/\beta = 0.5$

21.6 Acknowledgments

The author is very grateful to Professor Robert C. Rogers for his excellent work as Ph.D. thesis advisor at Virginia Tech-USA. We thank as well the Mathematics Department of that institution for the constant financial support received during our doctoral program.

21.7 Conclusion

In this chapter we develop duality principles for models in ferromagnetism met in reference [44], for example. The dual variational formulations here presented are convex (in fact concave) and the results are obtained through standard tools of convex analysis. It is important to emphasize that in some situations (specially the hard cases), the minima may not be attained through the primal approaches, so that the minimizers of the dual formulations reflect the average behavior of minimizing sequences for the primal problems, as weak cluster points of such sequences.

Chapter 22
The Generalized Method of Lines Applied to Fluid Mechanics

22.1 Introduction

In this chapter we develop solutions for the Navier–Stokes system through the generalized method of lines. The main reference for this chapter is R. Temam [66]. At this point we describe the system in question.

Consider $\Omega \subset \mathbb{R}^2$ an open, bounded, and connected set, whose internal boundary is denoted by Γ_0 and external one is denoted by Γ_1. Denoting by $u : \Omega \to \mathbb{R}$ the field of velocity in direction x of the Cartesian system (x, y), by $v : \Omega \to \mathbb{R}$ the velocity field in the direction y, by $p : \Omega \to \mathbb{R}$ the pressure field, so that $P = p/\rho$, where ρ is the constant fluid density, ν is the viscosity coefficient, and g is the gravity constant, the Navier–Stokes PDE system is expressed by

$$\begin{cases} \nu\nabla^2 u - u\partial_x u - v\partial_y u - \partial_x P + g_x = 0, & \text{in } \Omega, \\ \nu\nabla^2 v - u\partial_x v - v\partial_y v - \partial_y P + g_y = 0, & \text{in } \Omega, \\ \partial_x u + \partial_y v = 0, & \text{in } \Omega, \end{cases} \tag{22.1}$$

$$\begin{cases} u = v = 0, & \text{on } \Gamma_0, \\ u = u_\infty,\ v = 0,\ P = P_\infty, & \text{on } \Gamma_1 \end{cases} \tag{22.2}$$

In principle we look for solutions $(u, v, P) \in W^{2,2}(\Omega) \times W^{2,2}(\Omega) \times W^{1,2}(\Omega)$ despite the fact that less regular solutions are also possible specially concerning the weak formulation.

22.2 On the Solution of Steady-State Euler Equation

Through the next result we obtain a linear system whose solution also solves the steady-state Euler system.

F. Botelho, *Functional Analysis and Applied Optimization in Banach Spaces: Applications to Non-Convex Variational Models*, DOI 10.1007/978-3-319-06074-3_22,

Theorem 22.2.1. *A solution for the Euler system below indicated, that is,*

$$\begin{cases} -u\partial_x u - v\partial_y u - \partial_x P + g_x = 0, & in\ \Omega, \\ -u\partial_x v - v\partial_y v - \partial_y P + g_y = 0, & in\ \Omega, \\ \partial_x u + \partial_y v = 0, & in\ \Omega, \end{cases} \tag{22.3}$$

with the boundary conditions

$$\{\mathbf{u}\cdot\boldsymbol{n} = 0, \ \ on\ \Gamma, \tag{22.4}$$

where $\mathbf{u} = (u,v)$, *is given by*

$$\begin{cases} u = \partial_x w_0, \\ v = \partial_y w_0, \end{cases} \tag{22.5}$$

where w_0 *is a solution of the equation*

$$\begin{cases} \nabla^2 w_0 = 0 & in\ \Omega, \\ \nabla w_0 \cdot \boldsymbol{n} = 0, & on\ \Gamma. \end{cases} \tag{22.6}$$

Proof. For $u = \partial_x w_0$ and $v = \partial_y w_0$ define

$$F = -(\partial_x w_0)^2/2 - (\partial_y w_0)^2/2 = -u^2/2 - v^2/2.$$

The continuity equation in the system (22.3) stands for

$$\nabla^2 w_0 = 0 \text{ in } \Omega. \tag{22.7}$$

The first two equations in (22.3) correspond to

$$\partial_x F - \partial_x P + g_x = 0, \text{ in } \Omega, \tag{22.8}$$

and

$$\partial_y F - \partial_y P + g_y = 0, \text{ in } \Omega, \tag{22.9}$$

which may be solved in P for the particular w_0 solution of (22.7) with corresponding boundary conditions. Since

$$\partial_y(\partial_x F + g_x) = \partial_x(\partial_y F + g_y) = \partial_{xy} P,$$

it is clear that (22.8) and (22.9) have a solution in P with the proper boundary conditions above described.

22.3 The Generalized Method of Lines for the Navier–Stokes System

In this section we develop the solution for the Navier–Stokes system through the generalized method of lines. About such a method see Chap. 15 for more details.

Consider

$$\Omega = \{(r,\theta) \mid 1 \le r \le 2,\ 0 \le \theta \le 2\pi\},$$
$$\partial\Omega_0 = \{(1,\theta) \mid 0 \le \theta \le 2\pi\},$$

and

$$\partial\Omega_1 = \{(2,\theta) \mid 0 \le \theta \le 2\pi\}.$$

First for the boundary conditions

$$u = 0,\ v = 0,\ P = P_0(x) \text{ on } \partial\Omega_0,$$
$$u = u_f(x),\ v = v_f(x),\ P = P_f(x) \text{ on } \partial\Omega_1,$$

we have obtained the following general expressions for the $n-th$ lines (for an appropriate approximate system of equations in polar coordinates):

$$\begin{aligned} u_n(x) = {} & a_1[n]\cos(x)P_0(x) + a_2[n]\cos[x]P_f(x) + a_3[n]u_f(x) \\ & + a_4[n]\cos(x)u_f(x)^2 + a_5[n]\sin(x)u_f(x)v_f(x) \\ & + a_6[n]\sin(x)P_0'(x) + a_7[n]\sin(x)P_f'(x) + a_8[n]\sin(x)u_f(x)u_f'(x) \\ & + a_9[n]\cos(x)v_f(x)u_f'(x) + a_{10}[n]u_f''(x) \end{aligned}$$

$$\begin{aligned} v_n(x) = {} & b_1[n]P_0(x)\sin[x] + b_2[n]P_f(x)\sin(x) + b_3[n]v_f(x) \\ & + b_4[n]\cos[x]u_f(x)v_f(x) + b_5[n]\sin(x)v_f(x)^2 \\ & + b_6[n]\cos(x)P_0'(x) + b_7[n]\cos(x)P_f'(x) + b_8[n]\sin(x)u_f(x)v_f'(x) \\ & + b_9[n]\cos(x)v_f(x)v_f'(x) + b_{10}[n]v_f''(x) \end{aligned}$$

$$\begin{aligned} P_n(x) = {} & c_1[n]P_0(x) + c_2[n]P_f(x) + c_3[n]\cos(x)^2u_f(x)^2 \\ & + c_4[n]\cos(x)\sin(x)u_f(x)v_f(x) + c_5[n]\sin(x)^2v_f(x)^2 \\ & + c_6[n]\cos(x)\sin(x)u_f(x)u_f'(x) + c_7[n]\cos(x)^2v_f(x)u_f'(x) \\ & + c_8[n]\sin(x)^2u_f(x)^2 + c_9[n]\sin(x)^2u_f(x)v_f'(x) \\ & + c_{10}[n]\cos(x)\sin(x)v_f(x)v_f'(x) + c_{11}[n]\cos(x)\sin(x)u_f'(x)v_f'(x) \\ & + c_{12}[n]\cos(x)^2v_f'(x)^2 + c_{13}[n]P_0''(x) + c_{14}[n]P_f''(x) \end{aligned}$$

Denoting

$$J(u,v,P) = \int_\Omega \left(\nu\nabla^2 u - u\partial_x u - v\partial_y u - \partial_x P\right)^2 \, d\Omega$$
$$+ \int_\Omega \left(\nu\nabla^2 v - u\partial_x v - v\partial_y v - \partial_y P\right)^2 \, d\Omega$$
$$+ \int_\Omega (\partial_x u + \partial_y v)^2 \, d\Omega, \tag{22.10}$$

the coefficients $\{a_i[n]\}, \{b_i[n]\}, \{c_i[n]\}$ may be obtained through the numerical minimization of $J(u,v,P)$.

22.3.1 The General Case for Specific Boundary Conditions

For the boundary conditions

$$u = 0, \;\; v = 0, \; P = P_0(x) \text{ on } \partial\Omega_0,$$

$$u = u_f(x), \; v = v_f(x), \; P = P_f(x) \text{ on } \partial\Omega_1,$$

where

$$\Omega = \{(r,\theta) : \mid r(\theta) \leq r \leq 2r(\theta)\}$$

where $r(\theta)$ is a smooth periodic function. We recall that the system in question, in function of the variables (t,θ) where $t = r/r(\theta)$, is given by

$$L(u) - ud_1(u) - vd_2(u) - d_1(P) = 0, \tag{22.11}$$

$$L(v) - ud_1(v) - vd_2(v) - d_2(P) = 0, \tag{22.12}$$

$$d_1(u) + d_2(v) = 0. \tag{22.13}$$

where

$$L(u)/f_0(\theta) = \frac{\partial^2 u}{\partial t^2} + \frac{1}{t} f_2(\theta) \frac{\partial u}{\partial t}$$
$$+ \frac{1}{t} f_3(\theta) \frac{\partial^2 u}{\partial \theta \partial t} + \frac{f_4(\theta)}{t^2} \frac{\partial^2 u}{\partial \theta^2} = 0, \tag{22.14}$$

in Ω. Here $f_0(\theta), f_2(\theta)$, $f_3(\theta)$, and $f_4(\theta)$ are known functions.

More specifically, denoting

$$f_1(\theta) = \frac{-r'(\theta)}{r(\theta)},$$

we have

$$f_0(\theta) = 1 + f_1(\theta)^2,$$

$$f_2(\theta) = 1 + \frac{f_1'(\theta)}{1 + f_1(\theta)^2},$$

$$f_3(\theta) = \frac{2f_1(\theta)}{1+f_1(\theta)^2},$$

and

$$f_4(\theta) = \frac{1}{1+f_1(\theta)^2}.$$

Also

$$d_1u/f_0(\theta) = f_5(\theta)\frac{\partial u}{\partial t} + (f_6(\theta)/t)\frac{\partial u}{\partial \theta},$$

$$d_2u/f_0(\theta) = f_7(\theta)\frac{\partial u}{\partial t} + (f_8(\theta)/t)\frac{\partial u}{\partial \theta},$$

where

$$f_5(\theta) = \cos(\theta)/r(\theta) + \sin(\theta)r'(\theta)/r^3(\theta),$$

$$f_6(\theta) = -\sin(\theta)/r(\theta),$$

$$f_7(\theta) = \sin(\theta)/r(\theta) - \cos(\theta)r'(\theta)/r^3(\theta),$$

$$f_8(\theta) = \cos(\theta)/r(\theta).$$

Observe that $t \in [1,2]$ in Ω.
From (22.11) and (22.12) we may write

$$d_1(L(u) - ud_1(u) - vd_2(u) - d_1(P)) + d_2(L(v) - ud_1(v) - vd_2(v) - d_2(P)) = 0, \quad (22.15)$$

From (22.13) we have

$$d_1[L(u)] + d_2[L(v)] = L(d_1(u) + d_2(v)) = 0,$$

and considering that

$$d_1(d_1(P)) + d_2(d_2(P)) = L(P),$$

from (22.15) we have

$$L(P) + d_1(u)^2 + d_2(v)^2 + 2d_2(u)d_1(v) = 0, \text{ in } \Omega.$$

Hence, in fact we solve the approximate system

$$L(u) - ud_1(u) - vd_2(u) - d_1(P) = 0,$$

$$L(v) - ud_1(v) - vd_2(v) - d_2(P) = 0,$$

$$L(P) + d_1(u)^2 + d_2(v)^2 + 2d_2(u)d_1(v) = 0, \text{ in } \Omega.$$

For the field of velocity u we have obtained the following expressions for the lines (here x stands for θ):

Line 1

$$\begin{aligned}u_1(x) = {} & 0.1u_f(x) + 0.045f_5(x)P_0(x) - 0.045f_5(x)P_f(x) \\ & +0.034f_2(x)u_f(x) - 0.0165f_5(x)u_f(x)^2 \\ & -0.0165f_7(x)u_f(x)v_f(x) - 0.023f_6(x)P_0'(x) \\ & -0.011f_6(x)P_f'(x) + 0.034f_3(x)u_f'(x) \\ & -0.005f_6(x)u_f(x)u_f'(x) - 0.005f_8(x)v_f(x)u_f'(x) \\ & +0.008f_4(x)u_f''(x)\end{aligned}$$

Line 2

$$\begin{aligned}u_2(x) = {} & 0.2u_f(x) + 0.080f_5(x)P_0(x) - 0.080f_5(x)P_f(x) \\ & +0.058f_2(x)u_f(x) - 0.032f_5(x)u_f(x)^2 \\ & -0.032f_7(x)u_f(x)v_f(x) - 0.037f_6(x)P_0'(x) \\ & -0.022f_6(x)P_f'(x) + 0.058f_3(x)u_f'(x) \\ & -0.010f_6(x)u_f(x)u_f'(x) - 0.010f_8(x)v_f(x)u_f'(x) \\ & +0.015f_4(x)u_f''(x)\end{aligned}$$

Line 3

$$\begin{aligned}u_3(x) = {} & 0.3u_f(x) + 0.105f_5(x)P_0(x) - 0.105f_5(x)P_f(x) \\ & +0.075f_2(x)u_f(x) - 0.045f_5(x)u_f(x)^2 \\ & -0.045f_7(x)u_f(x)v_f(x) - 0.044f_6(x)P_0'(x) \\ & -0.030f_6(x)P_f'(x) + 0.075f_3(x)u_f'(x) \\ & -0.015f_6(x)u_f(x)u_f'(x) - 0.015f_8(x)v_f(x)u_f'(x) \\ & +0.020f_4(x)u_f''(x)\end{aligned}$$

Line 4

$$\begin{aligned}u_4(x) = {} & 0.4u_f(x) + 0.120f_5(x)P_0(x) - 0.120f_5(x)P_f(x) \\ & +0.083f_2(x)u_f(x) - 0.056f_5(x)u_f(x)^2 \\ & -0.056f_7(x)u_f(x)v_f(x) - 0.047f_6(x)P_0'(x) \\ & -0.037f_6(x)P_f'(x) + 0.083f_3(x)u_f'(x) \\ & -0.019f_6(x)u_f(x)u_f'(x) - 0.019f_8(x)v_f(x)u_f'(x) \\ & +0.024f_4(x)u_f''(x)\end{aligned}$$

Line 5

$$\begin{aligned}u_5(x) = {} & 0.5u_f(x) + 0.125f_5(x)P_0(x) - 0.125f_5(x)P_f(x) \\ & +0.085f_2(x)u_f(x) - 0.062f_5(x)u_f(x)^2\end{aligned}$$

$$
\begin{aligned}
&-0.062f_7(x)u_f(x)v_f(x)-0.045f_6(x)P_0'(x)\\
&-0.040f_6(x)P_f'(x)+0.085f_3(x)u_f'(x)\\
&-0.022f_6(x)u_f(x)u_f'(x)-0.022f_8(x)v_f(x)u_f'(x)\\
&+0.026f_4(x)u_f''(x)
\end{aligned}
$$

Line 6

$$
\begin{aligned}
u_6(x) = {} & 0.6u_f(x)+0.120f_5(x)P_0(x)-0.120f_5(x)P_f(x)\\
&+0.080f_2(x)u_f(x)-0.064f_5(x)u_f(x)^2\\
&-0.064f_7(x)u_f(x)v_f(x)-0.039f_6(x)P_0'(x)\\
&-0.040f_6(x)P_f'(x)+0.080f_3(x)u_f'(x)\\
&-0.024f_6(x)u_f(x)u_f'(x)-0.024f_8(x)v_f(x)u_f'(x)\\
&+0.025f_4(x)u_f''(x)
\end{aligned}
$$

Line 7

$$
\begin{aligned}
u_7(x) = {} & 0.7u_f(x)+0.105f_5(x)P_0(x)-0.105f_5(x)P_f(x)\\
&+0.068f_2(x)u_f(x)-0.059f_5(x)u_f(x)^2\\
&-0.059f_7(x)u_f(x)v_f(x)-0.032f_6(x)P_0'(x)\\
&-0.037f_6(x)P_f'(x)+0.068f_3(x)u_f'(x)\\
&-0.023f_6(x)u_f(x)u_f'(x)-0.023f_8(x)v_f(x)u_f'(x)\\
&+0.023f_4(x)u_f''(x)
\end{aligned}
$$

Line 8

$$
\begin{aligned}
u_8(x) = {} & 0.8u_f(x)+0.080f_5(x)P_0(x)-0.080f_5(x)P_f(x)\\
&+0.051f_2(x)u_f(x)-0.048f_5(x)u_f(x)^2\\
&-0.048f_7(x)u_f(x)v_f(x)-0.022f_6(x)P_0'(x)\\
&-0.029f_6(x)P_f'(x)+0.051f_3(x)u_f'(x)\\
&-0.019f_6(x)u_f(x)u_f'(x)-0.019f_8(x)v_f(x)u_f'(x)\\
&+0.018f_4(x)u_f''(x)
\end{aligned}
$$

Line 9

$$
\begin{aligned}
u_9(x) = {} & 0.9u_f(x)+0.045f_5(x)P_0(x)-0.045f_5(x)P_f(x)\\
&+0.028f_2(x)u_f(x)-0.059f_5(x)u_f(x)^2\\
&-0.028f_7(x)u_f(x)v_f(x)-0.028f_6(x)P_0'(x)\\
&-0.011f_6(x)P_f'(x)+0.017f_3(x)u_f'(x)
\end{aligned}
$$

$$-0.012f_6(x)u_f(x)u_f'(x) - 0.012f_8(x)v_f(x)u_f'(x)$$
$$+0.010f_4(x)u_f''(x)$$

For the field of velocity v we have obtained for the following expressions for the lines:

Line 1

$$\begin{aligned} v_1(x) = {} & 0.1v_f(x) + 0.045f_7(x)P_0(x) - 0.045f_7(x)P_f(x) \\ & +0.034f_2(x)v_f(x) - 0.017f_5(x)u_f(x)v_f(x) \\ & -0.017f_7(x)v_f(x)^2 - 0.023f_8(x)P_0'(x) \\ & -0.011f_8(x)P_f'(x) + 0.034f_3(x)v_f'(x) \\ & -0.005f_6(x)u_f(x)v_f'(x) - 0.005f_8(x)v_f(x)v_f'(x) \\ & +0.008f_4(x)v_f''(x) \end{aligned}$$

Line 2

$$\begin{aligned} v_2(x) = {} & 0.2v_f(x) + 0.080f_7(x)P_0(x) - 0.080f_7(x)P_f(x) \\ & +0.058f_2(x)v_f(x) - 0.032f_5(x)u_f(x)v_f(x) \\ & -0.032f_7(x)v_f(x)^2 - 0.037f_8(x)P_0'(x) \\ & -0.022f_8(x)P_f'(x) + 0.058f_3(x)v_f'(x) \\ & -0.010f_6(x)u_f(x)v_f'(x) - 0.010f_8(x)v_f(x)v_f'(x) \\ & +0.015f_4(x)v_f''(x) \end{aligned}$$

Line 3

$$\begin{aligned} v_3(x) = {} & 0.3v_f(x) + 0.105f_7(x)P_0(x) - 0.105f_7(x)P_f(x) \\ & +0.075f_2(x)v_f(x) - 0.045f_5(x)u_f(x)v_f(x) \\ & -0.045f_7(x)v_f(x)^2 - 0.045f_8(x)P_0'(x) \\ & -0.030f_8(x)P_f'(x) + 0.075f_3(x)v_f'(x) \\ & -0.015f_6(x)u_f(x)v_f'(x) - 0.015f_8(x)v_f(x)v_f'(x) \\ & +0.020f_4(x)v_f''(x) \end{aligned}$$

Line 4

$$\begin{aligned} v_4(x) = {} & 0.4v_f(x) + 0.120f_7(x)P_0(x) - 0.120f_7(x)P_f(x) \\ & +0.083f_2(x)v_f(x) - 0.056f_5(x)u_f(x)v_f(x) \\ & -0.056f_7(x)v_f(x)^2 - 0.047f_8(x)P_0'(x) \\ & -0.037f_8(x)P_f'(x) + 0.083f_3(x)v_f'(x) \\ & -0.019f_6(x)u_f(x)v_f'(x) - 0.019f_8(x)v_f(x)v_f'(x) \\ & +0.024f_4(x)v_f''(x) \end{aligned}$$

Line 5

$$\begin{aligned} v_5(x) = {} & 0.5v_f(x) + 0.125f_7(x)P_0(x) - 0.125f_7(x)P_f(x) \\ & +0.085f_2(x)v_f(x) - 0.062f_5(x)u_f(x)v_f(x) \\ & -0.062f_7(x)v_f(x)^2 - 0.045f_8(x)P_0'(x) \\ & -0.040f_8(x)P_f'(x) + 0.085f_3(x)v_f'(x) \\ & -0.022f_6(x)u_f(x)v_f'(x) - 0.022f_8(x)v_f(x)v_f'(x) \\ & +0.026f_4(x)v_f''(x) \end{aligned}$$

Line 6

$$\begin{aligned} v_6(x) = {} & 0.6v_f(x) + 0.120f_7(x)P_0(x) - 0.120f_7(x)P_f(x) \\ & +0.068f_2(x)v_f(x) - 0.064f_5(x)u_f(x)v_f(x) \\ & -0.064f_7(x)v_f(x)^2 - 0.039f_8(x)P_0'(x) \\ & -0.040f_8(x)P_f'(x) + 0.080f_3(x)v_f'(x) \\ & -0.024f_6(x)u_f(x)v_f'(x) - 0.024f_8(x)v_f(x)v_f'(x) \\ & +0.026f_4(x)v_f''(x) \end{aligned}$$

Line 7

$$\begin{aligned} v_7(x) = {} & 0.7v_f(x) + 0.105f_7(x)P_0(x) - 0.105f_7(x)P_f(x) \\ & +0.068f_2(x)v_f(x) - 0.059f_5(x)u_f(x)v_f(x) \\ & -0.059f_7(x)v_f(x)^2 - 0.032f_8(x)P_0'(x) \\ & -0.037f_8(x)P_f'(x) + 0.068f_3(x)v_f'(x) \\ & -0.023f_6(x)u_f(x)v_f'(x) - 0.023f_8(x)v_f(x)v_f'(x) \\ & +0.023f_4(x)v_f''(x) \end{aligned}$$

Line 8

$$\begin{aligned} v_8(x) = {} & 0.8v_f(x) + 0.080f_7(x)P_0(x) - 0.080f_7(x)P_f(x) \\ & +0.051f_2(x)v_f(x) - 0.048f_5(x)u_f(x)v_f(x) \\ & -0.048f_7(x)v_f(x)^2 - 0.022f_8(x)P_0'(x) \\ & -0.029f_8(x)P_f'(x) + 0.051f_3(x)v_f'(x) \\ & -0.019f_6(x)u_f(x)v_f'(x) - 0.019f_8(x)v_f(x)v_f'(x) \\ & +0.018f_4(x)v_f''(x) \end{aligned}$$

Line 9

$$
\begin{aligned}
v_9(x) = {} & 0.9v_f(x) + 0.045f_7(x)P_0(x) - 0.045f_7(x)P_f(x) \\
& +0.028f_2(x)v_f(x) - 0.028f_5(x)u_f(x)v_f(x) \\
& -0.028f_7(x)v_f(x)^2 - 0.011f_8(x)P_0'(x) \\
& -0.017f_8(x)P_f'(x) + 0.028f_3(x)v_f'(x) \\
& -0.012f_6(x)u_f(x)v_f'(x) - 0.012f_8(x)v_f(x)v_f'(x) \\
& +0.010f_4(x)v_f''(x)
\end{aligned}
$$

Finally, for the field of pressure P, we have obtained the following lines:

Line 1

$$
\begin{aligned}
P_1(x) = {} & 0.9P_0(x) + 0.1P_f(x) - 0.034f_2(x)P_1(x) \\
& +0.034f_2(x)P_f(x) + 0.045f_5(x)^2 f_0(x)u_f(x)^2 \\
& +0.090f_5(x)f_7(x)f_0(x)u_f(x)v_f(x) + 0.045f_7(x)^2 f_0(x)v_f(x0^2 \\
& -0.034f_3(x)P_0'(x) + 0.034f_3(x)P_f'(x) \\
& +0.022f_5(x)f_6(x)f_0(x)u_f(x)u_f'(x) + 0.022f_5(x)f_8(x)f_0(x)v_f(x)u_f'(x) \\
& +0.003f_6(x)^2 f_0(x)u_f'(x)^2 + 0.022f_6(x)f_7(x)f_0(x)u_f(x)v_f'(x) \\
& +0.022f_7(x)f_8(x)f_0(x)v_f(x)v_f'(x) + 0.007f_6(x)f_8(x)f_0(x)u_f'(x)v_f'(x) \\
& +0.003f_8(x)^2 f_0(x)v_f'(x)^2 + 0.018f_4(x)P_0''(x) \\
& +0.008f_4(x)P_f''(x)
\end{aligned}
$$

Line 2

$$
\begin{aligned}
P_2(x) = {} & 0.8P_0(x) + 0.2P_f(x) - 0.058f_2(x)P_0(x) \\
& +0.058f_2(x)P_f(x) + 0.080f_5(x)^2 f_0(x)u_f(x)^2 \\
& +0.160f_5(x)f_7(x)f_0(x)u_f(x)v_f(x) + 0.080f_7(x)^2 f_0(x)v_f(x)^2 \\
& -0.058f_3(x)P_0'(x) + 0.058f_3(x)P_f'(x) \\
& +0.043f_5(x)f_6(x)f_0(x)u_f(x)u_f'(x) + 0.043f_5(x)f_8(x)f_0(x)v_f(x)u_f'(x) \\
& +0.007f_6(x)^2 f_0(x)u_f'(x)^2 + 0.043f_6(x)f_7(x)f_0(x)u_f(x)v_f'(x) \\
& +0.043f_7(x)f_8(x)f_0(x)v_f(x)v_f'(x) + 0.013f_6(x)f_8(x)f_0(x)u_f'(x)v_f'(x) \\
& +0.007f_8(x)^2 f_0(x)v_f'(x)^2 + 0.028f_4(x)P_0''(x) \\
& +0.014f_4(x)P_f''(x)
\end{aligned}
$$

Line 3

$$
\begin{aligned}
P_3(x) = {} & 0.7P_0(x) + 0.3P_f(x) - 0.075f_2(x)P_0(x) \\
& +0.075f_2(x)P_f(x) + 0.104f_5(x)^2 f_0(x)u_f(x)^2
\end{aligned}
$$

$$+0.210f_5(x)f_7(x)f_0(x)u_f(x)v_f(x)+0.105f_7(x)^2f_0(x)v_f(x)^2$$
$$-0.075f_3(x)P_0'(x)+0.075f_3(x)P_f'(x)$$
$$+0.060f_5(x)f_6(x)f_0(x)u_f(x)u_f'(x)+0.060f_5(x)f_8(x)f_0(x)v_f(x)u_f'(x)$$
$$+0.010f_6(x)^2f_0(x)u_f'(x)^2+0.060f_6(x)f_7(x)f_0(x)u_f(x)v_f'(x)$$
$$+0.060f_7(x)f_8(x)f_0(x)v_f(x)v_f'(x)+0.020f_6(x)f_8(x)f_0(x)u_f'(x)v_f'(x)$$
$$+0.010f_8(x)^2f_0(x)v_f'(x)^2+0.034f_4(x)P_0''(x)$$
$$+0.020f_4(x)P_f''(x)$$

Line 4

$$\begin{aligned}P_4(x) = {} & 0.6P_0(x)+0.4P_f(x)-0.083f_2(x)P_0(x)\\ & +0.083f_2(x)P_f(x)+0.120f_5(x)^2f_0(x)u_f(x)^2\\ & +0.240f_5(x)f_7(x)f_0(x)u_f(x)v_f(x)+0.120f_7(x)^2f_0(x)v_f(x)^2\\ & -0.083f_3(x)P_0'(x)+0.083f_3(x)P_f'(x)\\ & +0.073f_5(x)f_6(x)f_0(x)u_f(x)u_f'(x)+0.073f_5(x)f_8(x)f_0(x)v_f(x)u_f'(x)\\ & +0.012f_6(x)^2f_0(x)u_f'(x)^2+0.073f_6(x)f_7(x)f_0(x)u_f(x)v_f'(x)\\ & +0.073f_7(x)f_8(x)f_0(x)v_f(x)v_f'(x)+0.073f_6(x)f_8(x)f_0(x)u_f'(x)v_f'(x)\\ & +0.012f_8(x)^2f_0(x)v_f'(x)^2+0.035f_4(x)P_0''(x)\\ & +0.024f_4(x)P_f''(x)\end{aligned}$$

Line 5

$$\begin{aligned}P_5(x) = {} & 0.5P_0(x)+0.5P_f(x)-0.085f_2(x)P_0(x)\\ & +0.085f_2(x)P_f(x)+0.125f_5(x)^2f_0(x)u_f(x)^2\\ & +0.250f_5(x)f_7(x)f_0(x)u_f(x)v_f(x)+0.125f_7(x)^2f_0(x)v_f(x)^2\\ & -0.085f_3(x)P_0'(x)+0.085f_3(x)P_f'(x)\\ & +0.080f_5(x)f_6(x)f_0(x)u_f(x)u_f'(x)+0.080f_5(x)f_8(x)f_0(x)v_f(x)u_f'(x)\\ & +0.014f_6(x)^2f_0(x)u_f'(x)^2+0.080f_6(x)f_7(x)f_0(x)u_f(x)v_f'(x)\\ & +0.080f_7(x)f_8(x)f_0(x)v_f(x)v_f'(x)+0.028f_6(x)f_8(x)f_0(x)u_f'(x)v_f'(x)\\ & +0.014f_8(x)^2f_0(x)v_f'(x)^2+0.033f_4(x)P_0''(x)\\ & +0.026f_4(x)P_f''(x)\end{aligned}$$

Line 6

$$\begin{aligned}P_6(x) = {} & 0.4P_0(x)+0.6P_f(x)-0.080f_2(x)P_0(x)\\ & +0.080f_2(x)P_f(x)+0.120f_5(x)^2f_0(x)u_f(x)^2\\ & +0.240f_5(x)f_7(x)f_0(x)u_f(x)v_f(x)+0.120f_7(x)^2f_0(x)v_f(x)^2\end{aligned}$$

$$-0.080 f_3(x) P_0'(x) + 0.080 f_3(x) P_f'(x)$$
$$+0.081 f_5(x) f_6(x) f_0(x) u_f(x) u_f'(x) + 0.081 f_5(x) f_8(x) f_0(x) v_f(x) u_f'(x)$$
$$+0.015 f_6(x)^2 f_0(x) u_f'(x)^2 + 0.081 f_6(x) f_7(x) f_0(x) u_f(x) v_f'(x)$$
$$+0.081 f_7(x) f_8(x) f_0(x) v_f(x) v_f'(x) + 0.030 f_6(x) f_8(x) f_0(x) u_f'(x) v_f'(x)$$
$$+0.015 f_8(x)^2 f_0(x) v_f'(x)^2 + 0.028 f_4(x) P_0''(x)$$
$$+0.026 f_4(x) P_f''(x)$$

Line 7

$$P_7(x) = 0.3 P_0(x) + 0.7 P_f(x) - 0.068 f_2(x) P_0(x)$$
$$+0.068 f_2(x) P_f(x) + 0.105 f_5(x)^2 f_0(x) u_f(x)^2$$
$$+0.210 f_5(x) f_7(x) f_0(x) u_f(x) v_f(x) + 0.105 f_7(x)^2 f_0(x) v_f(x)^2$$
$$-0.068 f_3(x) P_0'(x) + 0.068 f_3(x) P_f'(x)$$
$$+0.073 f_5(x) f_6(x) f_0(x) u_f(x) u_f'(x) + 0.073 f_5(x) f_8(x) f_0(x) v_f(x) u_f'(x)$$
$$+0.014 f_6(x)^2 f_0(x) u_f'(x)^2 + 0.073 f_6(x) f_7(x) f_0(x) u_f(x) v_f'(x)$$
$$+0.073 f_7(x) f_8(x) f_0(x) v_f(x) v_f'(x) + 0.027 f_6(x) f_8(x) f_0(x) u_f'(x) v_f'(x)$$
$$+0.014 f_8(x)^2 f_0(x) v_f'(x)^2 + 0.022 f_4(x) P_0''(x)$$
$$+0.023 f_4(x) P_f''(x)$$

Line 8

$$P_8(x) = 0.2 P_0(x) + 0.8 P_f(x) - 0.051 f_2(x) P_0(x)$$
$$+0.051 f_2(x) P_f(x) + 0.080 f_5(x)^2 f_0(x) u_f(x)^2$$
$$+0.160 f_5(x) f_7(x) f_0(x) u_f(x) v_f(x) + 0.080 f_7(x)^2 f_0(x) v_f(x)^2$$
$$-0.051 f_3(x) P_0'(x) + 0.051 f_3(x) P_f'(x)$$
$$+0.058 f_5(x) f_6(x) f_0(x) u_f(x) u_f'(x) + 0.058 f_5(x) f_8(x) f_0(x) v_f(x) u_f'(x)$$
$$+0.011 f_6(x)^2 f_0(x) u_f'(x)^2 + 0.058 f_6(x) f_7(x) f_0(x) u_f(x) v_f'(x)$$
$$+0.058 f_7(x) f_8(x) f_0(x) v_f(x) v_f'(x) + 0.022 f_6(x) f_8(x) f_0(x) u_f'(x) v_f'(x)$$
$$+0.011 f_8(x)^2 f_0(x) v_f'(x)^2 + 0.015 f_4(x) P_0''(x)$$
$$+0.018 f_4(x) P_f''(x)$$

Line 9

$$P_9(x) = 0.1 P_0(x) + 0.9 P_f(x) - 0.028 f_2(x) P_0(x)$$
$$+0.028 f_2(x) P_f(x) + 0.045 f_5(x)^2 f_0(x) u_f(x)^2$$
$$+0.090 f_5(x) f_7(x) f_0(x) u_f(x) v_f(x) + 0.045 f_7(x)^2 f_0(x) v_f(x)^2$$
$$-0.028 f_3(x) P_0'(x) + 0.028 f_3(x) P_f'(x)$$

$$+0.034 f_5(x) f_6(x) f_0(x) u_f(x) u_f'(x) + 0.034 f_5(x) f_8(x) f_0(x) v_f(x) u_f'(x)$$
$$+0.007 f_6(x)^2 f_0(x) u_f'(x)^2 + 0.034 f_6(x) f_7(x) f_0(x) u_f(x) v_f'(x)$$
$$+0.034 f_7(x) f_8(x) f_0(x) v_f(x) v_f'(x) + 0.013 f_6(x) f_8(x) f_0(x) u_f'(x) v_f'(x)$$
$$+0.007 f_8(x)^2 f_0(x) v_f'(x)^2 + 0.008 f_4(x) P_0''(x)$$
$$+0.010 f_4(x) P_f''(x)$$

22.3.2 A Numerical Example

We consider for the cases $\nu = 1$ and $\nu = 0.01$,

$$\Omega = \{(r,\theta) \mid 1 \leq r \leq 2,\ 0 \leq \theta \leq 2\pi\},$$

$$\partial\Omega_0 = \{(1,\theta) \mid 0 \leq \theta \leq 2\pi\},$$

and

$$\partial\Omega_1 = \{(2,\theta) \mid 0 \leq \theta \leq 2\pi\}.$$

For the present example, the boundary conditions are

$$u = -3.0 \sin(\theta),\ \ v - 3.0 \cos(\theta),\ \ \text{on } \partial\Omega_0,$$
$$u = v = 0,\ P = 2.0 \text{ on } \partial\Omega_1,$$

Through the generalized method of lines, truncating the series up to the terms in d^2 where $d = 1/N$ is the mesh thickness concerning the discretization in r, the general expression for the velocity and pressure fields on the line n is given by (here x stands for θ):

$$u_n(x) = a_1[n]\cos(x) + a_2[n]\sin(x) + a_3[n]\cos(x)^3 + a_4[n]\cos(x)\sin(x)^2$$

$$v_n(x) = b_1[n]\cos(x) + b_2[n]\sin(x) + b_3[n]\sin(x)^3 + b_4[n]\cos(x)^2\sin(x)$$

$$P_n(x) = c_1[n] + c_2[n]\sin(x)^4 + c_3[n]\cos(x)^4 + c_4[n]\cos(x)^2\sin(x)^2.$$

We have plotted the field of velocity u, for lines $n = 1$, $n = 5$, $n = 10$, $n = 15$, and $n = 19$, for a mesh 20×20. Please see Figs. 22.1, 22.2, 22.3, 22.4, and 22.5 for the case $\nu = 1.0$.

For the case $\nu = 0.01$ see Figs. 22.6, 22.7, 22.8, 22.9, and 22.10. For all graphs, please consider units in x to be multiplied by $2\pi/20$.

Again denoting

$$J = \int_\Omega \left(\nu\nabla^2 u - u\partial_x u - v\partial_y u - \partial_x P\right)^2 \, d\Omega$$

$$+\int_{\Omega}\left(\nu\nabla^2 v - u\partial_x v - v\partial_y v - \partial_y P\right)^2 \, d\Omega$$
$$+\int_{\Omega}(\partial_x u + \partial_y v)^2 \, d\Omega, \tag{22.16}$$

the coefficients $\{a_i[n]\}, \{b_i[n]\}, \{c_i[n]\}$ has been obtained through the numerical minimization of J, so that for the mesh in question, we have obtained

$$J \approx 0.0665 \text{ for } \nu = 1.0,$$

$$J \approx 0.0437 \text{ for } \nu = 0.01.$$

In any case it seems we have got good qualitative first approximations for the concerned solutions.

22.4 Conclusion

In this chapter we develop solutions for two-dimensional examples of incompressible Navier–Stokes system. Such solutions are obtained through the generalized method of lines. The extension of results to $\mathbb{R}^3$, compressible and time-dependent cases, is planned for a future work.

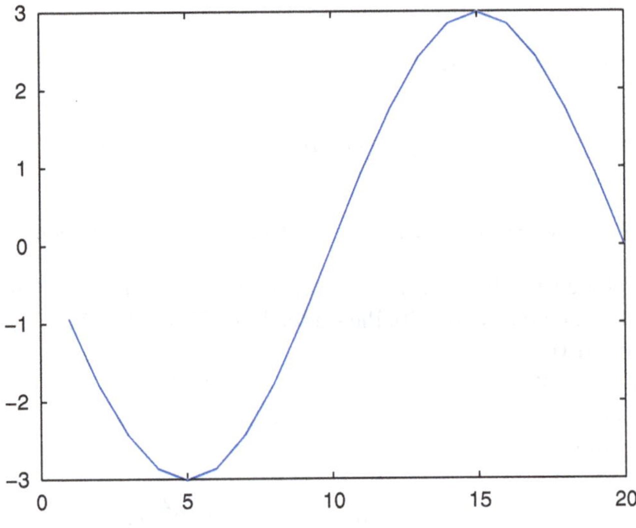

Fig. 22.1 Field of velocity $u_1(x)$-line n=1, case $\nu = 1.0$

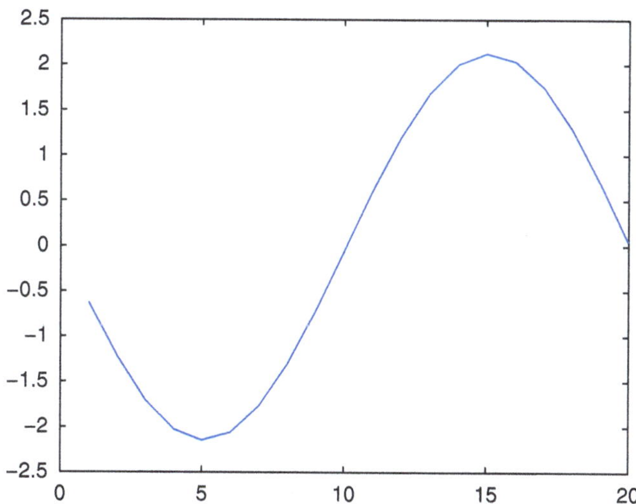

Fig. 22.2 Field of velocity $u_5(x)$-line n=5, case $\nu = 1.0$

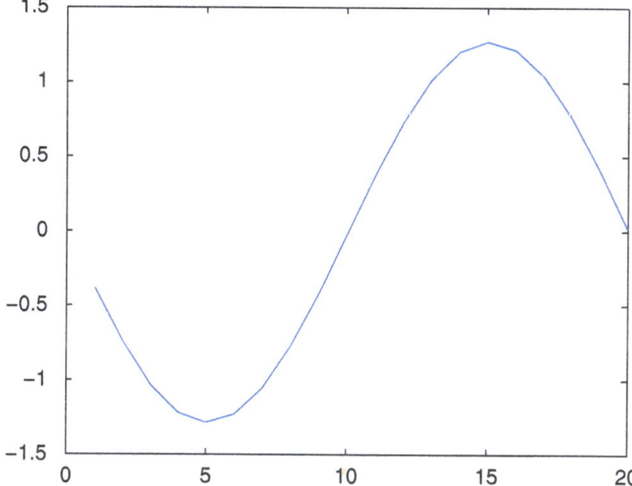

Fig. 22.3 Field of velocity $u_{10}(x)$-line n=10, case $\nu = 1.0$

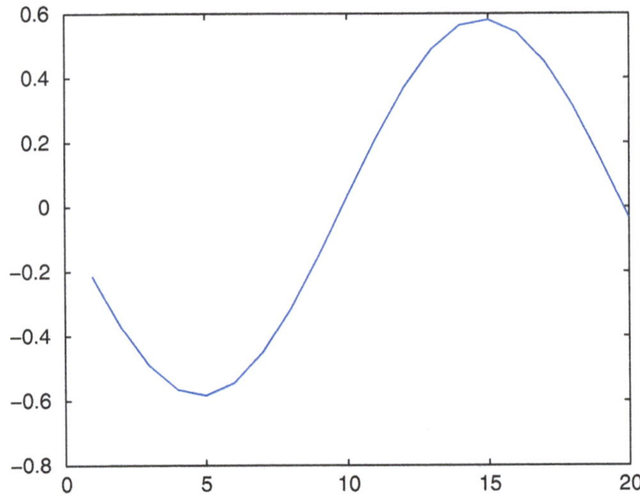

Fig. 22.4 Field of velocity $u_{15}(x)$-line n=15, case $\nu = 1.0$

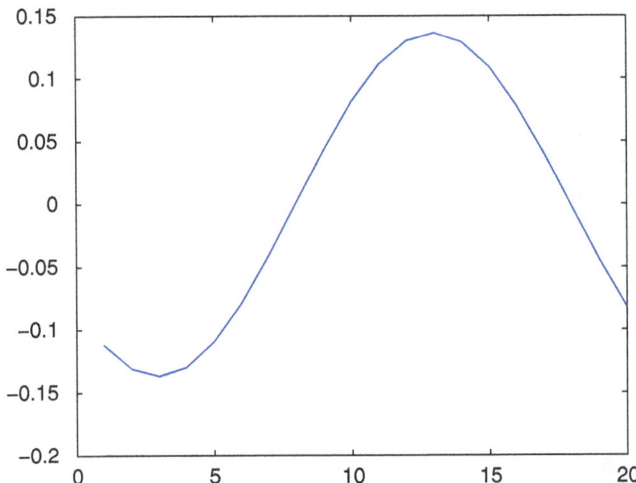

Fig. 22.5 Field of velocity $u_{19}(x)$-line n=19, case $\nu = 1.0$

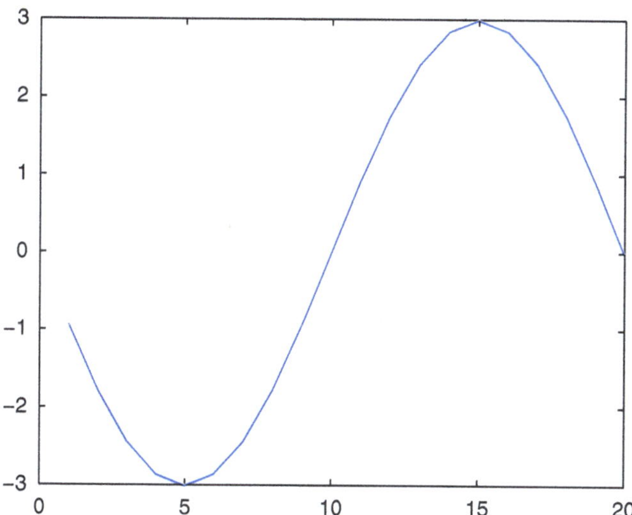

Fig. 22.6 Field of velocity $u_1(x)$-line n=1, case $\nu = 0.01$

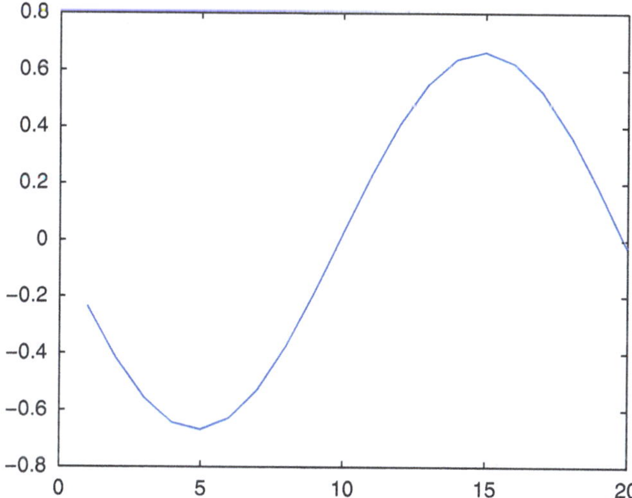

Fig. 22.7 Field of velocity $u_5(x)$-line n=5, case $\nu = 0.01$

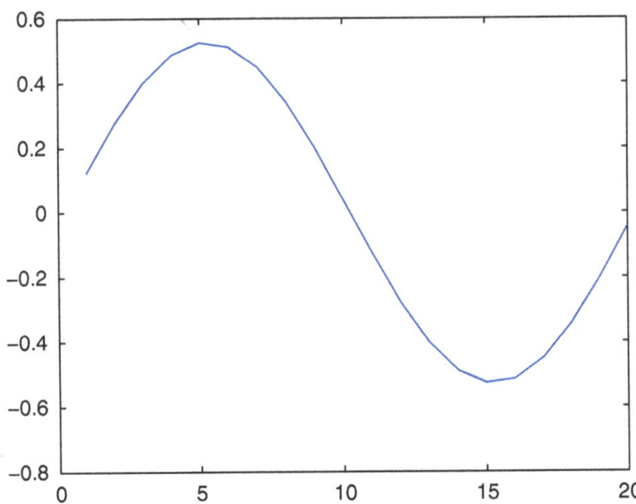

Fig. 22.8 Field of velocity $u_{10}(x)$-line n=10, case $\nu = 0.01$

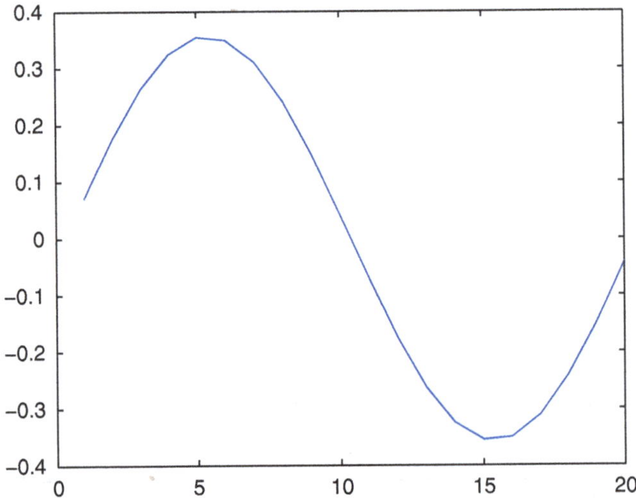

Fig. 22.9 Field of velocity $u_{15}(x)$-line n=15, case $\nu = 0.01$

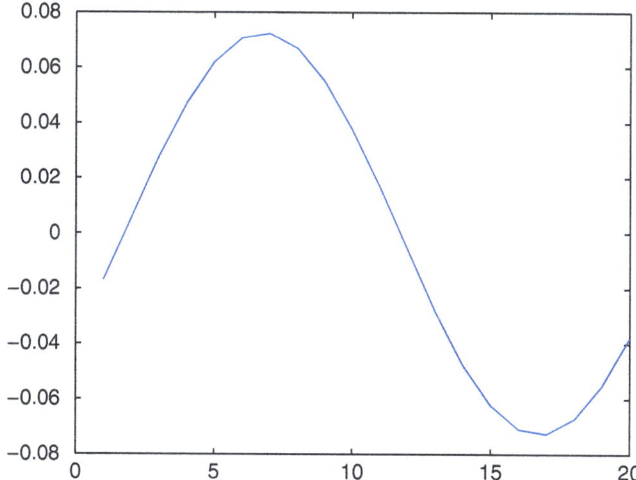

Fig. 22.10 Field of velocity $u_{19}(x)$-line n=19, case $\nu = 0.01$

Chapter 23
Duality Applied to the Optimal Control and Optimal Design of a Beam Model

23.1 Introduction

In this chapter, first we study duality for the optimal control concerning the energy minimization of a well-known model of beams. The duality principle developed includes a concave dual variational formulation suitable to obtain numerical results. For related results in optimization and convex analysis see [13, 14, 40, 47]. For details on the Sobolev spaces involved, see [1, 26]. We emphasize the dual problem always has a solution through which we may verify the optimality of the corresponding primal problem one. However in some situations the primal problem may not have a global minimum, so that in such cases, if there is no duality gap between the dual and primal problems, the dual formulation global maximum solution is a weak limit of minimizing sequences for the primal one. At this point we start to describe the primal problem.

Consider a straight beam represented by the set $\Omega = [0,l]$ where l is the beam length. Consider also the problem of minimizing the beam energy on a fixed interval $[0,T]$, under the equilibrium equations, that is, the problem of minimizing $J : U \to \mathbb{R}$, where

$$J(w,u) = \frac{1}{2}\int_0^T \int_0^l EI(w_{xx})^2\,dx\,dt + \frac{1}{2}\int_0^T \int_0^l \rho A(w_t)^2\,dx\,dt,$$

subject to

$$EIw_{xxxx} + \rho A w_{tt} + Cw_t + u(x,t)w_x - f(x,t) = 0, \text{ in } \Omega \times [0,T],$$

with the boundary conditions

$$w(0,t) = w_x(0,t) = 0, \text{ in } [0,T]$$

and the initial conditions

$$w(x,0) = w_1(x), \text{ and } w_t(x,0) = w_2(x) \text{ in } \Omega.$$

F. Botelho, *Functional Analysis and Applied Optimization in Banach Spaces: Applications to Non-Convex Variational Models*, DOI 10.1007/978-3-319-06074-3_23,

It is worth emphasizing that the boundary conditions refer to a clamped beam at $x = 0$ and free at $x = l$.

Here w denotes the field of vertical displacements, E is the Young modulus, I is a constant which depends on the cross-sectional geometry, ρ is the material density, A is cross-sectional area, and $C > 0$ is a constant which also depends on the type of material. We emphasize to assume E, I, ρ, A, C to be constant on $\Omega \times [0,T]$. Finally, $f \in L^2([0,T];L^2(\Omega))$ is an external dynamical load. The objective here is to obtain the control $u(x,t)$ which minimizes J, so that such a function satisfies the constraints:

$$-M_0 \leq u(x,t) \leq M_0, \text{ in } \Omega \times [0,T],$$

and

$$\int_0^l |u(x,t)|\,dx \leq c, \text{ in } [0,T],$$

where $M_0 \in \mathbb{R}$ and $0 < c < M_0 l$.

23.2 The Duality Principle

In this section we develop a dual variational formulation for the optimal control problem in question. Our main theoretical result is summarized by the next theorem in which we redefine the functional J without relabeling it.

Theorem 23.2.1. *Let $\varepsilon > 0$ be a small constant. Let $J : U \to \bar{\mathbb{R}} = \mathbb{R} \cup \{+\infty\}$ be redefined by*

$$J(w,u) = G(\Lambda w) + F(\Lambda w) + Ind(w,u),$$

where

$$\Lambda : U \to Y = [L^2([0,T];L^2(\Omega))]^3$$

is given by

$$\Lambda w = \{\Lambda_1 w, \Lambda_2 w, \Lambda_3 w\},$$

$$\Lambda_1 w = w_{xx},\ \Lambda_2 w = w_x,\ \Lambda_3 w = w_t,$$

$$\begin{aligned} G(\Lambda w) = {} & \frac{\hat{A}}{2}\int_0^T\int_0^l (w_{xx})^2\,dx\,dt + \frac{\hat{B}}{2}\int_0^T\int_0^l (w_t)^2\,dx\,dt \\ & -\frac{K}{2}\int_0^T\int_0^l (w_x)^2\,dx\,dt, \end{aligned} \tag{23.1}$$

$$\begin{aligned} F(\Lambda w) = {} & \frac{K}{2}\int_0^T\int_0^l (w_x)^2\,dx\,dt \\ & +\frac{\varepsilon}{2}\int_0^T\int_0^l (w_{xx})^2\,dx\,dt \\ & +\frac{\varepsilon}{2}\int_0^T\int_0^l (w_t)^2\,dx\,dt \end{aligned} \tag{23.2}$$

where $\hat{A} = EI - \varepsilon$, and $\hat{B} = \rho A - \varepsilon$.

Moreover,

$$Ind(w,u) = \begin{cases} 0, & \text{if } (w,u) \in B \\ +\infty, & \text{otherwise,} \end{cases} \tag{23.3}$$

$$B = \{(w,u) \in U \mid EIw_{xxxx} + \rho A w_{tt} + C w_t + u w_x - f = 0 \text{ in } \Omega \times [0,T]\},$$

$$U = \tilde{U} \times \tilde{B},$$

$$\tilde{U} = U_1 \cap U_2 \cap U_3,$$

$$\tilde{B} = B_1 \cap B_2,$$

$$U_1 = \{w \in L^2([0,T]; H^2(\Omega)) \mid w(0,t) = w_x(0,t) = 0, \text{ in } [0,T]\} \tag{23.4}$$

$$U_2 = \{w \in L^2([0,T]; H^2(\Omega)) \cap H^1([0,T]; L^2(\Omega)) \mid w(x,0) = w_1(x), \text{ and } w_t(x,0) = w_2(x), \text{ in } \Omega\}, \tag{23.5}$$

$$U_3 = \{w \in L^2([0,T]; H^2(\Omega)) \mid w_{xx}(l,t) = w_{xxx}(l,t) = 0, \text{ in } [0,T]\} \tag{23.6}$$

$$B_1 = \{u \in L^2([0,T]; L^2(\Omega)) \mid -M_0 \leq u(x,t) \leq M_0, \text{ in } \Omega \times [0,T]\}, \tag{23.7}$$

and

$$B_2 = \{u \in L^2([0,T]; L^2(\Omega)) \mid \int_0^l |u(x,t)|\, dx \leq c, \text{ in } [0,T]\}. \tag{23.8}$$

Also,

$$A^* = L^2([0,T]; L^2(\Omega)) \times L^2([0,T]; H^2(\Omega)), \tag{23.9}$$

and we assume that $K > 0$ is the largest constant such that

$$G(\Lambda w) \geq 0, \forall w \in \tilde{U}.$$

Under such hypotheses we have

$$\inf_{(w,u) \in U} \{J(w,u)\} \geq \sup_{(v^*,\lambda) \in A^*} \{-J^*(v^*,\lambda)\}, \tag{23.10}$$

where

$$J^*(v^*,\lambda) = (G \circ \Lambda)^*(-\Lambda^* v^*, \lambda) + F_1^*(v_1^* - EI\lambda_{xx}) + F_2^*(v_2^*, \lambda) + F_3^*(v_3^* + \rho A \lambda_t - C\lambda) + \langle \lambda, f \rangle_{L^2}, \tag{23.11}$$

$\lambda \in L^2([0,T];H^2(\Omega))$ is an appropriate Lagrange multiplier,

$$F_1(\Lambda_1 w) = \frac{\varepsilon}{2}\int_0^T\int_0^l (w_{xx})^2\,dx\,dt,$$

$$F_2(\Lambda_2 w,\lambda) = \inf_{u\in\tilde{B}}\left\{\int_0^T\int_0^l (\lambda u(x,t)w_x)\,dx\,dt + \frac{K}{2}\int_0^T\int_0^l (w_x)^2\,dx\,dt\right\},$$

and

$$F_3(\Lambda_3 w) = \frac{\varepsilon}{2}\int_0^T\int_0^l (w_t)^2\,dx\,dt.$$

Moreover,

$$\begin{aligned} F_1^*(v_1^* - EI\lambda_{xx}) &= \sup_{v_1\in L^2}\{\langle v_1, v_1^* - EI\lambda_{xx}\rangle_{L^2} - F_1(v_1)\} \\ &= \frac{1}{2\varepsilon}\int_0^T\int_0^l (v_1^* - EI\lambda_{xx})^2\,dx\,dt, \end{aligned} \tag{23.12}$$

$$\begin{aligned} F_2^*(v_2^*,\lambda) &= \sup_{v_2\in L^2}\{\langle v_2, v_2^*\rangle_{L^2} - F_2(v_2,\lambda)\} \\ &= \sup_{u\in\tilde{B}}\frac{1}{2K}\int_0^T\int_0^l (v_2^* - \lambda u(x,t))^2\,dx\,dt, \end{aligned} \tag{23.13}$$

$$\begin{aligned} F_3^*(v_3^* + \rho A\lambda_t - C\lambda) &= \sup_{v_3\in L^2}\{\langle v_3, v_3^* + \rho A\lambda_t - C\lambda\rangle_{L^2} - F_3(v_3)\} \\ &= \frac{1}{2\varepsilon}\int_0^T\int_0^l (v_3^* + \rho A\lambda_t - C\lambda)^2\,dx\,dt. \end{aligned}$$

Also,

$$\begin{aligned} &(G\circ\Lambda)^*(-\Lambda^* v^*,\lambda) \\ &= \sup_{w\in U}\left\{\langle \Lambda w, -v^*\rangle_Y - (G\circ\Lambda)(w)\right. \\ &\quad + \int_0^T \lambda(0,t)EIw_{xxx}(0,t)\,dt \\ &\quad - \int_0^T \lambda_x(0,t)EIw_{xx}(0,t)\,dt \\ &\quad - \int_0^l \lambda(x,T)\rho A w_t(x,T)\,dx \\ &\quad \left. + \int_0^l \lambda(x,0)\rho A w_2(x)\,dx\right\}. \end{aligned} \tag{23.14}$$

Under such assumptions there exists $(v_0^*, \lambda_0) \in A^*$ *such that*

$$-J^*(v_0^*, \lambda_0) = \max_{(v^*, \lambda) \in A^*} \{-J^*(v^*, \lambda)\},$$

so that (w_0, u_0) *such that*

$$w_0 = \frac{\partial (G \circ \Lambda)^*(-\Lambda^* v_0^*, \lambda_0)}{\partial w^*},$$

where $w^* = -\Lambda^* v^*$ *and*

$$F_2^*(v_{02}^*, \lambda_0) = \frac{1}{2K} \int_0^T \int_0^l (v_{02}^* - \lambda_0 u_0(x,t))^2 \, dx \, dt$$

are also such that

$$\begin{aligned}
&(G \circ \Lambda)(w_0) + F_1(\Lambda_1 w_0) + F_2^{**}(\Lambda_2 w_0, \lambda_0) \\
&\quad + F_3(\Lambda_3 w_0, \lambda_0) + \langle \lambda_0, EI w_{0xxxx} \rangle_{L^2} \\
&\quad + \langle \lambda_0, \rho A w_{0tt} + C w_{0t} \rangle_{L^2} - \langle \lambda_0, f \rangle_{L^2} \\
&= \min_{w \in \tilde{U}} \{ (G \circ \Lambda)(w) + F_1(\Lambda_1 w) + F_2^{**}(\Lambda_2 w, \lambda_0) \\
&\quad + F_3(\Lambda_3 w, \lambda_0) + \langle \lambda_0, EI w_{xxxx} \rangle_{L^2} \\
&\quad + \langle \lambda_0, \rho A w_{tt} + C w_t \rangle_{L^2} - \langle \lambda_0, f \rangle_{L^2} \} \\
&= \max_{(v^*, \lambda) \in A^*} \{-J^*(v^*, \lambda)\} = -J^*(v_0^*, \lambda_0),
\end{aligned} \tag{23.15}$$

where

$$\begin{aligned}
&(G \circ \Lambda)(w_0) + F_1(\Lambda_1 w_0) + F_2(\Lambda_2 w_0, \lambda_0) \\
&\quad + F_3(\Lambda_3 w_0, \lambda_0) + \langle \lambda_0, EI w_{0xxxx} \rangle_{L^2} \\
&\quad + \langle \lambda_0, \rho A w_{0tt} + C w_{0t} \rangle_{L^2} - \langle \lambda_0, f \rangle_{L^2} \\
&= (G \circ \Lambda)(w_0) + F(\Lambda w_0) \\
&\quad + \langle \lambda_0, EI w_{0xxxx} + \rho A w_{0tt} + C w_{0t} + u_0 w_{0x} - f \rangle_{L^2}.
\end{aligned}$$

Furthermore, we emphasize to denote $L^2(\Omega \times [0,T]) \equiv L^2$:

$$\langle \Lambda w, v^* \rangle_Y = \langle \Lambda_1 w, v_1^* \rangle_{L^2} + \langle \Lambda_2 w, v_2^* \rangle_{L^2} + \langle \Lambda_3 w, v_3^* \rangle_{L^2},$$

where

$$\langle g, h \rangle_{L^2} = \int_0^T \int_0^l g(x,t) h(x,t) \, dx \, dt$$

and

$$F^{**}(\Lambda_2 w, \lambda) = \sup_{v_2^* \in A^*} \{ \langle \Lambda_2 w, v_2^* \rangle_{L^2} - F_2^*(v_2^*, \lambda) \}.$$

Finally, if $K > 0$ *above specified is such that the optimal inclusion*

$$\Lambda_2\left(\frac{\partial(G\circ\Lambda)^*(-\Lambda^* v_0^*,\lambda_0)}{\partial w^*}\right)\in\partial_{v_2^*}F_2^*(v_{02}^*,\lambda_0) \tag{23.16}$$

stands for

$$\Lambda_2 w_0=\frac{\partial\tilde{F}_2^*(v_{02}^*,\lambda_0,u_0)}{\partial v_2^*},$$

where

$$w_0=\frac{\partial(G\circ\Lambda)^*(-\Lambda^* v_0^*,\lambda_0)}{\partial w^*},$$

and

$$\tilde{F}_2^*(v_2^*,\lambda,u)=\frac{1}{2K}\int_0^T\int_0^l (v_2^*-\lambda u(x,t))^2\,dx\,dt;$$

then

$$J(w_0,u_0)=\min_{(w,u)\in U}\{J(w,u)\}=\max_{(v^*,\lambda)\in A^*}\{-J^*(v^*,\lambda)\}=-J^*(v_0^*,\lambda_0).$$

Proof. Observe that

$$\begin{aligned}J(w,u)&=G(\Lambda w)+F(\Lambda w)+Ind(u,w)\\&\geq G(\Lambda w)+F(\Lambda w)\\&\quad+\langle\lambda,EIw_{xxxx}+\rho Aw_{tt}+Cw_t+uw_x-f\rangle_{L^2},\end{aligned} \tag{23.17}$$

$\forall(w,u)\in U,\ (v^*,\lambda)\in A^*$, so that

$$\begin{aligned}J(w,u)&=G(\Lambda w)+F(\Lambda w)+Ind(u,w)\\&\geq\langle\Lambda w,v^*\rangle_Y+G(\Lambda w)\\&\quad-\langle\Lambda_1 w,v_1^*\rangle_{L^2}+F_1(\Lambda_1 w)\\&\quad-\langle\Lambda_2 w,v_2^*\rangle_{L^2}+F_2(\Lambda_2 w,\lambda)\\&\quad-\langle\Lambda_3 w,v_3^*\rangle_{L^2}+F_3(\Lambda_3 w)\\&\quad+\langle\lambda_{xx},EIw_{xx}\rangle_{L^2}-\langle\lambda_t,\rho Aw_t\rangle_{L^2}\\&\quad+\langle\lambda,Cw_t\rangle_{L^2}-\int_0^T\lambda(0,t)EIw_{xxx}(0,t)\,dt\\&\quad+\int_0^T\lambda_x(0,t)EIw_{xx}(0,t)\,dt\\&\quad+\int_0^l\lambda(x,T)\rho Aw_t(x,T)\,dx\\&\quad-\int_0^l\lambda(x,0)\rho Aw_2(x)\,dx\\&\quad-\langle\lambda,f\rangle_{L^2},\end{aligned} \tag{23.18}$$

$\forall(w,u)\in U,(v^*,\lambda)\in A^*$. Thus,

$$
\begin{aligned}
J(w,u) \geq \inf_{w\in\tilde{U}} \Bigg\{ & \langle \Lambda w, v^* \rangle_Y + G(\Lambda w) \\
& - \int_0^T \lambda(0,t) EI w_{xxx}(0,t)\, dt \\
& + \int_0^T \lambda_x(0,t) EI w_{xx}(0,t)\, dt \\
& + \int_0^l \lambda(x,T) \rho A w_t(x,T)\, dx \\
& - \int_0^l \lambda(x,0) \rho A w_2(x)\, dx \Bigg\} \\
& + \inf_{v_1 \in L^2} \{ -\langle v_1, v_1^* - EI\lambda_{xx} \rangle_{L^2} + F_1(v_1) \} \\
& + \inf_{v_2 \in L^2} \{ -\langle v_2, v_2^* \rangle_{L^2} + F_2(v_2, \lambda) \} \\
& + \inf_{v_3 \in L^2} \{ -\langle v_3, v_3^* + \rho A \lambda_t - C\lambda \rangle_{L^2} + F_3(v_3) \} \\
& - \langle \lambda, f \rangle_{L^2},
\end{aligned} \tag{23.19}
$$

$\forall (w,u) \in U, (v^*, \lambda) \in A^*$.

Therefore,

$$
\begin{aligned}
J(w,u) \geq & -(G\circ\Lambda)^*(-\Lambda^* v^*, \lambda) - F_1^*(v_1^* - EI\lambda_{xx}) - F_2^*(v_2^*, \lambda) \\
& - F_3^*(v_3^* + \rho A \lambda_t - C\lambda) - \langle \lambda, f \rangle_{L^2},
\end{aligned} \tag{23.20}
$$

$\forall (w,u) \in U, (v^*, \lambda) \in A^*$.

Hence,

$$
\inf_{(w,u)\in U} \{J(w,u)\} \geq \sup_{(v^*,\lambda)\in A^*} \{-J^*(v^*, \lambda)\}. \tag{23.21}
$$

Since $-J^*(v^*, \lambda)$ is concave, coercive, continuous, and therefore weakly upper semicontinuous, from an application of the direct method of variations (considering it is a standard procedure, here we omit more details) there exists (v_0^*, λ_0) such that

$$
-J(v_0^*, \lambda_0) = \max_{(v^*,\lambda)\in A^*} \{-J^*(v^*, \lambda)\}.
$$

Such an optimal point is attained through the extremal equations:

$$
\Lambda_1 \left(\frac{\partial (G\circ\Lambda)^*(-\Lambda^* v_0^*, \lambda_0)}{\partial w^*} \right) - \frac{\partial F_1^*(v_{01}^* - EI\lambda_{0xx})}{\partial v_1^*} = \theta, \tag{23.22}
$$

$$
\Lambda_2 \left(\frac{\partial (G\circ\Lambda)^*(-\Lambda^* v_0^*, \lambda_0)}{\partial w^*} \right) \in \partial_{v_2^*} F_2^*(v_{02}^*, \lambda_0), \tag{23.23}
$$

$$
\Lambda_3 \left(\frac{\partial (G\circ\Lambda)^*(-\Lambda^* v_0^*, \lambda_0)}{\partial w^*} \right) - \frac{\partial F_3^*(v_{03}^* + \rho A \lambda_{0t} - C\lambda_0)}{\partial v_3^*} = \theta. \tag{23.24}
$$

Hence, for w_0 such that

$$w_0 = \frac{\partial (G \circ \Lambda)^*(-\Lambda^* v_0^*, \lambda_0)}{\partial w^*}, \tag{23.25}$$

we get

$$\Lambda_1 w_0 - \frac{\partial F_1^*(v_{01}^* - EI\lambda_{0xx})}{\partial v_1^*} = \theta, \tag{23.26}$$

$$\Lambda_2 w_0 \in \partial_{v_2^*} F_2^*(v_{02}^*, \lambda_0), \tag{23.27}$$

and

$$\Lambda_3 w_0 - \frac{\partial F_3^*(v_{03}^* + \rho A \lambda_{0t} - C\lambda_0)}{\partial v_3^*} = \theta. \tag{23.28}$$

Thus, from the last three relations, we obtain

$$F_1^*(v_{01}^* - EI\lambda_{0xx}) = \langle \Lambda_1 w_0, v_{01}^* - EI\lambda_{0xx}\rangle_{L^2} - F_1(\Lambda_1 w_0),$$

$$F_2^*(v_{02}^*, \lambda_0) = \langle \Lambda_2 w_0, v_{02}^*\rangle_{L^2} - F_2^{**}(\Lambda_2 w_0, \lambda_0),$$

and

$$F_3^*(v_{03}^* + \rho A\lambda_{0t} - C\lambda_0) = \langle \Lambda_3 w_0, v_{03}^* + \rho A \lambda_{0t} - C\lambda_0\rangle_{L^2} - F_3(\Lambda_3 w_0).$$

From (23.25), the extremal condition concerning the variation in λ and these last three equalities, we get

$$\begin{aligned} J^*(v_0^*, \lambda_0) &= (G\circ\Lambda)^*(-\Lambda^* v_0^*, \lambda_0) + F_1^*(v_{01}^* - EI\lambda_{0xx}) + F_2^*(v_{02}^*, \lambda_0) \\ &\quad + F_3^*(v_{03}^* + \rho A \lambda_{0t} - C\lambda_0) + \langle \lambda_0, f\rangle_{L^2} \\ &= -(G\circ\Lambda)(w_0) - F_1(\Lambda_1 w_0) - F_2^{**}(\Lambda_2 w_0, \lambda_0) \\ &\quad - \langle \lambda_0, EIw_{0xxxx}\rangle_{L^2} - \langle \lambda_0, \rho A w_{0tt} + C w_{0t}\rangle_{L^2} \\ &\quad - F_3(\Lambda_3 w_0) + \langle \lambda_0, f\rangle_{L^2}. \end{aligned} \tag{23.29}$$

Similarly as above, we may infer that

$$\begin{aligned} &\inf_{w\in\tilde{U}} \{G(\Lambda w) + F_1(\Lambda_1 w) + F_2^{**}(\Lambda_2 w, \lambda_0) \\ &\quad + F_3(\Lambda_3 w) + \langle \lambda_0, EIw_{xxxx} + \rho A w_{tt} + C w_t - f\rangle_{L^2}\} \\ &\geq -J^*(v^*, \lambda_0), \forall v^* \text{ such that } (v^*, \lambda_0) \in A^*. \end{aligned} \tag{23.30}$$

From this and (23.29) we obtain

$$\begin{aligned} &\min_{w\in\tilde{U}} \{G(\Lambda w) + F_1(\Lambda_1 w) + F_2^{**}(\Lambda_2 w, \lambda_0) \\ &\quad + F_3(\Lambda_3 w) + \langle \lambda_0, EIw_{xxxx} + \rho A w_{tt} + C w_t - f\rangle_{L^2}\} \\ &= G(\Lambda w_0) + F_1(\Lambda_1 w_0) + F_2^{**}(\Lambda_2 w_0, \lambda_0) \\ &\quad + F_3(\Lambda_3 w_0) + \langle \lambda_0, EI(w_0)_{xxxx} + \rho A (w_0)_{tt} + C(w_0)_t - f\rangle_{L^2} \\ &= -J^*(v_0^*, \lambda_0) = \max_{(v^*,\lambda)\in A^*} \{-J^*(v^*, \lambda)\}. \end{aligned} \tag{23.31}$$

Finally, if $K > 0$ above specified is such that

$$\Lambda_2 w_0 = \frac{\partial \tilde{F}_2^*(v_{02}^*, \lambda_0, u_0)}{\partial v_2^*},$$

where

$$\tilde{F}_2^*(v_2^*, \lambda, u) = \frac{1}{2K}\int_0^T\int_0^l (v_2^* - \lambda u(x,t))^2\, dx\, dt,$$

denoting

$$\tilde{F}_2(\Lambda_2 w, \lambda, u) = \int_0^T\int_0^l (\lambda u(x,t) w_x)\, dx\, dt + \frac{K}{2}\int_0^T\int_0^l (w_x)^2\, dx\, dt,$$

we have

$$\begin{aligned} F_2^*(v_{02}^*, \lambda_0) &= \tilde{F}_2^*(v_{02}^*, \lambda_0, u_0) = \langle \Lambda_2 w_0, v_{02}^* \rangle_{L^2} - \tilde{F}_2(\Lambda_2 w_0, \lambda_0, u_0) \\ &= \langle \Lambda_2 w_0, v_{02}^* \rangle_{L^2} - F_2(\Lambda_2 w_0, \lambda_0). \end{aligned} \tag{23.32}$$

Moreover, the variation in λ in the dual formulation gives us the extremal inclusion:

$$\begin{aligned} &EI\left(\frac{\partial F_1^*(v_{01}^* - EI\lambda_{0xx})}{\partial v_1^*}\right)_{xx} + \rho A\left(\frac{\partial F_3^*(v_{03}^* + \rho A\lambda_{0t} - C\lambda_0)}{\partial v_3^*}\right)_t \\ &+ C\frac{\partial F_3^*(v_{03}^* + \rho A\lambda_{0t} - C\lambda_0)}{\partial v_3^*} - f \in [\partial_{v_2^*} F_2^*(v_{02}^*, \lambda_0)](-u_0), \end{aligned} \tag{23.33}$$

so that

$$EIw_{0xxxx} + \rho A w_{0tt} + Cw_t + u_0 w_{0x} - f = 0, \text{ in } \Omega \times [0,T].$$

Hence, from this last equation, (23.29) and (23.32) we have

$$\begin{aligned} J(w_0, u_0) &= G(\Lambda w_0) + F(\Lambda w_0) + Ind(w_0, u_0) \\ &= G(\Lambda w_0) + F_1(\Lambda_1 w_0) \\ &\quad + \tilde{F}_2(\Lambda_2 w_0, \lambda_0, u_0) + F_3(\Lambda_3 w_0) \\ &\quad + \langle \lambda_0, EIw_{0xxxx} + \rho A w_{0tt} + Cw_{0t} + u_0 w_{0x} - f \rangle_{L^2} \\ &= -(G\circ\Lambda)^*(-\Lambda^* v_0^*, \lambda_0) - F_1^*(v_{01}^*) - \tilde{F}_2^*(v_{02}^*, \lambda_0, u_0) \\ &\quad - F_3^*(v_{03}^*) - \langle \lambda_0, f \rangle_{L^2} \\ &= -(G\circ\Lambda)^*(-\Lambda^* v_0^*, \lambda_0) - F_1^*(v_{01}^*) - F_2^*(v_{02}^*, \lambda_0) \\ &\quad - F_3^*(v_{03}^*) - \langle \lambda_0, f \rangle_{L^2} \\ &= -J^*(v_0^*, \lambda_0). \end{aligned} \tag{23.34}$$

From this and (23.21) we get

$$J(w_0, u_0) = \min_{(w,u)\in U}\{J(w,u)\} = \max_{(v^*,\lambda)\in A^*}\{-J^*(v^*,\lambda)\} = -J^*(v_0^*, \lambda_0).$$

The proof is complete.

23.3 Some Closely Related Simpler Examples with Numerical Results

Consider a straight beam with circular cross-sectional area given by $A(x)$, where $x \in [0,l]$, l being the beam length and $[0,l] = \Omega$ its axis. Suppose such a beam is simply supported, so that $w \in U$, where $w : \Omega \to \mathbb{R}$ is the field of vertical displacements and

$$U = \{w \in W^{2,2}(\Omega) \; : \; w(0) = w(l) = 0\}.$$

Also, the beam in question is assumed to be under a compressive axial load P applied at $x = l$. We shall look for the optimal distribution $A(x)$ which maximizes the buckling load P, where the following designed constraints must be satisfied:

$$\int_\Omega A(x)\,dx = V = cA_{max}l,$$

where $0 < c < 1$ and

$$0 < A_{min} \leq A(x) \leq A_{max}, \text{ in } \Omega.$$

Hence, our optimization problem translates into minimizing $-P$, subject to

$$c_0(A(x)^2 w_{,xx})_{,xx} + Pw_{,xx} = 0, \text{ in } \Omega,$$

where $c_0 > 0$ is an appropriate constant to be specified, so that P is such that

$$\frac{c_0}{2}\int_\Omega A(x)^2 w_{,xx}^2\,dx - \frac{P}{2}\int_\Omega w_{,x}^2\,dx \geq 0,$$

$\forall w \in U$. Furthermore, as above indicate, we must have

$$0 < A_{min} \leq A(x) \leq A_{max}, \text{ in } \Omega,$$

and

$$\int_\Omega A(x)\,dx = V = cA_{max}l,$$

where $0 < c < 1$.

Observe that from the concerned constraints

$$\frac{c_0}{2}\int_\Omega A(x)^2 w_{,xx}^2\,dx = \frac{P}{2}\int_\Omega w_{,x}^2\,dx,$$

so that through the appropriate constraint for the concerned eigenvalue problem, that is,

$$\int_\Omega w_{,x}^2\,dx = 1,$$

we get

$$\frac{c_0}{2}\int_\Omega A(x)^2 w_{,xx}^2\,dx = \frac{P}{2}.$$

Hence, we may define the above optimization problem by the minimization of

$$J(w,A) + Ind(w,P,A),$$

where

$$J(w,A) = \frac{-c_0}{2}\int_\Omega A(x)^2 w_{,xx}^2\,dx$$

and

$$Ind(w,P,A) = \begin{cases} 0, & \text{if } (w,P,A) \in A^*, \\ +\infty, & \text{otherwise,} \end{cases} \tag{23.35}$$

where

$$A^* = A_1 \cap A_2 \cap A_3,$$

$$\begin{aligned} A_1 = \{&(w,P,A) \in U \times \mathbb{R}^+ \times L^2(\Omega) \text{ such that} \\ &c_0(A(x)^2 w_{,xx}^2)_{,xx} - P w_{,xx} = 0, \text{ in } \Omega\} \end{aligned} \tag{23.36}$$

$$\begin{aligned} A_2 = \{&(P,A) \in \mathbb{R}^+ \times L^2(\Omega) \text{ such that} \\ &\tilde{J}(w,P,A) \geq 0, \forall w \in U\} \end{aligned} \tag{23.37}$$

where

$$\tilde{J}(w,P,A) = \frac{c_0}{2}\int_\Omega A(x)^2 w_{,xx}^2\,dx - \frac{P}{2}\int_\Omega w_{,x}^2\,dx.$$

Finally,

$$A_3 = \left\{ w \in U \mid \int_\Omega (w_{,x})^2\,dx = 1 \right\}.$$

At this point, denoting

$$G(w_{,xx},A) = \frac{c_0}{2}\int_\Omega A(x)^2 (w_{,xx})^2\,dx,$$

we define the extended functional $J_\lambda(w,P,A)$ by

$$\begin{aligned} J_\lambda(w,P,A) = &-G(w_{,xx},A) \\ &+\langle \lambda, (c_0 A(x)^2 w_{,xx})_{xx} + P w_{,xx} \rangle_{L^2} \end{aligned} \tag{23.38}$$

where λ is an appropriate Lagrange multiplier.

Observe that

$$\begin{aligned} J_\lambda(w,P,A) = &\langle w_{,xx}, v^* \rangle_{L^2} - G(w_{,xx},A) \\ &-\langle w_{,xx}, v^* \rangle_{L^2} + \langle \lambda, c_0(A(x)^2 w_{,xx})_{xx} + P w_{,xx} \rangle_{L^2}, \end{aligned} \tag{23.39}$$

so that

$$J_\lambda(w,P,A) \leq \sup_{v \in L^2}\{\langle v, v^* \rangle_{L^2} - G(v,A)\} + \sup_{w \in U}\{-\langle w_{,xx}, v^* \rangle_{L^2} + \langle \lambda, c_0(A(x)^2 w_{,xx})_{xx} + P w_{,xx} \rangle_{L^2}\}, \quad (23.40)$$

and therefore

$$J_\lambda(w,P,A) \leq G^*(v^*,A) + Ind_1(v^*,\lambda,P,A),$$

where

$$G^*(v^*,A) = \frac{1}{2c_0}\int_\Omega \frac{(v^*)^2}{A(x)^2}\,dx,$$

and

$$Ind_1(v^*,\lambda,P,A) = \begin{cases} 0, & \text{if } (v^*,\lambda,P,A) \in B^*, \\ +\infty, & \text{otherwise,} \end{cases} \quad (23.41)$$

where

$$\begin{aligned} B^* = \{&(v^*,\lambda,P,A) \in L^2 \times L^2 \times \mathbb{R}^+ \times L^2 \mid \\ &v^*_{,xx} - c_0(A(x)^2\lambda_{xx})_{,xx} + P w_{xx} = 0, \text{ in } \Omega, \\ &\text{and } v^*(0) = v^*(l) = 0\}. \end{aligned} \quad (23.42)$$

Summarizing the partial duality principle obtained, we have

$$J_\lambda(w,P,A) \leq \inf_{v^* \in L^2}\{G^*(v^*,A) + Ind_1(v^*,\lambda,P,A)\}.$$

Having this inequality in mind, we suggest the following algorithm to get critical points relating the original problem. It is worth emphasizing we have not formally proven its convergence:

1. Set $k = 1$ and choose $\tilde{w}_1^0 \in U$ such that
$$\int_\Omega [(\tilde{w}_1^0)_{,x}]^2\,dx = 1.$$
2. Set $A_1 = cA_{max}$.
3. Set $n = 1$ and $\tilde{w}_k^1 = \tilde{w}_k^0$.
4. Calculate $w_k^n \in U$ by solving the equation
$$c_0(A_k(x)^2(w_k^n)_{,xx})_{,xx} + (\tilde{w}_k^n)_{,xx} = 0, \text{ in } \Omega.$$
5. Define
$$\tilde{w}_k^{n+1} = w_k^n / S_k^n,$$
where
$$S_k^n = \sqrt{\int_\Omega [(w_k^n)_{,x}]^2\,dx}.$$
6. Set $n \to n+1$ and go to step (4), up to the satisfaction of an appropriate convergence criterion.

7. Define

$$\tilde{w}^0_{k+1} = \lim_{n\to\infty} \tilde{w}^n_k,$$

$$P_k = \lim_{n\to\infty} \frac{1}{S^n_k}.$$

8. Define

$$v^* = c_0 A_k(x)(\tilde{w}^0_{k+1})_{xx},$$

and obtain $A_{k+1}(x)$ by

$$A_{k+1}(x) = argmin_{A\in C^*}\{G^*(v^*,A)\},$$

where

$$C^* = C_1 \cap C_2,$$

$$C_1 = \{A \in L^2 \mid \int_\Omega A(x)\,dx = V = cA_{max}l\},$$

and

$$C_2 = \{A \in L^2 \mid 0 < A_{min} \leq A(x) \leq A_{max}, \text{ in } \Omega\}.$$

9. Set $k \to k+1$ and go to step (3), up to the satisfaction of an appropriate convergence criterion.

23.3.1 Numerical Results

We present numerical results $l = 1.0$, $c_0 = 10^5$, $c = 0.7$, $A_{min}/\alpha = 0.3$, and $A_{max}/\alpha = 1.0$ for an appropriate $\alpha > 0$. Here units refer to the international system.

We have obtained the buckling load $P = 6.0777 \cdot 10^5$, and for the optimal $A(x)$, see Fig. 23.1. The eigenvalue $P_1 = 4.8320 \cdot 10^5$ corresponds to $A(x) = cA_{max}$. Observe that $P > P_1$ as expected. Anyway, we have obtained just a critical point; at this point, we are not able to guarantee global optimality.

23.3.2 A Dynamical Case

In this section we develop analysis for a beam model dynamics, similarly as in the last section. Specifically, we consider the motion of a beam on an interval $[0,T]$. The beam model in question is the same as in the last section, so that the dynamical equation is given by

$$c_0(A(x)^2 w(x,t)_{,xx})_{xx} + \rho A(x) w(x,t)_{,tt} = 0, \text{ in } \Omega = [0,l],$$

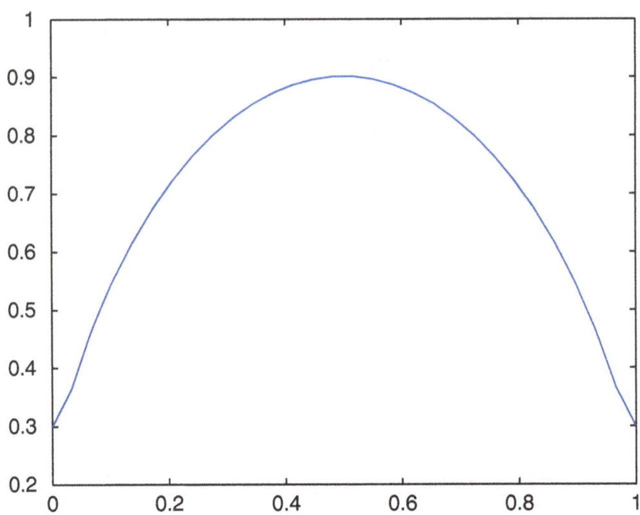

Fig. 23.1 Optimal distribution of area $A(x)/\alpha$, intending to maximize the buckling load

where ρ denotes the beam density and $w(x,t) : \Omega \times [0,T] \to \mathbb{R}$ denotes the field of vertical displacements. The motion results from proper initial conditions not yet specified.

For the last equation we look for a solution of the form

$$w(x,t) = e^{i\omega t} u(x),$$

where ω is the first natural frequency.

Replacing such a solution in the last equation we get

$$e^{i\omega t}\left(c_0(A(x)^2 u_{,xx})_{xx} - \omega^2 \rho A(x) u(x)\right) = 0, \text{ in } \Omega,$$

so that

$$(c_0(A(x)^2 u_{,xx})_{xx} - \omega^2 \rho A(x) u(x) = 0, \text{ in } \Omega. \tag{23.43}$$

At this point we consider the problem of finding $A(x)$ which maximizes the fundamental frequency ω, subject to (23.43):

$$\int_\Omega u^2 dx = 1,$$

$$\frac{1}{2}\int_\Omega c_o A(x)^2 (u_{,xx})^2 \, dx - \frac{\omega^2}{2}\int_\Omega \rho A(x) u^2 \, dx \geq 0,$$

$\forall u \in U$.

Moreover, the following design constraints must be satisfied:

$$\int_\Omega A(x) \, dx = V = cA_{max} l,$$

and

$$0 < A_{min} \leq A(x) \leq A_{max}, \text{ in } \Omega.$$

This problem is mathematically similar to the previous one, related to the maximization of the buckling load.

Thus, similarly as in the last section, we define $J_\lambda(u, A, \omega)$ by

$$\begin{aligned} J_\lambda(u, \omega, A) &= -G(u_{,xx}, A) \\ &\quad + \langle \lambda, (c_0 A(x)^2 u_{,xx})_{xx} - \omega^2 \rho A(x) u \rangle_{L^2} \end{aligned} \tag{23.44}$$

where

$$G(u_{,xx}, A) = \frac{c_0}{2} \int_\Omega A(x)^2 (u_{,xx})^2 \, dx,$$

and λ is an appropriate Lagrange multiplier.

Observe that

$$\begin{aligned} J_\lambda(w, \omega, A) &= \langle u_{,xx}, v^* \rangle_{L^2} - G(u_{,xx}, A) \\ &\quad - \langle u_{,xx}, v^* \rangle_{L^2} + \langle \lambda, (c_0 A(x)^2 u_{,xx})_{xx} - \omega^2 \rho u \rangle_{L^2}, \end{aligned} \tag{23.45}$$

so that

$$\begin{aligned} J_\lambda(w, \omega, A) &\leq \sup_{v \subset L^2} \{ \langle v, v^* \rangle_{L^2} - G(v, A) \\ &\quad + \sup_{w \in U} \{ -\langle u_{,xx}, v^* \rangle_{L^2} + \langle \lambda, c_0 (A(x)^2 u_{,xx})_{xx} - \omega^2 \rho A(x) u \rangle_{L^2} \}, \end{aligned}$$

and therefore

$$J_\lambda(w, P, A) \leq G^*(v^*, A) + Ind_2(v^*, \lambda, P, A),$$

where

$$G^*(v^*, A) = \frac{1}{2c_0} \int_\Omega \frac{(v^*)^2}{A(x)^2} \, dx,$$

and

$$Ind_1(v^*, \lambda, \omega, A) = \begin{cases} 0, & \text{if } (v^*, \lambda, \omega, A) \in B^*, \\ +\infty, & \text{otherwise,} \end{cases} \tag{23.46}$$

where

$$\begin{aligned} B^* = \{ &(v^*, \lambda, \omega, A) \in L^2 \times L^2 \times \mathbb{R}^+ \times L^2 \mid \\ &v^*_{,xx} - c_0 (A(x)^2 \lambda_{xx})_{,xx} - \omega^2 \rho A(x) \lambda = 0, \text{ in } \Omega, \\ &\text{and } v^*(0) = v^*(l) = 0 \}. \end{aligned} \tag{23.47}$$

Summarizing the partial duality principle obtained, we have

$$J_\lambda(w, \omega, A) \leq \inf_{v^* \in L^2} \{ G^*(v^*, A) + Ind_1(v^*, \lambda, \omega, A) \}.$$

Having such an inequality in mind, we develop an algorithm to obtain critical points, similar to that of the previous sections (we do not give the details here).

For the same constraints in $A(x)$ (in particular $c = 0.7$) as for the previous example, again for $l = 1$, $c_0 = 10^5$, and $\rho = 10$, we obtain the optimal $\omega^2 = 7.3885 \cdot 10^5$. For the optimal $A(x)$ see Fig. 23.2.

For the case $A(x) = cA_{max}$ we have obtained $\omega_1^2 = 6.8070 \cdot 10^5$. Units refer to the international system.

Observe that the optimal $\omega > \omega_1$ as naturally expected. Anyway, we emphasize to have calculated just a critical point. Again at this point we cannot guarantee global optimality.

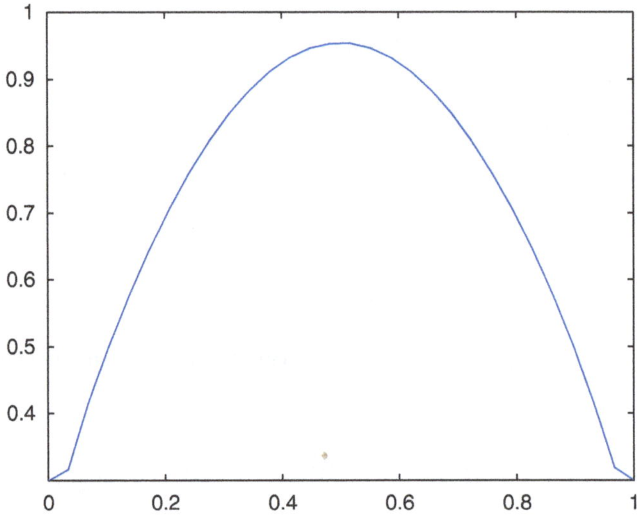

Fig. 23.2 Optimal distribution of area $A(x)/\alpha$, intending to maximize ω

23.4 Conclusion

In this chapter we develop a concave dual variational formulation for the optimal control of a well-known beam model. In practice, the results may be applied to the energy minimization with piezoelectric actuators, through which the beam vibration may be controlled (see [49] for related results). In a second step, we study the optimal design for this same beam model, with the objective of maximizing its buckling load and fundamental frequency, respectively. In both cases the numerical results obtained are consistent with the problem physics. Finally, we emphasize the approach here developed may be applied to many other situations, such as for plate and shell models for example.

Errata

Functional Analysis and Applied Optimization in Banach Spaces

Fabio Botelho

Department of Mathematics and Statistics, Federal University of Pelotas, Pelotas, RS-Brazil

F. Botelho, *Functional Analysis and Applied Optimization in Banach Spaces: Applications to Non-Convex Variational Models*, DOI 10.1007/978-3-319-06074-3,

---

DOI 10.1007/978-3-319-06074-3_24

The paperback and online versions of the book contain some errors, and the corrections to these versions are given on the following pages.

The online version of the original book can be found at http://dx.doi.org/10.1007/978-3-319-06074-3

F. Botelho, *Functional Analysis and Applied Optimization in Banach Spaces: Applications to Non-Convex Variational Models*, DOI 10.1007/978-3-319-06074-3_24,

1
Topological Vector Spaces

page 4, in the proof of Proposition 1.2.9, in the last two lines of this page, please replace the text part

"2. A,B closed implies that A^c and B^c are open, and by Definition 1.2.5, $A^c \cup B^c$ is open, so that $A \cap B = (A^c \cup B^c)^c$ is closed."

by

"2. A,B closed implies that A^c and B^c are open, and by Definition 1.2.5, $A^c \cap B^c$ is open, so that $A \cup B = (A^c \cap B^c)^c$ is closed."

page 12, in fact Definition 1.5.8 is not a definition, it is a proposition. Please replace

"Definition 1.5.8. Let (U,d) be a metric space. The set σ of all open sets, defined through the last definition, is indeed a topology for (U,d)."

by

"Proposition 1.5.8. Let (U,d) be a metric space. The set σ of all open sets, defined through the last definition, is indeed a topology for (U,d)."

page 32, at the bottom of this page, please replace the text part

"Thus, from the parallelogram law, we have

$$\begin{aligned} \|m_i - m_j\|_H^2 &= \|m_i - u - (m_j - u)\|_H^2 \\ &= 2\|m_i - u\|_H^2 + 2\|m_j - u\|_H^2 \\ &\quad -2\| - 2u + m_i + m_j\|_H^2 \\ &= 2\|m_i - u\|_H^2 + 2\|m_j - u\|_H^2 \\ &\quad -4\| - u + (m_i + m_j)/2\|_H^2 \\ &\to 2d^2 + 2d^2 - 4d^2 = 0, \text{ as } i,j \to +\infty. \end{aligned} \tag{1.46}$$

"

by

"Thus, from the parallelogram law, we have

$$\begin{aligned} \|m_i - m_j\|_H^2 &= \|m_i - u - (m_j - u)\|_H^2 \\ &= 2\|m_i - u\|_H^2 + 2\|m_j - u\|_H^2 \\ &\quad -\| - 2u + m_i + m_j\|_H^2 \\ &= 2\|m_i - u\|_H^2 + 2\|m_j - u\|_H^2 \\ &\quad -4\| - u + (m_i + m_j)/2\|_H^2 \\ &\to 2d^2 + 2d^2 - 4d^2 = 0, \text{ as } i,j \to +\infty. \end{aligned} \tag{1.46}$$

"

2
The Hahn–Banach Theorems and Weak Topologies

page 42, at line 10 from the bottom of such a page, inside the proof of Theorem 2.2.1, please replace the text part

"Clearly $e_\alpha \prec e$ so each linearly ordered set..."

by

"Clearly $e_\alpha \prec e$, $\forall \alpha \in A$, so each linearly ordered set..."

3
Topics on Linear Operators

page 57, at the beginning of section 3.1, first line, please replace

"First we recall that the set of all bounded linear operators, denoted by $\mathscr{L}(U,Y)$, is a Banach space with the norm..."

by

"Let U,Y be Banach spaces. First we recall that the set of all bounded linear operators from U into Y, denoted by $\mathscr{L}(U,Y)$, is a Banach space with the norm..."
page 59, in the proof of Theorem 3.2.3, please replace the text part

" 2. Observe that

$$(u,Av)_H = (A^*u,v)_H = (u,A^{**}v)_H, \forall u,v \in H.$$

"

by

"2. Observe that

$$(u,Av)_H = \overline{(Av,u)_H} = \overline{(v,A^*u)_H} = (A^*u,v)_H = (u,A^{**}v)_H, \forall u,v \in H.$$

"

page 63, in Theorem 3.2.7, at line 4 of this Theorem, please replace

"

$$(A^*)^{-1} = \{v^* \in Y^* \,:\, A^*v^* \in S^{\oplus}\}.$$

"

by

"

$$(A^*)^{-1}(S^{\oplus}) = \{v^* \in Y^* \,:\, A^*v^* \in S^{\oplus}\}.$$

"

page 68, in the 7 last lines of this page, please replace the text part

"Finally, we may write,

$$\begin{aligned}(ABu,u)_H &= \|A\|(A_1Bu,u)_H \\ &= \|A\|(BA_1u,u)_H \\ &= \|A\|(B\lim_{n\ldots}\sum_j = 1^n A_j^2u,u)_H \\ &= \|A\|\lim_{n\ldots}\sum_j = 1^n(BA_j^2u,u)_H \\ &= \|A\|\lim_{n\ldots}\sum_j = 1^n(BA_ju,BA_ju)_H \\ &\geq 0. \end{aligned} \tag{3.11}$$

"

by

"Finally, since $B \geq \theta$, we may write,

$$\begin{aligned}(ABu,u)_H &= \|A\|(A_1Bu,u)_H \\ &= \|A\|(BA_1u,u)_H \\ &= \|A\|(B\lim_{n\to\infty}\sum_{j=1}^n A_j^2u,u)_H \\ &= \|A\|\lim_{n\to\infty}\sum_{j=1}^n(BA_j^2u,u)_H \\ &= \|A\|\lim_{n\to\infty}\sum_{j=1}^n(B(A_ju),(A_ju))_H \\ &\geq 0. \end{aligned} \tag{3.11}$$

"

page 76, at line 1 at the top of the page, please replace the text part

"Thus

$$(u,(A-\lambda I)v))_H = 0, \forall u \in H,$$

so that

$$((A^*-\overline{\lambda}I)u,v)_H, \forall u \in H.$$

"

by

"Thus

$$(u,(A-\lambda I)v))_H = 0, \forall u \in H,$$

so that

$$((A^* - \overline{\lambda} I)u, v)_H = 0, \forall u \in H.$$

"

8
The Lebesgue and Sobolev Spaces

page 200, about line 13 from the page bottom, please replace the text part

"Observe that defining $\psi(x) = y$ from the continuity of ψ^{-1}, there exists $r_1 > 0$ such that

$$\psi^{-1}(B^+_{r_1}(y_0)) \subset \Omega \cap B_r(x_0),$$

"

by

"Observe that defining $\psi(x) = y$ from the continuity of ψ^{-1}, there exists $r_1 > 0$ such that

$$\psi^{-1}(B^+_{r_1}(y_0)) \subset \overline{\Omega} \cap B_r(x_0),$$

"

page 209, in the two last lines of this page and turning to page 210, please replace the text part

"Let $1 \leq p_1 \leq n$ such that

$$\hat{r} = np_1/(n - p_1).$$

Thus we have that

$$np/(n - jp) = np_1/(n - p_1),$$

so that

$$p_1 = np/(n - (j-1)p),$$

so that by above and the last theorem:"

by

"Let $1 \leq p_1 < n$ such that

$$\hat{r} = np_1/(n - p_1).$$

Thus we have that

$$np/(n - jp) = np_1/(n - p_1),$$

so that

$$p_1 = np/(n - (j-1)p),$$

so that by above and Theorem 8.4.15:"

page 210, about line 12 from the top, please replace the text part

"Hence $p_2 = np/(n-(j-2)p)$ so that by the last theorem..."

by

"Hence $p_2 = np/(n-(j-2)p)$ so that by the Theorem 8.4.15..."

page 210, about line 7 from the bottom of this page, please replace the text part

"Since $n > p$ we have that $n \geq 2$ so that $1 \leq p_1 < n$. From the last theorem we obtain..."

by

"Since $n > p$ we have that $n \geq 2$ so that $1 \leq p_1 < n$. From the Theorem 8.4.15 we obtain..."

page 211, line 3 from the top of the page, please replace the text part

"Finally, if $r_1 = \max\{n, p\} = p \geq n$, Define p_1 such that

$$r_1 = p = \frac{np_1}{n - p_1}$$

that is,

$$p_1 = \frac{np}{n+p} \leq p,$$

so that by last theorem

$$\|u\|_\infty \leq \|u\|_{m-j,r_1,\Omega} \leq C_5\|u\|_{m-(j-1),p_1,\Omega} \leq C_6\|u\|_{m,p,\Omega}.$$

This completes the proof."

by

"Finally, if $r_1 = \max\{n, p\} = p = n$, Define p_1 such that

$$r_1 = p = \frac{np_1}{n - p_1}$$

that is,

$$p_1 = \frac{np}{n+p} < n = p,$$

so that by Theorem 8.4.15

$$\|u\|_\infty \leq \|u\|_{m-j,r_1,\Omega} \leq C_5\|u\|_{m-(j-1),p_1,\Omega} \leq C_6\|u\|_{m,p,\Omega}.$$

This completes the proof."

9
Basic Concepts on the Calculus of Variations

page 234, in the last 3 lines of the proof of Theorem 9.5.9, please replace the text part

"Hence,

$$F(u_0) \leq F(u_0 + \varepsilon\varphi), \forall\, \varepsilon,\ \varphi \text{ such that } |\varepsilon| < r,\ \|\varphi\|_U < 1.$$

The proof is complete."

by

"Hence,

$$F(u_0) \leq F(u_0 + \varepsilon\varphi), \forall\, \varepsilon,\ \varphi \text{ such that } |\varepsilon| < \min\{r, 1\},\ \|\varphi\|_U < 1.$$

The proof is complete."

page 236, in Definition 9.7.1, please replace

"Definition 9.7.1. We say that $u \in \hat{C}^1([a,b];\mathbb{R}^N)$ if $u : [a,b] \to \mathbb{R}^N$ is continuous in $[a,b]$, and Du is continuous except on a finite set of points in $[a,b]$."

by

"Definition 9.7.1. We say that $u \in \hat{C}^1([a,b];\mathbb{R}^N)$ if $u : [a,b] \to \mathbb{R}^N$ is Lipschitz continuous in $[a,b]$, and Du is continuous except on a finite set of points in $[a,b]$. The points in which Du is not continuous are said to be the corner points of u. Hence, if $x_0 \in (a,b)$ is a corner point of u, we have

$$u'(x_0+) \equiv \lim_{h\to 0^+} u'(x_0+h) \neq \lim_{h\to 0^-} u'(x_0+h) \equiv u'(x_0-).$$

"

page 240, about at line 11 from the top, please replace the text part,

"Let $\lambda \in (0,\lambda_0)$ and

$$\phi \in C_c((-1,1);\mathbb{R})$$

"

by

"Let $\lambda \in (0,\lambda_0)$ and

$$\phi \in C_c^1((-1,1),\mathbb{R})$$

"

page 242, about line 14 from the top, please replace the text part

"the second and third terms in (9.20) are of $o(1)$ where

$$\lim_{\lambda \to 0^+} o(1)/\lambda = 0,$$

"

by

"the second and third terms in (9.20) are of $o(1)$ where

$$\lim_{\lambda \to 0^+} o(1) = \lim_{\lambda \to 0^+} o(\lambda)/\lambda = 0,$$

"

page 247, first line from the top, please replace the text part

"Define

$$\phi(\varepsilon) = F(x, \tilde{u}_\varepsilon, \tilde{u}'_\varepsilon(x)).$$

Thus ϕ has a local minimum at 0, so that $\phi'(0) = 0$, that is

$$\frac{d(F(x, \tilde{u}_\varepsilon, \tilde{u}'_\varepsilon(x)))}{d\varepsilon}|_{\varepsilon=0} = 0.$$

"

by

"Define

$$\phi(\varepsilon) = F(\tilde{u}_\varepsilon).$$

Thus ϕ has a local minimum at 0, so that $\phi'(0) = 0$, that is

$$\frac{d(F(\tilde{u}_\varepsilon))}{d\varepsilon}|_{\varepsilon=0} = 0.$$

"

page 247, at the last line of the page, please replace the text part

"for some $c_1 \in \mathbb{R}^N$."

by

"for some $c_1 \in \mathbb{R}$."

page 247, about line 9 from the bottom of page, please replace

"From

$$\frac{dF(\tilde{u}_\varepsilon)}{d\varepsilon}|_{\varepsilon=0},$$

"

by

"From

$$\frac{dF(\tilde{u}_\varepsilon)}{d\varepsilon}|_{\varepsilon=0} = 0,$$

"

page 248, in the last two lines of Proposition 9.11.1, please replace

"then u is a extremal of F which satisfies the following natural boundary conditions:

$$n_\alpha f_{\xi^i_\alpha}(x, u(x)\nabla u(x)) = 0, \text{ a.e. on } \Gamma_1, \forall i \in \{1,...,N\}.$$

"

by

"then u is a extremal of F which satisfies the following natural boundary conditions:

$$n_\alpha f_{\xi^i_\alpha}(x, u(x)\nabla u(x)) = 0, \text{ a.e. on } \Gamma_1, \forall i \in \{1,...,N\}.$$

Here $\{n_\alpha\}$ denotes the outward normal to $\partial\Omega$."

10
Basic Concepts on Convex Analysis

page 252, at line 4 from the top, in Definition 10.1.4, please replace

"Definition 10.1.4. Let U be a Banach space. Consider the weak topology $\sigma(U,U^*)$ and let $F : U \to \mathbb{R} \cup \{+\infty\}$. Such a function is said to be weakly lower semi-continuous if $\forall\lambda$ such that $\lambda < F(u)$, there exists a weak neighborhood $V_\lambda(u) \in \sigma(U,U^*)$ such that

$$F(v) > \lambda, \ \forall v \in V_\lambda(u).$$

"

by

"Definition 10.1.4. Let U be a Banach space. Consider the weak topology $\sigma(U,U^*)$ and let $F : U \to \mathbb{R} \cup \{+\infty\}$. Such a function is said to be weakly lower semi-continuous at $u \in U$ if $\forall\lambda$ such that $\lambda < F(u)$, there exists a weak neighborhood $V_\lambda(u) \in \sigma(U,U^*)$ of u such that

$$F(v) > \lambda, \ \forall v \in V_\lambda(u).$$

Also, if F is weakly lower semi-continuous at each $u \in U$, it is said to be weakly lower semi-continuous."

page 252, in Theorem 10.1.5, please replace (item 5)

"5.

$$\liminf_{v \rightharpoonup u} F(v) \geq F(u), \forall u \in U.$$

"

by

"5.

$$\liminf_{v \rightharpoonup u} F(v) \geq F(u), \forall u \in U,$$

where

$$\liminf_{v \rightharpoonup u} F(v) \equiv \sup_{V(u) \in \sigma(U,U^*)} \inf_{v \in V(u)} F(v).$$

"

page 253, at about line 9 from the top, right before the end of the proof o Theorem 10.1.5, please replace the text part

"Finally assume that

$$\liminf_{v \rightharpoonup u} F(v) \geq F(u).$$

Let $\lambda < F(u)$. Thus there exists a weak neighborhood $V(u)$ such that $F(v) \geq F(u) > \lambda, \forall v \in V(u)$. The proof is complete."

by

"Finally, for $u \in U$ assume that

$$\liminf_{v \rightharpoonup u} F(v) \geq F(u).$$

Let $\lambda < F(u)$. Define $\varepsilon = (F(u) - \lambda)/2$, thus there exists a weak neighborhood $V(u)$ such that $F(v) \geq F(u) - \varepsilon > \lambda, \forall v \in V(u)$. The proof is complete."

page 265, at about line 14 from the top, please replace the text part

"so that

$$\sup_{v \in B} \inf_{u \in A} L(u,v) \leq \sup_{vInB} L(u,v), \forall u \in A,$$

and hence..."

by

"so that

$$\sup_{v \in B} \inf_{u \in A} L(u,v) \leq \sup_{v \in B} L(u,v), \forall u \in A,$$

and hence..."

page 281, in equation (10.181), please replace

"...and hence

$$\varepsilon d(u_{n+m}, u_n) \leq \sum_{i=1}^{m} d(u_{n+i}, u_{n+i-1}) \leq F(u_n) - F(u_{n+m}).$$

"

by

"...and hence

$$\varepsilon d(u_{n+m}, u_n) \leq \varepsilon \sum_{i=1}^{m} d(u_{n+i}, u_{n+i-1}) \leq F(u_n) - F(u_{n+m}).$$

"

11 Constrained Variational Optimization

page 287, at line 2 of Definition 11.1.2, please replace the text part

"In particular $u \succeq \theta$ if and only if $u \in C$."

By

"In particular $u \geq \theta$ if and only if $u \in P$."

page 287, near the end of line 2 in the proof of Proposition 11.1.3, please replace the text part

"Since P is a cone we must have $\langle p, u^* \rangle_U \geq 0$; otherwise we would have $\langle u, u^* \rangle > \langle \alpha p, u^* \rangle_U$ for some $\alpha > 0$"

by

"Since P is a cone, for any fixed $p \in P$ we must have $\langle p, u^* \rangle_U \geq 0$; otherwise we would have $\langle u, u^* \rangle > \langle \alpha p, u^* \rangle_U$ for some $\alpha > 0$"

page 290, in Theorem 11.1.8 please replace equation (11.23)

"

$$F(u_0) + \langle G(u_0), z_0^* \rangle_Z =$$
$$\inf\{F(u) \mid u \in \Omega \text{ and } G(u) \leq G(u_0)\}.$$

"

by

“

$$F(u_0) = \inf\{F(u) \mid u \in \Omega \text{ and } G(u) \leq G(u_0)\}.$$

”

page 300, in Theorem 11.1, at the line 4 of this theorem, please replace the text part

“Furthermore, assume $u_0 \in U$ is a point of local minimum for $F : U \to \mathbb{R}$ subject to $G_1(u) = \theta$ and $G_2(u_0) \leq \theta$, where...”

by

“Furthermore, assume $u_0 \in U$ is a point of local minimum for $F : U \to \mathbb{R}$ subject to $G_1(u) = \theta$ and $G_2(u) \leq \theta$, where...”

page 304, in Theorem 11.2, at the line 4 of this theorem, please replace the text part

“Furthermore, assume $u_0 \in U$ is a point of local minimum for $F : U \to \mathbb{R}$ subject to $G_1(u) = \theta$ and $G_2(u_0) \leq \theta$, where...”

by

“Furthermore, assume $u_0 \in U$ is a point of local minimum for $F : U \to \mathbb{R}$ subject to $G_1(u) = \theta$ and $G_2(u) \leq \theta$, where...”

page 307, at section 11.7 at the top of the page, please replace the text part

“Now we recall a classical definition, namely, the Banach fixed theorem also known as the contraction mapping theorem.”

by

“Now we recall a classical result, namely, the Banach fixed theorem also known as the contraction mapping theorem.”

13 Duality Applied to a Plate Model

page 347, in equation (13.10), please replace

“

$$\tilde{F}^*(z^*) - G^*(z^*) > 0, \forall z^* \in Y_0^* \text{ such that } z^* \neq \theta, \tag{13.10}$$

”

by

“

$$\tilde{F}^*(z^*) - G^*(\theta, z^*) > 0, \forall z^* \in Y_0^* \text{ such that } z^* \neq \theta, \tag{13.10}$$

”

16
More on Duality and Computation for the Ginzburg–Landau System

page 420, about 6 lines from the page bottom, please replace the text part

"where

$$\tilde{\alpha}_1 = a_0\alpha_0 + d_0\tilde{\alpha}\alpha_1,$$

"

by

"where

$$\tilde{\alpha}_1 = a_0\alpha_1 + d_0\tilde{\alpha}\alpha_1,$$

"

page 422, at the top of this page, please replace the text part,

"Moreover,

$$\begin{aligned} C_{i+1} &= fn_i + gC_i \\ &= f(\alpha_{i+1}T_{i+1} + \beta_{i+1}) \\ &\quad + g(\hat{\alpha}_i T_i + \hat{\beta}_i) \\ &= f(\alpha_{i+1}T_{i+1} + \beta_{i+1}) \\ &\quad + g(\hat{\alpha}_i(\eta_i T_i + \xi_i) + \hat{\beta}_i) \\ &= \hat{\alpha}_{i+1}T_{i+1} + \hat{\beta}_{i+1}, \end{aligned} \tag{16.82}$$

"

by

"Moreover,

$$\begin{aligned} C_{i+1} &= fn_i + gC_i \\ &= f(\alpha_{i+1}T_{i+1} + \beta_{i+1}) \\ &\quad + g(\hat{\alpha}_i T_i + \hat{\beta}_i) \\ &= f(\alpha_{i+1}T_{i+1} + \beta_{i+1}) \\ &\quad + g(\hat{\alpha}_i(\eta_i T_{i+1} + \xi_i) + \hat{\beta}_i) \\ &= \hat{\alpha}_{i+1}T_{i+1} + \hat{\beta}_{i+1}, \end{aligned} \tag{16.82}$$

"

22

The Generalized Method of Lines Applied to Fluid Mechanics

page 517, in the line 4 at 22.1 Introduction, please replace the paragraph

"Consider $\Omega \subset \mathbb{R}^2$ an open, bounded and connected set, whose the internal boundary is denoted by Γ_0 and, the external one is denoted by Γ_1. Denoting by $u : \Omega \to \mathbb{R}$ the field of velocity in direction x of the Cartesian system (x,y), by $v : \Omega \to \mathbb{R}$, the velocity field in the direction y, by $p : \Omega \to \mathbb{R}$, the pressure field, so that $P = p/\rho$, where ρ is the constant fluid density, ν is the viscosity coefficient and, g is the gravity constant, the Navier-Stokes PDE system is expressed by:"

by

"Consider $\Omega \subset \mathbb{R}^2$ an open, bounded and connected set, whose the internal boundary is denoted by Γ_0 and, the external one is denoted by Γ_1. We assume Γ_0 and Γ_1 to be $\hat{C}_1$ (Lipschitz continuous). Denoting by $u : \Omega \to \mathbb{R}$ the field of fluid velocity in direction x of the cartesian system (x,y), by $v : \Omega \to \mathbb{R}$, the fluid velocity field in the direction y, by $p : \Omega \to \mathbb{R}$, the fluid pressure field, so that $P = p/\rho$, where ρ is the constant fluid density, ν is the viscosity coefficient and, g is the gravity constant, the Navier-Stokes PDE system is expressed by:"

page 518, in Theorem 22.2.1, at the line 5 from top, please replace the text part

"with the boundary conditions

$$\left\{ \mathbf{u} \cdot \mathbf{n} = 0, \text{ on } \Gamma, \right. \tag{22.4}$$

where $\mathbf{u} = (u, v)$, is given by

$$\begin{cases} u = \partial_x w_0, \\ v = \partial_y w_0, \end{cases} \tag{22.5}$$

where w_0 is a solution of the equation

$$\begin{cases} \nabla^2 w_0 = 0 & \text{in } \Omega, \\ \nabla w_0 \cdot \mathbf{n} = 0, & \text{on } \Gamma. \end{cases} \tag{22.6}$$

"

by

"with the boundary conditions

$$\begin{cases} \mathbf{u}\cdot\mathbf{n}=0, & \text{on } \Gamma_0, \\ \mathbf{u}\cdot\mathbf{n}=h\in L^2(\Gamma_1), & \text{on } \Gamma_1, \\ P=P_\infty & \text{on } \Gamma_1, \end{cases} \tag{22.4}$$

where $\mathbf{u}=(u,v)$, is given by

$$\begin{cases} u=\partial_x w_0, \\ v=\partial_y w_0, \end{cases} \tag{22.5}$$

where w_0 is a solution of the equation

$$\begin{cases} \nabla^2 w_0=0 & \text{in } \Omega, \\ \nabla w_0\cdot\mathbf{n}=0, & \text{on } \Gamma_0 \\ \nabla w_0\cdot\mathbf{n}=h, & \text{on } \Gamma_1,. \end{cases} \tag{22.6}$$

"

page 520, right after equation (22.13), please replace

"where

$$\begin{aligned} L(u)/f_0(\theta) = {} & \frac{\partial^2 u}{\partial t^2}+\frac{1}{t}f_2(\theta)\frac{\partial u}{\partial t} \\ & +\frac{1}{t}f_3(\theta)\frac{\partial^2 u}{\partial\theta\partial t}+\frac{f_4(\theta)}{t^2}\frac{\partial^2 u}{\partial\theta^2}=0, \end{aligned} \tag{22.14}$$

"

by

"where we have considered $\nu=1$, $g_x=g_y=0$, and

$$\begin{aligned} L(u)\left(r(\theta)^2/f_0(\theta)\right) = {} & \frac{\partial^2 u}{\partial t^2}+\frac{1}{t}f_2(\theta)\frac{\partial u}{\partial t} \\ & +\frac{1}{t}f_3(\theta)\frac{\partial^2 u}{\partial\theta\partial t}+\frac{f_4(\theta)}{t^2}\frac{\partial^2 u}{\partial\theta^2}, \end{aligned} \tag{22.14}$$

"

page 521, at line 4 from the top of this page, please replace

"Also

$$d_1u/f_0(\theta) = f_5(\theta)\frac{\partial u}{\partial t} + (f_6(\theta)/t)\frac{\partial u}{\partial \theta},$$

$$d_2u/f_0(\theta) = f_7(\theta)\frac{\partial u}{\partial t} + (f_8(\theta)/t)\frac{\partial u}{\partial \theta},$$

where

$$f_5(\theta) = \cos(\theta)/r(\theta) + \sin(\theta)r'(\theta)/r^3(\theta),$$

$$f_6(\theta) = -\sin(\theta)/r(\theta),$$

$$f_7(\theta) = \sin(\theta)/r(\theta) - \cos(\theta)r'(\theta)/r^3(\theta),$$

$$f_8(\theta) = \cos(\theta)/r(\theta).$$

"

by

"Also

$$d_1u = f_5(\theta)\frac{\partial u}{\partial t} + (f_6(\theta)/t)\frac{\partial u}{\partial \theta},$$

$$d_2u = f_7(\theta)\frac{\partial u}{\partial t} + (f_8(\theta)/t)\frac{\partial u}{\partial \theta},$$

where

$$f_5(\theta) = \cos(\theta)/r(\theta) + \sin(\theta)r'(\theta)/r^2(\theta),$$

$$f_6(\theta) = -\sin(\theta)/r(\theta),$$

$$f_7(\theta) = \sin(\theta)/r(\theta) - \cos(\theta)r'(\theta)/r^2(\theta),$$

$$f_8(\theta) = \cos(\theta)/r(\theta).$$

"

23

Duality Applied to the Optimal Control and Optimal Design of a Beam Model

page 546, about at line 8 from the top, please replace the text part

"... the buckling load *P*, where the following designed constraints must be satisfied:"

by

"... the buckling load *P*, where the following design constraints must be satisfied:"

page 546, at about line 16 from the top, please replace the text part

“Furthermore, as above indicate,...”

by

“Furthermore, as above indicated,...”

References

1. R.A. Adams, *Sobolev Spaces* (Academic Press, New York, 1975)
2. R.A. Adams, J.F. Fournier, *Sobolev Spaces*, 2nd edn. (Elsevier, Oxford, UK, 2003)
3. G. Allaire, *Shape Optimization by the Homogenization Method* (Springer, New York, 2002)
4. J.F. Annet, *Superconductivity, Superfluids and Condensates*. Oxford Master Series in Condensed Matter Physics, Oxford University Press, New YorK (2010)
5. H. Attouch, G. Buttazzo, G. Michaille, *Variational Analysis in Sobolev and BV Spaces*. MPS-SIAM Series in Optimization (SIAM, Philadelphia, 2006)
6. G. Bachman, L. Narici, *Functional Analysis* (Dover Publications, New York, 2000)
7. J.M. Ball, R.D. James, Fine mixtures as minimizers of energy. Arch. Ration. Mech. Anal. **100**, 15–52 (1987)
8. M.P. Bendsoe, O. Sigmund, *Topology Optimization, Theory Methods and Applications* (Springer, Berlin, New York 2003)
9. F. Bethuel, H. Brezis, F. Helein, *Ginzburg-Landau Vortices* (Birkhäuser, Basel, 1994)
10. F. Botelho, *Variational Convex Analysis*. Ph.D. Thesis (Virginia Tech, Blacksburg, 2009)
11. F. Botelho, Dual variational formulations for a non-linear model of plates. J. Convex Anal. **17**(1), 131–158 (2010)
12. F. Botelho, Existence of solution for the Ginzburg-Landau system, a related optimal control problem and its computation by the generalized method of lines. Appl. Math. Comput. **218**, 11976–11989 (2012)
13. F. Botelho, *Topics on Functional Analysis, Calculus of Variations and Duality* (Academic Publications, Sofia, 2011)
14. F. Botelho, On duality principles for scalar and vectorial multi-well variational problems. Nonlinear Anal. **75**, 1904–1918 (2012)
15. F. Botelho On the Lagrange multiplier theorem in Banach spaces, Comput. Appl. Math. **32**, 135–144 (2013)
16. H. Brezis, *Analyse Fonctionnelle* (Masson, Paris, 1987)
17. L.D. Carr, C.W. Clark, W.P. Reinhardt, Stationary solutions for the one-dimensional nonlinear Schrödinger equation. I- case of repulsive nonlinearity. Phys. Rev. A **62**, 063610 (2000)
18. I.V. Chenchiah, K. Bhattacharya, The relaxation of two-well energies with possibly unequal moduli. Arch. Rational Mech. Anal. **187**, 409–479 (2008)
19. M. Chipot, *Approximation and Oscillations*. Microstructure and Phase Transition, The IMA Volumes in Mathematics and Applications, Oxford, UK, vol. 54, 1993, pp. 27–38
20. R. Choksi, M.A. Peletier, J.F. Williams, On the phase diagram for microphase separation of diblock copolymers: an approach via a nonlocal Cahn-Hilliard functional. SIAM J. Appl. Math. **69**(6), 1712–1738 (2009)
21. P. Ciarlet, *Mathematical Elasticity*, vol. I, Three Dimensional Elasticity (North Holland, Elsevier, 1988)
22. P. Ciarlet, *Mathematical Elasticity*, vol. II Theory of Plates (North Holland, Elsevier, 1997)

F. Botelho, *Functional Analysis and Applied Optimization in Banach Spaces: Applications to Non-Convex Variational Models*, DOI 10.1007/978-3-319-06074-3,

23. P. Ciarlet, *Mathematical Elasticity*, vol. III – Theory of Shells (North Holland, Elsevier, 2000)
24. B. Dacorogna, *Direct Methods in the Calculus of Variations* (Springer, New York, 1989)
25. I. Ekeland, R. Temam, *Convex Analysis and Variational Problems* (Elsevier-North Holland, Amsterdam 1976).
26. L.C. Evans, *Partial Differential Equations*, Graduate Studies in Mathematics, vol. 19 (American Mathematical Society, Providence, Rhode Island, 1998)
27. U. Fidalgo and P. Pedregal: *A General Lower Bound for the Relaxation of an Optimal Design Problem with a General Quadratic Cost Functional and a General Linear State Equation*, Journal of Convex Analysis 19, number 1, 281–294 (2012)
28. N.B. Firoozye, R.V. Khon, *Geometric Parameters and the Relaxation for Multiwell Energies*. Microstructure and Phase Transition, the IMA volumes in mathematics and applications, vol. 54, 85–110 (Oxford, UK, 1993)
29. I. Fonseca, G. Leoni, *Modern Methods in the Calculus of Variations, L^p Spaces* (Springer, New York, 2007)
30. A. Galka, J.J. Telega, Duality and the complementary energy principle for a class of geometrically non-linear structures. Part I. Five parameter shell model Arch. Mech. **47**, 677–698; Part II. Anomalous dual variational principles for compressed elastic beams. Arch. Mech. **47**, 699–724 (1995)
31. D.Y. Gao, G. Strang, Geometric nonlinearity: potential energy, complementary energy and the gap function. Q. J. Appl. Math. **47**, 487–504 (1989a)
32. D.Y. Gao, On the extreme variational principles for non-linear elastic plates. Q. Appl. Math. **XLVIII**(2), 361–370 (1990)
33. D.Y. Gao, Pure complementary energy principle and triality theory in finite elasticity. Mech. Res. Comm. **26**(1), 31–37 (1999)
34. D.Y. Gao, General analytic solutions and complementary variational principles for large deformation nonsmooth mechanics. Meccanica **34**, 169–198 (1999)
35. D.Y. Gao, Finite deformation beam models and triality theory in dynamical post-buckling analysis. Int. J. Non Linear Mech. **35**, 103–131 (2000)
36. D.Y. Gao, *Duality Principles in Nonconvex Systems, Theory, Methods and Applications* (Kluwer, Dordrecht, 2000)
37. M. Giaquinta, S. Hildebrandt, **Calculus of Variations I**. A Series of Comprehensive Studies in Mathematics, vol. 310 (Springer, Berlin, 1996)
38. M. Giaquinta, S. Hildebrandt, Calculus of Variations II. A Series of Comprehensive Studies in Mathematics, vol. 311 (Springer, Berlin, 1996)
39. E. Giusti, *Direct Methods in the Calculus of Variations* (World Scientific, Singapore, reprint, 2005)
40. K. Ito, K. Kunisch, *Lagrange Multiplier Approach to Variational Problems and Applications*. Advances in Design and Control (SIAM, Philadelphia, 2008)
41. A. Izmailov, M. Solodov, *Otimização*, vol. 2 (IMPA, Rio de Janeiro, 2007)
42. A. Izmailov, M. Solodov, *Otimização*, vol. 1, 2nd edn. (IMPA, Rio de Janeiro, 2009)
43. R.D. James, D. Kinderlehrer, Frustration in ferromagnetic materials. Continuum Mech. Thermodyn. **2**, 215–239 (1990)
44. H. Kronmuller, M. Fahnle, *Micromagnetism and the Microstructure of Ferromagnetic Solids* (Cambridge University Press, Cambridge, UK, 2003)
45. L.D. Landau, E.M. Lifschits, *Course of Theoretical Physics, Vol. 5- Statistical Physics, Part 1* (Butterworth-Heinemann, Elsevier, reprint, 2008)
46. E.M. Lifschits, L.P. Pitaevskii, *Course of Theoretical Physics, Vol. 9- Statistical Physics, Part 2* (Butterworth-Heinemann, Elsevier, reprint, 2002)
47. D.G. Luenberger, *Optimization by Vector Space Methods* (Wiley, New York, 1969)
48. G.W. Milton, *Theory of composites*. Cambridge Monographs on Applied and Computational Mathematics (Cambridge University Press, Cambridge 2002)
49. A. Molter, O.A.A. Silveira, J. Fonseca, V. Bottega, *Simultaneous Piezoelectric Actuator and Sensor Placement Optimization and Control Design of Manipulators with Flexible Links Using SDRE Method*, Mathematical Problems in Engineering, Hindawi, 2010, pp. 1–23

50. P. Pedregal, *Parametrized measures and variational principles*, Progress in Nonlinear Differential Equations and Their Applications, vol. 30 (Birkhauser, Basel, 1997)
51. P. Pedregal and B. Yan, *On Two Dimensional Ferromagnetism*. Proceedings of the Royal Society of Edinburgh, 139A, 575–594 (2009)
52. M. Reed, B. Simon, *Methods of Modern Mathematical Physics, Volume I, Functional Analysis* (Reprint Elsevier, Singapore, 2003)
53. S.M. Robinson, Strongly regular generalized equations. Math. Oper. Res. **5**, 43–62 (1980)
54. R.T. Rockafellar, *Convex Analysis* (Princeton University Press, Princeton, New Jersey, 1970)
55. R.C. Rogers, Nonlocal variational problems in nonlinear electromagneto-elastostatics, SIAM J. Math. Anal. **19**(6), 1329–1347 (1988)
56. R.C. Rogers, A nonlocal model of the exchange energy in ferromagnet materials. J. Integr. Equat. Appl. **3**(1), 85–127 (1991)
57. W. Rudin, *Real and Complex Analysis*, 3rd edn. (McGraw-Hill, New York, 1987)
58. W. Rudin, *Functional Analysis*, 2nd edn. (McGraw-Hill, New York, 1991)
59. H. Royden, *Real Analysis*, 3rd edn. (Prentice Hall India, New Delhi, 2006)
60. O. Sigmund, A 99 line topology optimization code written in Matlab. Struc. Muldisc. Optim. **21**, 120–127 (2001)
61. J.J.A. Silva, A. Alvin, M. Vilhena, C.Z. Petersen, B. Bodmann, *On a Closed form Solution of the Point Kinetics Equations with Reactivity Feedback of Temperature*. International Nuclear Atlantic Conference- INAC- ABEN, Belo Horizonte, MG, Brazil, October 24–28, 2011, ISBN: 978 -85-99141-04-05
62. E.M. Stein, R. Shakarchi, *Real Analysis*, Princeton Lectures in Analysis III, (Princeton University Press, 2005)
63. J.C. Strikwerda, *Finite Difference Schemes and Partial Differential Equations*, 2d edn. SIAM (Philadelphia, 2004)
64. D.R.S. Talbot, J.R. Willis, Bounds for the effective constitutive relation of a nonlinear composite. Proc. R. Soc. Lond. **460**, 2705–2723 (2004)
65. J.J. Telega, On the complementary energy principle in non-linear elasticity. Part I: Von Karman plates and three dimensional solids, C.R. Acad. Sci. Paris, Serie II **308**, 1193–1198; Part II: Linear elastic solid and non-convex boundary condition. Minimax approach, C.R. Acad. Sci. Paris, Serie II **308**, 1313–1317 (1989)
66. R. Temam, *Navier-Stokes Equations* (AMS, Chelsea, reprint 2001)
67. J.F. Toland, A duality principle for non-convex optimisation and the calculus of variations. Arch. Rath. Mech. Anal. **71**(1), 41–61 (1979)
68. J.L. Troutman, *Variational Calculus and Optimal Control*, 2nd edn. (Springer, New York, 1996)

Index

F. Botelho, *Functional Analysis and Applied Optimization in Banach Spaces: Applications to Non-Convex Variational Models*, DOI 10.1007/978-3-319-06074-3,

Zeitfracht Medien GmbH
Ferdinand-Jühlke-Straße 7
99095 Erfurt, Deutschland
produktsicherheit@kolibri360.de